MATH TO THE MAX

1,200 PRACTICE QUESTIONS TO MAXIMIZE YOUR MATH POWER

LEARNING EXPRESS®

NEW YORK

Copyright © 2008 LearningExpress, LLC.

All rights reserved under International and Pan-American Copyright Conventions. Published in the United States by LearningExpress, LLC, New York.

Library of Congress Cataloging-in-Publication Data:
Math to the max: 1,200 practice questions to maximize your math power.
    p.   cm.
  ISBN 978-1-57685-703-8 (1-57685-703-4)
  1. Mathematics—Problems, exercises, etc.   I. LearningExpress (Organization)
  QA43.M2987   2008
  510.76—dc22

                                                                            2008012378

Printed in the United States of America

9 8 7 6 5 4 3 2 1

First Edition

For more information or to place an order, contact LearningExpress at:
  2 Rector Street
  26th Floor
  New York, NY 10006

Or visit us at:
  www.learnatest.com

Contents

CONTENTS

Introduction

O MATTER WHAT GRADE you may be in right now, or even if you are out of school, you can always improve your math skills. As with any skill, practice keeps your strengths sharp and helps you improve in the areas where you might struggle. This book will give you plenty of practice—1,200 questions of practice.

We'll start with basic operations involving integers, fractions, decimals, and percents. If you are already strong in these areas, you'll fly right through the first few chapters. Or, you might want to skip ahead to Chapter 4, where we'll begin to review algebra. In addition to hundreds of practice questions, this book will lead you through examples of different types of problems in each chapter, step-by-step, so you can understand how to solve each type. We'll also highlight some tips that may help you solve problems, as well as some pitfalls to avoid. Math can be tricky, so we'll show you how to avoid common mistakes. You'll also find many different kinds of practice questions for a single skill. For example, after showing that the area of a square is equal to the length of one side squared, you'll find questions that ask you to find the area of a square given the length of one side, and you'll also find questions that ask you to find the length of a side given the area, or to find the area given the perimeter. By asking questions in different ways and by combining different skills, this book will get you in prime mathematical shape to handle all kinds of exams.

You can read any chapter on its own for a review of one particular subject, but some chapters do build on others. If you are having difficulty solving algebraic equations (covered in Chapter 5), be sure to read about algebraic expressions in Chapter 4 first. Adding and subtracting like terms, as well as multiplying and dividing terms, are covered there, and reviewing this material might make solving algebraic equations easier.

After covering algebra, we'll move on to geometry. Some geometry problems might feel like algebra problems, because you are looking for a missing side of a triangle, or you need to manipulate a formula to find a missing value. This is why the algebra chapters come first. You may have an easier time handling certain geometry questions after reviewing algebra.

Keep plenty of scrap paper and a calculator on hand. Some questions involve a lot of work, and some questions might be easier to solve if you draw a picture. A calculator can make converting fractions to decimals or finding the square root of a large number easier.

Everyone finds some area of math to be difficult, so don't hesitate to reread a chapter. Sometimes one chapter might make more sense after reading other chapters, or the beginning of a chapter might click for you after reading the entire chapter. Algebraic expressions and equations are reviewed in Chapters 4 and 5 respectively, but if you have trouble working with exponents and radicals, we delve deeper into those topics in Chapter 6. After reading Chapter 6, the problems in Chapters 4 and 5 might be easier to solve.

This book is a guide to math that is taught in middle school, but this same math is used in high school and college. Algebra may be brand-new to you, but someone else might pick up this book to study for the SAT, GRE, or GMAT. If you hang on to this book, you may find yourself rereading it 10 years from now!

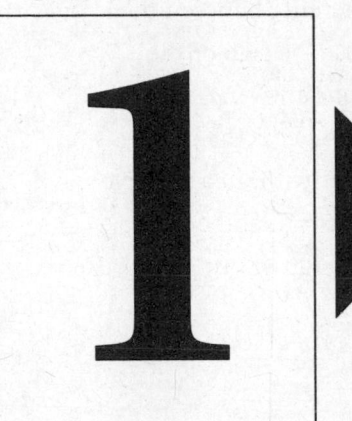

Working with Whole Numbers and Integers

THE SET OF **whole numbers** consists of all of the numbers 0 and greater that have no fractional part. This set is written as {0, 1, 2, 3, . . .}. These are the numbers we use every day.

The set of **integers** is made up of the whole numbers, their negatives, and zero. This set is written as {. . ., −3, −2, −1, 0, 1, 2, 3, . . .}. In this chapter, we review the four basic operations (addition, subtraction, multiplication, and division) with integers.

▶ Tips for Working with Integers

Addition

Signed numbers the same? Find the *sum* and use the same sign. Signed numbers different? Find the *difference* and use the sign of the larger number. (The larger number is the one whose value without a positive or negative sign is greater.)

Addition is commutative. That is, you can add numbers in any order and the result will be the same. As an example, **3 + 5 = 5 + 3**, or **–2 + –1 = –1 + –2**.

Subtraction

Change the operation sign to addition, change the sign of the number following the operation sign, and then follow the rules for addition.

Multiplication/Division

Signs the same? Multiply or divide and give the result a *positive* sign. Signs different? Multiply or divide and give the result a *negative* sign.

Multiplication is commutative. You can multiply terms in any order and the result will be the same. For example: **(2 × 5 × 7) = (2 × 7 × 5) = (5 × 2 × 7) = (5 × 7 × 2)** and so on.

Example

6 + 5 =

The positive numbers 6 and 5 can be added to each other without any worry about signs: 6 + 5 = 11.

Example

6 + –5 =

When the signs of two integers are different, we find the difference between the two numbers and use the sign of the larger number for our answer. As stated earlier, the larger number is the one whose value is greater without a positive or negative sign. 6 – 5 = 1, and because 6 is greater than 5, the sign of our answer is positive: 6 + –5 = 1.

Example

–6 + 5 =

Again, because the signs of the two integers are different, find the difference: 6 – 5 = 1. 6 is greater than 5, and the sign of 6 in this example is negative, so the sign of our answer is negative: –6 + 5 = –1.

Example

–6 – 5 =

The signs of these two integers are the same, so we add and keep the sign of the two integers: 6 + 5 = 11, so –6 – 5 = –11.

Example

6 – –5 =

Check the tips. To subtract an integer from another integer, change the operation to addition and change the sign of the number following the operation. 6 – –5 becomes 6 + 5; 6 + 5 = 11, so 6 – –5 = 11.

TIP

Here's another way to look at adding and subtracting signed numbers: When you have two signs in a row, combine the two signs into one. If the signs are the same, combine them into the addition symbol (+). If the signs are different, combine them into the subtraction symbol (–).

$6 + +5 = 6 + 5 = 11$

$6 + -5 = 6 - 5 = 1$

$6 - +5 = 6 - 5 = 1$

$6 - -5 = 6 + 5 = 11$

Basic Operations and Their Parts

The two numbers in an addition problem that are being added are called the **addends**. The result of an addition problem is called the **sum**. In the problem $2 + 4 = 6$, 2 and 4 are addends and 6 is their sum.

In a subtraction problem, the number being subtracted is called the **subtrahend**, the number from which we are subtracting is called the **minuend**, and the result of a subtraction problem is called the **difference**. In the problem $8 - 1 = 7$, 8 is the minuend, 1 is the subtrahend, and 7 is the difference.

The two numbers in a multiplication problem that are being multiplied are called the **factors**. The result of a multiplication problem is called the **product**. In the problem $7 \times 3 = 21$, 7 and 3 are the factors and 21 is their product.

In a division problem, the number being divided is called the **dividend**, the number that we are dividing by is called the **divisor**, and the result of a division problem is called the **quotient**. In the problem $35 \div 5 = 7$, 35 is the dividend, 5 is the divisor, and 7 is the quotient.

Example

$6 \times 5 =$

To multiply two numbers of the same sign, we multiply as we usually would and make the sign of our answer positive: $6 \times 5 = 30$.

Example

$-6 \times -5 =$

These two integers also have the same sign as each other, so we multiply and make the sign of our answer positive: $-6 \times -5 = 30$.

Example

$-6 \times 5 =$

$6 \times -5 =$

To multiply two numbers of different signs, we multiply as we usually would and make the sign of our answer negative: $-6 \times 5 = -30$, and $6 \times -5 = -30$.

Example

$24 \div 3 =$

Just as with multiplication, if the signs of both integers in a division problem are the same, we make the sign of our answer positive: $24 \div 3 = 8$.

Example

$24 \div -3 =$

Because the signs of the two integers in this division problem are different, we make the sign of our answer negative: $24 \div -3 = -8$.

Sometimes a problem will contain more than one operation. These operations must be performed in a certain order:

P Do operations inside *parentheses* (or other delimiters, such as brackets).

E Evaluate terms with *exponents*.

M D Do *multiplication* and *division* in order from left to right.

A S *Add* and *subtract* terms in order from left to right.

The acronym PEMDAS can help you remember the order of operations.

Example

$4 + 20 \div 4 =$

Division comes before addition in the order of operations, so we begin by dividing: $20 \div 4 = 5$. The problem is now $4 + 5$: $4 + 5 = 9$.

Example

$(4 + 20) \div 4 =$

This problem looks similar, but there are now parentheses around $4 + 20$. Because parentheses come before division in the order of operations, we add first: $4 + 20 = 24$. The problem is now $24 \div 4$: $24 \div 4 = 6$.

▶ **Practice Questions**

Evaluate the following expressions.

1. $27 + -5$ 22

2. $-18 + -20 - 16$ -54

3. $-15 - -7$ -8

4. $33 + -16$ 17

5. $8 + -4 - 12$ -8

6. $38 \div -2 + 9$ -10

7. $-25 \times -3 + 15 \times -5$ 0

8. $-5 \times -9 \times -2$ -90

9. $24 \times -8 + 2$ -190

10. $2 \times -3 \times -7$ 42

11. $-15 + 5 + -11$ -21

12. $(49 \div 7) - (48 \div -4)$ 19

13. $3 + -7 - 14 + 5$ -13

14. $-(5 \times 3) + (12 \div -4)$ -18

TIP

Addition: Words such as *total*, *sum*, *altogether*, *increase*, and *combine* can signal addition.

Subtraction: Words such as *difference*, *less than*, *more than*, and *decrease* can signal subtraction.

Multiplication: Words such as *product*, *times*, and *each* can signal multiplication.

Division: Words such as *quotient*, *share*, and *each* can signal division.

15. $(-18 \div 2) - (6 \times -3) = -9$

16. $23 + (64 \div -16) = 19$

17. $2^3 - (-4)^2 = -8$

18. $(3 - 5)^3 + (18 \div 6)^2 = 1$

19. $21 + (11 + -8)^3 = 48$

20. $(3^2 + 6) \div (-24 \div 8) = -5$

Now let's try some word problems. When working with a word problem, find the key words that signal which operation to use.

Example

Cindy has $56 and Nancy has $12. How much money do they have altogether?

The key word *altogether* signals addition: $56 + $12 = $68.

Example

Sherry has 400 safety pins. If she makes ten equal piles of safety pins, how many pins are in each pile?

Each can signal multiplication or division. In this problem, Sherry has divided her total number of safety pins into ten equal piles, so division is the operation we must use: 400 ÷ 10 = 40.

▶ **Practice Questions**

21. A scuba diver descends 80 feet, rises 25 feet, descends 12 feet, and then rises 52 feet where he will do a safety stop for five minutes before surfacing. At what depth did he do his safety stop?

22. A digital thermometer records the daily high and low temperatures. The high for the day was +5° C. The low was −11° C. What was the difference between the day's high and low temperatures?

23. A checkbook balance sheet shows an initial balance for the month of $300. During the month, checks were written in the amounts of $25, $82, $213, and $97. Deposits were made into the account in the amounts of $84 and $116. What was the balance at the end of the month?

24. A gambler begins playing a slot machine with $10 in quarters in her coin bucket. She plays 15 quarters before winning a jackpot of 50 quarters. She then plays 20 more quarters in the same machine before walking away. How many quarters does she now have in her coin bucket?

25. A glider is towed to an altitude of 2,000 feet above the ground before being released by the tow plane. The glider loses 450 feet of altitude before finding an updraft that lifts it 1,750 feet. What is the glider's altitude now?

26. Bonnie has twice as many cousins as Robert. George has five cousins, which is nine less than Bonnie has. How many cousins does Robert have?
   a. 17
   b. 22
   c. 4
   d. 7

27. Oscar sold two glasses of milk for every five sodas he sold. If he sold ten glasses of milk, how many sodas did he sell?
   a. 45
   b. 20
   c. 25
   d. 10

28. Justin earned scores of 85%, 92%, and 95% on his science tests. What does he need to earn on his next science test to have an average (arithmetic mean) of 93%?
  a. 93%
  b. 100%
  c. 85%
  d. 96%

29. Brad's class collected 270 cans of food. They boxed them in boxes of 30 cans each. How many boxes did they need?
  a. 240
  b. 10
  c. 9
  d. 5

30. Joey participated in a danceathon. His team started dancing at 10 A.M. on Friday and stopped at 6 P.M. on Saturday. How many hours did Joey's team dance?
  a. 52
  b. 56
  c. 30
  d. 32

31. Which expression has an answer of 18?
  a. $2 \times 5 + 4$
  b. $2 \times (4 + 5)$
  c. $5 \times (2 + 4)$
  d. $4 \times 2 + 5$

32. What is the square root of 64?
  a. 16
  b. 128
  c. 32
  d. 8

33. What is the prime factorization of 84?
  a. 42×2
  b. $2 \times 2 \times 4 \times 6$
  c. $2 \times 7 \times 6$
  d. $2 \times 2 \times 3 \times 7$

34. What is 2^5?
  a. 10
  b. 64
  c. 32
  d. 16

35. The low temperature in Anchorage, Alaska, today was −8° F. The low temperature in Los Angeles, California, was 63° F. What is the difference in the two low temperatures?
  a. 55°
  b. 71°
  c. 61°
  d. 14°

36. The Robin's Nest Nursing Home had a fund-raising goal of $9,500. By the end of the fund-raiser, they had exceeded their goal by $2,100. How much did they raise?
  a. $7,400
  b. $13,600
  c. $10,600
  d. $11,600

37. Mount Everest is 29,028 feet high. Mount Kilimanjaro is 19,340 feet high. How much taller is Mount Everest?
  a. 9,688 feet
  b. 10,328 feet
  c. 11,347 feet
  d. 6,288 feet

38. Mrs. Farrell's class has 26 students. Only 21 were present on Monday. How many were absent?
  a. 15
  b. 5
  c. 4
  d. 16

39. Lucy's youth group raised $1,569 for charity. They decided to split the money evenly among three charities. How much will each charity receive?
  a. $784.50
  b. $423.00
  c. $523.00
  d. $341.00

40. Jason made nine two-point baskets and three three-point baskets in Friday's basketball game. He did not score any other points. How many points did he score?
  a. 21
  b. 12
  c. 24
  d. 27

41. Susan traveled 114 miles in two hours. If she keeps going at the same rate, how long will it take her to go the remaining 285 miles of her trip?
  a. 5 hours
  b. 3 hours
  c. 7 hours
  d. 4 hours

42. A flight from Pittsburgh to Los Angeles took five hours and covered 3,060 miles. What was the plane's average speed?
  a. 545 mph
  b. 615 mph
  c. 515 mph
  d. 612 mph

43. Larry purchased three pairs of pants for $18 each and five shirts for $24 each. How much did Larry spend?
  a. $42
  b. $54
  c. $174
  d. $186

44. The temperature at 6 P.M. was 31° F. By midnight, it had dropped 40° F. What was the temperature at midnight?
  a. 9° F
  b. −9° F
  c. −11° F
  d. 0° F

45. Jody's English quiz scores are 56%, 93%, 72%, 89%, and 87%. What is the median of her scores?
  a. 72%
  b. 87%
  c. 56%
  d. 85.6%

46. What is the greatest common factor of 24 and 64?
  a. 8
  b. 4
  c. 12
  d. 36

47. Twelve coworkers go out for lunch together and order three pizzas. Each pizza is cut into eight slices. If each person gets the same number of slices, how many slices will each person get?

   a. 4

   b. 3

   c. 5

   d. 2

48. The value of a computer is depreciated evenly over five years for tax purposes (meaning that every year the computer is worth less by the same amount and that it is worth $0 after five years). If a business paid $2,400 for a computer, how much will it have depreciated after two years?

   a. $480

   b. $1,200

   c. $820

   d. $960

49. Steve earned a 96% on his first math test, a 74% on his second test, and an 85% on his third test. What is his test average?

   a. 91%

   b. 85%

   c. 87%

   d. 82%

50. A national park keeps track of how many people per car enter the park. Today, 57 cars had four people, 61 cars had two people, 9 cars had one person, and 5 cars had five people. What is the average number of people per car? Round to the nearest whole number.

   a. 2

   b. 3

   c. 4

   d. 5

51. A large pipe dispenses 750 gallons of water in 50 seconds. At this rate, how long will it take to dispense 330 gallons?

   a. 14 seconds

   b. 33 seconds

   c. 22 seconds

   d. 27 seconds

52. The light on a lighthouse blinks 45 times a minute. How long will it take the light to blink 405 times?

   a. 11 minutes

   b. 4 minutes

   c. 9 minutes

   d. 6 minutes

53. Wendy has five pairs of pants and seven shirts. How many different outfits can she make with these items (each outfit consists of one pair of pants and one shirt)?

   a. 12

   b. 24

   c. 35

   d. 21

54. The operator of an amusement park game kept track of how many tries it took participants to win the game. Here is the data from the first ten people:

2, 6, 3, 4, 6, 2, 8, 4, 3, 5

What is the median number of tries it took these participants to win the game?

   a. 8

   b. 6

   c. 4

   d. 2

55. Danny is a contestant on a TV game show. If he gets a question right, the points for that question are added to his score. If he gets a question wrong, the points for that question are subtracted from his score. Danny currently has 200 points. If he gets a 300-point question wrong, what will his score be?
   a. −100
   b. 0
   c. −200
   d. 100

56. The Ravens played 25 home games this year. They had 11 losses and 2 ties. How many games did they win?
   a. 12
   b. 13
   c. 14
   d. 11

57. The temperature at midnight was 4° F. By 2 A.M., it had dropped 9° F. What was the temperature at 2 A.M.?
   a. 13° F
   b. −5° F
   c. −4° F
   d. 0° F

58. Find the next number in the following pattern.
   320, 160, 80, 40, . . .
   a. 35
   b. 30
   c. 10
   d. 20

59. Which expression is equal to 5?
   a. $(1 + 2)^2$
   b. $9 - 2^2$
   c. $11 - 10 \times 5$
   d. $45 \div 3 \times 3$

60. Each week Jaime saves $25. How long will it take her to save $350?
   a. 12 weeks
   b. 14 weeks
   c. 16 weeks
   d. 18 weeks

61. Ashley's car insurance costs her $115 per month. How much does it cost her per year?
   a. $1,150
   b. $1,380
   c. $980
   d. $1,055

TERMS TO REVIEW

whole numbers

integers

addends

sum

subtrahend

minuend

difference

factors

product

dividend

divisor

quotient

▶ Answers

1. The signs of the terms are different, so find the difference of the values: $27 - 5 = 22$. The sign of the larger term is positive, so the sign of the result is positive: $27 + -5 = +22$.

2. Change the subtraction sign to addition by changing the sign of the number that follows it: $-18 + -20 + (-16)$. Because all the signs are negative, add the absolute value of the numbers: $18 + 20 + 16 = 54$. Because the signs were negative, the result is negative: $-18 + -20 + -16 = -54$. The simplified result of the numeric expression is: $-18 - 20 - 16 = -54$.

3. Change the subtraction sign to addition by changing the sign of the number that follows it: $-15 + 7$. Signs different? Subtract the absolute value of the numbers: $15 - 7 = 8$. Give the result the sign of the larger term: $-15 + 7 = -8$. The simplified expression is: $-15 - 7 = -8$.

4. Signs different? Subtract the value of the numbers: $33 - 16 = 17$. Give the result the sign of the larger term: $33 + -16 = +17$.

5. Change the subtraction sign to addition by changing the sign of the number that follows it: $8 + -4 + -12$. With three terms, first group like terms and add: $8 + (-4 + -12)$. Signs the same? Add the value of the terms and give the result the same sign: $(-4 + -12) = -16$. Substitute the result into the first expression: $8 + (-16)$. Signs different? Subtract the value of the numbers: $16 - 8 = 8$. Give the result the sign of the larger term: $8 + (-16) = -8$. The simplified result of the numeric expression is: $8 + -4 - 12 = -8$.

6. First divide. Signs different? Divide and give the result the negative sign: $(38 \div -2) = -19$. Substitute the result into the expression: $(-19) + 9$. Signs different? Subtract the value of the numbers: $19 - 9 = 10$. Give the result the sign of the term with the larger value: $(-19) + 9 = -10$. The simplified result of the numeric expression is: $38 \div -2 + 9 = -10$.

7. First perform the multiplications. Signs the same? Multiply the terms and give the result a positive sign: $-25 \times -3 = +75$. Signs different? Multiply the terms and give the result a negative sign: $15 \times -5 = -75$. Now substitute the results into the original expression: $(+75) + (-75)$. Signs different? Subtract the value of the numbers: $75 - 75 = 0$. The simplified result of the numeric expression is as follows: $-25 \times -3 + 15 \times -5 = 0$.

8. Because all the operators are multiplication, you could group any two terms and the result would be the same. Let's group the first two terms: $(-5 \times -9) \times -2$. Signs the same? Multiply the terms and give the result a positive sign: $5 \times 9 = +45$. Now substitute the result into the original expression: $+45 \times -2$. Signs different? Multiply the terms and give the result a negative sign: $+45 \times -2 = -90$. The simplified result of the numeric expression is: $-5 \times -9 \times -2 = -90$.

9. Group the terms being multiplied and evaluate: $(24 \times -8) + 2$. Signs different? Multiply the terms and give the result a negative sign: $24 \times -8 = -192$. Substitute: $(-192) + 2$. Signs different? Subtract the value of the terms: $192 - 2 = 190$. Give the result the sign of the term with the larger value: $(-192) + 2 = -190$. The simplified result of the numeric expression is: $24 \times -8 + 2 = -190$.

10. Because all the operators are multiplication, you could group any two terms and the result would be the same. Let's group the last two terms: $2 \times (-3 \times -7)$. Signs the same? Multiply the terms and give the result a positive sign: $(-3 \times -7) = +21$. Substitute: $2 \times (+21)$. Signs the same? Multiply the terms and give the result a positive sign: $2 \times (+21) = +42$. The simplified result of the numeric expression is: $2 \times -3 \times -7 = +42$.

11. Because all the operators are addition, you could group any two terms and the result would be the same. Or you could just work from left to right: $(-15 + 5) + -11$. Signs different? Subtract the value of the numbers: $15 - 5 = 10$. Give the result the sign of the term with the larger value: $(-15 + 5) = -10$. Substitute: $(-10) + -11$. Signs the same? Add the value of the terms and give the result the same sign: $10 + 11 = 21$; $(-10) + -11 = -21$. The simplified result of the numeric expression is: $-15 + 5 + -11 = -21$.

12. First evaluate the expressions within the parentheses: $49 \div 7 = 7$. Signs different? Divide and give the result a negative sign: $48 \div -4 = -12$. Substitute into the original expression: $(7) - (-12)$. Change the subtraction sign to addition by changing the sign of the number that follows it: $7 + +12$. Signs the same? Add the value of the terms and give the result the same sign: $7 + +12 = +19$. The simplified result of the numeric expression is: $(49 \div 7) - (48 \div -4) = +19$.

13. Change the subtraction sign to addition by changing the sign of the number that follows it: $3 + -7 + -14 + 5$. Now perform additions from left to right: $(3 + -7) + -14 + 5$. Signs different? Subtract the value of the numbers and give the result the sign of the higher value number: $7 - 3 = 4$; $3 + -7 = -4$. Substitute: $(-4) + -14 + 5$. Add from left to right: $(-4 + -14) + 5$. Signs the same? Add the value of the terms and give the result the same sign: $-4 + -14 = -18$. Substitute: $(-18) + 5$. Signs different? Subtract the value of the numbers and give the result the sign of the higher-value number: $18 - 5 = 13$; $(-18) + 5 = -13$. The simplified result of the numeric expression is: $3 + -7 + -14 + 5 = -13$.

14. First evaluate the expressions within the parentheses: $5 \times 3 = 15$. Signs different? Divide and give the result a negative sign: $12 \div -4 = -3$. Substitute the values into the original expression: $-(15) + (-3)$. Signs the same? Add the value of the terms and give the result the same sign: $15 + 3 = 18$; $-(15) + (-3) = -18$. The simplified result of the numeric expression is: $-(5 \times 3) + (12 \div -4) = -18$.

15. First evaluate the expressions within the parentheses: $(-18 \div 2)$. Signs different? Divide the value of the terms and give the result a negative sign: $18 \div 2 = 9$; $(-18 \div 2 = -9)$. Signs different? Multiply the term values and give the result a negative sign: (6×-3); $6 \times 3 = 18$; $(6 \times -3) = -18$. Substitute the values into the original expression: $(-9) - (-18)$. Change subtraction to addition and change the sign of the term that follows: $(-9) + (+18)$. Signs different? Subtract the value of the numbers and give the result the sign of the higher-value number: $18 - 9 = 9$; $(-9) + (+18) = +9$. The simplified result of the numeric expression is: $(-18 \div 2) - (6 \times -3) = +9$.

16. Evaluate the expressions within the parentheses: $(64 \div -16)$. Signs different? Divide and give the result a negative sign: $64 \div 16 = 4$; $(64 \div -16 = -4)$. Substitute the value into the original expression: $23 + (-4)$. Signs different? Subtract the value of the numbers and give the result the sign of the higher-value number: $23 - 4 = 19$; $23 + (-4) = +19$. The simplified result of the numeric expression is: $23 + (64 \div -16) = +19$.

17. The order of operations tells us to evaluate the terms with exponents first: $2^3 = 2 \times 2 \times 2 = 8$; $(-4)^2 = (-4) \times (-4)$. Signs the same? Multiply the terms and give the result a positive sign: $4 \times 4 = 16$; $(-4)^2 = +16$. Substitute the values of terms with exponents into the original expression: $2^3 - (-4)^2 = (8) - (+16)$. Change subtraction to addition and change the sign of the term that follows: $8 + -16$. Signs different? Subtract the value of the numbers and give the result the sign of the higher-value number: $16 - 8 = 8$; $8 + -16 = -8$. The simplified result of the numeric expression is: $2^3 - (-4)^2 = -8$.

18. First evaluate the expressions within the parentheses: $3 - 5$. Change subtraction to addition and change the sign of the term that follows: $3 + (-5)$. Signs different? Subtract the value of the numbers and give the result the sign of the higher-value number: $5 - 3 = 2$; $3 - 5 = -2$; $18 \div 6 = 3$. Substitute the values of the expressions in parentheses into the original expression: $(-2)^3 + (3)^2$. Evaluate the terms with exponents: $(-2)^3 = -2 \times -2 \times -2$; $(-2 \times -2) \times -2 = (+4) \times -2$. Signs different? Multiply the value of the terms and give the result a negative sign: $(+4) \times -2 = -8$; $(3)^2 = 3 \times 3 = 9$. Substitute the values into the expression: $(-2)^3 + (3)^2 = -8 + 9$. Signs different? Subtract the value of the numbers and give the result the sign of the higher-value number: $9 - 8 = +1$. The simplified result of the numeric expression is: $(3 - 5)^3 + (18 \div 6)^2 = +1$.

19. First evaluate the expression within the parentheses: $11 + -8$. Signs different? Subtract the value of the numbers and give the result the sign of the higher-value number: $11 - 8 = 3$; $11 + -8 = +3$. Substitute the value into the expression: $21 + (+3)^3$. Evaluate the term with the exponent: $(+3)^3 = 3^3 = 27$. Substitute the value into the expression: $21 + (27) = 48$. The simplified result of the numeric expression is: $21 + (11 + -8)^3 = 48$.

20. First evaluate the expressions within the parentheses: $3^2 + 6 = (9) + 6 = 15$. Signs different? Divide and give the result the negative sign: $-24 \div 8 = -3$. Substitute values into the original expression: $(15) \div (-3)$. Signs different? Divide the value of the terms and give the result a negative sign: $15 \div 3 = 5$; $(15) \div (-3) = -5$. The simplified result of the numeric expression is: $(3^2 + 6) \div (-24 \div 8) = -5$.

21. If you think of distance above sea level as a positive number, then you must think of going below sea level as a negative number. Going up is in the positive direction, while going down is in the negative direction. Give all the descending distances a negative sign and the ascending distances a positive sign. The resulting numerical expression would be: $-80 + +25 + -12 + +52$. Because addition is commutative, you can associate like-signed numbers: $(-80 + -12) + (+25 + +52)$. Evaluate the numerical expression in each set of parentheses: $-80 + -12 = -92$; $+25 + +52 = +77$. Substitute the values into the numerical expression: $(-92) + (+77)$. Signs different? Subtract the value of the numbers and give the result the sign of the higher-value number: $92 - 77 = 15$. The diver took his rest stop at -15 feet.

22. You could simply figure that $+5°$ C is 5° above zero and $-11°$ C is 11° below. So the difference is the total of $5° + 11° = 16°$. Or you could find the difference between $+5°$ and $-11°$. That would be represented by the following equation: $+5° - -11° = +5° + +11° = +16°$.

23. You can consider that balances and deposits are positive signed numbers, while checks are deductions, represented by negative signed numbers. An expression to represent the activity during the month would be: $300 + -25 + -82 + -213 + -97 + +84 + +116$. Because addition is commutative, you can associate like-signed numbers: $(300 + +84 + +116) + (-25 + -82 + -213 + -97)$. Evaluate the numbers within each set of parentheses: $300 + +84 + +116 = +500$; $-25 + -82 + -213 + -97 = -417$. Substitute the values into the revised expression: $(+500) + (-417) = +83$. The balance at the end of the month would be $83.

24. You first figure out how many quarters she starts with. Four quarters per dollar gives you $4 \times 10 = 40$ quarters. You can write an expression that represents the quarters in the bucket and the quarters added and subtracted. In chronological order, the expression would be: $40 - 15 + 50 - 20$. Change all operation signs to addition and the sign of the number that follows: $40 + -15 + 50 + -20$. Because addition is commutative, you can associate like-signed numbers: $(40 + 50) + (-15 + -20)$. Use the rules for adding integers with like signs: $40 + 50 = 90$; $-15 + -20 = -35$. Substitute into the revised expression: $(90) + (-35)$. Signs different? Subtract the value of the numbers and give the result the sign of the higher-value number: $90 - 35 = 55$. The simplified result of the numeric expression is: $40 + -15 + 50 + -20 = 55$.

25. As in problem 21, ascending is a positive number while descending is a negative number. You can assume ground level is the zero point. An expression that represents the problem is: $+2,000 + -450 + +1,750$. Because addition is commutative, you can associate like-signed numbers: $(+2,000 + +1,750) + -450$. Evaluate the expression in the parentheses: $+2,000 + +1,750 = +3,750$. Substitute into the revised equation: $(+3,750) + -450$. Signs different? Subtract the value of the numbers and give the result the sign of the higher-value number: $3,750 - 450 = 3,300$. The simplified result of the numeric expression is: $(+3,750) + -450 = +3,300$.

26. **d.** Work backwards to find the solution. George has five cousins, which is nine less than Bonnie has; therefore, Bonnie has 14 cousins. Bonnie has twice as many as Robert has, so half of 14 is 7. Robert has seven cousins.

27. **c.** Set up a proportion with $\frac{\text{milk}}{\text{soda}} \cdot \frac{2}{5} = \frac{10}{x}$. Cross multiply and solve: $(5)(10) = 2x$. Divide both sides by 2: $\frac{50}{2} = \frac{2x}{2}$; $x = 25$ sodas.

28. **b.** To earn an average of 93% on four tests, the sum of those four tests must be $(93)(4)$ or 372. The sum of the first three tests is $85 + 92 + 95 = 272$. The difference between the needed sum of four tests and the sum of the first three tests is 100. He needs a 100 to earn a 93% average.

29. **c.** To find the number of boxes needed, you should divide the number of cans by 30: $270 \div 30 = 9$ boxes.

30. **d.** From 10 A.M. Friday to 10 A.M. Saturday is 24 hours. Then, from 10 A.M. Saturday to 6 P.M. Saturday is another 8 hours. Together, that makes 32 hours.

31. **b.** Use the order of operations and try each option. The first option results in 14 because $2 \times 5 = 10$, then $10 + 4 = 14$. This does not work. The second option does result in 18. The numbers in parentheses are added first and result in 9, which is then multiplied by 2 to get a final answer of 18. Choice **c** does not work because the operation in parentheses is done first, yielding 6, which is then multiplied by 5 to get a result of 30. Choice **d** does not work because the multiplication is done first, yielding 8, which is added to 5 for a final answer of 13.

32. **d.** To find the square root ($\sqrt{}$), you ask yourself, "What number multiplied by itself gives me 64?" $8 \times 8 = 64$; therefore, 8 is the square root of 64.

33. **d.** This is the only answer choice that has only *prime* numbers. A prime number is a number with two and only two distinct factors. In choice **a**, 42 is not prime. In choice **b**, 4 and 6 are not prime. In choice **c**, 6 is not prime.

34. **c.** $2^5 = 2 \times 2 \times 2 \times 2 \times 2 = 32$.

35. **b.** Visualize a number line. The distance from -8 to 0 is 8. Then, the distance from 0 to 63 is 63. Add the two distances together to get 71; $63 + 8 = 71$.

36. **d.** *Exceeded* means "gone above." Therefore, if they exceeded their goal of \$9,500 by \$2,100, they went over their goal by \$2,100; $\$9,500 + \$2,100 = \$11,600$. If you chose **a**, you subtracted \$2,100 from \$9,500 instead of adding the two numbers.

37. **a.** Subtract Mount Kilimanjaro's height from Mount Everest's height: $29,028 - 19,340 = 9,688$. If you chose **b**, you did not borrow correctly when subtracting.

38. **b.** Subtract the number of students present from the total number in the class to determine how many students are missing: $26 - 21 = 5$.

39. **c.** Divide the money raised by 3 to find the amount each charity will receive: $\$1,569 \div 3 = \$523$.

40. **d.** Find the number of points scored on two-point baskets by multiplying 2×9; 18 points were scored on two-point baskets. Find the number of points scored on three-point baskets by multiplying 3×3; 9 points were scored on three-point baskets. The total number of points is the sum of these two totals: $18 + 9 = 27$.

41. **a.** Find the rate at which Susan is traveling by dividing her distance by time: $114 \div 2 = 57$ mph. To find out how long it will take her to travel 285 miles, divide her distance by her rate: $285 \div 57 = 5$ hours.

42. **d.** Divide the miles by the time to find the rate: $3,060 \div 5 = 612$ mph.

43. **c.** He spent $54 on pants ($3 \times \$18 = \$54$) and $120 on shirts ($5 \times \$24 = \$120$). Altogether he spent $174 ($\$54 + \$120 = \$174$). If you chose **a**, you calculated the cost of *one* pair of pants plus *one* shirt instead of *three* pairs of pants and *five* shirts.

44. **b.** Visualize a number line. The drop from 31° to 0° is 31°. There are still nine more degrees to drop. They will be below zero, so −9° F is the temperature at midnight.

45. **b.** To find the median, first put the numbers in order from least to greatest: 56, 72, 87, 89, 93. The middle number, 87, is the median. If you chose **a**, you forgot to put the numbers in order before finding the middle number.

46. **a.** List the factors of 24 and 64. The largest factor that they have in common is the greatest common factor.
Factors of 24: 1, 2, 3, 4, 6, 8, 12, 24
Factors of 64: 1, 2, 4, 8, 16, 32, 64
The largest number that appears in both lists is 8.

47. **d.** Find the total number of slices by multiplying 3 by 8 ($3 \times 8 = 24$). There are 24 slices to be shared among 12 coworkers. Divide the number of slices by the number of people to find the number of slices per person: $24 \div 12 = 2$ slices per person.

48. **d.** Find how much it depreciates over one year by dividing the cost by 5: $\$2,400 \div 5 = \$480$. Multiply this by 2 for two years: $\$480 \times 2 = \$960$. It will have depreciated $960.

49. **b.** Add the test grades ($96 + 74 + 85 = 255$) and divide the sum by the number of tests ($255 \div 3 = 85$). The average is 85%.

50. **b.** Find the total number of people and the total number of cars. Then, divide the total people by the total cars.

People:	$57 \times 4 =$	228
	$61 \times 2 =$	122
	$9 \times 1 =$	9
	$5 \times 5 =$	25
	Total	384 people

Cars: $57 + 61 + 9 + 5 = 132$
$384 \div 132 = 2.9$, which is rounded up to 3 people because 2.9 is closer to 3 than it is to 2.

51. **c.** Find the number of gallons per second by dividing 750 by 50 ($750 \div 50 = 15$ gallons per second). Divide 330 gallons by 15 to find how many seconds it will take ($330 \div 15 = 22$ seconds). It will take 22 seconds.

52. **c.** Divide 405 by 45 to get 9 minutes.

53. **c.** Multiply the number of choices for each item to find the number of outfits ($5 \times 7 = 35$). There are 35 combinations.

54. **c.** First, put the numbers in order from least to greatest, and then find the middle of the set.

2, 2, 3, 3, 4, 4, 5, 6, 6, 8

The middle (median) is the average (mean) of the 5th and 6th data items. The mean of 4 and 4 is 4, so the median is 4.

55. **a.** $200 - 300 = -100$ points.

56. **a.** Thirteen games are accounted for with the losses and ties ($11 + 2 = 13$). The remainder of the 25 games were won. Subtract to find the games won: $25 - 13 = 12$ games won.

57. **b.** If the temperature is only 4° and drops 9°, it goes below zero. It drops 4° to zero and another 5° to $-5°$ F.

```
   5°
   4°
   3°
   2°
   1°
   0°
  −1°
  −2°
  −3°
  −4°
  −5°
  −6°
```

58. **d.** Each number is divided by 2 to find the next number: $40 \div 2 = 20$, so 20 is the next number.

59. **b.** $9 - 2^2 = 9 - 4 = 5$. The correct order of operations must be used here. PEMDAS tells you that you should do the operations in the following order: **p**arentheses, **e**xponents, **m**ultiplication, **d**ivision, **a**ddition, and **s**ubtraction—all left to right.

a is $(1 + 2)^2 = (3)^2 = 9$;

c is $11 - 10 \times 5 = 11 - 50 = -39$;

d is $45 \div 3 \times 3 = 15 \times 3 = 45$.

60. **b.** Divide \$350 by \$25; $350 \div 25 = 14$ weeks.

61. **b.** Multiply \$115 by 12 because there are 12 months in a year; $\$115 \times 12 = \$1,380$ per year.

Conversions, Part I

RHODA SPENDS 80 MINUTES washing her clothes and 100 minutes drying them. How much time did she spend washing and drying? We could write our answer as 180 minutes (the sum of 80 and 100), or we could convert the minutes to hours and write our answer as three hours. In this chapter, we look at converting numbers from one form to another.

Use the following chart to help you convert from one measure of time to another.

Time Conversion

60 seconds = 1 minute
60 minutes = 1 hour
24 hours = 1 day
7 days = 1 week
365 days = 52 weeks = 1 year
10 years = 1 decade
100 years = 10 decades = 1 century
10 centuries = 1 millennium

To go from a small unit of time to a larger unit of time, use division. To go from a large unit of time to a smaller unit of time, use multiplication.

Example

120 hours = _____ days

Because days are a larger unit of measure, divide the total number of hours (120) by the number of hours in a day (24): 120 ÷ 24 = 5; 120 hours = 5 days.

Example

54 decades = _____ years

Because years are a smaller unit of measure, multiply the total number of decades (54) by the number of years in a decade (10): 54 × 10 = 540; 54 decades = 540 years.

▶ Practice Questions

62. It took Kaitlyn two hours to finish her homework. How many minutes did it take her to finish her homework?
   a. 90 minutes
   b. 60 minutes
   c. 100 minutes
   d. 120 minutes
   e. 150 minutes

63. Michael takes four minutes to shave each morning. How many seconds does Michael spend shaving each morning?
   a. 240 seconds
   b. 64 seconds
   c. 200 seconds
   d. 120 seconds
   e. 400 seconds

64. Bill worked for a steel manufacturer for three decades. How many years did Bill work for the steel manufacturer?
   a. 15 years
   b. 60 years
   c. 30 years
   d. 6 years
   e. 12 years

65. How many minutes are there in 12 hours?
   a. 24 minutes
   b. 1,440 minutes
   c. 36 minutes
   d. 1,200 minutes
   e. 720 minutes

66. Millie's ancestors first arrived in the United States exactly two centuries ago. How many years ago did Millie's ancestors arrive in the United States?
   a. 200 years
   b. 50 years
   c. 20 years
   d. 100 years
   e. 30 years

67. Rita is eight decades old. How many years old is Rita?
   a. 40 years old
   b. 16 years old
   c. 64 years old
   d. 48 years old
   e. 80 years old

68. Lucy spent 240 minutes working on her art project. How many hours did she spend on the project?
   a. 14,400 hours
   b. 24 hours
   c. 4 hours
   d. 2 hours
   e. 6 hours

69. How many days comprise two years (assume that neither year is a leap year)?
   a. 200 days
   b. 400 days
   c. 624 days
   d. 730 days
   e. 750 days

70. Jessica is 17 years old today. How many months old is she?
   a. 204 months
   b. 255 months
   c. 300 months
   d. 221 months
   e. 272 months

71. Lou just completed 48 months of service in the Army. How many years was Lou in the Army?
   a. 12 years
   b. 4 years
   c. 6 years
   d. 10 years
   e. 15 years

72. How many years are there in five centuries?
   a. 35 years
   b. 125 years
   c. 250 years
   d. 400 years
   e. 500 years

73. How many hours are there in one week?
   a. 168 hours
   b. 120 hours
   c. 84 hours
   d. 144 hours
   e. 96 hours

74. Marty has lived in his house for 156 weeks. How many years has he lived in his house?
   a. 2 years
   b. 5 years
   c. 3 years
   d. 11 years
   e. 9 years

75. How many hours are in five days?
   a. 60 hours
   b. 100 hours
   c. 120 hours
   d. 240 hours
   e. 600 hours

76. 2,520 seconds is equivalent to how many minutes?
   a. 84 minutes
   b. 42 minutes
   c. 151,200 minutes
   d. 126 minutes
   e. 252 minutes

77. A scientist has spent the past two decades studying a specific type of frog. How many years has the scientist spent studying this frog?
   a. 10 years
   b. 12 years
   c. 15 years
   d. 20 years
   e. 30 years

78. Josh slept 540 minutes last night. How many hours did he sleep?
   a. 9 hours
   b. 8 hours
   c. 7 hours
   d. 6 hours
   e. 5 hours

79. How many seconds are in two hours?
   a. 120 seconds
   b. 240 seconds
   c. 2,400 seconds
   d. 72,000 seconds
   e. 7,200 seconds

80. Betsy and Tim have been married 30 years. How many decades have they been married?
   a. 300 decades
   b. 3 decades
   c. 10 decades
   d. 6 decades
   e. 600 decades

81. Jacob is 7,200 minutes old. How many days old is Jacob?
   a. 120 days
   b. 5 days
   c. 24 days
   d. 36 days
   e. 2 days

82. How many decades are in five centuries?
   a. 500 decades
   b. 100 decades
   c. 40 decades
   d. 50 decades
   e. 25 decades

83. Today is Rosa's seventh birthday. How many months old is Rosa?
   a. 84 months
   b. 70 months
   c. 74 months
   d. 90 months
   e. 104 months

84. Kyra's biology class is 55 minutes long. How many seconds long is the class?
   a. 3,000 seconds
   b. 4,400 seconds
   c. 5,400 seconds
   d. 8,600 seconds
   e. 3,300 seconds

85. How many months are there in six decades?
   a. 60 months
   b. 720 months
   c. 48 months
   d. 660 months
   e. 960 months

86. Two hundred and forty months have passed since Guy started his current job. How many decades has Guy worked at his current job?

a. 20 decades

b. 4 decades

c. 2 decades

d. 10 decades

e. 12 decades

87. Peter is going on vacation in exactly five weeks. How many days until Peter goes on vacation?

a. 45 days

b. 30 days

c. 35 days

d. 40 days

e. 50 days

88. How many seconds are there in nine hours?

a. 32,400 seconds

b. 540 seconds

c. 60 seconds

d. 44,000 seconds

e. 960 seconds

89. Wendy took out a car loan for 60 months. The loan is for how many years?

a. 6 years

b. 5 years

c. 4 years

d. 3 years

e. 2 years

90. Lois spent a total of four hours working on her science project. How many minutes did Lois spend on the project?

a. 48 minutes

b. 96 minutes

c. 108 minutes

d. 220 minutes

e. 240 minutes

91. Two centuries have passed since Joan's house was built. How many months have passed since Joan's house was built?

a. 200 months

b. 1,200 months

c. 2,400 months

d. 8,000 months

e. 10,400 months

92. Sherman took his pulse for 10 seconds and counted 11 beats. What is Sherman's pulse rate in beats per minute?

a. 210 beats per minute

b. 110 beats per minute

c. 66 beats per minute

d. 84 beats per minute

Use the following chart to help convert from one unit of money to another.

Money Conversion

5 pennies = 1 nickel
10 pennies = 2 nickels = 1 dime
25 pennies = 5 nickels = 1 quarter
100 pennies = 20 nickels = 10 dimes = 1 dollar

Example

How many pennies are in five dimes?

Multiply the number of pennies in one dime, 10, by the number of dimes, 5: $10 \times 5 = 50$. There are 50 pennies in five dimes.

As with time conversions, sometimes we may need to use more than one step to convert from one unit of measure to another.

Example

How many nickels can Dylan receive in exchange for 15 quarters?

First, multiply the number of pennies in one quarter, 25, by the number of quarters, 15: $25 \times 15 = 375$ pennies. Since there are five pennies in one nickel, divide the number of pennies by 5: $375 \div 5 = 75$. Dylan receives 75 nickels for 15 quarters.

TIP

Many conversion problems can be solved more than one way. For instance, we could have found the number of nickels in a quarter by dividing the number of pennies in a quarter, 25, by 5 as there are five pennies in a nickel: $25 \div 5 = 5$. Then, we could have multiplied the number of nickels in a quarter by the number of quarters: $5 \times 15 = 75$. Either way, there are 75 nickels in 15 quarters.

▶ **Practice Questions**

93. Two quarters are equivalent to how many dimes?
   a. 2
   b. 4
   c. 5
   d. 6
   e. 10

94. How many quarters are there in $10?
   a. 80
   b. 40
   c. 10
   d. 100
   e. 250

95. How many pennies are there in a half dollar?
   a. 2
   b. 10
   c. 25
   d. 50
   e. 100

96. A $100 bill is equal to how many $5 bills?
   a. 20
   b. 100
   c. 40
   d. 5
   e. 10

97. Three dimes and four nickels are equal to how many pennies?
   a. 7
   b. 34
   c. 70
   d. 12
   e. 50

98. Marco needs quarters for the laundry machine. How many quarters will he receive if he puts a $5 bill into the change machine?
   a. 25
   b. 20
   c. 5
   d. 4
   e. 100

99. How many dimes are there in $20?
   a. 20
   b. 200
   c. 2,000
   d. 100
   e. 1,000

100. How many dimes are equal to six quarters?
   a. 16
   b. 60
   c. 10
   d. 12
   e. 15

101. Khai's meal cost $7. He gives the waitress a $20 bill and asks for the change in all $1 bills. How many $1 bills does he receive?
   a. 5
   b. 7
   c. 2
   d. 13
   e. 17

102. Sixteen half-dollar coins are equal to how many dollars?
   a. $8
   b. $50
   c. $16
   d. $80
   e. $100

103. How many nickels are there in three quarters?
   a. 5
   b. 25
   c. 15
   d. 6
   e. 50

104. Anna has 20 dimes. How many dollars is this?
   a. $1.00
   b. $2.00
   c. $5.00
   d. $10.00
   e. $20.00

105. The value of 100 quarters is equal to how many dollars?
   a. $2.50
   b. $10.00
   c. $25.00
   d. $1.00
   e. $50.00

106. Sandra used the following chart to keep track of her tips. How much money did she make in tips?

Currency	Amount of Each Currency
$1 bills	7
quarters	10
dimes	25
nickels	4

   a. $12.00
   b. $15.50
   c. $15.20
   d. $12.20
   e. $10.50

107. How many dimes are there in four half-dollars?
   a. 4
   b. 10
   c. 40
   d. 200
   e. 20

108. Jack has four $50 bills. This is equal to how many $1 bills?

a. 200

b. 4

c. 100

d. 2,000

e. 150

109. Five quarters is equal to how many nickels?

a. 5

b. 25

c. 50

d. 125

e. 20

110. One penny, one nickel, one dime, and one quarter are equal to how much money?

a. $0.28

b. $0.46

c. $0.41

d. $0.81

e. $1.01

111. How many quarters are there in $12?

a. 12

b. 120

c. 24

d. 84

e. 48

112. Sonya buys a lamp that cost $25. She hands the clerk a $100 bill and asks for her change to be only in $5 bills. How many $5 bills does she receive as change?

a. 20

b. 15

c. 50

d. 25

e. 5

113. How many pennies are there in the value of a $20 bill?

a. 200

b. 2,000

c. 20

d. 20,000

e. 1,000

114. A parking meter allows 15 minutes for every quarter. If Hung puts in $2 worth of quarters, how much time does he get on the meter?

a. 30 minutes

b. 45 minutes

c. 1 hour

d. 2 hours

e. 4 hours

115. How many nickels are there in five half-dollars?

a. 5

b. 100

c. 25

d. 50

e. 250

116. Luis won 50 nickels from the slot machine. How many dollars is this?

a. $0.50

b. $5

c. $25

d. $15

e. $2.50

117. How many quarters are needed to equal the value of 35 dimes?

a. 5

b. 10

c. 35

d. 12

e. 14

Given the conversion rate of a foreign currency, we can convert from U.S. dollars to a foreign currency, from a foreign currency to U.S. dollars, or from one foreign currency to another.

Example

If there are 1.5 dollars in one euro, how many dollars are in 14 euros?

Multiply the number of dollars in one euro, 1.5, by the number of euros, 14: $1.5 \times 14 = 21$. There are 21 dollars in 14 euros.

▶ **Practice Questions**

118. If $1 is equal to 11 pesos, how many dollars are equal to 143 pesos?
  a. $11
  b. $13
  c. $14
  d. $143
  e. $1,573

119. Twenty dollars is equal to how many pesos if $1 equals 11 pesos?
  a. 55
  b. 181
  c. 1.81
  d. 210
  e. 220

120. If a book costs 13 Canadian dollars, how many U.S. dollars is this? Assume that the current exchange rate is $1.30 Canadian for each U.S. dollar.
  a. $10.00
  b. $20.00
  c. $13.00
  d. $13.30
  e. $16.90

121. A British tourist needs to change 63 British pounds into U.S. dollars. If the exchange rate is 1 pound for every $1.80, how many dollars will he receive?
  a. $28.50
  b. $35.00
  c. $113.40
  d. $180.00
  e. $65.80

122. Thirty dollars is equal to how many yen if $1 equals 107 yen?
  a. 3.56
  b. 28
  c. 3,000
  d. 3,210
  e. 2,100

123. How many Canadian dollars are equal to 20 U.S. dollars if one U.S. dollar equals 1.3 Canadian dollars?
  a. 23
  b. 26
  c. 2.6
  d. 15.40
  e. 10.80

There are many other units of measure—units of weight, length, volume, and more—but first, we'll review how to work with fractions and decimals, since some of those conversions can involve numbers that aren't whole. We'll do some more conversions in Chapter 17.

▶ **Answers**

62. **d.** Multiply the number of minutes in an hour by the given number of hours. There are 60 minutes in an hour. Therefore, there are 120 minutes in two hours: 2 hours × 60 minutes = 120 minutes.

63. **a.** Multiply the number of seconds in a minute by the given number of minutes. There are 60 seconds in one minute. Therefore there are 240 seconds in four minutes: 4 minutes × 60 seconds = 240 seconds.

64. **c.** Multiply the number of years in a decade by the given number of decades. There are ten years in a decade, so three decades is 30 years: 3 decades × 10 years = 30 years.

65. **e.** Multiply the number of minutes in an hour by the given number of hours. There are 60 minutes in each hour. Therefore, there are 720 minutes in 12 hours: 12 hours × 60 minutes = 720 minutes.

66. **a.** Multiply the number of years in a century by the given number of centuries. There are 100 years in a century, so two centuries is 200 years: 2 centuries × 100 years = 200 years.

67. **e.** Multiply the number of years in a decade by the given number of decades. A decade is ten years. Eight decades is therefore 80 years: 8 decades × 10 years = 80 years.

68. **c.** Divide the total number of minutes by the number of minutes in an hour. There are 60 minutes in an hour. There are four hours in 240 minutes: 240 minutes ÷ 60 minutes = 4 hours.

69. **d.** Multiply the number of days in a year by the given number of years. A year is 365 days, so two years is 730 days: 365 days × 2 years = 730 days.

70. **a.** Multiply the number of months in a year by the given number of years. There are 12 months in a year, so there are 204 months in 17 years: 17 years × 12 months = 204 months.

71. **b.** Divide the total number of months by the number of months in a year. Each year is 12 months, so there are four years in 48 months: 48 months ÷ 12 months = 4 years.

72. **e.** Multiply the number of years in a century by the given number of centuries. There are 100 years in a century. Therefore, there are 500 years in five centuries: 5 centuries × 100 years = 500 years.

73. **a.** Multiply the number of days in one week by the number of hours in each day. First, there are seven days in a week. Next, there are 24 hours in each of those days. Therefore, there are 168 hours in a week: 7 days × 24 hours = 168 hours.

74. **c.** Divide the total number of weeks by the number of weeks in a year. Each year is made up of 52 weeks, so there are three years in 156 weeks: 156 weeks ÷ 52 weeks = 3 years.

75. **c.** Multiply the number of hours in a day by the given number of days. There are 24 hours in each day. There are 120 hours in five days: 5 days × 24 hours = 120 hours.

76. **b.** Divide the total number of seconds by the number of seconds in a minute. There are 60 seconds in a minute, so there are 2,520 seconds in 42 minutes: 2,520 seconds ÷ 60 seconds = 42 minutes.

77. **d.** Multiply the number of years in a decade by the given number of decades. A decade is ten years, so two decades is 20 years: 10 years × 2 decades = 20 years.

78. a. Divide the total number of minutes by the number of minutes in an hour. An hour is 60 minutes, so 540 minutes is nine hours: 540 minutes ÷ 60 minutes = 9 hours.

79. e. First, find the number of minutes in two hours. There are 60 minutes in one hour, so there are 120 minutes in two hours. Next, find the number of seconds in those 120 minutes. There are 60 seconds in a minute. Therefore, there are 7,200 seconds in 120 minutes: 120 minutes × 60 seconds = 7,200 seconds.

80. b. Divide the total number of years by the number of years in a decade. There are ten years in a decade; 30 years is three decades: 30 years ÷ 10 years = 3 decades.

81. b. You must change minutes to hours, then hours to days. First, there are 60 minutes in an hour; 7,200 minutes is 120 hours (7,200 minutes ÷ 60 minutes = 120 hours). Next, there are 24 hours in a day, so there are 120 hours in five days: 120 hours ÷ 24 hours = 5 days.

82. d. First, determine how many decades are in one century. There are 100 years in a century, and ten years in a decade. There are ten decades in a century (100 years ÷ 10 years = 10 decades). Therefore, there are 50 decades in five centuries: 5 centuries × 10 decades = 50 decades.

83. a. Multiply the number of months in a year by the given number of years. There are 12 months in a year. There are 84 months in 7 years: 7 years × 12 months = 84 months.

84. e. Multiply the number of seconds in a minute by the given number of minutes. A minute is 60 seconds, so 55 minutes is 3,300 seconds: 55 minutes × 60 seconds = 3,300 seconds.

85. b. First, determine how many years are in six decades. Since there are ten years in a decade, there are 60 years in six decades (6 decades × 10 years = 60 years). Next, multiply this number of years by the number of months in one year. Because there are 12 months in one year, there are 720 months in 60 years: 60 years × 12 months = 720 months.

86. c. You must convert the months to years and then the years to decades. First, divide the total number of months by the number of months in one year. There are 12 months in a year. Therefore, there are 20 years in 240 months (240 months ÷ 12 months = 20 years). Now, divide this total number of years by the number of years in one decade. There are ten years in a decade, so there are two decades in 20 years: 20 years ÷ 10 years = 2 decades.

87. c. Multiply the number of days in a week by the given number of weeks. There are seven days in a week, so there are 35 days in five weeks: 5 weeks × 7 days = 35 days.

88. a. First convert hours to minutes, and then change minutes to seconds. There are 60 minutes in an hour. Therefore, there are 540 minutes in nine hours (9 hours × 60 minutes = 540 minutes). Next, there are 60 seconds in a minute. Therefore, there are 32,400 seconds in 540 minutes: 540 minutes × 60 seconds = 32,400 seconds.

89. b. Divide the total number of months by the number of months in a year. There are 12 months in a year, so 60 months is five years: 60 months ÷ 12 months = 5 years.

90. e. Multiply the number of minutes in an hour by the given number of hours. There are 60 minutes in an hour, so four hours is 240 minutes: 4 hours × 60 minutes = 240 minutes.

91. c. You must convert centuries to years, then years to months. First, multiply the total number of centuries by the number of years in a century, so there are 100 years in a century. There are 200 years in two centuries (2 centuries × 100 years = 200 years). Now, multiply this total number of years by the number of months in a year. There are 12 months in a year, so there are 2,400 months in 200 years: 200 years × 12 months = 2,400 months.

92. c. A 10-second count is $\frac{1}{6}$ of a minute. To find the number of beats per minute, multiply the beat in 10 seconds by 6: 11 × 6 = 66 beats per minute.

93. c. Each quarter is worth $0.25, so two quarters are equal to $0.50 (2 × 0.25 = 0.50). Each dime is worth $0.10: $0.50 ÷ $0.10 = 5 dimes.

94. b. Each quarter is $0.25, so divide $10 by this amount to get the number of quarters: $10 ÷ $0.25 = 40 quarters.

95. d. A half-dollar is worth 50 cents. Each penny is a cent, so there are 50 pennies.

96. a. $100 ÷ $5 = 20. Therefore, 20 $5 bills are equal to $100.

97. e. Three dimes equal 30 cents (3 × 10 cents); four nickels equal 20 cents (4 × 5 cents). Add these two values together to get the total: 30 + 20 = 50 cents. Therefore, this is equal to 50 pennies.

98. b. A quarter is worth $0.25. There are four quarters in every dollar (4 × 0.25 = 1). Multiply 4 times the dollar amount to get the number of quarters: 4 × $5 = 20 quarters.

99. b. Each dime is worth $0.10, so there are ten dimes in every dollar (10 × 0.10 = 1). Multiply 10 by the dollar amount to get the number of dimes: 10 × $20 = 200 dimes.

100. e. Each quarter is worth $0.25, so six quarters equal $1.50 (6 × 0.25 = 1.50). Each dime is worth $0.10. Divide $1.50 by $0.10 to get the number of dimes: 1.50 ÷ 0.10 = 15 dimes.

101. d. The first step is to figure out how much change he should receive; $20 − $7 = $13. Since he wants all $1 bills, he will receive 13 bills.

102. a. A half-dollar is equal to $0.50: 16 × $0.50 = $8.00

103. c. Each quarter is worth 25 cents; 3 × 25 cents = 75 cents. Each nickel is worth 5 cents. Divide 75 cents by 5 cents to get the number of nickels: 75 ÷ 5 = 15 nickels.

104. b. Each dime is $0.10: 20 dimes × $0.10 = $2.00.

105. c. There are four quarters in every dollar: 100 quarters ÷ 4 quarters per dollar = $25.

106. d. First, calculate the dollar amount of each form of currency: $1 bills: $1 × 7 = $7. Quarters: $0.25 × 10 = $2.50. Dimes: $0.10 × 25 = $2.50. Nickels: $0.05 × 4 = $0.20. Next, add up all of the dollar values to get the total: $7 + $2.50 + $2.50 + $0.20 = $12.20.

107. e. Each half-dollar is worth $0.50; $0.50 × 4 = $2.00. Since each dime is worth $0.10, divide $2.00 by this value to get the number of dimes: $2.00 ÷ $0.10 = 20 dimes.

108. a. Multiply $50 by the number of bills: $50 × 4 = $200. Therefore, this is equal to 200 $1 bills.

109. b. A quarter is equal to $0.25: $0.25 × 5 = $1.25; $1.25 ÷ $0.05 (the value of a nickel) = 25 nickels.

110. c. Add up the amounts of each coin: $0.01 (penny) + $0.05 (nickel) + $0.10 (dime) + $0.25 (quarter) = $0.41.

111. e. There are four quarters in every dollar; $12 × 4 quarters per dollar = 48 quarters.

112. b. The first step is to figure out how much change she needs back: $100 − $25 (the cost of the lamp) = $75. Next, calculate how many $5 bills total $75: $75 ÷ $5 = 15 $5 bills.

113. b. There are 100 pennies in every dollar; $20 × 100 pennies/dollar = 2,000 pennies. A short-cut to this multiplication is to multiply the whole numbers (2 × 1) and then add the number of zeros at the end. There are three zeros (1 from the 20, two from the 100), so the total is 2,000.

114. d. There are four quarters in every dollar; $2 × 4 quarters/dollar = 8 quarters. Every quarter allows 15 minutes. Four quarters × 15 minutes/quarter = 120 minutes, which is equal to two hours (60 minutes per hour).

115. d. A half-dollar is equal to $0.50; 5 × $0.50 = $2.50. Divide this amount by the value of one nickel ($0.05) to find the total number of nickels; $2.50 ÷ $0.05 = 50 nickels.

116. e. Each nickel is worth $0.05; 50 nickels × $0.05/nickel = $2.50.

117. e. 35 dimes × $0.10 per dime = $3.50; $3.50 ÷ $0.25 per quarter = 14 quarters. An alternate solution would be to figure out that there are four quarters per dollar, so 12 quarters would equal $3 and another two quarters would add the extra 50 cents. Therefore, a total of 14 quarters (12 + 2 = 14) would equal $3.50.

118. b. Divide; there are 11 pesos for each dollar, so 143 pesos are worth $13: 143 ÷ 11 = $13.

119. e. Multiply: $20 × 11 pesos per dollar = 220 pesos. Note that the dollar units cancel out to leave the answer in units of pesos.

120. a. Divide; there are $1.30 Canadian for each $1 U.S., so $13 C is worth $10 U.S. Basically, divide 13 by 1.3 to get the U.S. dollar amount.

121. c. Multiply; 63 pounds × $1.80 per pound = $113.40. The units of pounds cancel out so that you are left with the dollar value.

122. d. Multiply; $30 × 107 yen per dollar = 3,210 yen. The dollar units cancel out so that the answer is in units of yen.

123. b. Multiply; 20 U.S. dollars × 1.3 Canadian dollars per U.S. dollar = 26 Canadian dollars (20 × 1.3 = 26).

CHAPTER

3 ▶ Fractions, Decimals, and Percents

WE USE INTEGERS to represent whole numbers, their negatives, and zero. We can represent parts of a whole using fractions, decimals, and percents. In this chapter, we look at comparing, adding, subtracting, multiplying, and dividing fractions and decimals, and we'll learn how to find the percent of a number, as well as percent increase and percent decrease.

Fractions

A **fraction** represents a part of a whole. A fraction itself is a division statement. The top number of the fraction, the **numerator**, is divided by the bottom number of the fraction, the **denominator**. In the fraction $\frac{2}{8}$, the numerator is 2 and the denominator is 8. If two fractions have the same denominator, then they are **like fractions**. If two fractions have different denominators, then they are **unlike fractions**.

Comparing Fractions

The fractions $\frac{1}{4}$ and $\frac{3}{4}$ are like fractions. Both have a denominator of 4. Like fractions are easy to compare: The fraction with the greater numerator is the greater fraction. The fractions $\frac{1}{4}$ and $\frac{2}{3}$ are unlike fractions, since they do not have the same denominator. To compare two unlike fractions, find a common denominator for the fractions and turn them into like fractions.

Example
Which is greater, $\frac{1}{4}$ or $\frac{2}{3}$?

We can find a common denominator for two fractions, such as $\frac{1}{4}$ and $\frac{2}{3}$, by finding the least common multiple of the denominators. The **least common multiple** of two numbers is the smallest number that is a multiple of both numbers. To find the least common multiple, list the first few multiples of each number. In this example, list the multiples of 4 and 3:

multiples of 3: 3, 6, 9, 12, 15, 18, 21, 24, . . .

multiples of 4: 4, 8, 12, 16, 20, 24, 28, . . .

The least common multiple of 3 and 4 is 12, because it is the smallest multiple that both 3 and 4 have in common. Now we must convert both fractions to a numerator over the common denominator, 12. To write $\frac{1}{4}$ as a number over 12, we must multiply the original denominator, 4, by 3. Therefore, we must also multiply the nu-

---

TIP

Here are some tips for comparing fractions:

- If two fractions have the same numerator but different denominators, the fraction with the *smaller* denominator is the *bigger* fraction. For instance, $\frac{1}{2} > \frac{1}{3}$. Check by finding common denominators: $\frac{1}{2} = \frac{3}{6}$, and $\frac{1}{3} = \frac{2}{6}$. $\frac{1}{2}$ is greater than $\frac{1}{3}$. $\frac{1}{2}$ is also greater than $\frac{1}{4}$, $\frac{1}{5}$, and $\frac{1}{6}$.

- If the denominator of one or both of the fractions is a prime number and the other denominator is not a multiple of it, the least common denominator will be the product of the two denominators. For instance, if you are comparing $\frac{1}{5}$ and $\frac{3}{8}$, the least common denominator will be 40 (5 × 8), since 5 is a prime number and 8 is not a multiple of 5. Check by listing the multiples of each:

  multiples of 5: 5, 10, 15, 20, 25, 30, 35, 40, . . .

  multiples of 8: 8, 16, 32, 40, . . .

- If two fractions are positive and one of them is improper while the other is proper, the improper fraction is greater—you don't even have to find common denominators to determine which is greater. All positive improper fractions are equal to 1 or more, while all positive proper fractions are equal to less than 1.

merator by 3, so that the value of the fraction does not change: $\frac{1}{4} = \frac{3}{12}$. In the same way, to write $\frac{2}{3}$ as a number over 12, we must multiply the original denominator, 3, by 4. Therefore, we must also multiply the numerator by 4: $\frac{2}{3} = \frac{8}{12}$. Now that the fractons have common denominators, we can compare them: $3 < 8$, $\frac{3}{12} < \frac{8}{12}$, and $\frac{1}{4} < \frac{2}{3}$.

► Reducing Fractions

We can make fractions smaller, or simpler, by reducing them. In order to reduce a fraction to its simplest form, we must divide the numerator and the denominator by the largest number that is a factor of both the numerator and the denominator. This number is called the greatest common factor. The **greatest common factor** of two numbers is the largest number that divides evenly into both numbers.

Example
Reduce $\frac{12}{36}$.

Begin by listing the factors of each number:

factors of 12: 1, 2, 3, 4, 6, 12

factors of 36: 1, 2, 3, 4, 6, 9, 12, 18, 36

Although 12 and 36 have many common factors, 12 is the greatest common factor. We can reduce $\frac{12}{36}$ to its simplest form if we divide its numerator and denominator by 12: $12 \div 12 = 1$ and $36 \div 12 = 3$. Therefore, $\frac{12}{36} = \frac{1}{3}$. $\frac{1}{3}$ is $\frac{12}{36}$ in its simplest form.

► Adding and Subtracting Fractions

To add two like fractions, add their numerators and keep the denominator. If the fractions are unlike, convert them to like fractions before adding. To subtract a like fraction from another like fraction, subtract the second fraction from the first and keep the denominator. If the fractions are unlike, convert them to like fractions before subtracting.

Example
$\frac{3}{5} + \frac{1}{6}$

Find a common denominator for these fractions.

multiples of 5: 5, 10, 15, 20, 25, 30, 35, . . .

multiples of 6: 6, 12, 18, 24, 30, 36, 42, . . .

The least common multiple of 5 and 6 is 30. Convert $\frac{3}{5}$ to a numerator over the common denominator, 30. $30 \div 5 = 6$, so we must multiply the numerator of $\frac{3}{5}$ by 6: $3 \times 6 = 18$. Therefore, $\frac{3}{5} = \frac{18}{30}$. Convert $\frac{1}{6}$ in the same way. $30 \div 6 = 5$, so multiply the numerator of $\frac{1}{6}$ by 5: $1 \times 5 = 5$. Therefore, $\frac{1}{6} = \frac{5}{30}$. Now we have like fractions: $\frac{18}{30} + \frac{5}{30}$. Add the numerators and keep the denominator: $18 + 5 = 23$, so $\frac{3}{5} + \frac{1}{6} = \frac{18}{30} + \frac{5}{30} = \frac{23}{30}$.

Sometimes, the sum of two fractions is greater than 1. When this happens, we write our answer as a mixed number. A **mixed number** has two parts: a whole number part and a fraction part. The number $1\frac{1}{2}$ is a mixed number. Mixed numbers, like **improper fractions**, which are fractions whose numerators are greater than or equal to their denominators, have a value that is greater than or equal to 1, or less than or equal to negative 1. The improper fraction $\frac{3}{2}$ is equal to the mixed number $1\frac{1}{2}$. We convert improper fractions to mixed numbers by dividing the numerator by the denominator. The remainder of that division is written as a new numerator over the denominator of the improper fraction.

Example

$9\frac{2}{3} + 15\frac{7}{15}$

First, add the whole numbers: $9 + 15 = 24$. Next, add the fractions. The least common denominator of 3 and 15 is 15. $\frac{2}{3} = \frac{10}{15}$. Now, we can add the fractions: $\frac{10}{15} + \frac{7}{15} = \frac{17}{15}$. Since 17 divided by 15 is 1 with 2 left over, $\frac{17}{15} = 1\frac{2}{15}$. Finally, add this mixed number to the sum of the whole numbers: $24 + 1\frac{2}{15} = 25\frac{2}{15}$.

Example

$\frac{7}{8} - \frac{7}{12}$

The least common multiple of 8 and 12 is 24. Convert $\frac{7}{8}$ to a number over 24. $24 \div 8 = 3$, so we must multiply the numerator and denominator of $\frac{7}{8}$ by 3: $\frac{7}{8} = \frac{21}{24}$. Convert $\frac{7}{12}$ to a number over 24. $24 \div 12 = 2$, so we must multiply the numerator and denominator of $\frac{7}{12}$ by 2: $\frac{7}{12} = \frac{14}{24}$. Now we have like fractions: $\frac{21}{24} - \frac{14}{24}$. Find the difference between 21 and 14 and keep the denominator: $21 - 14 = 7$, so $\frac{7}{8} - \frac{7}{12} = \frac{21}{24} - \frac{14}{24} = \frac{7}{24}$.

▶ Multiplying and Dividing Fractions

To multiply two fractions, whether like or unlike, multiply the numerators, and then multiply the denominators. To divide two fractions, whether like or unlike, take the reciprocal of the divisor and then multiply the fractions. To find the **reciprocal** of a fraction, switch the numerator and the denominator. Convert mixed numbers to improper fractions before multiplying or dividing them.

Example

$\frac{7}{10} \times \frac{5}{6}$

Multiply the numerators: $7 \times 5 = 35$. Multiply the denominators: $10 \times 6 = 60$. The product of $\frac{7}{10}$ and $\frac{5}{6}$ is $\frac{35}{60}$. The greatest common factor of 35 and 60 is 5, so $\frac{35}{60}$ reduces to $\frac{7}{12}$.

TIP

You've seen how to reduce a single fraction by dividing the numerator and denominator of that fraction by the same number. When multiplying fractions, we can divide the numerator of one fraction and the denominator of the other fraction by the same number. Look again at the preceding example. Before multiplying, we can divide the numerator of $\frac{5}{6}$ by 5 and the denominator of $\frac{7}{10}$ by 5, since 5 is the greatest common factor of 5 and 10. The problem now becomes $\frac{7}{2} \times \frac{1}{6}$, which is easier to multiply. The product of these fractions is $\frac{7}{12}$, which is the same as the product of $\frac{7}{10} \times \frac{5}{6}$—and, the answer is already in simplest form!

Example

$\frac{5}{21} \div \frac{10}{18}$

The reciprocal of $\frac{10}{18}$ is $\frac{18}{10}$, so $\frac{5}{21} \div \frac{10}{18}$ becomes $\frac{5}{21} \times \frac{18}{10}$. Now that it is a multiplication problem, we can simplify these fractions. By dividing the numerator of $\frac{5}{21}$ and the denominator of $\frac{18}{10}$ by 5, and by dividing the denominator of $\frac{5}{21}$ and the numerator of $\frac{18}{10}$ by 3, $\frac{5}{21} \times \frac{18}{10}$ becomes $\frac{1}{7} \times \frac{6}{2}$. $\frac{6}{2} = 3$, so now the problem becomes $\frac{1}{7} \times 3$, which is equal to $\frac{3}{7}$.

▶ Practice Questions

Label each pair of fractions as like or unlike.

124. $\frac{7}{8}$ and $\frac{8}{7}$

125. $\frac{5}{6}$ and $\frac{6}{6}$

126. $\frac{3}{5}$ and $\frac{8}{5}$

Place the proper symbol ($>$, $<$, or $=$) between each pair of fractions.

127. $\frac{9}{12}$ _____ $\frac{3}{4}$

128. $\frac{7}{8}$ _____ $\frac{5}{6}$

129. $\frac{1}{2}$ _____ $\frac{6}{10}$

130. $\frac{3}{5}$ _____ $\frac{4}{7}$

131. $\frac{11}{11}$ _____ $\frac{21}{21}$

Reduce each of the following fractions to its simplest form.

132. $\frac{10}{16}$

133. $\frac{12}{20}$

134. $\frac{24}{27}$

135. $\frac{30}{48}$

136. $\frac{56}{72}$

For each question, find the sum.

137. $\frac{3}{8} + \frac{4}{8}$

138. $\frac{3}{7} + \frac{3}{8}$

139. $\frac{4}{10} + \frac{6}{15}$

140. $9\frac{7}{11} + 10\frac{15}{11}$

141. $7\frac{5}{6} + 7\frac{4}{5}$

For each question, find the difference.

142. $\frac{11}{14} - \frac{5}{14}$

143. $\frac{5}{7} - \frac{4}{9}$

144. $\frac{17}{20} - \frac{2}{3}$

145. $8\frac{2}{3} - 5\frac{1}{2}$

146. $11\frac{7}{12} - 4\frac{3}{8}$

For each question, find the product.

147. $\frac{7}{10} \times \frac{6}{7}$

148. $\frac{9}{16} \times \frac{12}{18}$

149. $\frac{8}{13} \times \frac{5}{8} \times \frac{1}{5}$

150. $7\frac{1}{2} \times 9\frac{4}{9}$

151. $1\frac{11}{12} \times 4\frac{4}{5}$

For each question, find the quotient.

152. $\frac{1}{6} \div \frac{5}{8}$

153. $\frac{48}{18} \div \frac{18}{3}$

154. $\frac{20}{36} \div \frac{20}{27}$

155. $15\frac{5}{6} \div 3\frac{3}{8}$

156. $8\frac{9}{10} \div 5\frac{7}{12}$

Answer the following word problems.

157. Justin read $\frac{1}{8}$ of a book the first day, $\frac{1}{3}$ the second day, and $\frac{1}{4}$ the third day. On the fourth day he finished the book. What part of the book did Justin read on the fourth day?

a. $\frac{2}{5}$

b. $\frac{3}{8}$

c. $\frac{7}{24}$

d. $\frac{17}{24}$

158. During a commercial break in the Super Bowl, there were three half-minute commercials and five quarter-minute commercials. How many minutes was the commercial break?

a. $2\frac{3}{4}$

b. $\frac{3}{4}$

c. $3\frac{1}{2}$

d. 5

159. The Cheshire Senior Center is hosting a bingo night; $2,400 in prize money will be given away. The first winner of the night will receive $\frac{1}{3}$ of the money. The next ten winners will receive $\frac{1}{10}$ of the remaining amount. How many dollars will each of the ten winners receive?

a. $240

b. $800

c. $160

d. $200

160. Lori has half a pizza left over from dinner last night. For breakfast, she eats one-third of the leftover pizza. What fraction of the original pizza remains after Lori eats breakfast?

a. $\frac{1}{4}$

b. $\frac{1}{6}$

c. $\frac{1}{3}$

d. $\frac{3}{8}$

▶ Decimals

We can also represent parts of a whole and mixed numbers as decimals. A **decimal** is a number that is written using one or more of ten digits: 0, 1, 2, 3, 4, 5, 6, 7, 8, or 9. Let's look at comparing, rounding, adding, subtracting, multiplying, and dividing decimals.

▶ Comparing Decimals

We compare the values of decimals by comparing their corresponding digits, place by place, from left to right. The decimal with the larger digit is the larger number. If both decimals have the same digit in the same place, move one place to the right and compare again.

Example

8.52376 versus 8.5276

First, line up the decimal points and add a trailing zero to the second number:

8.52376

8.52760

Next, compare the digits from left to right. The numbers have the same digits in the ones, tenths, and hundredths places, so we must move to the thousandths place. The first number has a 3 in the thousandths place and the second number has a 7 in the thousandths place. Since 7 is greater than 3, 8.5276 is greater than 8.52376.

▶ Rounding Decimals

Rounding is the process of taking a number and making it less precise by removing one or more digits from the end of the number, replacing those digits with zeros if necessary. To round a number to a place, look at the digit to the immediate right. If that digit is 5 or greater, we round up. If that digit is less than 5, we round down.

Example

Round 574.299199 to the nearest thousandth.

The digit to the immediate right of the thousandths place is the digit in the ten-thousandths place, 1. Since 1 is less than 5, we round down. Notice that even though the thousandths digit itself is greater than 5, and even though the digits to the right of the ten-thousandths place are greater than 9, we still round down, because we are rounding to the thousandths place. The thousandths digit, 9, stays the same and the digits to the right become zero. 574.299199 to the nearest thousandth is 574.299.

▶ Adding and Subtracting Decimals

To add two decimals, line up the decimal points, add trailing zeros to keep the columns even, and add just as you would add whole numbers. Subtract decimals in the same way—line up the decimal points, add trailing zeros, and subtract.

Example

145.785 + 2.3

The first addend has six digits: three to the left of the decimal point and three to the right. The second addend has only two digits, one to the left of the decimal point and one to the right. Keep your columns straight and place two trailing zeros on the end of 2.3. Add column by column and carry the decimal point down into your answer.

```
  145.785
+   2.300
---------
  148.085
```

Example

602.107 − 74.87

Line up the decimal points. Place a trailing zero on 74.87, subtract, and carry down the decimal point:

$$
\begin{array}{r}
602.107 \\
-74.870 \\
\hline
527.237
\end{array}
$$

▶ Multiplying Decimals

Multiply two decimals just as you would two whole numbers, but be careful where you place the decimal point in your answer. Count the number of digits to the right of the decimal points in both factors. That sum is the number of digits to the right of the decimal point in your product.

Example

1.2534 × 24.05

The first factor has four digits to the right of the decimal point, and the second factor has two digits to the right of the decimal point. Our product will have six digits to the right of the decimal point:

$$
\begin{array}{r}
1.2534 \\
\times\,24.05 \\
\hline
62670 \\
0 \\
50136 \\
25068 \\
\hline
30.144270
\end{array}
$$

Now that we've placed our decimal point with six digits to its right, we can remove the trailing zero. The product of 1.2534 and 24.05 is 30.14427.

▶ Dividing Decimals

Before you can divide a decimal by a decimal, you must convert the divisor into a whole number. To do this, you must shift the decimal point to the right. In order to keep the value of the equation the same, every time you shift the decimal point of the divisor, you must also shift the decimal point of the dividend.

Example

62.314 ÷ 12.5

The divisor has one digit to the right of the decimal point. Shift the decimal point in both the divisor and the dividend one place to the right: 623.14 ÷ 125. The problem has become a decimal divided by a whole number. Place the decimal point in the quotient and divide:

$$
\begin{array}{r}
4.90512 \\
125{\overline{\smash{\big)}\,623.14000}} \\
\underline{500} \\
1231 \\
\underline{1125} \\
64 \\
\underline{0} \\
640 \\
\underline{625} \\
150 \\
\underline{125} \\
250 \\
\underline{250} \\
0
\end{array}
$$

TIP

We never actually divide a decimal by another decimal. After shifting the decimal point of the divisor, we divide a whole number into either a decimal or another whole number.

▶ Practice Questions

Choose the larger number.

161. 11.43781 versus 11.43718

162. 0.00051 versus 0.0005

Round these numbers to the nearest hundredth.

163. 874.561

164. 0.005

Round this number to the nearest thousandth.

165. 560.5454

Add the following numbers.

166. 74.65722 + 1,045

167. 874 + 8.74

168. 7.91 + 6,327.7

169. 19.159 + 207.44

170. 62.3906 + 58.906

Subtract.

171. 22 − 14.989

172. 571 − 46.4

173. 18.45 − 9

174. 38.766 − 9.558

175. 144.323 − 112.705

Multiply.

176. 4.4 × 1.3

177. 61.8 × 1.22

178. 107.4 × 0.631

179. 7.461 × 1.55

180. 9.241 × 2.477

Divide. If necessary, round your answers to the nearest ten-thousandth.

181. 9.4277 ÷ 6

182. 123.95 ÷ 9

183. 13.2 ÷ 4.4

184. 6.39 ÷ 0.18

185. 524.475 ÷ 3.5

Solve the following word problems.

186. Mike, Dan, Ed, and Sy played together on a baseball team. Mike's batting average was .349, Dan's was 0.2, Ed's was 0.35, and Sy's was .299. Who had the highest batting average?
   a. Mike
   b. Dan
   c. Ed
   d. Sy

187. Kenny used a micrometer to measure the thickness of a piece of construction paper. The paper measured halfway between 0.24 millimeter and 0.25 millimeter. What is the thickness of the paper?

    a. 0.05 millimeter

    b. 0.245 millimeter

    c. 0.255 millimeter

    d. 0.3 millimeter

188. Brian's 100-yard dash time was 2.68 seconds more than the school record. His time was 13.4 seconds. What is the school record?

    a. 10.72 seconds

    b. 11.28 seconds

    c. 10.78 seconds

    d. 16.08 seconds

189. Bill traveled 117 miles in 2.25 hours. What was his average speed?

    a. 26.3 miles per hour

    b. 5.2 miles per hour

    c. 46 miles per hour

    d. 52 miles per hour

▶ Percents

A **percent** is a ratio that represents a part to a whole as a number out of 100. Fractions are used to represent a part to a whole, but fractions can have many different denominators. A percent is like a fraction whose denominator is always 100. The value 36% is read as "36 percent" and represents 36 out of 100.

▶ Converting Percents to Decimals and Fractions

We can rewrite a percent as a decimal by removing the percent symbol and moving the decimal point two places to the left. We can rewrite a percent as a fraction by removing the percent symbol and placing the value over 100.

Example

Write 28% as a decimal and as a fraction.

Remove the percent symbol from 28% and move the decimal point two places to the left: 28% = 0.28. To write 28% as a fraction, remove the percent symbol and place 28 over 100. 28% = $\frac{28}{100}$, which reduces to $\frac{7}{25}$.

PITFALL

If you move the decimal point two places to the left of a single-digit number, you will first have to add a zero to the left of the number. For example, to write 2% as a decimal, you must add a leading zero to 2%: 02%. Now you can move the decimal point two places to the left, and 2% becomes .02.

We can convert from a decimal back to a percent by moving the decimal point two places to the right and adding the percent symbol. We can convert a fraction to a decimal by dividing the numerator by the denominator. We can then write that decimal as a percent.

Example

Write 5.6103 as a percent and as a mixed number.

Percents can be 100 or greater. After moving the decimal point two places to the right, we find that 5.6103 = 561.03%. The whole number part of 5.6103 is 5 and the fraction part is $\frac{6,103}{10,000}$, which means that $5.6103 = 5\frac{6,103}{10,000}$.

Example

Write $\frac{3}{4}$ as a decimal and as a percent.

Divide the numerator, 3, by the denominator, 4: $3 \div 4 = 0.75$. $\frac{3}{4} = 0.75$. Move the decimal point two places to the right and add the percent symbol: 0.75 = 75%.

▶ Working with Percents

To find the percent of a number, write the percent as a decimal and multiply it by the number. To represent a part of a number as a percent, divide the part by the whole and write the answer as a percent.

Example

What is 23% of 54?

23% = 0.23; 0.23 × 54 = 12.42, so 23% of 54 is 12.42.

Example

What percent is 30 of 96?

Divide 30 by 96: 30 ÷ 96 = 0.3125. Move the decimal point two places to the right and add the percent sign: 0.3125 = 31.25%, so 30 is 31.25% of 96.

▶ Percent Increase and Percent Decrease

Percent increase is the difference between an original value and a new value divided by the original value. When finding a percent increase, the original value is subtracted from the new value, since the new value is larger than the original value.

Example

Find the percent increase from 12 to 30.

Subtract the original value from the new value and divide by the original value:

30 − 12 = 18; 18 ÷ 12 = 1.5 = 150%.

Percent decrease, like percent increase, is the difference between an original value and a new value divided by the original value. However, when finding a percent decrease, the new value is subtracted from the original value, since the original value is larger than the new value.

TIP

The percent increase from 15 to 18 is 20%, but the percent decrease from 18 to 15 is only 16.67%. How can that be? When finding the percent increase from 15 to 18, our original value is 15, so we divide the difference between the values by 15. When we find the percent decrease from 18 to 15, our original value is 18, so we divide the difference between the values by 18. As long as the two values are different (and if they are the same, then there is no percent increase or percent decrease), the percent increase from the first to the second will always be different than the percent decrease from the second to the first.

Example

Find the percent decrease from 18 to 15.

Subtract the new value from the original value and divide by the original value: $18 - 15 = 3$; $3 \div 18 = 0.1667$, to the nearest ten-thousandth; $0.1667 = 16.67\%$.

▶ **Practice Questions**

Write each percent as a decimal and as a fraction.

190. 24%

191. 80%

192. 11%

193. 97%

194. 6%

Write each decimal as a percent and as a fraction or mixed number.

195. 0.47

196. 0.9

197. 0.133

198. 1.52

199. 9.01

Write each fraction or mixed number as a decimal and as a percent.

200. $\frac{8}{10}$

201. $\frac{22}{64}$

202. $\frac{3}{19}$

203. $7\frac{17}{30}$

204. $87\frac{53}{80}$

Find the given percent of each number.

205. 16% of 64

206. 55% of 250

207. 87% of 133

Answer the following questions.

208. What percent is 52 of 64?

209. What percent of 125 is 82?

210. What percent is 65 of 90?

Find the percent increase from each original value to its new value.

211. 20 to 26

212. 48 to 66

213. 33 to 44

214. 45 to 135

Find the percent decrease from each original value to its new value.

215. 64 to 56

216. 34 to 17

217. 98 to 0

218. 7 to −28

Solve the following word problems.

219. Michael scored 260 points during his junior year on the school basketball team. He scored 25% more points during his senior year. How many points did he score during his senior year?
a. 195
b. 65
c. 325
d. 345

220. Joey has 30 pages to read for history class tonight. He decided that he would take a break when he finished reading 70% of the pages assigned. How many pages must he read before he takes a break?
a. 7
b. 21
c. 9
d. 18

221. The Dow Jones Industrial Average fell 2% today. The Dow began the day at 10,600. What was the Dow at the end of the day after the 2% drop?
a. 10,400
b. 10,812
c. 10,388
d. 7,800

222. Peter was 60 inches tall on his thirteenth birthday. By the time he turned 15, his height had increased 15%. How tall was Peter when he turned 15?
a. 75 inches
b. 69 inches
c. 72 inches
d. 71 inches

TERMS TO REVIEW

fraction

numerator

denominator

like fractions

unlike fractions

least common multiple

greatest common factor

mixed number

improper fractions

reciprocal

decimal

rounding

percent

percent increase

percent decrease

FORMULAS TO REVIEW

percent increase: (new value – original value) ÷ (original value)

percent decrease: (original value – new value) ÷ (original value)

▶ **Answers**

124. The denominators of these fractions are different, so these fractions are unlike.

125. The denominators of these fractions are the same, so these fractions are like.

126. Even though the second fraction is improper, the denominators of these fractions are the same, so these fractions are like.

127. The least common denominator for $\frac{9}{12}$ and $\frac{3}{4}$ is 12. $\frac{3}{4} = \frac{9}{12}$ (both the numerator and denominator are 3 times bigger).

128. The least common denominator for $\frac{7}{8}$ and $\frac{5}{6}$ is 24. $\frac{7}{8} = \frac{21}{24}$ (both the numerator and denominator are 3 times bigger) and $\frac{5}{6} = \frac{20}{24}$ (both the numerator and denominator are 4 times bigger). Since 21 is greater than 20, $\frac{21}{24} > \frac{20}{24}$, so $\frac{7}{8} > \frac{5}{6}$.

129. The least common denominator for $\frac{1}{2}$ and $\frac{6}{10}$ is 10. $\frac{1}{2} = \frac{5}{10}$ (both the numerator and denominator are 5 times bigger). Since 5 is less than 6, $\frac{5}{10} < \frac{6}{10}$, so $\frac{1}{2} < \frac{6}{10}$.

130. The least common denominator for $\frac{3}{5}$ and $\frac{4}{7}$ is 35. $\frac{3}{5} = \frac{21}{35}$ (both the numerator and denominator are 7 times bigger), and $\frac{4}{7} = \frac{20}{35}$ (both the numerator and denominator are 5 times bigger). Since 21 is greater than 20, $\frac{21}{35} > \frac{20}{35}$, so $\frac{3}{5} > \frac{4}{7}$.

131. Any number over itself is equal to 1, so both fractions, $\frac{11}{11}$ and $\frac{21}{21}$, are equal to 1, and $\frac{11}{11} = \frac{21}{21}$.

132. First, list the factors of the numerator and the denominator. The factors of 10 are 1, 2, 5, and 10, and the factors of 16 are 1, 2, 4, 8, and 16. The greatest common factor is 2. $10 \div 2 = 5$ and $16 \div 2 = 8$, so $\frac{10}{16}$ reduces to $\frac{5}{8}$.

133. List the factors of the numerator and the denominator. The factors of 12 are 1, 2, 3, 4, 6, and 12, and the factors of 20 are 1, 2, 4, 5, 10, and 20. The greatest common factor is 4. $12 \div 4 = 3$ and $20 \div 4 = 5$, so $\frac{12}{20}$ reduces to $\frac{3}{5}$.

134. List the factors of the numerator and the denominator. The factors of 24 are 1, 2, 3, 4, 6, 8, 12, and 24, and the factors of 27 are 1, 3, 9, and 27. The greatest common factor is 3. $24 \div 3 = 8$ and $27 \div 3 = 9$, so $\frac{24}{27}$ reduces to $\frac{8}{9}$.

135. List the factors of the numerator and the denominator. The factors of 30 are 1, 2, 3, 5, 6, 10, 15, and 30, and the factors of 48 are 1, 2, 3, 4, 6, 8, 12, 16, 24, and 48. The greatest common factor is 6. $30 \div 6 = 5$ and $48 \div 6 = 8$, so $\frac{30}{48}$ reduces to $\frac{5}{8}$.

136. List the factors of the numerator and the denominator. The factors of 56 are 1, 2, 4, 7, 8, 14, 28, and 56, and the factors of 72 are 1, 2, 3, 4, 6, 8, 9, 12, 18, 24, 36, and 72. The greatest common factor is 8. $56 \div 8 = 7$ and $72 \div 8 = 9$, so $\frac{56}{72}$ reduces to $\frac{7}{9}$.

137. The denominators of these fractions are 8, so the denominator of our answer will be 8. The sum of the numerators is $3 + 4 = 7$, so $\frac{3}{8} + \frac{4}{8} = \frac{7}{8}$.

138. The least common multiple of 7 and 8 is 56. Multiply the numerator and denominator of $\frac{3}{7}$ by 8: $\frac{3}{7} = \frac{24}{56}$. Multiply the numerator and denominator of $\frac{3}{8}$ by 7: $\frac{3}{8} = \frac{21}{56}$. The sum of the numerators is $24 + 21 = 45$. $\frac{3}{7} + \frac{3}{8} = \frac{24}{56} + \frac{21}{56} = \frac{45}{56}$.

139. The least common multiple of 10 and 15 is 30. Multiply the numerator and denominator of $\frac{4}{10}$ by 3: $\frac{4}{10} = \frac{12}{30}$. Multiply the numerator and denominator of $\frac{6}{15}$ by 2: $\frac{6}{15} = \frac{12}{30}$. The sum of the numerators is $12 + 12 = 24$. $\frac{4}{10} + \frac{6}{15} = \frac{12}{30} + \frac{12}{30} = \frac{24}{30}$. The greatest common factor of 24 and 30 is 6. $24 \div 6 = 4$ and $30 \div 6 = 5$. $\frac{24}{30} = \frac{4}{5}$, so $\frac{4}{10} + \frac{6}{15} = \frac{4}{5}$.

140. Add the whole numbers: $9 + 10 = 19$. Add the fractions: $\frac{7}{11} + \frac{15}{11} = \frac{22}{11}$. Convert $\frac{22}{11}$ to a whole number: $\frac{22}{11} = 2$. Add it to the sum of the whole numbers: $19 + 2 = 21$.

141. Add the whole numbers: $7 + 7 = 14$. To add the fractions, find common denominators. The least common denominator of 6 and 5 is 30. $\frac{5}{6} = \frac{25}{30}$ and $\frac{4}{5} = \frac{24}{30}$. $\frac{25}{30} + \frac{24}{30} = \frac{49}{30}$. Convert $\frac{49}{30}$ to a mixed number: $\frac{49}{30} = 1\frac{19}{30}$. Add it to the sum of the whole numbers: $14 + 1\frac{19}{30} = 15\frac{19}{30}$.

142. The denominators of these fractions are both 14, so the denominator of our answer will be 14. Since $11 - 5 = 6$, $\frac{11}{14} - \frac{5}{14} = \frac{6}{14}$, which reduces to $\frac{3}{7}$.

143. The least common multiple of 7 and 9 is 63. Multiply the numerator and denominator of $\frac{5}{7}$ by 9: $\frac{5}{7} = \frac{45}{63}$. Multiply the numerator and denominator of $\frac{4}{9}$ by 7: $\frac{4}{9} = \frac{28}{63}$. Subtract the second numerator from the first: $45 - 28 = 17$. $\frac{5}{7} - \frac{4}{9} = \frac{45}{63} - \frac{28}{63} = \frac{17}{63}$.

144. The least common multiple of 20 and 3 is 60. Multiply the numerator and denominator of $\frac{17}{20}$ by 3: $\frac{17}{20} = \frac{51}{60}$. Multiply the numerator and denominator of $\frac{2}{3}$ by 20: $\frac{2}{3} = \frac{40}{60}$. Subtract the second numerator from the first: $51 - 40 = 11$. $\frac{17}{20} - \frac{2}{3} = \frac{51}{60} - \frac{40}{60} = \frac{11}{60}$.

145. Subtract the whole numbers: $8 - 5 = 3$. To subtract the fractions, the least common multiple of 3 and 2 is 6, so convert both fractions to a number over 6. $\frac{2}{3} = \frac{4}{6}$ and $\frac{1}{2} = \frac{3}{6}$. Subtract the fractions: $\frac{4}{6} - \frac{3}{6} = \frac{1}{6}$. $8\frac{2}{3} - 5\frac{1}{2} = 3\frac{1}{6}$.

146. Subtract the whole numbers: $11 - 4 = 7$. To subtract the fractions, the least common multiple of 12 and 8 is 24, so convert both fractions to a number over 24. $\frac{7}{12} = \frac{14}{24}$ and $\frac{3}{8} = \frac{9}{24}$. Subtract the fractions: $\frac{14}{24} - \frac{9}{24} = \frac{5}{24}$. $11\frac{7}{12} - 4\frac{3}{8} = 7\frac{5}{24}$.

147. $7 \times 6 = 42$ and $10 \times 7 = 70$, so $\frac{7}{10} \times \frac{6}{7} = \frac{42}{70}$, which reduces to $\frac{3}{5}$. We also could have canceled the 7 in the numerator of $\frac{7}{10}$ with the 7 in the denominator of $\frac{6}{7}$, and divided the denominator of $\frac{7}{10}$ and the numerator of $\frac{6}{7}$ by 2 before multiplying.

148. $9 \times 12 = 108$ and $16 \times 18 = 288$, so $\frac{9}{16} \times \frac{12}{18} = \frac{108}{288}$, which reduces to $\frac{3}{8}$. We also could have divided the denominator of $\frac{9}{16}$ and the numerator of $\frac{12}{18}$ by 4 and divided the numerator of $\frac{9}{16}$ and the denominator of $\frac{12}{18}$ by 9 before multiplying.

149. $8 \times 5 \times 1 = 40$ and $13 \times 8 \times 5 = 520$, so $\frac{8}{13} \times \frac{5}{8} \times \frac{1}{5} = \frac{40}{520}$, which reduces to $\frac{1}{13}$. We also could have canceled the 8 in the numerator of $\frac{8}{13}$ with the 8 in the denominator of $\frac{5}{8}$ and canceled the 5 in the numerator of $\frac{5}{8}$ with the 5 in the denominator of $\frac{1}{5}$ before multiplying.

150. Convert both mixed numbers to improper fractions. $7\frac{1}{2} = \frac{15}{2}$ and $9\frac{4}{9} = \frac{85}{9}$. Divide the 15 in the first fraction and the 9 in the second fraction by 3. The problem becomes $\frac{5}{2} \times \frac{85}{3}$. $5 \times 85 = 425$ and $2 \times 3 = 6$. 425 divided by 6 is 70 with 5 left over: $70\frac{5}{6}$.

151. Convert both mixed numbers to improper fractions. $1\frac{11}{12} = \frac{23}{12}$ and $4\frac{4}{5} = \frac{24}{5}$. Divide the 12 in the first fraction and the 24 in the second fraction by 12. The problem becomes $23 \times \frac{2}{5}$. $23 \times 2 = 46$. 46 divided by 5 is 9 with 1 left over: $9\frac{1}{5}$.

152. The reciprocal of the divisor, $\frac{5}{8}$, is $\frac{8}{5}$. $\frac{1}{6} \div \frac{5}{8}$ becomes $\frac{1}{6} \times \frac{8}{5}$. Divide the 6 in $\frac{1}{6}$ and the 8 in $\frac{8}{5}$ by 2, and the problem becomes $\frac{1}{3} \times \frac{4}{5}$. Multiply the numerators and the denominators. $1 \times 4 = 4$ and $3 \times 5 = 15$, so $\frac{1}{3} \times \frac{4}{5}$ and $\frac{1}{6} \div \frac{5}{8}$ both equal $\frac{4}{15}$.

153. The reciprocal of the divisor, $\frac{18}{3}$, is $\frac{3}{18}$. $\frac{48}{18} \div \frac{18}{3}$ becomes $\frac{48}{18} \times \frac{3}{18}$. Divide the 18 in $\frac{48}{18}$ and the 3 in $\frac{3}{18}$ by 3, and the problem becomes $\frac{48}{6} \times \frac{1}{18}$, or $8 \times \frac{1}{18}$, which is equal to $\frac{8}{18}$. Simplify $\frac{8}{18}$ by dividing the numerator and the denominator by 2: $\frac{8}{18} = \frac{4}{9}$.

154. The reciprocal of the divisor, $\frac{20}{27}$, is $\frac{27}{20}$. $\frac{20}{36} \div \frac{20}{27}$ becomes $\frac{20}{36} \times \frac{27}{20}$. Cancel the 20 in $\frac{20}{36}$ with the 20 in $\frac{27}{20}$ and divide the 36 in $\frac{20}{36}$ and the 27 in $\frac{27}{20}$ by 9, and the problem becomes $\frac{1}{4} \times \frac{3}{1}$. Multiply the numerators and the denominators. $1 \times 3 = 3$ and $4 \times 1 = 4$, so $\frac{1}{4} \times \frac{3}{1}$ and $\frac{20}{36} \div \frac{20}{27}$ both equal $\frac{3}{4}$.

155. Convert both mixed numbers to improper fractions. $15\frac{5}{6} = \frac{95}{6}$ and $3\frac{3}{8} = \frac{27}{8}$. The reciprocal of the divisor, $\frac{27}{8}$, is $\frac{8}{27}$. The problem is now $\frac{95}{6} \times \frac{8}{27}$. Divide the 6 in the first fraction and the 8 in the second fraction by 2. The problem is now $\frac{95}{3} \times \frac{4}{27} = \frac{380}{81}$. 380 divided by 81 is 4 with 56 left over, so $\frac{380}{81} = 4\frac{56}{81}$.

156. Convert both mixed numbers to improper fractions. $8\frac{9}{10} = \frac{89}{10}$ and $5\frac{7}{12} = \frac{67}{12}$. The reciprocal of the divisor, $\frac{67}{12}$, is $\frac{12}{67}$. The problem is now $\frac{89}{10} \times \frac{12}{67}$. Divide the 10 in the first fraction and the 12 in the second fraction by 2. The problem is now $\frac{89}{5} \times \frac{6}{67} = \frac{534}{335}$. 534 divided by 335 is 1 with 199 left over, so $\frac{534}{335} = 1\frac{199}{335}$.

157. c. First, find the fraction of the book that he has read by adding the three fractions using a common denominator of 24: $\frac{3}{24} + \frac{8}{24} + \frac{6}{24} = \frac{17}{24}$. Subtract the fraction of the book he has read from one whole, using a common denominator of 24; $\frac{24}{24} - \frac{17}{24} = \frac{7}{24}$. If you chose **d**, you found the fraction of the book that Justin had already read.

158. a. First, multiply 3 by $\frac{1}{2}$ to find the time taken by the three half-minute commercials; $3 \times \frac{1}{2} = \frac{3}{2}$. Then, multiply $\frac{1}{4}$ by 5 to find the time taken by the five quarter-minute commercials; $\frac{1}{4} \times 5 = \frac{5}{4}$. Add the two times together to find the total commercial time. Use a common denominator of 4; $\frac{6}{4} + \frac{5}{4} = \frac{11}{4}$, which simplifies to $2\frac{3}{4}$ minutes. If you chose **b**, you only calculated for *one* commercial of each length rather than *three* half-minute commercials and *five* quarter-minute commercials.

159. c. The prize money is divided into tenths after the first third has been paid out. Find one-third of $2,400 by dividing $2,400 by 3; $800 is paid to the first winner, leaving $1,600 for the next ten winners to split evenly ($2,400 − $800 = $1,600). Divide $1,600 by 10 to find how much each of the ten winners will receive: $1,600 ÷ 10 = $160. Each winner will receive $160.

160. c. Refer to the drawing. If half is broken into thirds, each third is one-sixth of the whole. Therefore, she has $\frac{2}{6}$ or $\frac{1}{3}$ of the pizza left.

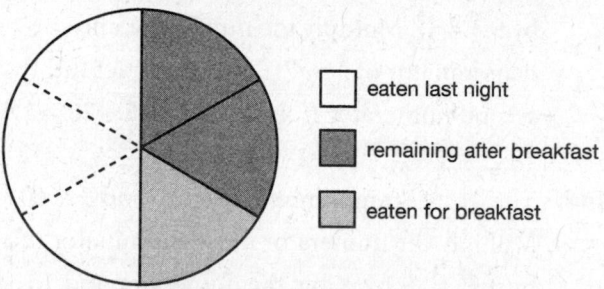

eaten last night

remaining after breakfast

eaten for breakfast

161. The numbers have the same digits in the tens, ones, tenths, hundredths, and thousandths places. The first number has an 8 in the ten-thousandths place and the second number has a 1 in the ten-thousandths place, so the first number is bigger. 11.43781 is greater than 11.43718.

162. Add a trailing zero to the second number. The numbers have the same digits in the ones, tenths, hundredths, thousandths, and ten-thousandths places. The first number has a 1 in the hundred-thousandths place and the second number has a trailing zero in the hundred-thousandths place, so the first number is bigger. 0.00051 is greater than 0.0005.

163. The digit to the immediate right of the hundredths place is the digit in the thousandths place, 1. Because 1 is less than 5, we round down. The hundredths digit stays at 6, and the digit to the right becomes zero. 874.561 to the nearest hundredth is 874.56.

164. The digit to the immediate right of the hundredths place is the digit in the thousandths place, 5. We round up when the digit is 5 or greater. The hundredths digit increases by 1 to 1, and the digit to the right becomes zero. 0.005 to the nearest hundredth is 0.010, or 0.01.

165. The digit to the immediate right of the thousandths place is the digit in the ten-thousandths place, 4. Because 4 is less than 5, we round down. The thousandths digit stays at 5, and the digit to the right becomes zero. 560.5454 to the nearest thousandth is 560.545.

166. 74.65722 + 1,045.00000 = 1,119.65722. There are five digits to the right of the decimal point in 74.65722, and five digits to the right of the decimal point in the answer.

167. 874 + 8.74 = 882.74. There are two digits to the right of the decimal point in 8.74, and two digits to the right of the decimal point in the answer.

168. 6,335.61

169. 226.599

170. 121.2966

171. 7.011. There are three digits to the right of the decimal point in 14.989, and three digits to the right of the decimal point in the answer.

172. 524.6. There is one digit to the right of the decimal point in 46.4, and one digit to the right of the decimal point in the answer.

173. 9.45

174. 29.208

175. 31.618

176. 4.4 has one digit to the right of the decimal point and 1.3 has one digit to the right of the decimal point, so our answer has two digits to the right of the decimal point: 5.72.

177. 61.8 has one digit to the right of the decimal point and 1.22 has two digits to the right of the decimal point, so our answer has three digits to the right of the decimal point: 75.396.

178. 107.4 has one digit to the right of the decimal point and 0.631 has three digits to the right of the decimal point, so our answer has four digits to the right of the decimal point: 67.7694.

179. 7.461 has three digits to the right of the decimal point and 1.55 has two digits to the right of the decimal point, so our answer has five digits to the right of the decimal point: 11.56455.

180. 9.241 has three digits to the right of the decimal point and 2.477 has three digits to the right of the decimal point, so our answer has six digits to the right of the decimal point: 22.889957.

181. Add one zero to the dividend: 9.42770 ÷ 6 = 1.57128. Since the hundred-thousandths digit of 1.57128 is an 8, 1.57128 rounds up to 1.5713.

$$
\begin{array}{r}
1.57128 \\
6\overline{)9.42770} \\
\underline{6} \\
34 \\
\underline{30} \\
42 \\
\underline{42} \\
07 \\
\underline{6} \\
17 \\
\underline{12} \\
50 \\
\underline{48} \\
2
\end{array}
$$

182. Add three zeros to the dividend. Since the hundred-thousandths digit of 13.77222 is a 2, 13.77222 rounds down to 13.7722.

$$
\begin{array}{r}
13.77222 \\
9\overline{)123.95000} \\
\underline{9} \\
33 \\
\underline{27} \\
69 \\
\underline{63} \\
65 \\
\underline{63} \\
20 \\
\underline{18} \\
20 \\
\underline{18} \\
20 \\
\underline{18} \\
2
\end{array}
$$

183. Shift the decimal point of both the divisor and the dividend one place to the right. 13.2 ÷ 4.4 becomes 132 ÷ 44:

$$
\begin{array}{r}
3 \\
44\overline{)132} \\
\underline{132} \\
0
\end{array}
$$

184. Shift the decimal point of both the divisor and the dividend two places to the right.
6.39 ÷ 0.18 becomes 639 ÷ 18:

$$
\begin{array}{r}
35.5 \\
18\overline{)639.0} \\
\underline{54} \\
99 \\
\underline{90} \\
90 \\
\underline{90} \\
0
\end{array}
$$

185. Shift the decimal point of both the divisor and the dividend one place to the right. 524.475 ÷ 3.5 becomes 5,244.75 ÷ 35:

$$
\begin{array}{r}
149.85 \\
35\overline{)5244.75} \\
\underline{35} \\
174 \\
\underline{140} \\
344 \\
\underline{315} \\
297 \\
\underline{280} \\
175 \\
\underline{175} \\
0
\end{array}
$$

186. c. If you add zeros to the end of Dan's and Ed's averages to make them have the traditional three decimal places, it will be easy to compare the batting averages. The four averages are: .349, .200, .350, and .299; .350 is the largest.

187. b. Find the difference between 0.24 millimeter and 0.25 millimeter by subtracting: 0.25 − 0.24 = 0.01 millimeter. Half of this is 0.01 ÷ 2 = 0.005. Add to 0.24 to get 0.245 millimeter.

188. a. The school record is less than Brian's time. Therefore, 2.68 must be subtracted from 13.4; 13.4 − 2.68 = 10.72. To subtract decimals, line up the numbers vertically so that the decimal points are aligned. Since 13.4 has one less decimal place than 2.68, you must add a zero after the 4 (13.40) before subtracting. After you have done this, subtract normally. If you chose **d**, you added instead of subtracted.

189. d. Divide the distance, 117 miles, by 2.25 hours after shifting the decimal point of both the divisor and the dividend two places to the right: $11,700 \div 225 = 52$. The rate is 52 miles per hour.

190. $24\% = 0.24 = \frac{24}{100} = \frac{6}{25}$

191. $80\% = 0.80$ or $0.8 = \frac{8}{10} = \frac{4}{5}$

192. $11\% = 0.11 = \frac{11}{100}$.

193. $97\% = 0.97 = \frac{97}{100}$

194. $6\% = 0.06 = \frac{6}{100} = \frac{3}{50}$

195. $0.47 = 47\% = \frac{47}{100}$

196. $0.9 = 90\% = \frac{90}{100} = \frac{9}{10}$

197. $0.133 = 13.3\% = \frac{133}{1,000}$

198. $1.52 = 152\% = 1\frac{52}{100} = 1\frac{13}{25}$

199. $9.01 = 901\% = 9\frac{1}{100}$

200. $8 \div 10 = 0.8 = 80\%$

201. $22 \div 64 = 0.34375 = 34.375\%$

202. $3 \div 19 = 0.158$, to the nearest thousandth; $0.158 = 15.8\%$

203. $17 \div 30 = 0.567$, to the nearest thousandth. Write the whole number, 7, to the left of the fraction and the decimal point. $7\frac{17}{30}$ is approximately 7.567, which equals 756.7%.

204. $53 \div 80 = 0.6625$. Write the whole number, 87, to the left of the fraction and the decimal point. $87\frac{53}{80} = 87.6625 = 8,766.25\%$.

205. $16\% = 0.16$, $0.16 \times 64 = 10.24$

206. $55\% = 0.55$, $0.55 \times 250 = 137.5$

207. $87\% = 0.87$, $0.87 \times 133 = 115.71$

208. $52 \div 64 = 0.8125 = 81.25\%$

209. $82 \div 125 = 0.656 = 65.6\%$

210. $65 \div 90 = 0.7222$ to the nearest ten-thousandth; $0.7222 = 72.22\%$

211. $26 - 20 = 6$, $6 \div 20 = 0.3 = 30\%$

212. $66 - 48 = 18$, $18 \div 48 = 0.375 = 37.5\%$

213. $44 - 33 = 11$, $11 \div 33 = 0.3333$ to the nearest ten-thousandth; $0.3333 = 33.33\%$

214. $135 - 45 = 90$, $90 \div 45 = 2 = 200\%$

215. $64 - 56 = 8$, $8 \div 64 = 0.125 = 12.5\%$

216. $34 - 17 = 17$, $17 \div 34 = 0.5 = 50\%$

217. $98 - 0 = 98$, $98 \div 98 = 1 = 100\%$

218. $7 - (-28) = 35$, $35 \div 7 = 5 = 500\%$

219. c. If the number of points is increased by 25%, the number of points in his senior year is 125% of the number of points in his junior year (100% + 25% = 125%). To find 125% of the number of points in his junior year, multiply the junior year points by the decimal equivalent of 125%; $260 \times 1.25 = 325$. If you chose **a**, you calculated what his points would be if he scored 25% *less* than he did in his junior year.

220. b. To find 70% of 30, you must multiply 30 by the decimal equivalent of 70% (0.70); $30 \times 0.70 = 21$. If you chose **c**, you calculated how many pages he has left to read after his break.

221. c. The Dow lost 2%, so it is worth 98% of what it was worth at the beginning of the day (100% − 2% = 98%). To find 98% of 10,600, multiply 10,600 by the decimal equivalent of 98%: $10,600 \times 0.98 = 10,388$.

222. b. Find 15% of 60 inches and add it to 60 inches. Find 15% by multiplying 60 by the decimal equivalent of 15% (0.15); $60 \times 0.15 = 9$. Add 9 inches to 60 inches to get 69 inches.

4 ▶ Algebraic Expressions

WE'VE REVIEWED HOW to work with integers, fractions, and decimals—adding, subtracting, multiplying, and dividing them. We have also reviewed how to follow the order of operations to evaluate an expression that contains more than one operation. Now, we'll look at how to perform those same operations on algebraic expressions.

▶ Evaluating Algebraic Expressions

A letter or symbol that takes the place of a number in an expression or an equation is called a **variable**. That's what algebra is all about—variables. A real number expression might be $4 + 7$, and an algebraic expression could be $4 + x$. An **algebraic expression** is an expression that contains one or more variables.

Example

$x - 6$

This is an algebraic expression because 6 is subtracted from x, a variable that represents a number.

Example

$(7x^2 + 5) - 3y$

This is a more complex algebraic expression. The variable x is squared, and then multiplied by 7. The result of that is added to 5, and then finally, 3 times the variable y is subtracted from that sum.

If we know the value of a variable in an algebraic expression, we can replace the variable with the value. Let's look again at that first example.

Example

$x - 6$

If we are told that the value of x is 7, we can replace x with 7 and the expression becomes $7 - 6$. Now that we have two real numbers, we can evaluate the expression: $7 - 6 = 1$. We say that when $x = 7$, $x - 6 = 1$.

What if x is equal to 2? Substitute 2 for x in the expression: $2 - 6 = -4$. When $x = 2$, $x - 6 = -4$.

Example

What is the value of $(7x^2 + 5) - 3y$ when $x = 1$ and $y = -1$?

Substitute 1 for x and -1 for y: $[7(1)^2 + 5] - 3(-1)$. Remembering the order of operations, evaluate parentheses (or in this case brackets) first: $7(1)^2 + 5$. Exponents come before multiplication or addition in the order of operations, so we begin by squaring 1. The expression is now $(7 + 5) - 3(-1)$. Now handle the operation in parentheses: $7 + 5 = 12$. The expression has become $12 - 3(-1)$. Multiplication comes before subtraction, so we multiply 3 by -1. $3 \times -1 = -3$, and our expression has been simplified to $12 - (-3)$. Replace the two negative signs with a positive sign, and $12 + 3 = 15$. When $x = 1$ and $y = -1$, $(7x^2 + 5) - 3y = 15$.

Sometimes a variable appears in an expression more than once. Be sure to replace every occurrence of the variable with its real value.

Example

If $a = 3$, what is the value of $4a + 3a^2$?

Replace both instances of a with 3: $4(3) + 3(3^2) = 4(3) + 3(9) = 12 + 27 = 39$.

▶ Practice Questions

Evaluate the following algebraic expressions when

$$a = 3$$
$$b = -5$$
$$x = 6$$
$$y = \frac{1}{2}$$
$$z = -8$$

223. $4a + z$

224. $3x \div z$

225. $2ax - z$

226. $5ab + xy$

227. $4b^2 - az$

228. $7x \div 2yz$

229. $bx + z \div y$

230. $6y - 2ab$

231. $a(b + z)^2$

232. $2(a^2 + 2y) \div b$

233. $a^3 + 24y - 3b$

234. $-2x - b + az$

235. $5z^2 - 2z + 2$

236. $5xy \div 2b$

237. $7x + \frac{12}{x} - z$

238. $2b^2 \div y$

239. $bx(z + 3)$

240. $6y(z \div y) + 3ab$

241. $2bx \div (z - b)$

242. $12ab \div y$

243. $y[(\frac{x}{2} - 3) - 4a]$

244. $10b^3 - 4b^2$

245. $8y(a^3 - 2y)$

246. $z^2 - 4a^2y$

247. $3x^2b(5a - 3b)$

▶ Combining Like Terms

Sometimes, we can simplify an algebraic expression by combining one or more of the terms in the expression. A **term** is an element in the expression that is subtracted, added, multiplied, or divided (or, in a single-term expression, an element that stands alone). In the expression $x - 6$, x and 6 are both terms. In the expression $4a + 3a^2$, $4a$ and $3a^2$ are both terms, but 4 is not considered a term because it is multiplied by a. In the same way, a is not a term, either, but $4a$, the product of 4 and a, is a term.

If two terms have the same base, then the two terms can be combined using addition or subtraction. These kinds of terms are called **like terms**. In the expression $4a + 3a^2$, $4a$ and $3a^2$ have different bases. The base of $4a$ is a, or a^1, and the base of $3a^2$ is a^2. These terms cannot be combined because they are **unlike terms**.

PITFALL

The language we use can be tricky. *Unlike terms*, such as $4a$ and $3a^2$, cannot be combined into a single term using addition or subtraction because they have different bases. However, it would be wrong to say that $4a$ and $3a^2$ cannot be added together. We can add $4a$ and $3a^2$, and we show that addition as $4a + 3a^2$. It's just that we can't do anything more. Any two terms can be added, but we can only *perform the addition and combine them* if they are *like terms*.

What makes two algebraic terms like terms? First, they must contain the same variable. Second, the exponent(s) of the variable(s) in each term must be the same. The terms $2x^2$ and $3x^2$ are like terms because both have the same base: x^2. However, $2x^2$ and

$3x^3$ are not like terms. Even though they both have the same variable in their base, x, that variable is raised to the second power in the first term and raised to the third power in the second term. These terms are not alike and cannot be combined.

The term $3x^2$ has a base of x^2 and a coefficient of 3. The number that is multiplied by the base of a term is called a **coefficient**. If a term appears to have no coefficient, then the coefficient is 1. The coefficient of b^4 is 1, because $1b^4 = b^4$.

▶ Adding Like Terms

To add two like terms, add the coefficients of the terms and keep the bases.

Example

$6y + 9y$

The coefficient of $6y$ is 6 and the coefficient of $9y$ is 9; $6 + 9 = 15$, so $6y + 9y = 15y$. What would be the sum of $6y$ and $9z$? That's a trick question: We can't combine $6y$ and $9z$, because they are unlike terms.

▶ Subtracting Like Terms

To subtract two like terms, subtract the coefficient of the second term (the subtrahend) from the coefficient of the first term and keep the bases.

Example

$8ab^3c^2 - 5ab^3c^2$

Subtract the coefficient of the second term from the coefficient of the first term: $8 - 5 = 3$, so $8ab^3c^2 - 5ab^3c^2 = 3ab^3c^2$.

▶ Multiplying Terms

We don't need like terms in order to multiply two terms. To multiply two terms, multiply their coefficients and multiply their bases. How do we multiply the bases? After all, the bases are variables, not numbers. We simply list the different variables next to each other, to show that they are being multiplied. If two variables are the same, we multiply them by adding their exponents.

Example

$(ab)(b^2)$

To multiply ab by b^2, list a next to b (since these variables are different) and add the exponent of b in the first term to the exponent of b in the second term. In the first term, b is raised to the first power, and in the second term, it is raised to the second power. Since $1 + 2 = 3$, $(b)(b^2) = b^3$, and $(ab)(b^2) = ab^3$.

Example

$(12x^3yz^4)(2xy^2z^2)$

First, multiply the coefficients: $(12)(2) = 24$. Next, multiply the variables: $(x^3)(x) = x^4$, $(y)(y^2) = y^3$, and $(z^4)(z^2) = z^6$. $(12x^3yz^4)(2xy^2z^2) = 24x^4y^3z^6$. The coefficient is listed right alongside the variables, which are listed right alongside each other, to show that they are all being multiplied together.

PITFALL

Don't forget about multiplication. When you see a number next to a variable, even though no multiplication symbol is shown, that number (the coefficient) is being multiplied by the variable: $3x^2$ means $3 \times x^2$.

▶ Dividing Terms

Just as with multiplication, we can divide two terms that are like or unlike. Divide the coefficient of the dividend by the coefficient of the divisor. If the same variable appears in both the dividend and the divisor, subtract the exponent of the variable in the divisor from the exponent of that same variable in the dividend.

Example

$(60a^2b^6c) \div (20a^5b^3)$

First, divide the coefficients: $60 \div 20 = 3$. Next, divide the variables: $2 - 5 = -3$, so $(a^2) \div (a^5) = a^{-3}$; $6 - 3 = 3$, so $(b^6) \div (b^3) = b^3$. Since there is no c in the divisor, we carry c right into our answer. $(60a^2b^6c) \div (20a^5b^3) = 3a^{-3}b^3c$.

TIP

Why did we carry c from the dividend right into our answer in the preceding example? You can think of that two ways: c divided by 1 is c, since any number divided by 1 is itself, or you can think of c as c^1 divided by c^0. Subtracting exponents, $1 - 0 = 1$, so $(c^1) \div (c^0) = c^1$, or c. That's why, when a variable appears in the dividend but not the divisor of an algebraic expression, we carry it right into our simplified result.

▶ Tips for Combining Like Terms

Distributive Property of Multiplication

The distributive property of multiplication tells you how to multiply the terms inside parentheses by the term outside the parentheses. Study the following general and specific examples.

$$a(b + c) = ab + ac$$
$$a(b - c) = ab - ac$$
$$(b + c)a = ba + ca$$
$$4(6 + 3) = 4 \times 6 + 4 \times 3 = 24 + 12 = 36$$
$$(-5 + 8)3 = -5 \times 3 + 8 \times 3 = -15 + 24 = 9$$
$$7(10 + 3) = 7 \times 10 + 7 \times 3 = 70 + 21 = 91$$
$$3(x + 2y) = 3 \times x + 3 \times 2y = 3x + 6y$$
$$a(b - 5d) = a \times b - a \times 5d = ab - 5ad$$

Numerical examples of the commutative properties for addition and multiplication were given in the "Tips for Working with Integers" section in Chapter 1. Now look at the following examples:

Commutative Property of Addition

$$a + b = b + a$$

This equation reminds us that terms being combined by addition can change their location (commute), but the value of the expression remains the same.

Commutative Property of Multiplication

$$x \times y = y \times x$$

This equation reminds us that the order in which we multiply expressions can change without changing the value of the result.

Associative Property of Addition

$$(q + r) + s = q + (r + s)$$

This equation reminds us that when you are performing a series of additions of terms, you can associate any term with any other and the result will be the same.

Associative Property of Multiplication

$$(d \times e) \times f = d \times (e \times f)$$

This equation reminds us that you can multiply three or more terms in any order without changing the value of the result.

Identity Property of Addition

$$n + 0 = n$$

Identity Property of Multiplication

$$n \times 1 = n$$

Term Equivalents

$$x = 1 \times x$$

For purposes of combining like terms, a variable by itself is understood to mean one of that term.

$$n = +n$$

A term without a sign in front of it is considered to be positive.

$$a + -b = a - +b = a - b$$

Adding a negative term is the same as subtracting a positive term. Look at the expressions on either side of the equal signs. Which one looks simpler? Of course, it's the last, $a - b$. Clarity is valued in mathematics. Writing expressions as simply as possible is always appreciated.

While it may not seem relevant yet, as you go through the practice exercises, you will see how each of these properties will come into play as we simplify algebraic expressions by combining like terms.

Here is one more example, and then we'll be ready to practice.

Example

$$[5g - 2(4g)](3g^3h^2)$$

Begin with the first set of delimiters. $[5g - 2(4g)] = (5g - 8g) = -3g$. The expression is now $-3g(3g^3h^2)$, which equals $-9g^4h^2$.

▶ Practice Questions

Simplify the following expressions by combining like terms.

248. $5a + 2a + 7a$

249. $7a + 6b + 3a$

250. $4x + 2y - x + 3y$

251. $27 - 3m + 12 - 5m$

252. $7h + 6 + 2w - 3 + h$

253. $4(x + 2y) + 2(x + y)$

254. $3(2a + 3b) + 7(a - b)$

255. $11(4m + 5) + 3(-3m + 8)$

256. $64 + 5(n - 8) + 12n - 24$

257. $4(x + y - 4) + 6(2 - 3y)$

258. $-7(a + b) + 12a - 16b$

259. $14 + 9(2w + 7) - 2(6 - w)$

260. $8s - 3r + 5(2r - s)$

261. $6(3m - 12) - 4(9m + 8)$

262. $5(15 - 2j) + 11(7j - 3)$

263. $a(a + 4) + 3a^2 - 2a + 10$

264. $8(x - 7) + x(2 - x)$

265. $3r^2 + r(2 - r) + 6(r + 4)$

266. $2x - x(5 + y) + 3xy + 4(2x - y)$

267. $-7(c - 2d) + 21c - 3(d - 5)$

268. $5(3x - y) + x(5 + 2y) - 4(3 + x)$

269. $6(m - 3n) + 3m(n + 5) - 2n(3 - m)$

270. $9(2x - t) + 23xt + x(-4 + 5t)$

271. $4[2a(a + 3) + 6(4 - a)] + 5a^2$

272. $8(2a - b - 3c) + 3(2a - b) - 4(6 - b)$

▶ Translating Words into Algebraic Expressions

We can take a phrase or sentence and convert it into an algebraic expression.

Example

four less than twice x

"Twice x" means "two times x," or $2x$.
Four less than that is $2x - 4$. Four less than twice x in algebraic terms is $2x - 4$.

Example

If apples cost z cents and bananas cost w cents, what is the cost of four apples and five bananas?

If the cost of one apple is z cents, then the cost of four apples is four times that amount, or $4z$. In the same way, the cost of five bananas is five times the cost of

one banana, or $5w$. The cost of four apples and five bananas is $4z + 5w$.

▶ Practice Questions

273. Assume that the number of hours Katie spent practicing soccer is represented by x. Michael practiced four hours less than two times the number of hours that Katie practiced. How long did Michael practice?
   a. $2x - 4$
   b. $2x + 4$
   c. $2x + 8$
   d. $4x + 4$

274. Patrick gets paid three dollars less than four times what Kevin gets paid. If the number of dollars that Kevin gets paid is represented by x, what does Patrick get paid?
   a. $3 - 4x$
   b. $3x - 4$
   c. $4x - 3$
   d. $4 - 3x$

275. If the expression $9y - 5$ represents a certain number, which of the following could NOT be the translation?
   a. five less than nine times y
   b. five less than the sum of 9 and y
   c. the difference between $9y$ and 5
   d. the product of 9 and y, decreased by 5

276. Susan starts work at 4:00 P.M. and Dee starts at 5:00 P.M. They both finish at the same time. If Susan works x hours, how many hours does Dee work?
   a. $x + 1$
   b. $x - 1$
   c. x
   d. $2x$

277. Frederick bought 11 books that cost d dollars each. What is the total cost of the books?

    a. $d + 11$

    b. $d - 11$

    c. $11d$

    d. $\frac{d}{11}$

278. There are m months in a year, w weeks in a month, and d days in a week. How many days are there in a year?

    a. mwd

    b. $m + w + d$

    c. $\frac{mw}{d}$

    d. $d + \frac{w}{d}$

279. Carlie received x dollars each hour she spent babysitting. She babysat a total of h hours. She then gave half of the money to a friend who had stopped by to help her. How much money did Carlie have after she had paid her friend?

    a. $\frac{hx}{2}$

    b. $\frac{x}{2} + h$

    c. $\frac{h}{2} + x$

    d. $2hx$

280. A long distance call costs x cents for the first minute and y cents for each additional minute. How much would a ten-minute call cost?

    a. $10xy$

    b. $x + 10y$

    c. $\frac{xy}{10}$

    d. $x + 9y$

281. Melissa is four times as old as Jim. Pat is five years older than Melissa. If Jim is y years old, how old is Pat?

    a. $4y + 5$

    b. $5y + 4$

    c. $4 \times 5y$

    d. $y + 5$

282. Sally gets paid x dollars per hour for a 40-hour workweek and y dollars for each hour she works over 40 hours. How much did Sally earn if she worked 48 hours?

    a. $48xy$

    b. $40y + 8x$

    c. $40x + 8y$

    d. $48x + 48y$

283. Eduardo is combining two six-inch pieces of wood with a piece that measures four inches. How many total inches of wood does he have?

    a. 10 inches

    b. 16 inches

    c. 8 inches

    d. 12 inches

284. Mary has $2 in her pocket. She does yard work for four different neighbors and earns $3 per yard. She then spends $2 on a soda. How much money does she have left?

    a. $18

    b. $10

    c. $12

    d. $14

285. Ten is decreased by four times the quantity of eight minus three. One is then added to that result. What is the final answer?

    a. -5

    b. -9

    c. 31

    d. -8

286. Eight is added to the quantity three minus the sum of negative seven and positive six. This answer is then multiplied by three. What is the result?

    a. 30

    b. -21

    c. 36

    d. 57

287. John and Charlie have a total of 80 dollars. John has x dollars. How many dollars does Charlie have?

a. 80

b. $80 + x$

c. $80 - x$

d. $x - 80$

288. Marty used the following mathematical statement to show he could change an expression and still get the same answer on both sides:

$$10 \times (6 \times 5) = (10 \times 6) \times 5$$

Which mathematical property did Marty use?

a. identity property of multiplication

b. commutative property of multiplication

c. distributive property of multiplication over addition

d. associative property of multiplication

289. Tori was asked to give an example of the commutative property of addition. Which of the following choices would be correct?

a. $3 + (4 + 6) = (3 + 4) + 6$

b. $3(4 + 6) = 3(4) + 3(6)$

c. $3 + 4 = 4 + 3$

d. $3 + 0 = 3$

290. Jake needed to find the perimeter of an equilateral triangle whose sides measure $x + 4$ cm each. Jake realized that he could multiply $3(x + 4) = 3x + 12$ to find the total perimeter in terms of x. Which property did he use to multiply?

a. associative property of addition

b. distributive property of multiplication over addition

c. commutative property of multiplication

d. inverse property of addition

291. Noel rode $3x$ miles on his bike and Jamie rode $5x$ miles on hers. In terms of x, what is the total number of miles they rode?

a. $15x$ miles

b. $15x^2$ miles

c. $8x$ miles

d. $8x^2$ miles

292. If the areas of two sections of a garden are $6a + 2$ and $5a$, what is the difference between the areas of the two sections in terms of a?

a. $a - 2$

b. $3a + 2$

c. $a + 2$

d. $11a - 2$

293. Laura has a rectangular garden whose width is x^3 and whose length is x^4. In terms of x, what is the area of her garden?

a. $2x^7$

b. x^7

c. x^{12}

d. $2x^{12}$

294. Jamestown High School has a soccer field whose dimensions can be expressed as $7y^2$ and $3xy$. What is the area of this field in terms of x and y?

a. $10xy^2$

b. $10xy^3$

c. $21xy^3$

d. $21xy^2$

295. The area of a parallelogram is x^8. If the base is x^4, what is the height in terms of x?

a. x^4

b. x^2

c. x^{12}

d. x^{32}

296. What is the quotient of $3d^4$ divided by $12d^6$?

   a. $4d^2$

   b. $4d^8$

   c. $\frac{1}{4d^8}$

   d. $\frac{1}{4d^2}$

297. The product of $6x^2$ and $4xy^2$ is divided by $3x^3y$. What is the simplified expression?

   a. $8y$

   b. $\frac{4y}{x}$

   c. $4y$

   d. $\frac{8y}{x}$

298. If the side of a square can be expressed as a^2b^3, what is the area of the square in simplified form?

   a. a^4b^5

   b. a^4b^6

   c. a^2b^6

   d. a^2b^5

299. If $3x^2$ is multiplied by the quantity $2x^3y$ raised to the fourth power, what would this expression simplify to?

   a. $48x^{14}y^4$

   b. $1{,}296x^{16}y^4$

   c. $6x^9y^4$

   d. $6x^{14}y^4$

300. Sara's bedroom is in the shape of a rectangle. The dimensions are $2x$ and $4x + 5$. What is the area of Sara's bedroom?

   a. $18x$

   b. $18x^2$

   c. $8x^2 + 5x$

   d. $8x^2 + 10x$

301. Express the product of $-9p^3r$ and the quantity $5p - 6r$ in simplified form.

   a. $-4p^4r - 15p^3r^2$

   b. $-45p^4r + 54p^3r^2$

   c. $-45p^4r - 6r$

   d. $-45p^3r + 54p^3r^2$

302. A number, x, increased by 3 is multiplied by the same number, x, increased by 4. What is the product of the two numbers in terms of x?

   a. $x^2 + 7$

   b. $x^2 + 12$

   c. $x^2 + 7x + 12$

   d. $x^2 + x + 7$

303. The length of Kara's rectangular patio can be expressed as $2x - 1$ and the width can be expressed as $x + 6$. In terms of x, what is the area of her patio?

   a. $2x^2 + 13x - 6$

   b. $2x^2 - 6$

   c. $2x^2 - 5x - 6$

   d. $2x^2 + 11x - 6$

TERMS TO REVIEW

variable

algebraic expression

term

like terms

unlike terms

coefficient

▶ **Answers**

223. Substitute the values for the variables into the expression: $4(3) + (-8)$. Order of operations tells you to multiply first: $4(3) = 12$. Substitute: $(12) + (-8)$. Signs different? Subtract the values of the numbers and give the result the sign of the larger value (no sign means +): $12 - 8 = 4$. The value of the expression is: $4a + z = 4$.

224. Substitute the values for the variables into the expression: $3(6) \div (-8)$. PEMDAS: Multiply the first term: $3(6) = 18$. Substitute: $(18) \div (-8)$. Signs different? Divide and give the result the negative sign $(-2\frac{1}{4})$: $18 \div 8 = 2\frac{2}{8} = 2\frac{1}{4}$. The value of the expression is: $3x \div z = -2\frac{1}{4}$ or -2.25.

225. Substitute the values for the variables into the expression. $2(3)(6) - (-8)$. Multiply the factors of the first term: $2(3)(6) = 36$. Substitute: $(36) - (-8)$. Change the operator to addition and change the sign of the number that follows: $(36) + (8)$. Signs the same? Add the values of the terms and give the result the same sign: $36 + 8 = 44$. The simplified value of the expression is: $2ax - z = 44$.

226. Substitute the values for the variables into the expression: $5(3)(-5) + (6)(\frac{1}{2})$. Evaluate the first term of the expression: $5(3)(-5) = 15(-5)$. Signs different? Multiply the terms and give the result a negative sign: $15(-5) = -75$. Evaluate the second term of the expression: $(6)(\frac{1}{2}) = 3$. Substitute the equivalent values into the original expression: $(-75) + (3)$. Signs different? Subtract the values of the numbers and give the result the sign of the larger value: $(-75) + (3) = -72$. The simplified value of the expression is: $5ab + xy = -72$.

227. Substitute the values for the variables into the expression: $4(-5)^2 - (3)(-8)$. PEMDAS: Evaluate the term with the exponent: $(-5)^2 = (-5) \times (-5)$. Signs the same? Multiply the terms, and the result is positive: $5 \times 5 = 25$. Substitute: $4(25) = 100$. Now evaluate the other term: $(3)(-8)$. Signs different? Multiply and give the result a negative sign: $3 \times -8 = -24$. Substitute the equivalent values into the original expression: $(100) - (-24)$. Change the operator to addition and change the sign of the number that follows: $100 + 24 = 124$. The simplified value of the expression is: $4b^2 - az = 124$.

228. Substitute the values for the variables into the expression: $7(6) \div 2(\frac{1}{2})(-8)$. PEMDAS: Multiply the terms in the expression: $7 \times 6 = 42 [2(\frac{1}{2})](-8) = (1)(-8) = -8$. Substitute the equivalent values into the original expression: $(42) \div (-8)$. Signs different? Divide and give the result a negative sign: $(42) \div (-8) = -5.25$. The simplified value of the expression is: $7x \div 2yz = -5.25$.

229. Substitute the values for the variables into the expression: $(-5)(6) + (-8) \div (\frac{1}{2})$. Group terms using order of operations: $(-5)(6) + [(-8) \div (\frac{1}{2})]$. PEMDAS: Multiply or divide the terms in the expression: $(-5)(6)$. Signs different? Multiply and give the result a negative sign: $(-5)(6) = -30$. Consider the second term: $(-8) \div (\frac{1}{2})$. Signs different? Divide and give the result a negative sign. To divide by a fraction, you multiply by the reciprocal: $8 \div \frac{1}{2} = 8 \times \frac{2}{1} = 8 \times 2 = 16(-8) \div (\frac{1}{2}) = -16$. Substitute the equivalent values into the original expression: $(-30) + (-16)$. Signs the same? Add the value of the terms and give the result the same sign: $30 + 16 = 46(-30) + (-16) = -46$. The simplified value of the expression is: $bx + z \div y = -46$.

230. Substitute the values for the variables into the expression: $6(\frac{1}{2}) - 2(3)(-5)$. Evaluate the terms on either side of the subtraction sign: $6(\frac{1}{2}) = 3$; $2(3)(-5) = 2 \times 3 \times -5$. Positive times positive is positive. Positive times negative is negative: $6 \times -5 = -30$. Substitute the equivalent values into the original expression: $(3) - (-30)$. Change the operator to addition and change the sign of the number that follows: $3 - (-30) = 3 + 30$; $3 + 30 = 33$. The simplified value of the expression is: $6y - 2ab = 33$.

231. Substitute the values for the variables into the expression: $3[(-5) + (-8)]^2$. PEMDAS: You must add the terms inside the parentheses first: $(-5) + (-8)$. Signs the same? Add the values of the terms and give the result the same sign: $(-5) + (-8) = -13$. Substitute into the original expression: $3(-13)^2$. Next you evaluate the term with the exponent: $(-13)^2 = -13 \times -13$. Signs the same? Multiply the terms, and the result is positive: $13 \times 13 = 169$. Substitute the equivalent values into the original expression: $3(169) = 507$. The simplified value of the expression is: $a(b + z)^2 = 507$.

232. Substitute the values for the variables into the expression: $2[(3)^2 + 2(\frac{1}{2})] \div (-5)$. Look first to evaluate the term inside the outer parentheses (brackets): $(3)^2 + 2(\frac{1}{2})$. The first term has an exponent. Evaluate it: $(3)^2 = 3 \times 3 = 9$. Evaluate the second term: $2(\frac{1}{2}) = 1$; $(3)^2 + 2(\frac{1}{2}) = 9 + 1 = 10$. Substitute into the original numerical expression: $2(10) \div (-5)$. Evaluate the first term: $2(10) = 20$. Substitute into the numerical expression: $(20) \div (-5)$. Signs different? Divide and give the result a negative sign: $(20) \div (-5) = -4$. The simplified value of the expression is: $2(a^2 + 2y) \div b = -4$.

233. Substitute the values for the variables into the expression: $(3)^3 + 24(\frac{1}{2}) - 3(-5)$. Evaluate the term with the exponent: $(3)^3 = 3 \times 3 \times 3 = 27$. Evaluate the second term: $24(\frac{1}{2}) = 12$. Evaluate the third term: $3(-5)$. Signs different? Multiply and give the result a negative sign: $3(-5) = -15$. Substitute the equivalent values into the original expression: $(27) + (12) - (-15)$. Change the subtraction to addition and change the sign of the number that follows: $(27) + (12) + (15)$. Signs the same? Add the values of the terms and give the result the same sign: $27 + 12 + 15 = 54$. The simplified value of the expression is: $a^3 + 24y - 3b = 54$.

234. Substitute the values for the variables into the expression: $-2(6) - (-5) + (3)(-8)$. Evaluate first and last terms. Positive times negative results in a negative: $-2(6) = -2 \times 6 = -12$ $(3)(-8) = 3 \times -8 = -24$. Substitute the equivalent values into the original expression: $(-12) - (-5) + (-24)$. Change the subtraction to addition and change the sign of the number that follows: $-12 + (5) + -24$. Commutative property of addition allows grouping of like signs: $(-12 + -24) + (5)$. Signs the same? Add the values of the terms and give the result the same sign: $-12 + -24 = -36$. Substitute: $(-36) + (5)$. Signs different? Subtract and give the result the sign of the higher-value number: $-36 + 5 = -31$. The simplified value of the expression is: $-2x - b + az = -31$.

235. Substitute the values for the variables into the expression: $5(-8)^2 - 2(-8) + 2$. PEMDAS: Evaluate the term with the exponent first: $(-8)^2 = (-8)(-8) = 64$. Substitute the value into the numerical expression: $5(64) - 2(-8) + 2$. PEMDAS: Evaluate terms with multiplication next: $5(64) = 320$; $2(-8) = -16$. Substitute the values into the numerical expression: $320 - (-16) + 2$. Change the subtraction to addition and change the sign of the number that follows: $320 + (16) + 2$. Add terms from left to right. All terms are positive, a result of addition: $320 + 16 + 2 = 338$. The simplified value of the expression is: $5z^2 - 2z + 2 = 338$.

236. Substitute the values for the variables into the expression: $5(6)(\frac{1}{2}) \div 2(-5)$. Consider the two terms on either side of the division sign. Evaluate the first term by multiplying: $5 \times 6 \times \frac{1}{2} = (5 \times 6) \times \frac{1}{2} = 15$. Evaluate the second term: $2(-5) = -10$. Substitute the values into the original numerical expression: $(15) \div (-10)$. Signs different? Divide and give the result a negative sign: $(15) \div (-10) = -1\frac{1}{2}$. The simplified value of the expression is: $5xy \div 2b = -1\frac{1}{2}$ or -1.5.

237. Substitute the values for the variables into the expression: $7(6) + \frac{12}{(6)} - (-8)$. Evaluate the first term: $7(6) = 7 \times 6 = 42$. Evaluate the second term: $\frac{12}{(6)} = 12 \div 6 = 2$. Substitute the values into the original numerical expression: $(42) + (2) - -8$. Change the subtraction to addition and change the sign of the number that follows: $42 + 2 + 8$. Add terms from left to right: $42 + 2 + 8 = 52$. The simplified value of the expression is: $7x + \frac{12}{x} - z = 52$.

238. Substitute the values for the variables into the expression: $2(-5)^2 \div \frac{1}{2}$. First, evaluate the term with the exponent: $2(-5)^2 = 2 \times (-5)(-5)$. Multiply from left to right: $[2 \times (-5)] \times (-5)$. Signs different? Multiply and give the result a negative sign: $2 \times (-5) = -10$. Signs the same? Multiply and the result is positive: $(-10) \times (-5) = 50$. Substitute the values into the original numerical expression: $(50) \div \frac{1}{2}$. Change division to multiplication and change the value to its reciprocal: $(50) \times 2 = 100$. The simplified value of the expression is: $2b^2 \div y = 100$.

239. Substitute the values for the variables into the expression: $(-5)(6)[(-8) + 3]$. First, evaluate the expression inside the outer parentheses (brackets): $(-8) + 3$. Signs different? Subtract and give the result the sign of the higher-value number: $(-8) + 3 = -5$. Substitute the result into the numerical expression: $(-5)(6)(-5)$. Multiply from left to right. Negative times positive equals negative: $-5 \times 6 = -30$. Signs the same? Multiply and the result is positive: $(-30) \times -5 = 150$. The simplified value of the expression is: $bx(z + 3) = 150$.

240. Substitute the values for the variables into the expression: $6(\frac{1}{2})(-8 \div \frac{1}{2}) + 3(3)(-5)$. First, evaluate the expression inside the parentheses. Division by a fraction is the same as multiplication by its reciprocal: $-8 \div \frac{1}{2} = -8 \times \frac{2}{1} = -16$. Substitute the result into the numerical expression: $6(\frac{1}{2})(-16) + 3(3)(-5)$. Evaluate the first term in the expression: $6(\frac{1}{2})(-16) = 6 \times \frac{1}{2} \times -16$; $3 \times -16 = -48$. Evaluate the second term in the expression: $3(3)(-5) = 3 \times 3 \times -5$; $9 \times -5 = -45$. Substitute the result into the numerical expression: $(-48) + (-45)$. Signs the same? Add the values of the terms and give the result the same sign: $-48 + -45 = -93$. The simplified value of the expression is: $6y(z \div y) + 3ab = -93$.

241. Substitute the values for the variables into the expression: $2(-5)(6) \div [(-8) - (-5)]$. First, evaluate the expression inside the outer parentheses (brackets): $(-8 - -5)$. Change the subtraction to addition and change the sign of the number that follows: $-8 + 5$. Signs different? Subtract and give the result the sign of the higher-value number: $-8 + 5 = -3$. Substitute the results into the numerical expression: $2(-5)(6) \div (-3)$. Multiply from left to right: $2 \times -5 \times 6 = -60$. Substitute the result into the numerical expression: $-60 \div -3$. Signs the same? Divide and the result is positive: $-60 \div -3 = 20$. The simplified value of the expression is: $2bx \div (z - b) = 20$.

242. Substitute the values for the variables into the expression: $12(3)(-5) \div (\frac{1}{2})$. Evaluate the first term. Multiply from left to right: $12 \times 3 \times -5 = 36 \times -5$. Signs different? Multiply the numbers and give the result a negative sign: $36 \times 5 = 180$; $36 \times -5 = -180$. Substitute the result into the numerical expression: $(-180) \div (\frac{1}{2})$. Division by a fraction is the same as multiplication by its reciprocal: $-180 \times (\frac{2}{1}) = -180 \times 2$. Signs different? Multiply numbers and give the result a negative sign: $-180 \times 2 = -360$. The simplified value of the expression is: $12ab \div y = -360$.

243. Substitute the values for the variables into the expression: $(\frac{1}{2})[(\frac{(6)}{2} - 3) - 4(3)]$. Evaluate the expression in the innermost parentheses: $(\frac{(6)}{2} - 3) = \frac{6}{2} - 3$. PEMDAS: Division before subtraction: $\frac{6}{2} - 3 = 3 - 3 = 0$. Substitute the result into the numerical expression: $(\frac{1}{2})[(0) - 4(3)]$. Evaluate the expression inside the brackets: $[0 - 4(3)] = 0 - 4 \times 3$. PEMDAS: Multiply before subtraction: $0 - 4 \times 3 = 0 - 12$. Change subtraction to addition and change the sign of the term that follows: $0 - 12 = 0 + -12 = -12$. Substitute the result into the numerical expression: $(\frac{1}{2})(-12) = \frac{1}{2} \times -12$. Signs different? Multiply numbers and give the result a negative sign: $\frac{1}{2} \times -12 = -6$. The simplified value of the expression is: $y[(\frac{x}{2} - 3) - 4a] = -6$.

244. Substitute the values for the variables into the expression: $10(-5)^3 - 4(-5)^2$. Evaluate the first term: $10(-5)^3 = 10 \times -5 \times -5 \times -5$. Multiply from left to right: $10 \times -5 = (-50)$; $(-50) \times -5 = 250$; $250 \times -5 = -1,250$. Evaluate the second term in the numerical expression: $4(-5)^2 = 4 \times -5 \times -5$. Multiply from left to right: $4 \times -5 = -20$; $-20 \times -5 = 100$. Substitute the results into the numerical expression: $-1,250 - 100$. Change subtraction to addition and change the sign of the term that follows: $-1,250 + -100$. Same signs? Add the values of the terms and give the result the same sign: $1,250 + 100 = 1,350$; $10(-5)^3 - 4(-5)^2 = -1,350$. The simplified value of the expression is: $10b^3 - 4b^2 = -1,350$.

245. Substitute the values for the variables into the expression: $8(\frac{1}{2})[(3)^3 - 2(\frac{1}{2})]$. Evaluate the expression in the outer parentheses (brackets): $[(3)^3 - 2(\frac{1}{2})] = (3^3) - (2 \times \frac{1}{2})$. Evaluate the first term. Multiply from left to right: $3^3 = 3 \times 3 \times 3 = 9 \times 3 = 27$. Evaluate the second term: $2 \times \frac{1}{2} = 1$. Substitute the results into the numerical expression in the parentheses: $(27) - (1)$. Subtract: $27 - 1 = 26$. Substitute the result into the original expression: $8(\frac{1}{2})(26)$. Multiply from left to right: $8 \times \frac{1}{2} = 4$; $4 \times 26 = 104$. The simplified value of the expression is: $8y(a^3 - 2y) = 104$.

246. Substitute the values for the variables into the expression: $(-8)^2 - 4(3)^2(\frac{1}{2})$. Evaluate the first term: $(-8)^2 = -8 \times -8$. Signs the same? Multiply and the result is positive: $-8 \times -8 = 64$. Evaluate the second term: $4(3)^2(\frac{1}{2}) = 4 \times 3 \times 3 \times \frac{1}{2}$. Multiply from left to right: $4 \times 3 = 12$; $12 \times 3 = 36$; $36 \times \frac{1}{2} = 18$. Substitute the results into the numerical expression: $(64) - (18)$. Yes, you can just subtract: $64 - 18 = 46$. The simplified value of the expression is: $z^2 - 4a^2y = 46$.

247. Substitute the values for the variables into the expression: $3(6)^2(-5)[5(3) - 3(-5)]$ PEMDAS: Evaluate the expression in the outer parentheses (brackets) first: $[5(3) - 3(-5)] = 5 \times 3 - 3 \times -5$; $5 \times 3 - 3 \times -5 = 15 - -15$. Change subtraction to addition and change the sign of the term that follows: $15 + 15 = 30$. Substitute the result into the numerical expression: $3(6)^2(-5)(30)$. PEMDAS: Evaluate terms with exponents next: $(6)^2 = 6 \times 6 = 36$. Substitute the result into the numerical expression: $3(36)(-5)(30)$. Multiply from left to right: $3(36) = 108$. Signs different? Multiply the values and give a negative sign: $(108) \times (-5) = -540$; $(-540) \times (30) = -16{,}200$; $3(6)^2(-5)[5(3) - 3(-5)] = -16{,}200$. The simplified value of the expression is: $3x^2b(5a - 3b) = -16{,}200$.

248. Use the associative property of addition: $(5a + 2a) + 7a$. Add like terms: $5a + 2a = 7a$. Substitute the results into the original expression: $(7a) + 7a$. Add like terms: $7a + 7a = 14a$. The simplified result of the algebraic expression is: $5a + 2a + 7a = 14a$.

249. Use the commutative property of addition to move like terms together: $7a + 3a + 6b$. Use the associative property for addition: $(7a + 3a) + 6b$. Add like terms: $(7a + 3a) = 10a$. Substitute: $(10a) + 6b$. The simplified result of the algebraic expression is: $7a + 6b + 3a = 10a + 6b$.

250. Change subtraction to addition and change the sign of the term that follows: $4x + 2y + (-x) + 3y$. Use the commutative property of addition to move like terms together: $4x + (-x) + 2y + 3y$. Use the associative property for addition: $(4x + -x) + (2y + 3y)$. Add like terms: $4x + -x = 3x$; $2y + 3y = 5y$. Substitute the results into the expression: $(4x + -x) + (2y + 3y) = (3x) + (5y)$. The simplified algebraic expression is: $4x + 2y - x + 3y = 3x + 5y$.

251. Change subtraction to addition and change the sign of the term that follows: $27 + -3m + 12 + -5m$. Use the commutative property of addition to put like terms together: $27 + 12 + -3m + -5m$. Use the associative property for addition: $(27 + 12) + (-3m + -5m)$. Add like terms: $27 + 12 = 39$; $-3m + -5m = -8m$. Substitute the results into the expression: $(27 + 12) + (-3m + -5m) = (39) + (-8m)$. Rewrite addition of a negative term as subtraction of a positive term by changing addition to subtraction and changing the sign of the following term: $(39) + (-8m) = 39 - 8m$. The simplified algebraic expression is: $27 - 3m + 12 - 5m = 39 - 8m$.

252. Change subtraction to addition and change the sign of the term that follows: $7h + 6 + 2w + (-3) + h$. Use the commutative property of addition to put like terms together: $7h + h + 2w + 6 + -3$. Use the associative property for addition: $(7h + h) + 2w + (6 + -3)$. Add like terms: $(7h + h) = 8h$; $(6 + -3) = 3$. Substitute the result into the expression: $(8h) + 2w + (3)$. The simplified algebraic expression is: $8h + 2w + 3$.

253. Use the distributive property of multiplication on the first expression: $4(x + 2y) = 4 \times x + 4 \times 2y$; $4x + 8y$. Use the distributive property of multiplication on the second expression: $2(x + y) = 2 \times x + 2 \times y$; $2x + 2y$. Substitute the results into the expression: $(4x + 8y) + (2x + 2y)$. Use the commutative property of addition to put like terms together: $4x + 2x + 8y + 2y$. Use the associative property for addition: $(4x + 2x) + (8y + 2y)$. Add like terms: $4x + 2x = 6x$; $8y + 2y = 10y$. Substitute the results into the expression: $(6x) + (10y)$. The simplified algebraic expression is: $6x + 10y$.

254. Use the distributive property of multiplication on the first term: $3(2a + 3b) = 3 \times 2a + 3 \times 3b$; $6a + 9b$. Use the distributive property of multiplication on the second term: $7(a - b) = 7 \times a - 7 \times b$; $7a - 7b$. Substitute the results into the expression: $(6a + 9b) + (7a - 7b)$; $6a + 9b + 7a - 7b$. Change subtraction to addition and change the sign of the term that follows: $6a + 9b + 7a + (-7b)$. Use the commutative property of addition to put like terms together: $6a + 7a + 9b + (-7b)$. Use the associative property for addition: $(6a + 7a) + (9b + -7b)$. Add like terms: $6a + 7a = 13a$. Signs different? Subtract the values of the terms: $9b + -7b = 2b$. Substitute the result into the expression: $(13a) + (2b)$. The simplified algebraic expression is: $13a + 2b$.

255. Use the distributive property of multiplication on the first term: $11(4m + 5) = 11 \times 4m + 11 \times 5$; $44m + 55$. Use the distributive property of multiplication on the second term: $3(-3m + 8) = 3 \times -3m + 3 \times 8$; $-9m + 24$. Substitute the result into the expression: $(44m + 55) + (-9m + 24)$; $44m + 55 + -9m + 24$. Use the commutative property of addition to put like terms together: $44m + -9m + 55 + 24$. Use the associative property for addition: $(44m + -9m) + (55 + 24)$. Add like terms: $44m + -9m = 35m$; $55 + 24 = 79$. Substitute the result into the expression: $(35m) + (79)$. The simplified algebraic expression is: $35m + 79$.

256. Use the distributive property of multiplication on the second term: $5(n - 8) = 5 \times n - 5 \times 8$; $5n - 40$. Substitute the result into the expression: $64 + (5n - 40) + 12n - 24$. Parentheses are no longer needed: $64 + 5n - 40 + 12n - 24$. Change subtraction to addition and change the sign of the term that follows: $64 + 5n + -40 + 12n + -24$. Use the commutative property of addition to put like terms together: $5n + 12n + 64 + -40 + -24$. Use the associative property for addition: $(5n + 12n) + (64 + -40 + -24)$. Add like terms: $5n + 12n = 17n$. Add like terms: $64 + -40 + -24 = 64 + (-40 + -24)$; $64 + -64 = 0$. Substitute the results into the expression: $(17n) + (0)$. The simplified algebraic expression is: $17n$.

257. Use the distributive property of multiplication on the first term: $4(x + y - 4) = 4 \times x + 4 \times y - 4 \times 4$; $4 \times x + 4 \times y - 4 \times 4 = 4x + 4y - 16$. Use the distributive property of multiplication on the second term: $6(2 - 3y) = 6 \times 2 - 6 \times 3y = 12 - 18y$. Substitute the results into the expression: $(4x + 4y - 16) + (12 - 18y)$. Parentheses are no longer needed: $4x + 4y - 16 + 12 - 18y$. Use the commutative property of addition to put like terms together: $4x + 4y - 18y - 16 + 12$. Change subtraction to addition and change the signs of the terms that follow: $4x + 4y + -18y + -16 + 12$. Use the associative property for addition: $4x + (4y + -18y) + (-16 + 12)$. Add like terms: $4y + -18y = -14y$; $-16 + 12 = 12 + -16 = -4$. Substitute the results into the expression: $4x + (-14y) + (-4)$. Rewrite addition of a negative term as subtraction of a positive term by changing addition to subtraction and changing the sign of the following term. The simplified algebraic expression is: $4x - 14y - 4$.

258. Use the distributive property of multiplication on the first term: $-7(a + b) = -7 \times a + -7 \times b$. Substitute the results into the expression: $(-7a + -7b) + 12a - 16b$. Parentheses are no longer needed: $-7a + -7b + 12a - 16b$. Use the commutative property of addition to put like terms together: $12a + -7a + -7b - 16b$. Change subtraction to addition and change the signs of the terms that follow: $12a + -7a + -7b + -16b$. Use the associative property for addition: $(12a + -7a) + (-7b + -16b)$. Add like terms: $12a + -7a = 5a$; $-7b + -16b = -23b$. Substitute the results into the expression: $(5a) + (-23b)$. Adding a negative term is the same as subtracting a positive term: $5a - 23b$.

259. Change subtraction to addition and change the signs of the terms that follow: $14 + 9(2w + 7) + -2(6 + -w)$. Use the distributive property of multiplication on the second term: $9(2w + 7) = 9 \times 2w + 9 \times 7$; $9 \times 2w + 9 \times 7 = 18w + 63$. Use the distributive property of multiplication on the third term: $-2(6 + -w) = -2 \times 6 + -2 \times -w$. Notice the result of multiplication for opposite and like-signed terms: $-2 \times 6 + -2 \times -w = -12 + 2w$. Substitute the results into the original expression: $14 + (18w + 63) + (-12 + 2w)$. Parentheses are no longer needed: $14 + 18w + 63 + -12 + 2w$. Use the commutative property of addition to put like terms together: $18w + 2w + 14 + 63 + -12$. Use the associative property for addition: $(18w + 2w) + (14 + 63 + -12)$. Add like terms: $18w + 2w = 20w$. Add from left to right: $14 + 63 + -12 = 77 + -12 = 65$. Substitute the results into the expression: $(20w) + (65)$. Parentheses are no longer needed: $20w + 65$.

260. Change subtraction to addition and change the signs of the terms that follow: $8s + -3r + 5(2r + -s)$. Use the distributive property of multiplication on the third term: $5(2r + -s) = 5 \times 2r + 5 \times -s$; $5 \times 2r + 5 \times -s = 10r + -5s$. Substitute the results into the expression: $8s + -3r + (10r + -5s)$. Parentheses are no longer needed: $8s + -3r + 10r + -5s$. Use the commutative property of addition to put like terms together: $8s + -5s + 10r + -3r$. Use the associative property for addition: $(8s + -5s) + (10r + -3r)$. Add like terms: $8s + -5s = 3s$; $10r + -3r = 7r$. Substitute the results into the expression: $(3s) + (7r)$. Parentheses are no longer needed: $3s + 7r$.

261. Change subtraction to addition and change the signs of the terms that follow: $6(3m + -12) + -4(9m + 8)$. Use the distributive property of multiplication on the first term: $6(3m + -12) = 6 \times 3m + 6 \times -12$; $6 \times 3m + 6 \times -12 = 18m + -72$. Use the distributive property of multiplication on the second term: $-4(9m + 8) = -4 \times 9m + -4 \times 8$; $-4 \times 9m + -4 \times 8 = -36m + -32$. Substitute the results into the expression: $(18m + -72) + (-36m + -32)$. Parentheses are no longer needed: $18m + -72 + -36m + -32$. Use the commutative property of addition to put like terms together: $18m + -36m + -72 + -32$. Use the associative property for addition: $(18m + -36m) + (-72 + -32)$. Add terms using the rules for terms with different signs: $18m + -36m = -18m$. Add terms using the rules for terms with the same signs: $-72 + -32 = -104$. Substitute the results into the expression: $(-18m) + (-104)$. Parentheses are no longer needed: $-18m + -104$. Adding a negative term is the same as subtracting a positive term: $-18m - +104$. Use the commutative property of addition: $-104 + -18m$. Change addition to subtraction and change the sign of the term that follows: $-104 - (+18m)$. The positive symbol in front of $18m$ can be dropped: $-104 - 18m$.

262. Change subtraction to addition and change the signs of the terms that follow: $5(15 + -2j) + 11(7j + -3)$. Use the distributive property of multiplication on the first term: $5(15 + -2j) = 5 \times 15 + 5 \times -2j$; $5 \times 15 + 5 \times -2j = 75 + -10j$. Use the distributive property of multiplication on the second term: $11(7j + -3) = 11 \times 7j + 11 \times -3$; $11 \times 7j + 11 \times -3 = 77j + -33$. Substitute the results into the expression: $(75 + -10j) + (77j + -33)$. Parentheses are no longer needed. $75 + -10j + 77j + -33$. Use the commutative property of addition to put like terms together: $77j + -10j + 75 + -33$. Use the associative property for addition: $(77j + -10j) + (75 + -33)$. Add terms using the rules for terms with different signs: $77j + -10j = 67j$. Add terms using the rules for terms with different signs: $75 + -33 = 42$. Substitute the results into the expression: $(67j) + (42)$. Parentheses are no longer needed: $67j + 42$.

263. Change subtraction to addition and change the signs of the terms that follow: $a(a + 4) + 3a^2 + -2a + 10$. Use the distributive property of multiplication on the first term: $a(a + 4) = a \times a + a \times 4$. Use the commutative property of multiplication for the second term: $a \times a + a \times 4 = a^2 + 4a$. Substitute the results into the expression: $(a^2 + 4a) + 3a^2 + -2a + 10$. Parentheses are no longer needed: $a^2 + 4a + 3a^2 + -2a + 10$. Use the commutative property of addition to put like terms together: $a^2 + 3a^2 + 4a + -2a + 10$. Use the associative property for addition: $(a^2 + 3a^2) + (4a + -2a) + 10$. Add the first term using the rules for terms with the same signs: $a^2 + 3a^2 = 4a^2$. Add the second term using the rules for terms with different signs: $4a + -2a = 2a$. Substitute the results into the expression: $(4a^2) + (2a) + 10$. Parentheses are no longer needed: $4a^2 + 2a + 10$.

264. Change subtraction to addition and change the signs of the terms that follow: $8(x + -7) + x(2 + -x)$. Use the distributive property of multiplication on the first term: $8(x + -7) = 8 \times x + 8 \times -7$; $8 \times x + 8 \times -7 = 8x + -56$. Use the distributive property of multiplication on the second term: $x(2 + -x) = x \times 2 + x \times -x$; $x \times 2 + x \times -x = 2x + -x^2$. Substitute the results into the expression: $(8x + -56) + 2x + -x^2$. Parentheses are no longer needed: $8x + -56 + 2x + -x^2$. Use the commutative property of addition to put like terms together: $8x + 2x + -56 + -x^2$. Use the associative property for addition: $(8x + 2x) + -56 + -x^2$. Add terms in parentheses using the rules for terms with the same signs: $(10x) + -56 + -x^2$. Change addition to subtraction and change the sign of the term that follows: $10x - (56) + -x^2$. Use the commutative property of addition to put the terms in exponential order: $-x^2 + 10x - 56$.

265. Change subtraction to addition and change the signs of the terms that follow: $3r^2 + r(2 + -r) + 6(r + 4)$. Use the distributive property of multiplication on the second term: $r(2 + -r) = r \times 2 + r \times -r$; $r \times 2 + r \times -r = 2r + -r^2$. Use the distributive property of multiplication on the third term: $6(r + 4) = 6 \times r + 6 \times 4$; $6 \times r + 6 \times 4 = 6r + 24$. Substitute the results into the expression: $3r^2 + (2r + -r^2) + (6r + 24)$. Remove the parentheses: $3r^2 + 2r + -r^2 + 6r + 24$. Use the commutative property of addition to put like terms together: $3r^2 + -r^2 + 2r + 6r + 24$. Use the associative property for addition: $(3r^2 + -r^2) + (2r + 6r) + 24$. Add the first term using the rules for terms with different signs: $3r^2 + -r^2 = 2r^2$. Add the second term using the rules for terms with the same signs: $2r + 6r = 8r$. Substitute the results into the expression: $(2r^2) + (8r) + 24$. Remove the parentheses: $2r^2 + 8r + 24$.

266. Change subtraction to addition and change the signs of the terms that follow: $2x + -x(5 + y) + 3xy + 4(2x + -y)$. Use the distributive property of multiplication on the second term: $-x(5 + y) = -x \times 5 + -x \times y$. Use the rules for multiplying signed terms: $-x \times 5 + -x \times y = -5x + -xy$. Use the distributive property of multiplication on the fourth term: $4(2x + -y) = 4 \times 2x + 4 \times -y$. Use the rules for multiplying signed terms: $4 \times 2x + 4 \times -y = 8x + -4y$. Substitute the results into the expression: $2x + (-5x + -xy) + 3xy + (8x + -4y)$. Remove the parentheses: $2x + -5x + -xy + 3xy + 8x + -4y$. Use the associative and commutative properties for addition: $(2x + -5x + 8x) + (-xy + 3xy) + (-4y)$. Add the first set of terms using the rules for terms with different signs: $2x + -5x + 8x = 5x$. Add the second set of terms using the rules for terms with different signs: $-xy + 3xy = 2xy$. Substitute the results into the expression: $(-5x) + (2xy) + (-4y)$. Remove the parentheses: $5x + 2xy - 4y$.

267. Change subtraction to addition and change the signs of the terms that follow: $-7(c + -2d) + 21c + -3(d + -5)$. Use the distributive property of multiplication on the first term: $-7(c + -2d) = -7 \times c + -7 \times -2d$. Use the rules for multiplying signed terms: $-7 \times c + -7 \times -2d = -7c + 14d$. Use the distributive property of multiplication on the third term: $-3(d + -5) = -3 \times d + -3 \times -5$. Use the rules for multiplying signed terms: $-3 \times d + -3 \times -5 = -3d + 15$. Substitute the results into the last expression: $(-7c + 14d) + 21c + (-3d + 15)$. Remove the parentheses: $-7c + 14d + 21c + -3d + 15$. Use the commutative property of addition to move terms together: $21c + -7c + 14d + -3d + 15$. Use the associative property for addition: $(21c + -7c) + (14d + -3d) + 15$. Combine like terms using addition rules for signed numbers: $(14c) + (11d) + 15$. Remove the parentheses: $14c + 11d + 15$.

268. Change subtraction to addition and change the signs of the terms that follow: $5(3x + -y) + x(5 + 2y) + -4(3 + x)$. Use the distributive property of multiplication on the first term: $5(3x + -y) = 5 \times 3x + 5 \times -y$. Use the rules for multiplying signed terms: $5 \times 3x + 5 \times -y = 15x + -5y$. Use the distributive property of multiplication on the second term: $x(5 + 2y) = x \times 5 + x \times 2y$. Use the rules for multiplying signed terms: $x \times 5 + x \times 2y = 5x + 2xy$. Use the distributive property of multiplication on the third term: $-4(3 + x) = -4 \times 3 + -4 \times x$. Use the rules for multiplying signed terms: $-4 \times 3 + -4 \times x = -12 + -4x$. Substitute the results into the original expression: $(15x + -5y) + (5x + 2xy) + (-12 + -4x)$. Remove the parentheses: $15x + -5y + 5x + 2xy + -12 + -4x$. Use the commutative property of addition to move like terms together. Use the associative property for addition: $(15x + 5x + -4x) + -5y + 2xy + -12$. Combine like terms using addition rules for signed numbers: $(16x) + -5y + 2xy + -12$. Adding a negative term is the same as subtracting a positive term: $(16x) - (5y) + 2xy - (12)$. Remove the parentheses: $16x - 5y + 2xy - 12$.

269. Change subtraction to addition and change the signs of the terms that follow: $6(m + -3n) + 3m(n + 5) + -2n(3 + -m)$. Use the distributive property of multiplication on the first term: $6(m + -3n) = 6 \times m + 6 \times -3n$. Use the rules for multiplying signed terms: $6 \times m + 6 \times -3n = 6m + -18n$. Use the distributive property of multiplication on the second term: $3m(n + 5) = 3m \times n + 3m \times 5$. Use the rules for multiplying signed terms: $3m \times n + 3m \times 5 = 3mn + 15m$. Use the distributive property of multiplication on the third term: $-2n(3 + -m) = -2n \times 3 + -2n \times -m$. Use the rules for multiplying signed terms: $-2n \times 3 + -2n \times -m = -6n + 2mn$. Substitute the results into the original expression: $(6m + -18n) + (3mn + 15m) + (-6n + 2mn)$. Remove the parentheses: $6m + -18n + 3mn + 15m + -6n + 2mn$. Use the commutative property of addition to move like terms together. Use the associative property for addition: $(6m + 15m) + (3mn + 2mn) + (-6n + -18n)$. Combine like terms using addition rules for signed numbers: $(21m) + (5mn) + (-24n)$. Adding a negative term is the same as subtracting a positive term: $21m + 5mn - 24n$.

270. Change subtraction to addition and change the signs of the terms that follow: $9(2x + -t) + 23xt + x(-4 + 5t)$. Use the distributive property of multiplication on the first term: $9(2x + -t) = 9 \times 2x + 9 \times -t$. Use the rules for multiplying signed terms: $9 \times 2x + 9 \times -t = 18x + -9t$. Use the distributive property of multiplication on the third term: $x(-4 + 5t) = x \times -4 + x \times 5t$. Use the rules for multiplying signed terms: $x \times -4 + x \times 5t = -4x + 5xt$. Substitute the results into the expression: $(18x + -9t) + 23xt + (-4x + 5xt)$. Remove the parentheses: $18x + -9t + 23xt + -4x + 5xt$. Use the commutative property of addition to move like terms together: $18x + -4x + -18t + 23xt + 5xt$. Use the associative property for addition: $(18x + -4x) + -9t + (23xt + 5xt)$. Combine like terms using addition rules for signed numbers: $(14x) + -9t + (28xt)$. Adding a negative term is the same as subtracting a positive term: $14x - 9t + 28xt$.

271. Change subtraction to addition and change the signs of the terms that follow: $4[2a(a + 3) + 6(4 + -a)] + 5a^2$. Simplify the terms inside the outer parentheses (brackets) first: $2a(a + 3) + 6(4 + -a)$. Use the distributive property of multiplication on the first term: $2a(a + 3) = 2a \times a + 2a \times 3$. Use the rules for multiplying signed terms: $2a \times a + 2a \times 3 = 2a^2 + 6a$. Use the distributive property of multiplication on the second term: $6(4 + -a) = 6 \times 4 + 6 \times -a$. Use the rules for multiplying signed terms: $6 \times 4 + 6 \times -a = 24 + -6a$. Substitute the results into the expression: $(2a^2 + 6a) + (24 + -6a)$. Remove the parentheses: $2a^2 + 6a + 24 + -6a$. Use the commutative property of addition: $2a^2 + 6a + -6a + 24$. Use the associative property for addition: $2a^2 + (6a + -6a) + 24$. Combine like terms using addition rules for signed numbers: $2a^2 + (0) + 24$. Use the identity property of addition: $2a^2 + 24$. Substitute the results into the expression: $4(2a^2 + 24) + 5a^2$. Use the distributive property of multiplication on the first term: $4 \times 2a^2 + 4 \times 24; 8a^2 + 96$. Substitute into the expression: $(8a^2 + 96) + 5a^2$. Remove the parentheses: $8a^2 + 96 + 5a^2$. Use the commutative property of addition: $8a^2 + 5a^2 + 96$. Use the associative property for addition: $(8a^2 + 5a^2) + 96$. Add like terms: $13a^2 + 96$.

272. Change subtraction to addition andchange the signs of the terms that follow: $8(2a + -b + -3c) + 3(2a + -b) + -4(6 + -b)$. Use the distributive property of multiplication on the first term: $8(2a + -b + -3c) = 8 \times 2a + 8 \times -b + 8 \times -3c$. Use the rules for multiplying signed terms: $8 \times 2a + 8 \times -b + 8 \times -3c = 16a + -8b + -24c$. Use the distributive property of multiplication on the second term: $3(2a + -b) = 3 \times 2a + 3 \times -b$. Use the rules for multiplying signed terms: $3 \times 2a + 3 \times -b = 6a + -3b$. Use the distributive property of multiplication on the third term: $-4(6 + -b) = -4 \times 6 + -4 \times -b$. Use the rules for multiplying signed terms: $-4 \times 6 + -4 \times -b = -24 + 4b$. Substitute the results into the expression: $(16a + -8b + -24c) + (6a + -3b) + (-24 + 4b)$. Remove the parentheses: $16a + -8b + -24c + 6a + -3b + -24 + 4b$. Use the commutative property of addition to move like terms together: $16a + 6a + -8b + -3b + 4b + -24c + -24$. Use the associative property for addition: $(16a + 6a) + (-8b + -3b + 4b) + -24c + -24$. Combine like terms using addition rules for signed numbers: $(22a) + (-7b) + -24c + -24$. Adding a negative term is the same as subtracting a positive term: $22a - 7b - 24c - 24$.

273. a. The translation of "two times the number of hours" is $2x$. Four hours less than $2x$ becomes $2x - 4$.

274. c. When the key words *less than* appear in a sentence, it means that you will subtract from the next part of the sentence, so the number that you subtract will appear at the end of the expression. "Four times a number" is equal to $4x$ in this problem. Three less than $4x$ is $4x - 3$.

275. b. Each one of the answer choices would translate to $9y - 5$ except for choice **b.** The word *sum* is a key word for addition, and $9y$ means "9 times y."

276. b. Because Susan starts one hour before Dee, Dee works for one less hour than Susan works; thus, $x - 1$.

277. c. Frederick would multiply the number of books, 11, by how much each one cost, d. For example, if each one of the books cost $10, he would multiply 11 times $10 and get $110. Therefore, the answer is $11d$.

278. a. In this problem, multiply d and w to get the total days in one month and then multiply that result by m, to get the total days in the year. This can be expressed as mwd, which means m times w times d.

279. a. To calculate the total she received, multiply x dollars per hour times h, the number of hours she worked. This becomes hx. Divide this amount by 2, because she gave half to her friend. Thus, $\frac{hx}{2}$ is how much money she has left.

280. d. The cost of the call is x cents plus y times the additional minutes. Because the call is ten minutes long, the caller will pay x cents for one minute and y cents for the other nine. Therefore the expression is $1x + 9y$, or $x + 9y$, as it is not necessary to write a 1 in front of a variable.

281. a. Start with Jim's age, y, since he appears to be the youngest. Melissa is four times as old as he is, so her age is $4y$. Pat is five years older than Melissa, so Pat's age would be Melissa's age, $4y$, plus another five years; thus, $4y + 5$.

282. c. Because she worked 48 hours, Sally will get paid her regular amount, x dollars per hour, for 40 hours and a different amount, y dollars per hour, for the additional eight hours. This becomes 40 times x plus eight times y, which translates to $40x + 8y$.

283. b. This problem translates to the expression $6 \times 2 + 4$. Using order of operations, do the multiplication first: $6 \times 2 = 12$; then add $12 + 4 = 16$ inches.

284. c. This translates to the expression $2 + 3 \times 4 - 2$. Using order of operations, multiply 3×4 first; the problem becomes $2 + 12 - 2$. Add and subtract the numbers in order from left to right: $2 + 12 = 14$; $14 - 2 = 12$.

285. b. This problem translates to the expression $10 - 4(8 - 3) + 1$. Using order of operations, do the operation inside the parentheses first: $10 - 4(5) + 1$. Multiplication is next, so multiply 4×5; the problem becomes $10 - 20 + 1$. Add and subtract in order from left to right: $10 - 20 = -10$; $-10 + 1 = -9$.

286. c. This problem translates to the expression $3\{[3 - (-7 + 6)] + 8\}$. When dealing with multiple grouping symbols, start from the innermost set and work your way out. Add and subtract in order from left to right inside the brackets. First, $-7 + 6 = -1$. Remember that subtraction is the same as adding the opposite, so $3 - (-1)$ is $3 + (+1) = 4$; the problem becomes $3(4 + 8)$. Multiply 3×12 to finish the problem: $3 \times 12 = 36$.

287. c. If the total amount for both is 80 dollars, then the amount for one person is 80 minus the amount of the other person. John has x dollars, so Charlie's amount is $80 - x$.

288. d. In the statement, the order of the numbers does not change; however, the grouping of the numbers in parentheses does. Each side, if simplified, results in an answer of 300, even though both sides look different. Changing the grouping in a problem like this is an example of the associative property of multiplication.

289. c. Choice **a** is an example of the associative property of addition, where changing the grouping of the numbers will still result in the same answer. Choice **b** is an example of the distributive property of multiplication over addition. Choice **d** is an example of the identity property of addition, where any number added to zero equals itself. Choice **c** is an example of the commutative property of addition, where we can change the order of the numbers that are being added and the result is always the same.

290. b. In the statement, 3 is being multiplied by the quantity in the parentheses, $x + 4$. The distributive property allows you to multiply $3 \times x$ and add it to 3×4, simplifying to $3x + 12$.

291. c. The terms $3x$ and $5x$ are like terms because they have exactly the same variable with the same exponent. Therefore, you just add the coefficients and keep the variable: $3x + 5x = 8x$.

292. c. Because the question asks for the *difference* between the areas, you need to subtract the expressions: $6a + 2 - 5a$. Subtract like terms: $6a - 5a + 2 = 1a + 2$; $1a = a$, so the simplified answer is $a + 2$.

293. b. Since the area of a rectangle is $A = length \times width$, multiply $(x^3)(x^4)$. When multiplying like bases, add the exponents: $x^{3+4} = x^7$.

294. c. Since the area of the soccer field would be found by the formula $A = length \times width$, multiply the dimensions together: $7y^2 \times 3xy$. Use the commutative property to arrange like variables and the coefficients next to each other: $7 \times 3 \times x \times y^2 \times y$. Multiply; remember that $y^2 \times y = y^2 \times y^1 = y^{2+1} = y^3$. The answer is $21xy^3$.

295. a. Since the area of a parallelogram is $A = base \times height$, then the area divided by the base would give you the height: $\frac{x^8}{x^4}$; when dividing like bases, subtract the exponents: $x^{8-4} = x^4$.

296. d. The key word *quotient* means division, so the problem becomes $\frac{3d^4}{12d^6}$. Divide the coefficients: $\frac{1d^4}{4d^6}$. When dividing like bases, subtract the exponents: $\frac{1d^{-2}}{4}$. A variable in the numerator with a negative exponent is equal to the same variable in the denominator with the opposite sign—in this case, a positive sign on the exponent: $\frac{1}{4d^2}$.

297. a. The translation of the question is $\frac{6x^2 \times 4xy^2}{3x^3y}$. The key word *product* tells you to multiply $6x^2$ and $4xy^2$. The result is then divided by $3x^3y$. Use the commutative property in the numerator to arrange like variables and the coefficients together: $\frac{6 \times 4x^2xy^2}{3x^3y}$. Multiply in the numerator. Remember that $x^2 \times x = x^2 \times x^1 = x^{2+1} = x^3$: $\frac{24x^3y^2}{3x^3y}$. Divide the coefficients; $24 \div 3 = 8$: $\frac{24x^3y^2}{3x^3y}$. Divide the variables by subtracting the exponents: $8x^{3-3}y^{2-1}$; simplify. Recall that anything to the zero power is equal to 1: $8x^0y^1 = 8y$.

298. b. Since the formula for the area of a square is $A = s^2$, then by substituting, $A = (a^2b^3)^2$. Multiply the outer exponent by each exponent inside the parentheses: $a^{2 \times 2}b^{3 \times 2}$. Simplify: a^4b^6.

299. a. The statement in the question would translate to $3x^2(2x^3y)^4$. The word *quantity* reminds you to put that part of the expression in parentheses. Evaluate the exponent by multiplying each number or variable inside the parentheses by the exponent outside the parentheses: $3x^2(2^4x^{3 \times 4}y^4)$; simplify: $3x^2(16x^{12}y^4)$. Multiply the coefficients and add the exponents of like variables: $3(16x^{2 + 12}y^4)$; simplify: $48x^{14}y^4$.

300. d. Since the area of a rectangle is $A = length \times width$, multiply the dimensions to find the area: $2x(4x + 5)$. Use the distributive property to multiply each term inside the parentheses by $2x$: $2x \times 4x + 2x \times 5$. Simplify by multiplying the coefficients of each term and adding the exponents of the like variables: $8x^2 + 10x$.

301. b. The translated expression would be $-9p^3r(5p - 6r)$. Remember that the key word *product* means multiply. Use the distributive property to multiply each term inside the parentheses by $-9p^3r$: $-9p^3r \times 5p - (-9p^3r) \times 6r$. Simplify by multiplying the coefficients of each term and adding the exponents of the like variables: $-9 \times 5p^{3 + 1}r - (-9 \times 6p^3r^{1 + 1})$. Simplify: $-45p^4r - (-54p^3r^2)$. Change subtraction to addition and change the sign of the following term to its opposite: $-45p^4r + (+54p^3r^2)$; this simplifies to: $-45p^4r + 54p^3r^2$.

302. c. The two numbers in terms of x would be $x + 3$ and $x + 4$ since *increased by* would tell you to add. *Product* tells you to multiply these two quantities: $(x + 3)(x + 4)$. Use **FOIL** (first terms of each binomial multiplied, outer terms of each multiplied, inner terms of each multiplied, and last terms of each binomial multiplied) to multiply the binomials: $(x \times x) + (4 \times x) + (3 \times x) + (3 \times 4)$; simplify each term: $x^2 + 4x + 3x + 12$. Combine like terms: $x^2 + 7x + 12$.

303. d. Since the area of a rectangle is $A = length \times width$, multiply the two expressions together: $(2x - 1)(x + 6)$. Use **FOIL** (first terms of each binomial multiplied, outer terms of each multiplied, inner terms of each multiplied, and last terms of each binomial multiplied) to multiply the binomials: $(2x \times x) + (2x \times 6) - (1 \times x) - (1 \times 6)$. Simplify: $2x^2 + 12x - x - 6$; combine like terms: $2x^2 + 11x - 6$.

Algebraic Equations

WE CAN USE OUR SKILLS at combining like terms and evaluating expressions to help us solve algebraic equations. While an algebraic expression is an *expression* containing one or more variables, an **algebraic equation** is an *equation* containing one or more variables. An equation is an expression with an equal sign.

Algebraic equations are at the heart of much of the math you will learn in high school and college. Even some geometry questions are as much algebra as they are geometry. This chapter is where you will build the foundation for most of the math you will learn in the future.

When solving any equation, there is one unbreakable rule: Always perform the same action on both sides of the equation. If you add 3 to one side of the equation, add 3 to the other side of the equation. If you divide one side of the equation by 20, you must divide the other side of the equation by 20.

▶ One-Step Equations

A one-step equation can be solved by performing a single operation. Usually, that one operation is addition, subtraction, multiplication, or division, but finding a square or square root might also help you solve an equation. We solve equations by using operations to isolate the variable on one side of the equation.

▶ Addition and Subtraction Equations

Addition and subtraction equations go together. To solve an addition equation, we must use subtraction. Take the number being added to the variable and subtract it from both sides of the equation to leave the variable alone on one side of the equation. To solve a subtraction equation, we must use addition. Take the number being subtracted from the variable and add it to both sides of the equation so that the variable remains alone on one side of the equation.

Example
$x + 5 = 8$

To solve this addition equation, subtract 5 from both sides of the equation. Remember, in order to keep an equation true, whatever we do to one side of the equation, we must do to the other side: $x + 5 - 5 = 8 - 5$, and $x = 3$.

Example
$w - 9 = 22$

To solve this subtraction equation, use addition. Add 9 to both sides of the equation: $w - 9 + 9 = 22 + 9$, and $w = 31$. Since $31 - 9 = 22$, our answer is correct.

▶ Multiplication and Division Equations

Multiplication and division equations also go together. To solve a multiplication equation, divide both sides of the equation by the coefficient of the variable. To solve a division equation, multiply both sides of the equation by the number that is dividing the variable.

Example
$5k = 35$

The coefficient of the variable is 5, so we must divide both sides of the equation by 5: $\left(\frac{5k}{5}\right) = \frac{35}{5}$, and $k = 7$.

Example
$\frac{j}{4} = 3$

The variable is being divided by 4, so we must multiply both sides of the equation by 4. $4\left(\frac{j}{4}\right) = 4(3)$, and $j = 12$.

TIP

When a variable is being multiplied or divided by a number, we can always multiply the variable by the reciprocal of that number. To find the reciprocal of a number, reverse the values of the numerator and denominator of the number. Remember, a whole number has a denominator of 1. Let's look again at the last two examples.

$5k = 35$

The reciprocal of 5, or $\frac{5}{1}$, is $\frac{1}{5}$. Multiply both sides of the equation by $\frac{1}{5}$: $\frac{1}{5}(5k) = \frac{1}{5}(35)$, and $k = 7$. Multiplying by $\frac{1}{5}$ is the same as dividing by 5.

$\frac{j}{4} = 3$

We could also write this problem as $\frac{1}{4}j = 3$. The reciprocal of $\frac{1}{4}$ is 4: $4(\frac{1}{4})j = 4(3)$, and $j = 12$.

▶ Word Problems

A word problem might describe a situation where a quantity is unknown. Represent that unknown with a variable (such as x), and then solve the equation for x.

Example
Tom and Jerry have $120. If Tom has $48, how much does Jerry have?

Let x represent the amount of money Jerry has. Since Tom has $48 and he and Jerry have a total of $120, $48 + x = $120. Solve the equation by subtracting $48 from both sides of the equation: $48 + x − $48 = $120 − $48, and $x = $72. Jerry has $72.

▶ Practice Questions

Solve the following equations.

304. $a + 21 = 32$

305. $x - 25 = 32$

306. $y + 17 = -12$

307. $b - 15 = 71$

308. $12 - c = -9$

309. $s - -4 = -1$

310. $a + 1\frac{3}{4} = 6\frac{1}{4}$

311. $b - \frac{5}{2} = -\frac{2}{3}$

312. $c - 4(\frac{1}{2} - 5) = 20$

313. $m + 2(5 - 24) = -76$

314. $2a = 24$

315. $4x = -20$

316. $-3y = 18$

317. $27b = 9$

318. $45r = -30$

319. $0.2c = 5.8$

320. $\frac{x}{7} = 16$

321. $\frac{y}{-4} = -12$

322. $\frac{2}{3}a = 54$

323. $\frac{8}{5}b = -56$

324. Jack paid \$21,000 for his new car. This was $\frac{7}{8}$ the suggested selling price of the car. What was the suggested selling price of the car?

325. After putting 324 teddy bears into packing crates, there were 54 crates filled with bears. If each crate contained the same number of bears, how many bears were in each packing crate?

326. Only 3% of turtle hatchlings will live to become breeding adults. How many turtles must have been born if the current number of breeding adults is 1,200?

327. This year, a farmer planted 300 acres of corn. This was 1.5 times as many acres as he planted last year. How many acres did he plant last year?

328. A business executive received a \$6,000 bonus check from her company at the end of the year. This was 5% of her annual salary. How much was her annual salary before receiving the bonus?

▶ Multistep Equations

Sometimes, more than one step is needed to get a variable alone on one side of the equation. We might need to add a number to both sides of an equation and then divide both sides of the equation by a number, or we may need to use both division and subtraction.

Example
$5u + 13 = 48$

In this example, the variable u is multiplied by 5, and then 13 is added to that product. Since there are multiplication and addition in this equation, we will need to use division and subtraction to solve the equation. First, subtract 13 from both sides of the equation.

$5u + 13 - 13 = 48 - 13$

$5u = 35$

Now we're looking at a one-step equation, and we already know how to solve these. Divide both sides of the equation by 5: $\frac{5u}{5} = \frac{35}{5}$, and $u = 7$.

TIP

In the preceding example, we needed to use division and subtraction. Why did we subtract first? In general, it is best to use addition and subtraction before multiplication or division when solving multistep equations. If we began by dividing, we would have had to divide three terms (instead of two) by 5, and that would have left us with fractions:

$u + \frac{13}{5} = \frac{48}{5}$

Our next step would have been to subtract $\frac{13}{5}$ from both sides of the equation. Subtracting 13 is a lot easier than subtracting $\frac{13}{5}$! By adding or subtracting first, you are less likely to have to work with either very large numbers (if you multiply first) or fractions (if you divide first).

TIP

No matter how many steps are required to isolate the variable, each step should reduce the number of terms on one or both sides of the equation.

Example

$\frac{q}{8} - 24 = -18$

Since there is division and subtraction in this equation, we will need to use multiplication and addition to solve the equation. Start with addition:

$\frac{q}{8} - 24 + 24 = -18 + 24$

$\frac{q}{8} = 6$

Now multiply: $8(\frac{q}{8}) = 8(6)$, and $q = 48$.

Let's check our answer. It becomes more important to check your work as the number of steps needed to solve an equation increases, because there are more places in which you could make a mistake. Substitute 48 for q in the original equation:

$\frac{48}{8} - 24 = -18$

$6 - 24 = -18$

$-18 = -18$

Our answer is correct.

▶ Practice Questions

Find the solutions to the following equations.

329. $4x + 7 = 11$

330. $13x + 21 = 60$

331. $3x - 8 = 16$

332. $5x - 6 = -26$

333. $\frac{x}{3} + 4 = 10$

334. $\frac{x}{7} - 5 = 1$

335. $39 = 3a - 9$

336. $4 = 4a + 20$

337. $10a + 5 = 7$

338. $0.3a + 0.25 = 1$

339. $\frac{2}{3}m + 8 = 20$

340. $9 = \frac{3}{4}m - 3$

341. $41 - 2m = 65$

342. $4m - 14 = 50$

343. $\frac{2m}{5} + 16 = 24$

344. $7m - 6 = -2.5$

345. $10s - 6 = 0$

346. $\frac{s}{4} + 2.7 = 3$

347. $8s - 7 = 41$

348. $-55 = 25 - s$

Find the solutions to the following word problems by letting a variable equal the unknown quantity, making an equation from the information given, and then solving the equation.

349. A farmer is raising a hog that weighed 20 pounds when he bought it. He expects it to gain 12 pounds per month. He will sell it when it weighs 200 pounds. How many months will it be before he will sell the animal?

350. Mary earns $1.50 less than twice Bill's hourly wage. Mary earns $12.50 per hour. What is Bill's hourly wage?

351. At year's end, a share of stock in Axon Corporation was worth $37. This was $8 less than three times its value at the beginning of the year. What was the price of a share of Axon stock at the beginning of the year?

352. By changing jobs, Jennifer increased her income to $4,000 more than one and half times her former salary. She earns $64,000 at her new job. What was her salary at her previous employment?

353. Twenty-five more girls than two-thirds the number of boys participate in interscholastic sports at a local high school. If the number of girls participating is 105, how many boys participate?

▶ Solving Multistep Equations by Combining Like Terms

In the preceding chapter, we learned to combine like terms to simplify and evaluate algebraic expressions. Now, we'll combine like terms to help us solve algebraic equations. So far, the examples we've seen have had only one variable in them, and only one occurrence of that variable. These next examples have one variable, but more than one occurrence of it.

Example
$3g + 8 = 5g - 4$

This equation has multiplication, addition, and subtraction. It also has a g term on both sides of the equation. We can start by subtracting 8 from both sides of the equation, or we could start by adding 4 to both sides of the equation. Another approach would be to start by combining the g terms. To do so, we can either sub-

tract $3g$ from both sides of the equation or subtract $5g$ from both sides of the equation. We have a lot of options, but no matter which we choose, we will arrive at the correct answer.

Let's begin by combining the g terms. We'll subtract $3g$ from both sides rather than subtracting $5g$ so that we don't have to work with negative numbers.

$3g + 8 - 3g = 5g - 4 - 3g$
$8 = 2g - 4$

Next, we can either subtract 8 from both sides or add 4 to both sides. If we subtract 8 from both sides, we won't have the g term alone on one side of the equation, which is our goal. We therefore add 4 to both sides of the equation:

$8 + 4 = 2g - 4 + 4$
$12 = 2g$

Now we're one step away from our answer. Divide both sides of the equation by 2: $\frac{12}{2} = \frac{2g}{2}$, and $g = 6$. Check the answer with substitution:

$3(6) + 8 = 5(6) - 4$
$18 + 8 = 30 - 4$
$26 = 26$

The two sides of the equation are equal, so our answer is correct.

Example
$3p + 5 - 6(p - 2) = 2(13 + p) - 4$

This example contains many operations. Begin by distributing the 6 on the left side of the equation and the 2 on the right side of the equation. This will give us a clearer idea of what we need to do.

$3p + 5 - 6p + 12 = 26 + 2p - 4$

Next, combine like terms on each side of the equation. On the left side, $3p - 6p = -3p$

and $5 + 12 = 17$. On the right side, $26 - 4 = 22$. The equation is now:

$-3p + 17 = 22 + 2p$

This looks a little like the preceding example. Combine the p terms. We can either add $3p$ to both sides of the equation or subtract $2p$ from both sides of the equation. It doesn't matter which we do, so let's add $3p$:

$-3p + 17 + 3p = 22 + 2p + 3p$

$17 = 22 + 5p$

To get the p term alone on the right side of the equation, subtract 22 from both sides:

$17 - 22 = 22 + 5p - 22$

$\quad\quad -5 = 5p$

We have just one step left now: Divide both sides by 5: $\frac{-5}{5} = \frac{5p}{5}$, so $p = -1$. The check proves our answer is true:

$3(-1) + 5 - 6[(-1) - 2] = 2[13 + (-1)] - 4$

$-3 + 5 - 6(-3) = 2(12) - 4$

$2 + 18 = 24 - 4$

$20 = 20$

Every equation we've seen so far has had exactly one solution. However, some equations can have infinitely many solutions—or no solutions at all. If the quantity on one side of the equation is identical to the quantity on the other side of the equation after like terms are combined and simplified, then the equation has infinite solutions. If the two quantities on either side of the equal sign are not in fact equal, the equation will have no solution.

Example

$5(2x + 4) = 10x + 20$

First, distribute the 5 on the left side of the equation. $5(2x + 4) = 10x + 20$, which is the same as the quantity on the right

side of the equation. If we subtract 20 from both sides of the equation, we are left with $10x = 10x$. Divide both sides by 10, and we have $x = x$; x will equal x for any value of x, so this equation has infinite solutions.

Example

$4(3x + 5) = 2(6x - 10)$

Begin by distributing the 4 on the left side of the equation and the 2 on the right side of the equation:

$4(3x + 5) = 12x + 20$

$2(6x - 10) = 12x - 20$

The equation is now $12x + 20 = 12x - 20$. If we subtract $12x$ from both sides of the equation, we are left with $20 = -20$, which is never true. This equation has no solution.

▶ Practice Questions

Find the solutions for the following equations.

354. $11x + 7 = 3x - 9$

355. $3x - 23 = 54 - 4x$

356. $5x + 3 + 6x = 10x + 9 - x$

357. $10x + 27 - 5x - 46 = 32 + 3x - 19$

358. $20x - 11 - 3x = 9x + 43$

359. $0.4 + 3x - 0.25 = 1.15 - 2x$

360. $2x + 17 - 1.2x = 10 - 0.2x + 11$

361. $2 + 6x - 0.2 = 5x + 2.1$

362. $0.4 + 3x - 0.25 = 1.15 - 2x$

363. $3x + 12 - 0.8x = 3.4 - 0.8x - 9.4$

364. $7(x + 2) + 1 = 3(x + 14) - 4x$

365. $4(4x + 3) = 6x - 28$

366. $13 - 8(x - 2) = 7(x + 4) + 46$

367. $13x + 3(3 - x) = -3(4 + 3x) - 2x$

368. $2(2x + 19) - 9x = 9(13 - x) + 21$

369. $12x - 4(x - 1) = 2(x - 2) + 16$

370. $2x + 1\frac{4}{5}x = 1 + 3x$

371. $\frac{5}{2}(x - 2) + 3x = 3(x + 2) - 10$

372. $6(\frac{1}{2}x + \frac{1}{2}) = 3(x + 1)$

373. $0.7(0.2x - 1) = 0.3(3 - 0.2x)$

374. $10(x + 2) + 7(1 - x) = 3(x + 9)$

375. $4(9 - x) = 2x - 6(x + 6)$

376. $5(2x + 3) - 9 = 14(x + 1)$

377. $7(x - 10) + 110 = 4(x - 25) + 7x$

378. $0.8(x + 20) - 4.5 = 0.7(5 + x) - 0.9x$

▶ Common Word Problems

Now that we know how to solve algebraic equations, let's look at a few common formulas:

> distance formula: $D = rt$ where r = rate and
> t = time
> simple interest: $I = prt$ where p = principal,
> r = interest rate, and t = time in years
> area of a trapezoid: $A = \frac{1}{2}h(b_1 + b_2)$ where
> h = height and b_1 and b_2 are the bases
> Fahrenheit/Celsius equivalence: $C = \frac{5}{9}(F - 32)$;
> $F = \frac{9}{5}C + 32$
> volume of a cylinder: $V = \pi r^2 h$ (let π = 3.14)
>   where r = radius and h = height
> surface area of a cylinder: $S = 2\pi r(r + h)$
>   where r = radius of the base and
> h = height of cylinder (let π = 3.14)

The distance formula consists of distance, rate, and time. Given any two of those quantities, we can find the third using multiplication or division. The simple interest formula is similar—given any three of the four values in that formula, you can find the fourth value using multiplication or division. The temperature formula may look complicated, but it's not—given the temperature in either Celsius or Fahrenheit, you can find the temperature in the other unit using multiplication and addition.

Example

If the temperature is 86 degrees Fahrenheit, what is the temperature in degrees Celsius?

Substitute 86 for F in the equation:

$C = \frac{5}{9}(86 - 32)$

$C = \frac{5}{9}(54)$

$C = 5(6)$

$C = 30$ degrees

► **Practice Questions**

379. An airplane flies at an average velocity of 350 miles per hour. A flight from Miami to Aruba takes three and a half hours. How far is it from Miami to Aruba?

380. A bicycle tour group planned to travel 68 miles between two New England towns. How long would the trip take if they averaged 17 miles per hour for the trip?

381. A hiking group wanted to travel along a segment of the Appalachian Trail for four days. They planned to hike six hours per day and wanted to complete a trail segment that was 60 miles long. What rate of speed would they have to average to complete the trail segment as planned?

382. At an interest rate of 6%, how much interest would $12,000 earn over two years?

383. Over a three-year period, the total interest paid on a $4,500 loan was $1,620. What was the interest rate?

384. In order to earn $1,000 in interest over two years at an annual rate of 4%, how much principal must be put into a savings account?

385. How long will it take for a $3,000 savings account to double its value at a simple interest rate of 10%?

386. You plan to visit a tropical island where the average daytime temperature is 40° Celsius. What would that temperature be on the Fahrenheit scale?

387. Jeff won't play golf if the temperature falls below 50° Fahrenheit. He is going to a country where the temperature is reported in Celsius. What would Jeff's low temperature limit be in that country? (Round your answer to the nearest degree.)

388. The temperature in Hillsville was 20° Celsius. What is the equivalent of this temperature in degrees Fahrenheit?
   a. 4°
   b. 43.1°
   c. 68°
   d. 132°

389. Peggy's town has an average temperature of 23° Fahrenheit in the winter. What is the average temperature on the Celsius scale?
   a. −16.2°
   b. 16.2°
   c. 5°
   d. −5°

390. Celine deposited $505 into her savings account. If the interest rate of the account is 5% per year, how much interest will she have made after four years?
   a. $252.50
   b. $606
   c. $10,100
   d. $101

391. A certain bank pays 3.4% interest per year for a certificate of deposit (CD). What is the total balance of an account after 18 months with an initial deposit of $1,250?
   a. $765
   b. $2,015
   c. $63.75
   d. $1,313.75

392. Joe took out a car loan for $12,000. He is paying $4,800 in interest at a rate of 8% per year. How many years will it take him to pay off the loan?

a. 5

b. 2.5

c. 8

d. 4

393. What is the annual interest rate on an account that earns $711 in simple interest over 36 months with an initial deposit of $7,900?

a. 4.3%

b. 3%

c. 30%

d. 4%

394. If 1 is added to the difference when $10x$ is subtracted from $-18x$, the result is 57. What is the value of x?

a. -2

b. -7

c. 2

d. 7

395. If 0.3 is added to 0.2 times the quantity $x - 3$, the result is 2.5. What is the value of x?

a. 1.7

b. 26

c. 14

d. 17

396. If twice the quantity $x + 6$ is divided by negative 7, the result is 6. Find the number.

a. -18

b. -27

c. -15

d. -0.5

397. The difference between six times the quantity $6x + 1$ and three times the quantity $x - 1$ is 108. What is the value of x?

a. $\frac{12}{11}$

b. $\frac{35}{11}$

c. 12

d. 3

398. Negative 4 is multiplied by the quantity $x + 8$. If $6x$ is then added to this, the result is $2x + 32$. What is the value of x?

a. no solution

b. identity

c. 0

d. 16

399. A telephone company charges $0.35 for the first minute of a phone call and $0.15 for each additional minute of the call. Which of the following represents the cost y of a phone call lasting x minutes?

a. $y = 0.15(x - 1) + 0.35$

b. $x = 0.15(y - 1) + 0.35$

c. $y = 0.15x + 0.35$

d. $x = 0.15y + 0.35$

400. A ride in a taxicab costs $1.25 for the first mile and $1.15 for each additional mile. Which of the following could be used to calculate the total cost y of a ride that is x miles?

a. $x = 1.25(y - 1) + 1.15$

b. $x = 1.15(y - 1) + 1.25$

c. $y = 1.25(x - 1) + 1.15$

d. $y = 1.15(x - 1) + 1.25$

401. Jackie invested money in two different accounts, one of which earned 12% interest per year and another that earned 15% interest per year. The amount invested at 15% was $100 more than twice the amount at 12%. How much was invested at 12% if the total annual interest earned was $855?

a. $4,100

b. $2,100

c. $2,000

d. $4,000

402. Kevin invested $4,000 in an account that earns 6% interest per year and $x in a different account that earns 8% interest per year. How much is invested at 8% if the total amount of interest earned annually is $405.50?

a. $2,075.00

b. $4,000.00

c. $2,068.75

d. $2,075.68

403. Megan bought x pounds of coffee that cost $3 per pound and 18 pounds of coffee at $2.50 per pound for the company picnic. Find the total number of pounds of coffee purchased if the average cost per pound of both types together is $2.85.

a. 42

b. 18

c. 63

d. 60

404. The student council bought two different types of candy for the school fair. They purchased 40 pounds of candy at $2.15 per pound and x pounds at $1.90 per pound. What is the total number of pounds they bought if the total amount of money spent on candy was $162?

a. 42

b. 40

c. 80

d. 52

405. The sum of the squares of two consecutive positive odd integers is 74. What is the value of the smaller integer?

a. 3

b. 7

c. 5

d. 11

406. If the difference between the squares of two consecutive integers is 15, find the larger integer.

a. 8

b. 7

c. 6

d. 9

TERM TO REVIEW

algebraic equation

▶ **Answers**

304. Subtract 21 from both sides of the equation. Associate like terms: $a + (21 - 21) = (32 - 21)$. Perform the numerical operation in the parentheses: $a + (0) = (11)$. Zero is the identity element for addition: $a = 11$.

305. Add 25 to each side of the equation: $x - 25 + 25 = 32 + 25$. Use the commutative property for addition: $x + 25 - 25 = 32 + 25$. Associate like terms: $x + (25 - 25) = (32 + 25)$. Perform the numerical operation in the parentheses: $x + (0) = (57)$. Zero is the identity element for addition: $x = 57$.

306. Subtract 17 from both sides of the equation: $y + 17 - 17 = -12 - 17$. Associate like terms: $y + (17 - 17) = (-12 - 17)$. Change subtraction to addition and change the sign of the term that follows: $y + (17 + -17) = (-12 + -17)$. Apply the rules for operating with signed numbers: $y + (0) = -29$. Zero is the identity element for addition: $y = -29$.

307. Change subtraction to addition and change the sign of the term that follows: $b + -15 = 71$. Add 15 to each side of the equation: $b + -15 + 15 = 71 + 15$. Associate like terms: $b + (-15 + 15) = (71 + 15)$. Apply the rules for operating with signed numbers: $b + (0) = 86$. Zero is the identity element for addition: $b = 86$.

308. Change subtraction to addition and change the sign of the term that follows: $12 + -c = -9$. Add c to each side of the equation: $12 + -c + c = -9 + c$. Associate like terms: $12 + (-c + c) = -9 + c$; $12 + (0) = -9 + c$; $12 = -9 + c$. Add 9 to each side of the equation: $9 + 12 = 9 + -9 + c$. Associate like terms: $(9 + 12) = (9 + -9) + c$. Apply the rules for operating with signed numbers: $(21) = (0) + c$. Zero is the identity element for addition: $21 = c$.

309. Change subtraction to addition and change the sign of the term that follows: $s + 4 = -1$. Add -4 to each side of the equation: $s + 4 + -4 = -1 + -4$. Associate like terms: $s + (4 + -4) = -1 + -4$. Apply the rules for operating with signed numbers: $s + (0) = (-5)$. Subtracting zero is the same as adding zero: $s = -5$.

310. Add $-1\frac{3}{4}$ to each side of the equation: $a + 1\frac{3}{4} + -1\frac{3}{4} = 6\frac{1}{4} + -1\frac{3}{4}$. Associate like terms: $a + (1\frac{3}{4} + -1\frac{3}{4}) = 6\frac{1}{4} + -1\frac{3}{4}$. Apply the rules for operating with signed numbers: $a + (0) = 4\frac{1}{2}$. Zero is the identity element for addition: $a = 4\frac{1}{2}$.

311. Change subtraction to addition and change the sign of the term that follows: $b + -\frac{5}{2} = -\frac{2}{3}$. Subtract $-\frac{5}{2}$ from both sides of the equation: $b + -\frac{5}{2} - -\frac{5}{2} = -\frac{2}{3} - -\frac{5}{2}$. Associate like terms: $b + (-\frac{5}{2} - -\frac{5}{2}) = -\frac{2}{3} - -\frac{5}{2}$. Change subtraction to addition and change the sign of the term that follows: $b + (-\frac{5}{2} + \frac{5}{2}) = -\frac{2}{3} + \frac{5}{2}$. Apply the rules for operating with signed numbers: $b + (0) = \frac{11}{6}$. Change the improper fraction to a mixed number: $b = 1\frac{5}{6}$.

312. Change subtraction to addition and change the sign of the term that follows: $c + -4(\frac{1}{2} + -5) = 20$. Use the distributive property of multiplication: $c + (-4 \times \frac{1}{2} + -4 \times -5) = 20$. Perform the operation in parentheses: $c + (-2 + 20) = 20$. Apply the rules for operating with signed numbers: $c + (18) = 20$. Subtract 18 from both sides of the equation: $c + 18 - 18 = 20 - 18$. Associate like terms: $c + (18 - 18) = 20 - 18$. Perform the operation in parentheses: $c + 0 = 2$. Zero is the identity element for addition: $c = 2$.

313. Use the distributive property of multiplication: $m + (2 \times 5 - 2 \times 24) = -76$. The order of operations is to multiply first: $m + (10 - 48) = -76$. Add 38 to each side of the equation: $m + -38 + 38 = -76 + 38$. Associate like terms: $m + (-38 + 38) = -76 + 38$. Apply the rules for operating with signed numbers: $m + (0) = -38$; $m = -38$.

314. Divide both sides of the equation by 2: $2a \div 2 = 24 \div 2$; $a = 24 \div 2$. Apply the rules for operating with signed numbers: $\frac{2a}{2} = \frac{24}{2}$; $a = 12$. Another method is: $\frac{2a}{2} = \frac{24}{2}$; $a = 12$.

315. Divide both sides of the equation by 4: $4x \div 4 = -20 \div 4$; $x = -20 \div 4$. Apply the rules for operating with signed numbers: $x = -5$: Another look for this solution method is: $\frac{4x}{4} = -\frac{20}{4}$; $x = -5$.

316. Divide both sides of the equation by -3: $\frac{-3y}{-3} = \frac{18}{-3}$; $y = \frac{18}{-3}$. Apply the rules for operating with signed numbers: $y = -6$.

317. Divide both sides of the equation by 27: $\frac{27b}{27} = \frac{9}{27}$; $b = \frac{9}{27}$. Reduce fractions to their simplest form: $b = \frac{1}{3}$.

318. Divide both sides of the equation by 45: $\frac{45r}{45} = \frac{-30}{45}$; $r = \frac{-30}{45}$. Reduce fractions to their simplest form (common factor of 15): $r = \frac{-2}{3}$.

319. Divide both sides of the equation by 0.2: $\frac{0.2c}{0.2} = \frac{5.8}{0.2}$; $c = \frac{5.8}{0.2}$. Divide: $c = 29$.

320. Multiply both sides of the equation by 7: $7(\frac{x}{7}) = 7(16)$; $x = 7(16)$. Multiply: $x = 112$.

321. Multiply both sides of the equation by -4: $-4 \times \frac{y}{-4} = -4 \times -12$; $y = -4 \times -12$. Signs the same? Multiply and the result is positive: $y = 48$.

322. Divide both sides of the equation by $\frac{2}{3}$: $\frac{2}{3}a \div \frac{2}{3} = 54 \div \frac{2}{3}$; $a = 54 \div \frac{2}{3}$. Dividing by a fraction is the same as multiplying by its reciprocal: $a = 54 \times \frac{3}{2}$; $a = 81$.

323. Divide both sides of the equation by $\frac{8}{5}$: $\frac{8}{5}b \div \frac{8}{5} = -56 \div \frac{8}{5}$; $b = -56 \div \frac{8}{5}$. Dividing by a fraction is the same as multiplying by its reciprocal: $b = -56 \times \frac{5}{8}$. There are several ways to multiply fractions and whole numbers. Here's one: $b = \frac{-56}{1} \times \frac{5}{8}$; $b = \frac{-280}{8}$; $b = -35$.

324. Let x = the suggested selling price of the car. The first and second sentences tell you that $\frac{7}{8}$ of the suggested price = \$21,000. So your equation is: $\frac{7}{8}x = \$21,000$. Divide both sides of the equation by $\frac{7}{8}$: $\frac{7}{8}x \div \frac{7}{8} = 21,000 \div \frac{7}{8}$; $x = 21,000 \div \frac{7}{8}$. Dividing by a fraction is the same as multiplying by its reciprocal: $x = 21,000 \times \frac{8}{7}$; $x = \$24,000$.

325. Let b = the number of bears in each packing crate. The first sentence tells you that the number of packing crates (54) times the number of bears in each is equal to the total number of bears (324). Your equation is: $54b = 324$. Divide both sides of the equation by 54: $54b \div 54 = 324 \div 54$; $b = 324 \div 54$. Divide: $b = 6$.

326. Let t = the number of turtle hatchlings born. The first sentence tells you that only 3% survive to adulthood. Three percent of the turtles born is 1,200. Your equation will be: $(3\%)t = 1,200$. The decimal equivalent of 3% is 0.03, so the equation becomes: $0.03t = 1,200$. Divide both sides of the equation by 0.03: $0.03t \div 0.03 = 1,200 \div 0.03$; $t = 1,200 \div 0.03$. Divide: $t = 40,000$.

327. Let c = the number of acres planted last year. 1.5 times c is 300: $1.5c = 300$. Divide both sides of the equation by 1.5: $1.5c \div 1.5 = 300 \div 1.5$; $c = 300 \div 1.5$. Divide: $c = 200$.

328. Let d = her annual salary. Five percent of her salary equals her yearly bonus. Your equation will be: $(5\%)d = \$6{,}000$. The decimal equivalent of 5% is 0.05, so the equation becomes: $0.05d = 6{,}000$. Divide both sides of the equation by 0.05: $0.05d \div 0.05 = 6{,}000 \div 0.05$; $d = 6{,}000 \div 0.05$. Divide: $d = \$120{,}000$.

329. Subtract 7 from both sides of the equation: $4x + 7 - 7 = 11 - 7$. Associate like terms: $4x + (7 - 7) = (11 - 7)$. Perform numerical operations: $4x + (0) = (4)$. Zero is the identity element for addition: $4x = 4$. Divide both sides of the equation by 4: $4x \div 4 = 4 \div 4$; $x = 1$.

330. Subtract 21 from both sides of the equation: $13x + 21 - 21 = 60 - 21$. Associate like terms: $13x + (21 - 21) = (60 - 21)$. Perform numerical operations: $13x + (0) = (39)$. Zero is the identity element for addition: $13x = 39$. Divide both sides of the equation by 13: $13x \div 13 = 39 \div 13$; $x = 3$.

331. Add 8 to each side of the equation: $3x - 8 + 8 = 16 + 8$. Change subtraction to addition and change the sign of the term that follows: $3x + -8 + 8 = 16 + 8$. Associate like terms: $3x + (-8 + 8) = 16 + 8$. Perform numerical operations: $3x + (0) = 24$. Zero is the identity element for addition: $3x = 24$. Divide both sides of the equation by 3: $3x \div 3 = 24 \div 3$; $x = 8$.

332. Add 6 to each side of the equation: $5x - 6 + 6 = -26 + 6$. Change subtraction to addition and change the sign of the term that follows: $5x + -6 + 6 = -26 + 6$. Associate like terms: $5x + (-6 + 6) = -26 + 6$. Perform numerical operations: $5x + (0) = -20$. Zero is the identity element for addition: $5x = -20$. Divide both sides of the equation by 5: $5x \div 5 = -20 \div 5$; $x = -4$.

333. Subtract 4 from both sides of the equation: $\frac{x}{3} + 4 - 4 = 10 - 4$. Associate like terms: $\frac{x}{3} + (4 - 4) = 10 - 4$. Perform numerical operations: $\frac{x}{3} + (0) = 6$. Zero is the identity element for addition: $\frac{x}{3} = 6$. Multiply both sides of the equation by 3: $3(\frac{x}{3}) = 3(6)$; $x = 18$.

334. Add 5 to each side of the equation: $\frac{x}{7} - 5 + 5 = 1 + 5$. Change subtraction to addition and change the sign of the term that follows: $\frac{x}{7} + -5 + 5 = 1 + 5$. Associate like terms: $\frac{x}{7} + (-5 + 5) = 1 + 5$. Perform numerical operations: $\frac{x}{7} + (0) = 6$. Zero is the identity element for addition: $\frac{x}{7} = 6$. Multiply both sides of the equation by 7: $7(\frac{x}{7}) = 7(6)$; $x = 42$.

335. Add 9 to each side of the equation: $39 + 9 = 3a - 9 + 9$. Change subtraction to addition and change the sign of the term that follows: $39 + 9 = 3a + (-9 + 9)$. Associate like terms: $39 + 9 = 3a + (-9 + 9)$. Perform numerical operations: $48 = 3a + (0)$. Zero is the identity element for addition: $48 = 3a$. Divide both sides of the equation by 3: $48 \div 3 = 3a \div 3$; $16 = a$.

336. Subtract 20 from both sides of the equation: $4 - 20 = 4a + 20 - 20$. Associate like terms: $4 - 20 = 4a + (20 - 20)$. Perform numerical operations: $-16 = 4a + (0)$. Zero is the identity element for addition: $-16 = 4a$. Divide both sides of the equation by 4: $\frac{-16}{4} = \frac{4a}{4}$; $-4 = a$.

337. Subtract 5 from both sides of the equation: $10a + 5 - 5 = 7 - 5$. Associate like terms: $10a + (5 - 5) = 7 - 5$. Perform numerical operations: $10a + (0) = 2$. Zero is the identity element for addition: $10a = 2$. Divide both sides of the equation by 10: $\frac{10a}{10} = \frac{2}{10}$; $a = \frac{1}{5}$ or 0.2.

338. Subtract 0.25 from both sides of the equation: $0.3a + 0.25 - 0.25 = 1 - 0.25$. Associate like terms: $0.3a + (0.25 - 0.25) = 1 - 0.25$. Perform numerical operations: $0.3a + (0) = 0.75$. Zero is the identity element for addition: $0.3a = 0.75$. Divide both sides of the equation by 0.3: $\frac{0.3a}{0.3} = \frac{0.75}{0.3}$. Simplify the result: $a = 2.5$.

339. Subtract 8 from both sides of the equation: $\frac{2}{3}m + 8 - 8 = 20 - 8$. Associate like terms: $\frac{2}{3}m + (8 - 8) = 20 - 8$. Perform numerical operations: $\frac{2}{3}m + (0) = 12$. Zero is the identity element for addition: $\frac{2}{3}m = 12$. Multiply both sides of the equation by the reciprocal of $\frac{2}{3}$: $\frac{3}{2}(\frac{2}{3}m) = \frac{3}{2}(12)$; $m = 18$.

340. Add 3 to both sides of the equation: $9 + 3 = \frac{3}{4}m - 3 + 3$. Change subtraction to addition and change the sign of the term that follows: $9 + 3 = \frac{3}{4}m + -3 + 3$. Associate like terms: $9 + 3 = \frac{3}{4}m + (-3 + 3)$. Perform numerical operations: $12 = \frac{3}{4}m + (0)$. Zero is the identity element for addition: $12 = \frac{3}{4}m$. Multiply both sides of the equation by the reciprocal of $\frac{3}{4}$: $\frac{4}{3}(12) = \frac{4}{3}(\frac{3}{4}m)$; $16 = m$.

341. Subtract 41 from both sides of the equation: $41 - 41 - 2m = 65 - 41$. Associate like terms: $(41 - 41) - 2m = 65 - 41$. Perform numerical operations: $(0) - 2m = 24$. Zero is the identity element for addition: $-2m = 24$. You can change the subtraction to addition and change the sign of the term following to its opposite, which in this case is $-2m$: $-2m = 24$. Divide both sides of the equation by -2: $\frac{-2m}{-2} = \frac{24}{-2}$. Use the rules for operating with signed numbers: $m = -12$.

342. Add 14 to each side of the equation: $4m - 14 + 14 = 50 + 14$. Change subtraction to addition and change the sign of the term that follows: $4m + -14 + 14 = 50 + 14$. Associate like terms: $4m + (-14 + 14) = 50 + 14$. Perform numerical operations: $4m + (0) = 64$. Zero is the identity element for addition: $4m = 64$. Divide both sides of the equation by 4: $\frac{4m}{4} = \frac{64}{4}$; $m = 16$.

343. This equation presents a slightly different look. The variable in the numerator has a coefficient. There are two methods for solving. Subtract 16 from both sides of the equation: $\frac{2m}{5} + 16 - 16 = 24 - 16$. Associate like terms: $\frac{2m}{5} + (16 - 16) = 24 - 16$. Perform numerical operations: $\frac{2m}{5} + (0) = 8$. Zero is the identity element for addition: $\frac{2m}{5} = 8$. Method 1: Multiply both sides of the equation by 5: $5(\frac{2m}{5}) = 5(8)$. Use rules for multiplying whole numbers and fractions: $\frac{5}{1}(\frac{2m}{5}) = 40$; $\frac{5 \times 2m}{1 \times 5} = 40$; $\frac{10m}{5} = 40$; $2m = 40$. Divide both sides by 2: $\frac{2m}{2} = \frac{40}{2}$; $m = 20$. Method 2: You can recognize that: $\frac{2m}{5} = (\frac{2}{5})m$. Then you would multiply by the reciprocal of the coefficient: $\frac{5}{2}(\frac{2}{5})m = (\frac{5}{2})8$; $m = 20$.

344. Add 6 to each side of the equation: $7m - 6 + 6 = -2.5 + 6$. Change subtraction to addition and change the sign of the term that follows: $7m + -6 + 6 = -2.5 + 6$. Associate like terms: $7m + (-6 + 6) = -2.5 + 6$. Perform numerical operations: $7m + (0) = 3.5$. Divide both sides of the equation by 7: $\frac{7m}{7} = \frac{3.5}{7}$; $m = 0.5$.

345. Add 6 to each side of the equation: $10s - 6 + 6 = 0 + 6$. Change subtraction to addition and change the sign of the term that follows: $10s + -6 + 6 = 0 + 6$. Associate like terms: $10s + (-6 + 6) = 0 + 6$. Perform numerical operations: $10s + (0) = 6$. Divide both sides of the equation by 10: $\frac{10s}{10} = \frac{6}{10}$. Express the answer in the simplest form: $s = \frac{3}{5} = 0.6$.

346. Subtract 2.7 from both sides of the equation: $\frac{s}{4} + 2.7 - 2.7 = 3 - 2.7$. Associate like terms: $\frac{s}{4} + (2.7 - 2.7) = 3 - 2.7$. Perform numerical operations: $\frac{s}{4} + (0) = 0.3$. Multiply both sides of the equation by 4: $4(\frac{s}{4}) = 4(0.3)$; $s = 1.2$.

347. Add 7 to each side of the equation: $8s - 7 + 7 = 41 + 7$. Change subtraction to addition and change the sign of the term that follows: $8s + -7 + 7 = 41 + 7$. Associate like terms: $8s + (-7 + 7) = 41 + 7$. Perform numerical operations: $8s + (0) = 48$. Divide both sides of the equation by 8: $\frac{8s}{8} = \frac{48}{8}$. Express the answer in simplest form: $s = 6$.

348. Subtract 25 from both sides of the equation: $-55 - 25 = 25 - 25 - s$. Change subtraction to addition and change the sign of the term that follows: $-55 + -25 = 25 + -25 + -s$. Associate like terms: $(-55 + -25) = (25 + -25) + -s$. Perform numerical operations: $-80 = (0) + -s$. Zero is the identity element for addition: $-80 = -s$. You are to solve for s, but the term remaining is $-s$. If you multiply both sides by -1, you will be left with just s: $-1(-80) = -1(-s)$. Use the rules for operating with signed numbers: $80 = s$.

349. Let $x =$ the number of months. The number of months (x), times 12 (pounds per month), plus the starting weight (20), will be equal to 200 pounds. An equation that represents these words would be: $12x + 20 = 200$. Subtract 20 from both sides of the equation: $12x + 20 - 20 = 200 - 20$. Associate like terms: $12x + (20 - 20) = 200 - 20$. Perform numerical operations: $12x + (0) = 180$. Divide both sides of the equation by 12: $\frac{12x}{12} = \frac{180}{12}$; $x = 15$. The farmer would have to wait 15 months before selling his hog.

350. Let $x =$ Bill's hourly wage. Then $2x$ minus $1.50 is equal to Mary's hourly wage, $12.50. The equation representing the last statement would be: $2x - 1.50 = 12.50$. Add 1.50 to both sides of the equation: $2x - 1.50 + 1.50 = 12.50 + 1.50$. Perform numerical operations: $2x = 14.00$. Divide both sides of the equation by 2: $\frac{2x}{2} = \frac{14.00}{2}$; $x = 7.00$. Bill's hourly wage is $7.00 per hour.

351. Let $x =$ the share price at the beginning of the year. The statements tell us that if we multiply the share price at the beginning of the year by 3 and then subtract $8, it will equal $37. An equation that represents this amount is: $3x - 8 = 37$. Add 8 to both sides of the equation: $3x - 8 + 8 = 37 + 8$. Perform numerical operations: $3x = 45$. Divide both sides of the equation by 3: $\frac{3x}{3} = \frac{45}{3}$; $x = 15$. One share of Axon cost $15 at the beginning of the year.

352. Let $x =$ her previous salary. The statements tell us that $64,000 is equal to 1.5 times x plus $4,000. An algebraic equation to represent this statement is: $64,000 = 1.5x + 4,000$. Subtract 4,000 from both sides of the equation: $64,000 - 4,000 = 1.5x + 4,000 - 4,000$. Perform numerical operations: $60,000 = 1.5x$. Divide both sides of the equation by 1.5: $\frac{60,000}{1.5} = \frac{1.5x}{1.5}$; $40,000 = x$. Jennifer's former salary was $40,000 per year.

353. Let x = the number of boys who participate in interscholastic sports. The question tells us that $\frac{2}{3}$ the number of boys plus 25 is equal to the number of girls who participate, 105. An equation that represents this statement is: $\frac{2}{3}x + 25 = 105$. Subtract 25 from both sides of the equation: $\frac{2}{3}x + 25 - 25 = 105 - 25$. Perform numerical operations: $\frac{2}{3}x = 80$. Multiply by the reciprocal of $\frac{2}{3}$: $\frac{3}{2}(\frac{2}{3}x) = \frac{3}{2}(80)$; $x = 120$. The number of boys who participate is 120.

354. Subtract 7 from both sides of the equation: $11x + 7 - 7 = 3x - 9 - 7$. Simplify: $11x + (0) = 3x - 16$. Identity property of 0 for addition: $11x = 3x - 16$. Subtract $3x$ from both sides of the equation: $11x - 3x = 3x - 3x - 16$. Simplify: $8x = -16$. Divide both sides of the equation by 8: $\frac{8x}{8} = \frac{-16}{8}$. Simplify: $1x = -2$. The solution is: $x = -2$. Let's check the answer. Substitute -2 for x in the original equation: $11(-2) + 7 = 3(-2) - 9$. Simplify: $-22 + 7 = -6 - 9$; $-15 = -15$. The result is a true statement, so this answer is a correct solution.

355. Add 23 to both sides of the equation: $3x - 23 + 23 = 54 + 23 - 4x$. Simplify by combining like terms: $3x + 0 = 77 - 4x$. Identity property of 0 for addition: $3x = 77 - 4x$. Now add $4x$ to both sides: $3x + 4x = 77 - 4x + 4x$. Simplify: $7x = 77$. Divide both sides of the equation by 7: $\frac{7x}{7} = \frac{77}{7}$. Simplify: $x = 11$.

356. Use the commutative property of addition with like terms: $5x + 6x + 3 = 10x - x + 9$. Combine like terms on each side of the equation: $11x + 3 = 9x + 9$. Subtract 3 from both sides: $11x + 3 - 3 = 9x + 9 - 3$. Simplify: $11x = 9x + 6$. Now subtract $9x$ from both sides of the equation: $11x - 9x = 9x - 9x + 6$. Simplify: $2x = 6$. Divide both sides by 2: $\frac{2x}{2} = \frac{6}{2}$. Simplify: $x = 3$.

357. Use the commutative property of addition with like terms: $10x - 5x + 27 - 46 = 3x + 32 - 19$. Combine like terms on each side of the equation: $5x - 19 = 3x + 13$. Add 19 to both sides of the equation: $5x - 19 + 19 = 3x + 13 + 19$. Combine like terms on each side of the equation: $5x = 3x + 32$. Subtract $3x$ from both sides of the equation to isolate the variable on one side of the equation: $5x - 3x = 3x - 3x + 32$. Simplify: $2x = 32$. Divide both sides of the equation by 2: $\frac{2x}{2} = \frac{32}{2}$. Simplify. The solution is: $x = 16$. Let's check this answer. Substitute 16 for x in the original equation: $10(16) + 27 - 5(16) - 46 = 32 + 3(16) - 19$. Simplify by multiplying factors: $(160) + 27 - (80) - 46 = 32 + (48) - 19$. Parentheses are not needed. Add and subtract from left to right: $187 - 80 - 46 = 80 - 19$; $107 - 46 = 61$; $61 = 61$. This statement is true; therefore, the solution is correct.

358. Use the commutative property to move like terms: $20x - 3x - 11 = 9x + 43$. Combine like terms on each side of the equation: $17x - 11 = 9x + 43$. Add 11 to both sides of the equation: $17x + 11 - 11 = 9x + 43 + 11$. Simplify: $17x = 9x + 54$. Subtract $9x$ from both sides of the equation: $17x - 9x = 9x - 9x + 54$. Simplify: $8x = 54$. Divide both sides of the equation by 8: $\frac{8x}{8} = \frac{54}{8}$. Simplify the expression: $x = 6.75$.

359. Use the commutative property to move like terms: $3x + 0.4 - 0.25 = 1.15 - 2x$. Combine like terms on each side of the equation: $3x + 0.15 = 1.15 - 2x$. Subtract 0.15 from both sides of the equation: $3x + 0.15 - 0.15 = 1.15 - 0.15 - 2x$. Combine like terms on each side of the equation: $3x + (0) = 1 - 2x$. Identity property of 0 for addition: $3x = 1 - 2x$. Add $2x$ to both sides of the equation: $3x + 2x = 1 - 2x + 2x$. Simplify: $5x = 1$. Divide both sides of the equation by 5: $\frac{5x}{5} = \frac{1}{5}$. Simplify the expression: $x = \frac{1}{5}$.

360. Use the commutative property with like terms. $2x - 1.2x + 17 = 10 + 11 - 0.2x$. Combine like terms on each side of the equation: $0.8x + 17 = 21 - 0.2x$. Subtract 17 from both sides of the equation: $0.8x + 17 - 17 = 21 - 17 - 0.2x$. Combine like terms on each side of the equation: $0.8x = 4 - 0.2x$. Add $0.2x$ to both sides of the equation: $0.8x + 0.2x = 4 - 0.2x + 0.2x$. Combine like terms on each side of the equation: $1x = 4$. Use the identity property of multiplication: $x = 4$.

361. Use the commutative property with like terms: $6x + 2 - 0.2 = 5x + 2.1$. Combine like terms and simplify the expression: $6x + 1.8 = 5x + 2.1$. Subtract 1.8 from both sides of the equation: $6x + 1.8 - 1.8 = 5x + 2.1 - 1.8$. Combine like terms and simplify the expression: $6x = 5x + 0.3$. Subtract $5x$ from both sides of the equation: $6x - 5x = 5x - 5x + 0.3$. Combine like terms on each side of the equation. The solution is: $x = 0.3$. Let's try this a slightly different way. Instead of subtracting 1.8, try subtracting 2.1 from both sides of the equation: $6x + 1.8 - 2.1 = 5x + 2.1 - 2.1$. Combine like terms on each side of the equation: $6x - 0.3 = 5x$. Now subtract $6x$ from both sides of the equation: $6x - 6x - 0.3 = 5x - 6x$. Combine like terms on each side of the equation: $-0.3 = -x$. Now divide both sides by the coefficient of the variable x, which is -1: $\frac{-0.3}{-1} = \frac{-x}{-1}$. Dividing like signs results in a positive. Simplify: $0.3 = x$. The answer is the same. As long as you do proper math, you can manipulate an equation many ways and still get the correct solution.

362. Use the commutative property with like terms: $3x + 0.4 - 0.25 = 1.15 - 2x$. Associate like terms on each side of the equation: $3x + (0.4 - 0.25) = 1.15 - 2x$. Simplify the expression: $3x + (0.15) = 1.15 - 2x$. Remove the parentheses: $3x + 0.15 = 1.15 - 2x$. Add $2x$ to both sides of the equation: $3x + 2x + 0.15 = 1.15 - 2x + 2x$. Associate like terms: $(3x + 2x) + 0.15 = 1.15 + (2x - 2x)$. Simplify the expression: $5x + 0.15 = 1.15$. Subtract 0.15 from both sides of the equation: $5x + 0.15 - 0.15 = 1.15 - 0.15$. Combine like terms on each side of the equation: $5x + 0 = 1$. Identity property of 0 for addition: $5x = 1$. Divide both sides of the equation by 5: $\frac{5x}{5} = \frac{1}{5}$. Simplify the expression: $x = 0.2$.

363. Use the commutative property with like terms: $3x - 0.8x + 12 = 3.4 - 9.4 - 0.8x$. Associate like terms on each side of the equation: $(3x - 0.8x) + 12 = (3.4 - 9.4) - 0.8x$. Simplify the expression: $2.2x + 12 = -6 - 0.8x$. Subtract 12 from both sides of the equation: $2.2x + 12 - 12 = -6 - 12 - 0.8x$. Associate like terms on each side of the equation: $2.2x + (12 - 12) = (-6 - 12) - 0.8x$. Simplify the expression: $2.2x + (0) = (-18) - 0.8x$. Identity property of 0 for addition: $2.2x = -18 - 0.8x$. Add $0.8x$ to both sides of the equation: $2.2x + 0.8x = -18 + 0.8x - 0.8x$. Combine like terms on each side of the equation: $3x = -18$. Divide both sides of the equation by 3: $\frac{3x}{3} = \frac{-18}{3}$. Simplify the expression: $x = -6$.

364. Use the distributive property of multiplication: $7(x) + 7(2) + 1 = 3(x) + 3(14) - 4x$. Simplify the expression: $7x + 14 + 1 = 3x + 42 - 4x$. Use the commutative property with like terms: $7x + 14 + 1 = 3x - 4x + 42$. Combine like terms on each side of the equation: $7x + 15 = -1x + 42$. Subtract 15 from both sides of the equation: $7x + 15 - 15 = -1x + 42 - 15$. Combine like terms on each side of the equation: $7x = -1x + 27$. Add x to both sides of the equation: $7x + x = -1x + x + 27$. Combine like terms on each side of the equation: $8x = 27$. Divide both sides of the equation by 8: $\frac{8x}{8} = \frac{27}{8}$. Simplify the expression: $x = 3\frac{3}{8}$.

365. Use the distributive property of multiplication: $4(4x) + 4(3) = 6x - 28$. Simplify the expression: $16x + 12 = 6x - 28$. Subtract 12 from both sides of the equation: $16x + 12 - 12 = 6x - 28 - 12$. Combine like terms on each side of the equation: $16x = 6x - 40$. Subtract $6x$ from both sides of the equation: $16x - 6x = 6x - 6x - 40$. Simplify the expression: $10x = -40$. Divide both sides of the equation by 10: $\frac{10x}{10} = \frac{-40}{10}$. Simplify the expression: $x = -4$.

366. Use the distributive property of multiplication: $13 - 8(x) - 8(-2) = 7(x) + 7(4) + 46$. Simplify the expression: $13 - 8x + 16 = 7x + 28 + 46$. Use the commutative property with like terms: $13 + 16 - 8x = 7x + 28 + 46$. Combine like terms on each side of the equation: $29 - 8x = 7x + 74$. Add $8x$ to both sides of the equation: $29 - 8x + 8x = 7x + 8x + 74$. Combine like terms on each side of the equation: $29 = 15x + 74$. Subtract 74 from both sides of the equation: $29 - 74 = 15x + 74 - 74$. Combine like terms on each side of the equation: $-45 = 15x$. Divide both sides of the equation by 15: $\frac{-45}{15} = \frac{15x}{15}$. Simplify the expression: $-3 = x$.

367. Use the distributive property of multiplication: $13x + 3(3) - 3(x) = -3(4) - 3(3x) - 2x$. Simplify the expression: $13x + 9 - 3x = -12 - 9x - 2x$. Use the commutative property with like terms: $9 + 13x - 3x = -12 - 9x - 2x$. Combine like terms on each side of the equation: $9 + 10x = -12 - 11x$. Add $11x$ to both sides of the equation: $9 + 10x + 11x = -12 - 11x + 11x$. Combine like terms on each side of the equation: $9 + 21x = -12$. Subtract 9 from both sides of the equation: $9 - 9 + 21x = -12 - 9$. Combine like terms on each side of the equation: $21x = -21$. Divide both sides of the equation by 21: $\frac{21x}{21} = \frac{-21}{21}$. Simplify the expression: $x = -1$.

368. Use the distributive property of multiplication: $2(2x) + 2(19) - 9x = 9(13) - 9(x) + 21$. Simplify the expression: $4x + 38 - 9x = 117 - 9x + 21$. Use the commutative property with like terms: $38 + 4x - 9x = 117 + 21 - 9x$. Combine like terms on each side of the equation: $38 - 5x = 138 - 9x$. Subtract 38 from both sides of the equation: $38 - 38 - 5x = 138 - 38 - 9x$. Combine like terms on each side of the equation: $-5x = 100 - 9x$. Add $9x$ to both sides of the equation: $-5x + 9x = 100 - 9x + 9x$. Combine like terms on each side of the equation: $4x = 100$. Divide both sides of the equation by 4: $\frac{4x}{4} = \frac{100}{4}$. Simplify the expression. The solution is: $x = 25$. Now let's check the answer by substituting the solution into the original equation: $2[2(25) + 19] - 9(25) = 9[13 - (25)] + 21$. Simplify the expression. Use order of operations: $2(50 + 19) - 225 = 9(-12) + 21$; $2(69) - 225 = -108 + 21$; $138 - 225 = -87$; $-87 = -87$. The solution is correct.

369. Use the distributive property of multiplication: $12x - 4(x) - 4(-1) = 2(x) + 2(-2) + 16$. Simplify the expression: $12x - 4x + 4 = 2x - 4 + 16$. Combine like terms on each side of the equation: $8x + 4 = 2x + 12$. Subtract 4 from both sides of the equation: $8x + 4 - 4 = 2x + 12 - 4$. Simplify the expression: $8x = 2x + 8$. Subtract $2x$ from both sides of the equation: $8x - 2x = 2x - 2x + 8$. Simplify the expression: $6x = 8$. Divide both sides of the equation by 6: $\frac{6x}{6} = \frac{8}{6}$. Simplify the expression. The solution is: $x = \frac{4}{3}$ or $1\frac{1}{3}$. Now let's check the answer by substituting the solution into the original equation: $12(\frac{4}{3}) - 4[(\frac{4}{3}) - 1] = 2[(\frac{4}{3}) - 2] + 16$. Operate inside the parentheses first. Change whole numbers to fractional equivalents: $12(\frac{4}{3}) - 4(\frac{4}{3} - \frac{3}{3}) = 2(\frac{4}{3} - \frac{6}{3}) + 16$; $12(\frac{4}{3}) - 4(\frac{1}{3}) = 2(\frac{-2}{3}) + 16$. Multiply: $\frac{48}{3} - \frac{4}{3} = -\frac{4}{3} + 16$. Add $\frac{4}{3}$ to both sides of the equation: $\frac{48}{3} - \frac{4}{3} + \frac{4}{3} = -\frac{4}{3} + \frac{4}{3} + 16$. Simplify the expression: $16 = 16$. A true statement, so this solution is correct.

370. Simplify the equation by adding like terms: $3\frac{4}{5}x = 1 + 3x$. Subtract $3x$ from both sides of the equation: $3\frac{4}{5}x - 3x = 1 + 3x - 3x$. Combine like terms on each side of the equation: $\frac{4}{5}x = 1$. Divide both sides of the equation by $\frac{4}{5}$. Remember that division by a fraction is the same as multiplication by the reciprocal of the fraction. You can multiply by $\frac{5}{4}$: $\frac{5}{4}(\frac{4}{5}x) = \frac{5}{4}(1)$. Simplify the expression: $x = \frac{5}{4}$.

371. Use the distributive property on both sides: $\frac{5}{2}(x) - \frac{5}{2}(2) + 3x = 3(x) + 3(2) - 10$. Simplify the expression: $\frac{5}{2}x - 5 + 3x = 3x + 6 - 10$. Use the commutative property with like terms: $\frac{5}{2}x + 3x - 5 = 3x + 6 - 10$. A simple way to avoid having to operate with fractions is to multiply the equation by a factor that will eliminate the denominator. In this case, that would be a 2: $2(\frac{5}{2}x + 3x - 5) = 2(3x + 6 - 10)$. Use the distributive property: $2(\frac{5}{2}x) + 2(3x) - 2(5) = 2(3x) + 2(6) - 2(10)$. Simplify the expression: $5x + 6x - 10 = 6x + 12 - 20$. Combine like terms on each side of the equation: $11x - 10 = 6x - 8$. Add 10 to both sides of the equation: $11x - 10 + 10 = 6x - 8 + 10$. Simplify the expression: $11x = 6x + 2$. Subtract $6x$ from both sides of the equation: $11x - 6x = 6x - 6x + 2$. Combine like terms on each side of the equation: $5x = 2$. Divide both sides of the equation by 5: $\frac{5x}{5} = \frac{2}{5}$. Simplify the expression: $x = \frac{2}{5}$.

372. Use the distributive property on both sides: $6(\frac{1}{2}x) + 6(\frac{1}{2}) = 3(x) + 3(1)$. Simplify the expression: $3x + 3 = 3x + 3$. Subtract 3 from both sides of the equation: $3x + 3 - 3 = 3x + 3 - 3$. Simplify the expression: $3x = 3x$. Divide both sides of the equation by 3: $\frac{3x}{3} = \frac{3x}{3}$. Simplify the expression: $x = x$. An infinite number of solutions exists for this equation.

373. Use the distributive property on both sides: $0.7(0.2x) - 0.7(1) = 0.3(3) - 0.3(0.2x)$. Simplify the expression: $0.14x - 0.7 = 0.9 - 0.06x$. Add $0.06x$ to both sides of the equation: $0.14x + 0.06x - 0.7 = 0.9 - 0.06x + 0.06x$. Combine like terms on each side of the equation: $0.2x - 0.7 = 0.9$. Add 0.7 to both sides of the equation: $0.2x - 0.7 + 0.7 = 0.9 + 0.7$. Combine like terms on each side of the equation: $0.2x = 1.6$. Divide both sides of the equation by 0.2: $\frac{0.2x}{0.2} = \frac{1.6}{0.2}$. Simplify the expression: $x = 8$.

374. Use the distributive property of multiplication: $10(x) + 10(2) + 7(1) - 7(x) = 3(x) + 3(9)$. Simplify the expression: $10x + 20 + 7 - 7x = 3x + 27$. Use the commutative property with like terms: $10x - 7x + 20 + 7 = 3x + 27$. Combine like terms on each side of the equation: $3x + 27 = 3x + 27$. Look familiar? Subtract 27 from both sides: $3x + 27 - 27 = 3x + 27 - 27$. Simplify the expression: $3x = 3x$. Divide both sides of the equation by 3: $x = x$. The solutions for this equation are infinite.

375. Use the distributive property of multiplication: $4(9) - 4(x) = 2x - 6(x) - 6(6)$. Simplify the expression: $36 - 4x = 2x - 6x - 36$. Combine like terms: $36 - 4x = -4x - 36$. Add $4x$ to both sides of the equation: $36 - 4x + 4x = -4x + 4x - 36$. Combine like terms on each side of the equation: The two sides are not equal: $36 \neq -36$. *There is no solution for this equation.* Another way of saying this is to say that the solution for this equation is the null set.

376. Use the distributive property of multiplication: $5(2x) + 5(3) - 9 = 14(x) + 14$. Simplify the expression: $10x + 15 - 9 = 14x + 14$. Combine like terms: $10x + 6 = 14x + 14$. Subtract $10x$ from both sides of the equation: $10x - 10x + 6 = 14x - 10x + 14$. Combine like terms on each side of the equation: $6 = 4x + 14$. Subtract 14 from both sides of the equation: $6 - 14 = 4x + 14 - 14$. Combine like terms on each side of the equation: $-8 = 4x$. Divide both sides of the equation by 4: $\frac{-8}{4} = \frac{4x}{4}$. Simplify the expression: $-2 = x$.

377. Use the distributive property of multiplication: $7(x) - 7(10) + 110 = 4(x) - 4(25) + 7x$. Simplify the expression: $7x - 70 + 110 = 4x - 100 + 7x$. Combine like terms on each side of the equation: $7x + 40 = 11x - 100$. Subtract 40 from both sides of the equation: $7x + 40 - 40 = 11x - 100 - 40$. Combine like terms on each side of the equation: $7x = 11x - 140$. Subtract $11x$ from both sides of the equation: $7x - 11x = 11x - 11x - 140$. Combine like terms on each side of the equation: $-4x = -140$. Divide both sides of the equation by -4: $\frac{-4x}{-4} = \frac{-140}{-4}$. Simplify the expression: $x = 35$.

378. Look for opportunities to simplify your work. In this equation, many of the terms are in decimal form. If you multiply the equation by 10, it might be easier to solve: $10[0.8(x + 20) - 4.5] = 10[0.7(5 + x) - 0.9x]$. Use the distributive property of multiplication: $10[0.8(x + 20)] - 10(4.5) = 10[0.7(5 + x)] - 10(0.9x)$. Simplify the expression: $8(x + 20) - 45 = 7(5 + x) - 9x$. Use the distributive property again: $8(x) + 8(20) - 45 = 7(5) + 7(x) - 9x$. Simplify the expression: $8x + 160 - 45 = 35 + 7x - 9x$. Combine like terms on each side of the equation: $8x + 115 = 35 - 2x$. Subtract 115 from both sides of the equation: $8x + 115 - 115 = 35 - 115 - 2x$. Combine like terms on each side of the equation: $8x = -80 - 2x$. Add $2x$ to both sides of the equation: $8x + 2x = -80 - 2x + 2x$. Combine like terms on each side of the equation: $10x = -80$. Divide both sides of the equation by 10: $\frac{10x}{10} = \frac{-80}{10}$. Simplify the expression: $x = -8$.

379. Write the applicable formula: $D = rt$. List the values for the variables: $D = ?$; $r = 350$; $t = 3.5$. Substitute the values for the variables: $D = (350)(3.5)$. Simplify the expression: $D = 1,225$. Include the units: $D = 1,225$ miles.

380. Write the applicable formula: $D = rt$. List the values for the variables: $D = 68$; $r = 17$; $t = ?$. Substitute the values for the variables: $68 = (17)t$. Simplify the expression: $68 = 17t$. Divide both sides of the equation by 17: $\frac{68}{17} = \frac{17t}{17}$. Simplify the expression: $4 = t$. Include the units: $t = 4$ hours.

381. Write the applicable formula: $D = rt$. List the values for the variables: $D = 60$; $r = ?$. Calculate the total number of hours: 4 days × 6 hours per day = 24 hours; $t = 24$. Substitute the given values into the formula: $60 = r(24)$. Simplify the expression: $60 = 24r$. Divide both sides of the equation by 24: $\frac{60}{24} = \frac{24r}{24}$. Simplify the expression: $2.5 = r$. Include the units: $r = 2.5$ mph.

382. Write the applicable formula: $I = prt$. List the values for the variables: $I = ?$; $p = 12,000$; $r = 6\%$; $t = 2$. Substitute the given values into the formula: $I = (12,000)(0.06)(2)$. Simplify the expression: $I = 1,440$. Include the units: $I = \$1,440$.

383. Write the applicable formula: $I = prt$. List the values for the variables: $I = 1,620$; $p = 4,500$; $r = ?$; $t = 3$. Substitute the given values into the formula: $1,620 = (4,500)r(3)$. Simplify the expression: $1,620 = 13,500r$. Divide both sides of the equation by 13,500: $\frac{1,620}{13,500} = \frac{13,500r}{13,500}$. Simplify the expression: $0.12 = r$. Express as a percent: $r = 12\%$.

384. Write the applicable formula: $I = prt$. List the values for the variables: $I = 1,000$; $p = ?$; $r = 4\%$; $t = 2$. Substitute the given values into the formula: $1,000 = p(0.04)(2)$. Simplify the expression: $1,000 = 0.08p$. Divide both sides of the equation by 0.08: $\frac{1,000}{0.08} = \frac{0.08p}{0.08}$. Simplify the expression: $12,500 = p$. Include the units: $p = \$12,500$.

385. Write the applicable formula: $I = prt$. To double its value, the account would have to earn $3,000 in interest. List the values for the variables: $I = 3,000$; $p = 3,000$; $r = 10\%$; $t = ?$. Substitute the given values into the formula: $3,000 = (3,000)(0.10)t$. Simplify the expression: $3,000 = 300t$. Divide both sides of the equation by 300: $\frac{3,000}{300} = \frac{300t}{300}$. Simplify the expression: $10 = t$. Include the units: $t = 10$ years.

386. Write the applicable formula: $C = \frac{5}{9}(F - 32)$. List the values for the variables: $C = 40$; $F = ?$. Substitute the given values into the formula: $40 = \frac{5}{9}(F - 32)$. Multiply both sides of the equation by 9: $9(40) = 9[\frac{5}{9}(F - 32)]$. Simplify the expression: $360 = 5(F - 32)$; $360 = 5F - 5(32)$; $360 = 5F - 160$. Add 160 to both sides of the equation: $360 + 160 = 5F - 160 + 160$. Combine like terms on each side of the equation: $520 = 5F$. Divide both sides of the equation by 5: $\frac{520}{5} = \frac{5F}{5}$. Simplify the expression: $104 = F$.

387. Write the applicable formula: $C = \frac{5}{9}(F - 32)$. List the values for the variables: $C = ?$; $F = 50$. Substitute the given values into the formula: $C = \frac{5}{9}(50 - 32)$. Simplify the expression: $C = \frac{5}{9}(18)$; $C = \frac{5}{9}(\frac{18}{1})$; $C = 10$.

388. **c.** Use the formula $F = \frac{9}{5}C + 32$. Substitute the Celsius temperature of 20° for C in the formula. This results in the equation $F = \frac{9}{5}(20) + 32$. Following the order of operations, multiply $\frac{9}{5}$ and 20 to get 36. The final step is to add $36 + 32$ for an answer of 68°.

389. **d.** Use the formula $C = \frac{5}{9}(F - 32)$. Substitute the Fahrenheit temperature of 23° for F in the formula. This results in the equation $C = \frac{5}{9}(23 - 32)$. Following the order of operations, begin calculations inside the parentheses first and subtract 23 from 32 to get −9. Multiply $\frac{5}{9}$ times −9 to get an answer of −5°.

390. **d.** Using the simple interest formula, interest = principal × rate × time, or $I = prt$, substitute $p = \$505$, $r = 0.05$ (the interest rate as a decimal), and $t = 4$; $I = (505)(0.05)(4)$. Multiply to get a result of $I = \$101$.

391. **d.** Using the simple interest formula, interest = principal × rate × time, or $I = prt$, substitute $p = \$1,250$, $r = 0.034$ (the interest rate as a decimal), and $t = 1.5$ (18 months is equal to 1.5 years), $I = (1,250)(.034)(1.5)$. Multiply to get a result of $I = \$63.75$. To find the total amount in the account after 18 months, add the interest to the initial principal: $\$63.75 + \$1,250 = \$1,313.75$.

392. **a.** Using the simple interest formula, interest = principal × rate × time, or $I = prt$, substitute $I = \$4,800$, $p = \$12,000$, and $r = 0.08$ (the interest rate as a decimal); $4,800 = (12,000)(0.08)(t)$. Multiply 12,000 and 0.08 to get 960, so $4,800 = 960t$. Divide both sides by 960 to get $5 = t$. Therefore, the time is five years.

393. **b.** Using the simple interest formula, interest = principal × rate × time, or $I = prt$, substitute $I = \$711$, $p = \$7,900$, and $t = 3$ (36 months are equal to 3 years); $711 = (7,900)(r)(3)$. Multiply 7,900 and 3 on the right side to get a result of $711 = 23,700r$. Divide both sides by 23,700 to get $r = 0.03$, which is a decimal equal to 3%.

394. **a.** The statement "If 1 is added to the difference when $10x$ is subtracted from $-18x$, the result is 57" translates to the equation $-18x - 10x + 1 = 57$. Combine like terms on the left side of the equation: $-28x + 1 = 57$. Subtract 1 from both sides of the equation: $-28x + 1 - 1 = 57 - 1$. Divide each side of the equation by -28: $\frac{-28x}{-28} = \frac{56}{-28}$. The variable is now alone: $x = -2$.

395. **c.** The statement "If 0.3 is added to 0.2 times the quantity $x - 3$, the result is 2.5" translates to the equation $0.2(x - 3) + 0.3 = 2.5$. Remember to use parentheses for the expression when the words *the quantity* are used. Use the distributive property on the left side of the equation: $0.2x - 0.6 + 0.3 = 2.5$. Combine like terms on the left side of the equation: $0.2x + -0.3 = 2.5$. Add 0.3 to both sides of the equation: $0.2x + -0.3 + 0.3 = 2.5 + 0.3$. Simplify: $0.2x = 2.8$. Divide both sides by 0.2: $\frac{0.2x}{0.2} = \frac{2.8}{0.2}$. The variable is now alone: $x = 14$.

396. **b.** Let $x =$ the number. The sentence "If twice the quantity $x + 6$ is divided by negative 7, the result is 6" translates to $\frac{2(x+6)}{-7} = 6$. Remember to use parentheses for the expression when the words *the quantity* are used. There are different ways to approach solving this problem. Method 1: Multiply both sides of the equation by -7: $-7 \times \frac{2(x+6)}{-7} = 6 \times -7$. This simplifies to: $2(x + 6) = -42$. Divide each side of the equation by 2: $\frac{2(x+6)}{2} = \frac{-42}{2}$. This simplifies to: $x + 6 = -21$. Subtract 6 from both sides of the equation: $x + 6 - 6 = -21 - 6$. The variable is now alone: $x = -27$. Method 2: Another way to look at the problem is to multiply each side by -7 in the first step to get: $2(x + 6) = -42$. Then use the distributive property on the left side: $2x + 12 = -42$. Subtract 12 from both sides of the equation: $2x + 12 - 12 = -42 - 12$. Simplify: $2x = -54$. Divide each side by 2: $\frac{2x}{2} = \frac{-54}{2}$. The variable is now alone: $x = -27$.

397. **d.** Translating the sentence "The difference between six times the quantity $6x + 1$ and three times the quantity $x - 1$ is 108" into symbolic form results in the equation: $6(6x + 1) - 3(x - 1) = 108$. Remember to use parentheses for the expression when the words *the quantity* are used. Perform the distributive property twice on the left side of the equation: $36x + 6 - 3x + 3 = 108$. Combine like terms on the left side of the equation: $33x + 9 = 108$. Subtract 9 from both sides of the equation: $33x + 9 - 9 = 108 - 9$. Simplify: $33x = 99$. Divide both sides of the equation by 33: $\frac{33x}{33} = \frac{99}{33}$. The variable is now alone: $x = 3$.

398. **a.** This problem translates to the equation $-4(x + 8) + 6x = 2x + 32$. Remember to use parentheses for the expression when the words *the quantity* are used. Use the distributive property on the left side of the equation: $-4x - 32 + 6x = 2x + 32$. Combine like terms on the left side of the equation: $2x - 32 = 2x + 32$. Subtract $2x$ from both sides of the equation: $2x - 2x - 32 = 2x - 2x + 32$. The two sides are not equal. There is no solution: $-32 \neq 32$.

399. **a.** Let x = the total minutes of the call. Therefore, $x - 1$ = the additional minutes of the call. This choice is correct because in order to calculate the cost, the charge is 35 cents plus 15 cents times the number of additional minutes. If y represents the total cost, then y equals 0.35 plus 0.15 times the quantity $x - 1$. This translates to $y = 0.35 + 0.15(x - 1)$ or $y = 0.15(x - 1) + 0.35$.

400. **d.** Let x = the total miles of the ride. Therefore, $x - 1$ = the additional miles of the ride. The correct equation takes $1.25 and adds it to $1.15 times the number of additional miles, $x - 1$. Translating, this becomes y (the total cost) $= 1.25 + 1.15(x - 1)$, which is the same equation as $y = 1.15(x - 1) + 1.25$.

401. **c.** Let x = the amount invested at 12% interest. Let y = the amount invested at 15% interest. Since the amount invested at 15% is $100 more then twice the amount at 12%, then $y = 2x + 100$. Since the total interest was $855, use the equation $0.12x + 0.15y = 855$. You have two equations with two variables. Use the second equation, $0.12x + 0.15y = 855$, and substitute $(2x + 100)$ for y: $0.12x + 0.15(2x + 100) = 855$. Use the distributive property: $0.12x + 0.3x + 15 = 855$. Combine like terms: $0.42x + 15 = 855$. Subtract 15 from both sides: $0.42x + 15 - 15 = 855 - 15$; simplify: $0.42x = 840$. Divide both sides by 0.42: $\frac{0.42x}{0.42} = \frac{840}{0.42}$. Therefore, $x = \$2,000$, which is the amount invested at 12% interest.

402. **c.** Let x = the amount invested at 8% interest. Since the total interest is $405.50, use the equation $0.06(4,000) + 0.08x = 405.50$. Simplify the multiplication: $240 + 0.08x = 405.50$. Subtract 240 from both sides: $240 - 240 + 0.8x = 405.50 - 240$; simplify: $0.08x = 165.50$. Divide both sides by 0.08: $\frac{0.08x}{0.08} = \frac{165.50}{0.08}$. Therefore, $x = \$2,068.75$, which is the amount invested at 8% interest.

403. **d.** Let x = the amount of coffee at \$3 per pound. Let y = the total amount of coffee purchased. If there are 18 pounds of coffee at \$2.50 per pound, then the total amount of coffee can be expressed as $y = x + 18$. Use the equation $3x + 2.50(18) = 2.85y$, because the average cost of the y pounds of coffee is \$2.85 per pound. To solve, substitute $y = x + 18$ into $3x + 2.50(18) = 2.85y$; $3x + 2.50(18) = 2.85(x + 18)$. Multiply on the left side and use the distributive property on the right side: $3x + 45 = 2.85x + 51.30$. Subtract $2.85x$ on both sides: $3x - 2.85x + 45 = 2.85x - 2.85x + 51.30$. Simplify: $0.15x + 45 = 51.30$. Subtract 45 from both sides: $0.15x + 45 - 45 = 51.30 - 45$. Simplify: $0.15x = 6.30$. Divide both sides by 0.15: $\frac{0.15x}{0.15} = \frac{6.30}{0.15}$; so $x = 42$ pounds, which is the amount of coffee that costs \$3 per pound. Therefore, the total amount of coffee is $42 + 18$, which is 60 pounds.

404. **c.** Let x = the amount of candy at \$1.90 per pound. Let y = the total number of pounds of candy purchased. If it is known that there are 40 pounds of candy at \$2.15 per pound, then the total amount of candy can be expressed as $y = x + 40$. Use the equation $1.90x + 2.15(40) = \$162$, because the total amount of money spent was \$162. Multiply on the left side: $1.90x + 86 = 162$. Subtract 86 from both sides: $1.90x + 86 - 86 = 162 - 86$. Simplify: $1.90x = 76$. Divide both sides by 1.90: $\frac{1.90x}{1.90} = \frac{76}{1.90}$; so $x = 40$ pounds, which is the amount of candy that costs \$1.90 per pound. Therefore, the total amount of candy is $40 + 40$, which is 80 pounds.

405. **c.** Let x = the lesser odd integer and let $x + 2$ = the greater odd integer. The translation of the sentence "The sum of the squares of two consecutive positive odd integers is 74" is the equation $x^2 + (x + 2)^2 = 74$. Multiply $(x + 2)^2$ out as $(x + 2)(x + 2)$ using the distributive property: $x^2 + (x^2 + 2x + 2x + 4) = 74$. Combine like terms on the left side of the equation: $2x^2 + 4x + 4 = 74$. Put the equation in standard form by subtracting 74 from both sides, and set it equal to zero: $2x^2 + 4x - 70 = 0$; factor the trinomial completely: $2(x^2 + 2x - 35) = 0$; $2(x - 5)(x + 7) = 0$. Set each factor equal to zero and solve: $2 \neq 0$ or $x - 5 = 0$ or $x + 7 = 0$; $x = 5$ or $x = -7$. Since you are looking for a positive integer, reject the solution of $x = -7$. Therefore, the smaller positive integer is 5.

406. **a.** Let x = the lesser integer and let $x + 1$ = the greater integer. The sentence "the difference between the squares of two consecutive integers is 15" can translate to the equation $(x + 1)^2 - x^2 = 15$. Multiply the binomial $(x + 1)^2$ as $(x + 1)(x + 1)$ using the distributive property: $x^2 + 1x + 1x + 1 - x^2 = 15$. Combine like terms: $2x + 1 = 15$; subtract 1 from both sides of the equation: $2x + 1 - 1 = 15 - 1$. Divide both sides by 2: $\frac{2x}{2} = \frac{14}{2}$. The variable is now alone: $x = 7$. Therefore, the larger consecutive integer is $x + 1 = 8$.

CHAPTER

6 ▶ Exponents and Radicals

▶ Exponents

You probably learned long ago that $3 \times 3 = 9$. When a number is multiplied by itself, we say that we are "squaring" that number. Three squared is nine. We can represent three squared (3×3) as 3^2. The small number above and to the right of the 3 is called an exponent. An **exponent** tells us how many of a number we must multiply by itself. The number 3 in this example is the base. The **base** is the number that is multiplied according to the value of the exponent.

Example

4^2

In this example, four is squared, or raised to the second power. Exponents are also referred to as "powers." Because the exponent is 2, we must multiply two fours: $4 \times 4 = 16$.

Example

4^3

In this example, four is raised to the third power. When a base is raised to the third power, we say that the base is "cubed." Because the exponent is 3, we must multiply three fours: $4 \times 4 \times 4 = 64$.

Example

6^5

As exponents get larger, we may need a calculator to find our answer. $6^5 = 6 \times 6 \times 6 \times 6 \times 6 = 7,776$.

TIP

Memorize the following common squares. Just as with addition, subtraction, multiplication, and division facts, knowing these squares will help you find larger squares faster:

$1^2 = 1$	$9^2 = 81$
$2^2 = 4$	$10^2 = 100$
$3^2 = 9$	$11^2 = 121$
$4^2 = 16$	$12^2 = 144$
$5^2 = 25$	$13^2 = 169$
$6^2 = 36$	$14^2 = 196$
$7^2 = 49$	$15^2 = 225$
$8^2 = 64$	$25^2 = 625$

▶ Negative Exponents

The preceding examples all raise bases to positive exponents. But what if an exponent is negative? Here's where it gets a little tricky: A negative exponent doesn't represent a negative number—it means we must rewrite the base with a positive exponent in the denominator of a fraction that has a numerator of 1.

Example

4^{-3}

If four were raised to positive 3 (4^3), we would multiply three fours (i.e., multiply four by itself twice). When four is raised to negative 3, we begin by rewriting four with an exponent of positive 3, in the denominator of a fraction that has a numerator of 1:

$4^{-3} = \frac{1}{4^3}$

Now we can multiply four by itself twice:

$4 \times 4 \times 4 = 64$, so $4^{-3} = \frac{1}{4^3} = \frac{1}{64}$.

Example

8^{-4}

$8^{-4} = \frac{1}{8^4} = \frac{1}{4,096}$

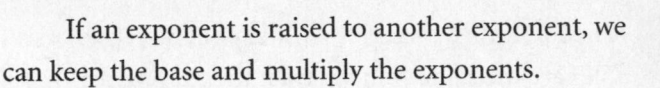
▶ Operations with Exponents

Now that we know how to evaluate exponents, here are a few shortcuts we can take to simplify problems with exponents.

If two identical bases are multiplied, the product is equal to the base raised to the sum of the exponents.

Example

$9^5 \times 9^7$

We can simplify $9^5 \times 9^7$ by adding 5 and 7: $5 + 7 = 12$, so $9^5 \times 9^7 = 9^{12}$.

Example

$11^8 \times 11^{73}$

Because $8 + 73 = 81$, $11^8 \times 11^{73} = 11^{81}$.

If two identical bases are divided, the product is equal to the base raised to the difference between the exponents.

Example

$10^8 \div 10^3$

We can simplify $10^8 \div 10^3$ by subtracting 3 from 8: $8 - 3 = 5$, so $10^8 \div 10^3 = 10^5$.

Example

$15^2 \div 15^9$

Because $2 - 9 = -7$, $15^2 \div 15^9 = 15^{-7}$.

PITFALL

Be careful with these rules. We can use them only if the bases are the same. We cannot simplify $6^9 \times 7^9$ or $8^2 \div 2^2$, because the bases are different. Also, these rules can be used only when the bases are multiplied or divided, not added or subtracted. We cannot simplify $9^5 + 9^7$ or $9^5 - 9^7$.

If an exponent is raised to another exponent, we can keep the base and multiply the exponents.

Example

$(2^3)^5$

The base 2 is being raised to the third power, which is then raised to the fifth power. We can simplify this by keeping the base and multiplying the exponents: $3 \times 5 = 15$, so $(2^3)^5 = 2^{15}$. We can check our answer by doing the multiplication:

$(2^3) = 8$; $8^5 = 8 \times 8 \times 8 \times 8 \times 8 = 32,768$

$2^{15} = 2 \times 2 \times 2 \times 2 \times 2 \times 2 \times 2 \times 2 \times 2 \times 2 \times 2 \times 2 \times 2 \times 2 \times 2 = 32,768$

Example

$(5^7)^6$

Because $7 \times 6 = 42$, $(5^7)^6 = 5^{42}$.

These same rules can be applied to algebraic expressions and equations.

▶ Tips for Working with Exponents

Add and subtract like terms:

$$3n + 5n = 8n$$
$$5x^2y - 3x^2y = 2x^2y$$

When multiplying variables with exponents, if the variables are the same, add the exponents and write the base only once:

$$(a^4)(a^3) = a^{(4+3)} = a^7$$
$$(x^2y^3)(axy^5) = ax^{(2+1)}y^{(3+5)} = ax^3y^8$$

When dividing variables with exponents, if the variables are the same, you subtract the exponents:

$$\frac{n^5}{n^2} = \frac{n \times n \times n \times n \times n}{n \times n} = n^{5-2} = n^3$$

If the exponent of a similar term in the denominator is larger than the one in the numerator, the exponent will have a negative sign:

$$\frac{2x^3}{x^4} = 2x^{-1}$$
$$\frac{n^5}{n^8} = n^{5-8} = n^{-3}$$

A negative numerator becomes positive when the variable is moved into the denominator.

$$2x^{-1} = 2\left(\frac{1}{x^1}\right) = \frac{2}{x}$$
$$n^{-3} = \frac{1}{n^3}$$

When the result of a division leaves an exponent of zero, the term raised to the power of zero equals 1:

$$z^0 = 1$$
$$3\frac{r^2}{r^2} = 3r^0 = 3(1) = 3$$

When a variable with an exponent is raised to a power, you multiply the exponent to form the new term:

$$(b^2)^3 = b^2 \times b^2 \times b^2 = b^{(2+2+2)} = b^6$$
$$(2x^2y)^2 = 2x^2y \times 2x^2y = 2 \times 2 \times x^2 \times x^2 \times y \times y$$
$$= 2^2 x^{2+2} y^{1+1} = 4x^4y^2$$

Remember order of operations: PEMDAS. Generally, list terms in order from highest power to lowest power.

▶ **Practice Questions**

Simplify the following expressions.

407. $5x^2 + 8x^2$

408. $5ab^4 - ab^4$

409. $9mn^3 + 8mn + 2mn^3$

410. $5c^2 + 3c - 2c^2 + 4 - 7c$

411. $3x^2 + 4ax - 8a^2 + 7x^2 - 2ax + 7a^2$

412. $(5n^2)(2n^5 - 2n^3 + 3n^7)$

413. $5xy \times 6xy + 7x^2y^2$

414. $(5a^2 \times 3ab) + 2a^3b$

415. $\frac{8xy^2}{2xy}$

416. $(4a^2)^3 + (2a^3)^2 - 11a^6$

417. $\frac{(3x)^3}{x^2 \times x^4}$

418. $\frac{(12s^2)(2s^4)}{3s^3}$

419. $\frac{7a^3b^5}{28ab^2}$

420. $(3xy^5)^2 - 11x^2y^2(4y^4)^2$

421. $\frac{2(3x^2y)^2(xy)^3}{3(xy)^2}$

422. $\frac{2x^2y^5}{x^5y^3}$

423. $4x^{-2}(3ax)^5$

424. $\dfrac{3x^{-2}}{x5} - \dfrac{2x}{x^8}$

425. $(5a^2x^3y)^3$

426. $\dfrac{24x^3}{(2x)^2} + \dfrac{3x^5}{x^4} - \dfrac{(3ax)^2}{a^2x}$

427. $(4x^2y)^3 + \dfrac{(2x^2y)^4}{2x^2y}$

428. $\dfrac{8ax^2}{(a^2x)^3}$

429. $(ab^2)^3 + 2b^2 - (4a)^3b^6$

430. $\dfrac{(4b)^2x^{-2}}{(2ab^2x)^2}$

431. $(2xy \times \frac{4}{x})^2 + \dfrac{9y^2}{(3y)^2}$

Often in math, we learn how one operation is the opposite of another. After learning how to add one number to another, we learn how to subtract. After learning how to multiply, we learn to divide. Radicals are the opposites of exponents.

▶ Radicals

An exponent tells us how many times to multiply a base by itself. A **radical** asks us to find the "root" of the number under the radical symbol. The number under the radical symbol is called the **radicand**. The type of root we're looking for is written inside the pointed part of the radical.

Example
$\sqrt[5]{7,776}$

This radical asks us to find the fifth root of 7,776. The radicand, found under the radical symbol, is 7,776, and the root is 5, found inside the pointed part of the radical.

But what is a root? In this example, the root is the number that must be multi-

plied by itself a specified number of times in order to equal 7,776. Look again at the third example in the "Exponents" section at the beginning of this chapter: $6^5 = 7,776$. When we multiply five sixes (i.e., we multiply six by itself four times), the result is 7,776. Six is the fifth root of 7,776.

That's how exponents and radicals go together: $5^4 = 625$ and $\sqrt[4]{625} = 5$. A base, raised to an exponent, yields a product. The root, to the degree of the exponent, of the radical of that product is equal to the base.

Example
$\sqrt[3]{27}$

What number, when multiplied by itself twice, is equal to 27? $2 \times 2 \times 2 = 8$, but $3 \times 3 \times 3 = 27$. Three is the third root of 27: $\sqrt[3]{27} = 3$.

TIP

Often, finding the root of a radical takes some trial and error if you don't have a calculator. Memorize these common square roots, and use them to help you estimate third roots and larger roots:

$\sqrt{1} = 1$	$\sqrt{81} = 9$
$\sqrt{4} = 2$	$\sqrt{100} = 10$
$\sqrt{9} = 3$	$\sqrt{121} = 11$
$\sqrt{16} = 4$	$\sqrt{144} = 12$
$\sqrt{25} = 5$	$\sqrt{169} = 13$
$\sqrt{36} = 6$	$\sqrt{196} = 14$
$\sqrt{49} = 7$	$\sqrt{225} = 15$
$\sqrt{64} = 8$	$\sqrt{625} = 25$

Compare this chart to the common squares chart at the beginning of the chapter. The square root of the square of a base is equal to that base, since taking the square root of a base is the inverse of squaring that base.

▶ Tips for Solving Radical Equations

- Squaring both sides of an equation is a valuable tool when solving radical equations. Use the following property: When a and b are algebraic expressions, if $a = b$, then $a^2 = b^2$.
- Isolate the radical on one side of an equation before using the squaring property.
- Squaring a radical results in the radical symbol disappearing, for example, $(\sqrt{x+5})^2 = x + 5$.
- For second-degree equations, which are equations in which at least one variable has an exponent of 2, use the radical sign on both sides of the equation to find a solution for the variable.

▶ Practice Questions

Solve the following radical equations. Watch for extraneous solutions (solutions that do not work).

432. $x^2 = 49$

433. $x^2 = 135$

434. $\sqrt{n} = 11$

435. $2\sqrt{a} = 24$

436. $\sqrt{2x} - 4 = 4$

437. $\sqrt{4x + 6} = 8$

438. $\sqrt{3x + 4} + 8 = 12$

439. $\sqrt{5x - 4} + 3 = 12$

440. $\sqrt{4x + 9} = -13$

441. $\sqrt{5x - 6} + 3 = 11$

442. $\sqrt{9 - x} + 14 = 25$

443. $3\sqrt{3x + 1} = 15$

444. $3\sqrt{-x} + 7 = 25$

445. $3 = 10 - \sqrt{100x - 1}$

446. $\sqrt{3x + 46} + 24 = 38$

447. $-7 = 10 - \sqrt{25x + 39}$

448. $3\sqrt{13x + 43} - 4 = 29$

449. $\dfrac{28}{\sqrt{5x + 1}} = 7$

450. $x = \sqrt{8 - 2x}$

451. $x = \sqrt{3x + 4}$

452. $x = \sqrt{x + 12}$

453. $x = \sqrt{7x - 1\,0}$

454. $\sqrt{4x + 3} = 2x$

455. $\sqrt{2 - \frac{7}{2}x} = x$

456. $x = \sqrt{\frac{3}{2}x + 10}$

457. The square of a positive number is 49. What is the number?

    a. $\sqrt{7}$

    b. −7

    c. 7 or −7

    d. 7

458. The square of a number added to 25 equals 10 times the number. What is the number?

    a. −5

    b. 10

    c. −10

    d. 5

459. The sum of the square of a number and 12 times the number is −27. What is the smaller possible value of this number?

    a. −3

    b. −9

    c. 3

    d. 9

TERMS TO REVIEW

exponent

base

radical

radicand

▶ **Answers**

407. Add like terms: $5x^2 + 8x^2 = 13x^2$.

408. Subtract like terms: $5ab^4 - ab^4 = 4ab^4$.

409. Use the commutative property of addition: $9mn^3 + 2mn^3 + 8mn$. Combine like terms: $11mn^3 + 8mn$.

410. Use the commutative property of addition: $5c^2 - 2c^2 + 3c - 7c + 4$. Combine like terms: $3c^2 - 4c + 4$.

411. Use the commutative property of addition: $3x^2 + 7x^2 + 4ax - 2ax - 8a^2 + 7a^2$. Combine like terms: $10x^2 + 2ax - a^2$.

412. Use the distributive property: $(5n^2)(2n^5) - (5n^2)(2n^3) + (5n^2)(3n^7)$. Use the commutative property of multiplication: $(5 \times 2 \times n^2 n^5) - (5 \times 2 \times n^2 n^3) + (5 \times 3 \times n^2 n^7)$. Add the exponents of the variables: $(10 \times n^{2+5}) - (10 \times n^{2+3}) + (15 \times n^{2+7})$; $10n^7 - 10n^5 + 15n^9$. Show the expression in decreasing exponential order: $15n^9 + 10n^7 - 10n^5$.

413. Use the commutative property of multiplication: $5 \times 6xxyy + 7x^2y^2$. When the same variables are multiplied, add the exponents of the variables: $(30x^2y^2) + 7x^2y^2$. Combine like terms: $37x^2y^2$.

414. Use the commutative property of multiplication: $(5 \times 3a^2ab) + 2a^3b$. When the same variables are multiplied, add the exponents of the variables: $15a^3b + 2a^3b$. Combine like terms: $17a^3b$.

415. Divide numerical terms: $\frac{8xy^2}{2xy} = \frac{4xy^2}{xy}$. When similar factors, or bases, are being divided, subtract the exponent in the denominator from the exponent in the numerator: $\frac{4xy^2}{xy} = 4x^{1-1}y^{2-1}$. Simplify: $4x^0y^1 = 4(1)y = 4y$.

416. Terms within parentheses are the base of the exponent outside the parentheses: $(4a^2)(4a^2)(4a^2) + (2a^3)(2a^3) - 11a^6$. Use the distributive property of multiplication: $(4 \times 4 \times 4)(a^2 \times a^2 \times a^2) + (2 \times 2)(a^3 \times a^3) - 11a^6$. When the same variables are multiplied, add the exponents of the variables: $64(a^{2+2+2}) + 4(a^{3+3}) - 11a^6$. Simplify: $64a^6 + 4a^6 - 11a^6$. Combine like terms: $57a^6$. Another way of solving this problem is to multiply the exponents of each factor inside the parentheses by the exponent outside the parentheses: $4^{(1)3}a^{(2)3} + 2^{(1)2}a^{(3)2} - 11a^6$. Simplify the expressions in the exponents: $4^3a^6 + 2^2a^6 - 11a^6$. Simplify terms: $64a^6 + 4a^6 - 11a^6$. Combine like terms: $57a^6$.

417. In the numerator, multiply the exponents of each factor inside the parentheses by the exponent outside the parentheses: $\frac{3^3x^3}{x^2 \times x^4}$. In the denominator, add the exponents of similar bases: $\frac{3^3x^3}{x^{2+4}} = \frac{27x^3}{x^6}$. When similar factors, or bases, are being divided, you subtract the exponent in the denominator from the exponent in the numerator: $27x^{3-6} = 27x^{-3}$. A base with a negative exponent in the numerator is equivalent to the same variable or base in the denominator with the inverse sign for the exponent: $27x^{-3} = \frac{27x}{3}$.

418. Use the commutative property of multiplication: $\frac{12 \times 2s^2 \times s^4}{3s^3}$. When similar factors, or bases, are multiplied, add the exponents of the variables: $\frac{12 \times 2s^{2+4}}{3s^3}$. Simplify the exponent and coefficients: $\frac{8s^6}{s^3}$. When similar factors, or bases, are being divided, subtract the exponent in the denominator from the exponent in the numerator: $8s^{6-3} = 8s^3$.

419. Divide out the common factor of 7 in the numerator and denominator: $\frac{1 \times 7\, a^3b^5}{4 \times 7\, ab^2} = \frac{1a^3b^5}{4ab^2}$. When similar factors, or bases, are being divided, subtract the exponent in the denominator from the exponent in the numerator: $\frac{1a^3b^5}{4ab^2} = \frac{1a^{3-1}b^{5-2}}{4}$. Simplify exponents: $\frac{1a^2b^3}{4}$. Separate the coefficient from the variable: $\frac{1a^2b^3}{4} = \frac{1}{4}a^2b^3$. Either expression is an acceptable answer.

420. Multiply the exponents of each factor inside the parentheses by the exponent outside the parentheses: $3^2x^2y^{10} - 11x^2y^2 4^2y^8$. Use the commutative property of multiplication: $3^2x^2y^{10} - 11 \times 4^2x^2y^2y^8$. When similar factors, or bases, are multiplied, add the exponents of the variables: $3^2x^2y^{10} - 11 \times 16 \times x^2y^{10}$. Evaluate numerical factors: $9x^2y^{10} - 176x^2y^{10}$. Combine like terms: $-167x^2y^{10}$.

421. Multiply the exponents of each factor inside the parentheses by the exponent outside the parentheses: $\frac{2(3^2x^4y^2)(x^3y^3)}{3(x^2y^2)}$. Use the commutative property of multiplication: $\frac{2 \times 9 \times x^4x^3y^2y^3}{x^2y^2}$. When similar factors, or bases, are multiplied, add the exponents of the variables: $\frac{2 \times 9 \times x^{4+3}y^{2+3}}{x^2y^2}$. Factor out like numerical terms in the fraction, and simplify exponents with operations: $\frac{6x^7y^5}{x^2y^2}$. When similar factors, or bases, are being divided, subtract the exponent in the denominator from the exponent in the numerator: $6x^{7-2}y^{5-2}$. Simplify the operations in the exponents: $6x^5y^3$.

422. When similar factors, or bases, are being divided, subtract the exponent in the denominator from the exponent in the numerator: $\frac{2x^2y^5}{x^5y^3} = 2x^{2-5}y^{5-3}$. Simplify the operations in the exponents: $2x^{2-5}y^{5-3} = 2x^{-3}y^2$. A base with a negative exponent in the numerator is equivalent to the same variable or base in the denominator with the inverse sign for the exponent: $2x^{-3}y^2 = \frac{2y^2}{x^3}$.

423. Multiply the exponents of each factor inside the parentheses by the exponent outside the parentheses: $4x^{-2}(2^3a^3x^3)$. Use the commutative property of multiplication: $4 \times 3^3 \times a^3x^3x^{-2}$. Evaluate numerical terms: $108a^3x^3x^{-2}$. When similar factors, or bases, are being multiplied, add the exponents: $108a^3x^{3-2}$. Simplify the operations in the exponents: $108a^3x$.

424. When similar factors, or bases, are being divided, subtract the exponent in the denominator from the exponent in the numerator: $3x^{-2-5} - 2x^{1-8}$. Simplify the operations in the exponents: $3x^{-7} - 2x^{-7}$. Subtract like terms: x^{-7}. A base with a negative exponent in the numerator is equivalent to the same variable or base in the denominator with the inverse sign for the exponent: $x^{-7} = \frac{1}{x^7}$.

425. Multiply the exponents of each factor inside the parentheses by the exponent outside the parentheses: $5^3a^6x^9y^3$. Evaluate the numerical coefficient: $125a^6x^9y^3$.

426. Multiply the exponents of each factor inside the parentheses by the exponent outside the parentheses: $\frac{24x^3}{2^2x^2} + \frac{3x^5}{x^4} - \frac{3^2a^2x^2}{a^2x}$. Evaluate the numerical coefficients and divide out common numerical factors in the terms: $\frac{6x^3}{x^2} + \frac{3x^5}{x^4} - \frac{9a^2x^2}{a^2x}$. When similar factors, or bases, are being divided, subtract the exponent in the denominator from the exponent in the numerator: $6x^{3-2} + 3x^{5-4} - 9a^{2-2}x^{2-1}$. Simplify the operations in the exponents: $6x^1 + 3x^1 - 9a^0x^1$. Any term to the power of zero equals 1: $6x + 3x - 9(1)x$. Combine like terms: $6x + 3x - 9x = 0x = 0$.

427. Multiply the exponents of each factor inside the parentheses by the exponent outside the parentheses: $4^3x^6y^3 + \frac{2^4x^8y^4}{2x^2y}$. When similar factors, or bases, are being divided, subtract the exponent in the denominator from the exponent in the numerator: $4^3x^6y^3 + 2^{4-1}x^{8-2}y^{4-1}$. Simplify the operations in the exponents: $4^3x^6y^3 + 2^3x^6y^3$. Evaluate the numerical coefficients: $64x^6y^3 + 8x^6y^3$. Add like terms: $72x^6y^3$.

428. Multiply the exponents of each factor inside the parentheses by the exponent outside the parentheses: $\frac{8ax^2}{a^6x^3}$. When similar factors, or bases, are being divided, subtract the exponent in the denominator from the exponent in the numerator: $8a^{1-6}x^{2-3}$. Simplify the operations in the exponents: $8a^{-5}x^{-1}$. A base with a negative exponent in the numerator is equivalent to the same variable or base in the denominator with the inverse sign for the exponent: $8a^{-5}x^{-1} = \frac{8}{a^5x^1} = \frac{8}{a^5x}$.

429. Multiply the exponents of each factor inside the parentheses by the exponent outside the parentheses: $a^3b^6 + 2b^2 - 4^3a^3b^6$. Evaluate the numerical coefficients: $a^3b^6 + 2b^2 - 64a^3b^6$. Use the commutative property of addition: $2b^2 - 64a^3b^6 + a^3b^6$. Combine like terms in the expression: $2b^2 - 63a^3b^6$.

430. Multiply the exponents of each factor inside the parentheses by the exponent outside the parentheses: $\frac{4^2 b^2 x^{-2}}{2^2 a^2 b^4 x^2}$. Evaluate the numerical coefficients: $\frac{16 b^2 x^{-2}}{4 a^2 b^4 x^2}$. Simplify the numerical factors in the numerator and the denominator: $\frac{4 b^2 x^{-2}}{a^2 b^4 x^2}$. When similar factors, or bases, are being divided, subtract the exponent in the denominator from the exponent in the numerator: $\frac{4 b^{2-4} x^{-2-2}}{a^2}$. Simplify the operations in the exponents: $\frac{4 b^{-2} x^{-4}}{a^2}$. A base with a negative exponent in the numerator is equivalent to the same variable or base in the denominator with the inverse sign for the exponent: $\frac{4}{a^2 b^2 x^4}$.

431. Multiply the exponents of each factor inside the parentheses by the exponent outside the parentheses: $(2xy)^2 \times (\frac{4}{x})^2 + \frac{9y^2}{3^2 y^2}$. Repeat the previous step: $(2^2 x^2 y^2)(\frac{4^2}{x^2}) + \frac{9y^2}{3^2 y^2}$. Evaluate numerical factors: $(4x^2 y^2)(\frac{16}{x^2}) + \frac{9y^2}{9y^2}$. The last term is equivalent to 1: $(4x^2 y^2)(\frac{16}{x^2}) + 1$. Multiply the fraction in the first term by the factor in the first term: $[\frac{(4x^2 y^2)16}{x^2}] + 1$. Use the commutative property of multiplication: $(\frac{4 \times 16 x^2 y^2}{x^2}) + 1$. Evaluate numerical factors: $\frac{64 x^2 y^2}{x^2} + 1$. Divide out the common factor of x^2 in the numerator and denominator: $64y^2 + 1$.

432. Use the radical sign on both sides of the equation: $\sqrt{x^2} = \sqrt{49}$. Show both solutions for the square root of 49: $x = \pm 7$. Check the first solution in the original equation: $(7)^2 = 49; 49 = 49$. Check the second solution in the original equation: $(-7)^2 = 49; 49 = 49$. Both solutions, $x = \pm 7$, check out.

433. Use the radical sign on both sides of the equation: $\sqrt{x^2} = \sqrt{135}$. Simplify the radical: $x = \sqrt{135} = \sqrt{9 \times 15} = \pm 3\sqrt{15}$. Check the first solution in the original equation: $(3\sqrt{15})^2 = 135; 3^2(\sqrt{15})^2 = 135; 9(15) = 135; 135 = 135$. Check the second solution in the original equation: $(-3\sqrt{15})^2 = 135; (-3)^2(\sqrt{15})^2 = 135; (9)(15) = 135; 135 = 135$; Both solutions, $x = \pm 3\sqrt{15}$, check out.

434. First, square both sides of the equation: $(\sqrt{n})^2 = 11^2$. Simplify both terms: $n = 121$. Check by substituting in the original equation: $\sqrt{121} = 11$. The original equation asks for only the positive root of n. So when you substitute 121 into the original equation, only the positive root $\sqrt{n} = 11$ is to be considered. $11 = 11$ checks out. Although this may seem trivial at this point, as the radical equations become more complex, this will become important.

435. Isolate the radical on one side of the equation. Divide both sides by 2: $\frac{2\sqrt{a}}{2} = \frac{24}{2}$. Simplify terms: $\sqrt{a} = 12$. Now square both sides of the equation: $(\sqrt{a})^2 = 12^2$. Simplify terms: $a = 144$. Check the solution in the original equation: $2\sqrt{144} = 24; 2(12) = 24$. The solution $a = 144$ checks out: $24 = 24$.

436. Begin by adding 4 to both sides to isolate the radical: $\sqrt{2x} - 4 + 4 = 4 + 4$. Combine like terms on each side: $\sqrt{2x} = 8$. Square both sides of the equation: $(\sqrt{2x})^2 = 8^2$. Simplify terms: $2x = 64$. Divide both sides by 2: $x = 32$. Check the solution in the original equation: $\sqrt{2(32)} - 4 = 4; \sqrt{64} - 4 = 4$. The solution $x = 32$ checks out: $8 - 4 = 4; 4 = 4$.

437. Square both sides of the equation: $(\sqrt{4x+6})^2 = 8^2$; $4x + 6 = 64$. Subtract 6 from both sides of the equation: $4x = 58$. Divide both sides by 4 and simplify: $x = \frac{58}{4} = 14.5$. Check the solution in the original equation: $\sqrt{4(14.5)+6} = 8$. Simplify terms: $\overline{64} = 8$. The solution $x = 14.5$ checks out: $8 = 8$.

438. Subtract 8 from both sides in order to isolate the radical: $\sqrt{3x+4} = 4$. Square both sides of the equation: $(\sqrt{3x+4})^2 = 4^2$. Simplify terms: $3x + 4 = 16$. Subtract 4 from both sides and divide by 3: $x = 4$. Check your solution in the original equation: $\sqrt{3(4)+4} + 8 = 12$. Simplify terms: $\sqrt{16} + 8 = 12$; $4 + 8 = 12$. The solution $x = 4$ checks out: $12 = 12$.

439. Subtract 3 from both sides of the equation, isolating the radical: $\sqrt{5x-4} = 9$. Square both sides of the equation: $(\sqrt{5x-4})^2 = 9^2$. Simplify terms on both sides: $5x - 4 = 81$. Add 4 to both sides and then divide by 5: $x = \frac{85}{5} = 17$. Check your solution in the original equation: $\sqrt{5(17)-4} + 3 = 12$. Simplify terms under the radical sign: $\sqrt{81} + 3 = 12$. Find the positive square root of 81: $9 + 3 = 12$. Simplify: $12 = 12$. The solution $x = 17$ checks out.

440. Square both sides of the equation: $(\sqrt{4x+9})^2 = (-13)^2$. Simplify terms on both sides of the equation: $4x + 9 = 169$. Subtract 9 from both sides and then divide by 4: $x = 40$. Substitute the solution in the original equation: $\sqrt{4(40+9)} = -13$. Simplify the expression under the radical sign: $\sqrt{169} = -13$. The radical sign calls for the positive square root: $13 \neq -13$. The solution does not check out. There is no solution for this equation.

441. Subtract 3 from both sides, isolating the radical: $\sqrt{5x-6} = 8$. Square both sides of the equation: $(\sqrt{5x-6})^2 = 8^2$. Simplify terms: $5x - 6 = 64$. Add 6 to both sides and divide the result by 5: $x = 14$. Check the solution in the original equation: $\sqrt{5(14)-6} + 3 = 11$. Simplify the expression under the radical: $\sqrt{64} + 3 = 11$. Find the positive square root of 64 and add 3: $8 + 3 = 11$. The solution $x = 14$ checks out: $11 = 11$.

442. Subtract 14 from both sides to isolate the radical: $\sqrt{9-x} = 11$. Now square both sides of the equation: $9 - x = 121$. Subtract 9 from both sides: $-x = 112$. Multiply both sides by negative 1 to solve for x: $x = -112$. Check the solution in the original equation: $\sqrt{9-(-112)} + 14 = 25$. Simplify the expression under the radical sign: $\sqrt{121} + 14 = 25$. The square root of 121 is 11. Add 14 and the solution $x = -112$ checks out: $25 = 25$.

443. To isolate the radical, divide both sides by 3: $\sqrt{3x+1} = 5$. Square both sides of the equation: $(\sqrt{3x+1})^2 = 5^2$. Simplify terms: $3x + 1 = 25$. Subtract 1 from both sides of the equation and divide by 3: $x = 8$. Check the solution in the original equation: $3\sqrt{3(8)+1} = 15$. Simplify the expression under the radical sign: $3\sqrt{25} = 15$. Multiply 3 by the positive root of 25: $3(5) = 15$. The solution $x = 8$ checks out: $15 = 15$.

444. Subtract 7 from both sides of the equation: $3\sqrt{-x} = 18$. Divide both sides by 3 to isolate the radical: $\sqrt{-x} = 6$. Square both sides of the equation: $(\sqrt{-x})^2 = 6^2$. Simplify terms: $-x = 36$. Multiply both sides by negative 1: $x = -36$. Check the solution in the original equation: $3\sqrt{-(-36)} + 7 = 25$. Simplify terms under the radical: $3\sqrt{36} + 7 = 25$. Use the positive square root of 36: $3(6) + 7 = 25$; $18 + 7 = 25$. The solution $x = -25$ checks out: $25 = 25$.

445. Add $\sqrt{100x - 1}$ to bothsides of the equation: $3 + \sqrt{100x - 1} = 10 - \sqrt{100x - 1} + \sqrt{100x - 1}$. Combine like terms and simplify the equation: $3 + \sqrt{100x - 1} = 10$. Now subtract 3 from both sides: $\sqrt{100x - 1} = 7$. Square both sides of the equation: $100x - 1 = 49$. Add 1 to both sides of the equation and divide by 100: $x = 0.5$. Check the solution in the original equation: $3 = 10 - \sqrt{100(0.5) - 1}$. Simplify the expression under the radical sign: $3 = 10 - \sqrt{49}$. The equation asks you to subtract the positive square root of 49 from 10: $3 = 10 - 7$. The solution $x = 0.5$ checks out: $3 = 3$.

446. Subtract 24 from both sides to isolate the radical: $\sqrt{3x + 46} = 14$. Square both sides of the equation: $3x + 46 = 196$. Subtract 46 from both sides of the equation and divide by 3: $x = 50$. Check the solution in the original equation: $\sqrt{3(50) + 46} + 24 = 38$. Simplify the expression under the radical sign: $\sqrt{196} + 24 = 38$; $14 + 24 = 38$. The solution $x = 50$ checks out: $38 = 38$.

447. Add $\sqrt{25x + 39}$ to both sides of the equation: $\sqrt{25x + 39} - 7 = 10 - \sqrt{25x + 39} + \sqrt{25x + 39}$. Combine like terms and simplify the equation: $\sqrt{25x + 39} - 7 = 10$. Add 7 to both sides of the equation: $\sqrt{25x + 39} = 17$. Square both sides: $25x + 39 = 289$. Subtract 39 from both sides and divide the result by 25: $x = 10$. Check the solution in the original equation: $-7 = 10 - \sqrt{25(10) + 39}$. Simplify the expression under the radical sign: $-7 = 10 - \sqrt{289}$. The equation asks you to subtract the positive square root of 289 from 10: $-7 = 10 - 17$. The solution $x = 10$ checks out: $-7 = -7$.

448. To isolate the radical on one side of the equation, add 4 to both sides and divide the result by 3: $\sqrt{13x + 43} = 11$. Square both sides of the equation: $13x + 43 = 121$. Subtract 43 from both sides and divide by 13: $x = 6$. Check the solution in the original equation: $3\sqrt{13(6) + 43} - 4 = 29$. Simplify the expression under the radical sign: $3\sqrt{121} - 4 = 29$. Evaluate the left side of the equation: $(11) - 4 = 29$. The solution $x = 6$ checks out: $29 = 29$.

449. To isolate the radical on one side of the equation, multiply both sides by $\sqrt{5x + 1}$: $28 = 7\sqrt{5x + 1}$. Divide both sides of the equation by 7: $4 = \sqrt{5x + 1}$. Square both sides of the equation: $16 = 5x + 1$. Subtract 1 from both sides and divide the result by 5: $3 = x$. Check the solution in the original equation: $\frac{28}{\sqrt{5(3) + 1}} = 7$. Simplify the expression under the radical sign: $\frac{28}{\sqrt{16}} = 7$. Divide the numerator by the positive square root of 16: $\frac{28}{4} = 7$. The solution $3 = x$ checks out: $7 = 7$.

450. The radical is alone on one side. Square both sides: $x^2 = 8 - 2x$. Transform the equation by putting all terms on one side: $x^2 + 2x - 8 = 0$. The result is a quadratic equation. Solve for x by factoring using the trinomial factor form and setting each factor equal to zero and solving for x. (For this and the following questions, refer to Chapter 10 for practice and tips for factoring quadratic equations.) It will be important to check each solution: $x^2 + 2x - 8 = (x + 4)(x - 2) = 0$. Let the first factor equal zero and solve for x: $x + 4 = 0$. Subtract 4 from both sides: $x = -4$. Check the solution in the original equation: $(-4) = \sqrt{8 - 2(-4)}$. Evaluate the expression under the radical sign: $-4 = \sqrt{16}$. The radical sign calls for a positive root: $-4 \neq 4$. Therefore, x cannot equal -4: $x = -4$ is an example of an *extraneous root*. Let the second factor equal zero and solve for x: $x - 2 = 0$. Subtract 2 from both sides: $x = 2$. Check the solution in the original equation: $(2) = \sqrt{8 - 2(2)}$. Evaluate the expression under the radical sign: $2 = \sqrt{4}$. The positive square root of 4 is 2: $2 = 2$. Therefore, the only solution for the equation is $x = 2$.

451. With the radical alone on one side of the equation, square both sides: $x^2 = 3x + 4$. The resulting quadratic equation may have up to two solutions. Put it into standard form and factor the equation using the trinomial factor form to find the solutions. Then check the solutions in the original equation: $x^2 - 3x - 4 = (x - 4)(x + 1) = 0$. Letting each factor equal zero and solving for x results in two possible solutions, $x = 4$ and/or -1. Check the first possible solution in the original equation: $(4) = \sqrt{3(4) + 4} = \sqrt{16} = 4$. The solution checks out. Now check the second possible solution in the original equation. $(-1) = \sqrt{3(-1) + 4} = \sqrt{1} = 1$; $-1 \neq 1$. Therefore, $x \neq -1$. $x = -1$ is an *extraneous root*. The only solution for the original equation is $x = 4$.

452. Square both sides of the equation. $x^2 = x + 12$. Subtract $(x + 12)$ from both sides of the equation: $x^2 - x - 12 = 0$. The resulting quadratic equation may have up to two solutions. Factor the equation to find the solutions and check in the original equation: $x^2 - x - 12 = (x - 4)(x + 3) = 0$. Let the first factor equal zero and solve for x. $x - 4 = 0$, so $x = 4$. Let the second factor equal zero and solve for x: $x + 3 = 0$, so $x = -3$. Check each solution in the original equation to rule out an extraneous solution: $(4) = \sqrt{(4) + 12}$. Simplify the expression under the radical sign: $4 = \sqrt{16}$. The solution $x = 4$ checks out. Now check the second possible solution in the original equation: $(-3) = \sqrt{(-3) + 12}$. You could simplify the expression under the radical sign to get the square root of 9. However, the radical sign indicates that the positive solution is called for, and the left side of the original equation when $x = -3$ is a negative number. So, $x = -3$ is *not* a solution.

453. Square both sides of the equation: $x^2 = 7x - 10$. Subtract $(7x - 10)$ from both sides of the equation: $x^2 - 7x + 10 = 0$. Factor the quadratic equation to find the solutions, and check each in the original equation to rule out any extraneous solution: $x^2 - 7x + 10 = (x - 5)(x - 2) = 0$. The first factor will give you the solution $x = 5$. The second factor will give the solution $x = 2$. Check the first solution for the quadratic equation in the original equation: $(5) = \sqrt{7(5) - 10}$. Simplify the expression under the radical sign: $5 = \sqrt{25}$ or $5 = 5$. The solution $x = 5$ is a solution to the original equation. Now check the second solution to the quadratic equation in the original: $(2) = \sqrt{7(2) - 10}$. Simplify the expression under the radical sign: $2 = \sqrt{4}$ or $2 = 2$. There are two solutions to the original equation: $x = 2$ and $x = 5$.

454. Square both sides of the radical equation: $4x + 3 = 4x^2$. Transform the equation into a quadratic equation: $4x^2 - 4x - 3 = 0$. Factor the result using the trinomial factor form: $4x^2 - 4x - 3 = (2x + 1)(2x - 3) = 0$. Let the first factor equal zero and solve for x: $2x + 1 = 0$, so $x = -0.5$. Let the second factor equal zero and solve for x: $2x - 3 = 0$, so $x = 1.5$. When you substitute -0.5 for x in the original equation, the result will be $\sqrt{1} = -1$. That cannot be true for the original equation, so $x \neq -0.5$. Substitute 1.5 for x in the original equation: $\sqrt{4(1.5) + 3} = 2(1.5)$. Simplify the terms on each side of the equal sign: $\sqrt{9} = 3$ or $3 = 3$. The only solution for the original equation is $x = 1.5$.

455. Square both sides of the equation: $2 - \frac{7}{2}x = x^2$. Subtract $(2 - \frac{7}{2}x)$ from both sides of the equation: $0 = x^2 + \frac{7}{2}x - 2$. Multiply both sides of the equation by 2 to simplify the fraction: $0 = 2x^2 + 7x - 4$. Factor using the trinomial factor form: $2x^2 + 7x - 4 = (2x - 1)(x + 4) = 0$. Let the first factor equal zero and solve for x: $2x - 1 = 0$, so $x = \frac{1}{2}$. Let the second factor equal zero and solve for x: $x + 4 = 0$, so $x = -4$. Check the first possible solution in the original equation: $\sqrt{2 - \frac{7}{2}\left(\frac{1}{2}\right)} = \left(\frac{1}{2}\right)$. Simplify the expression under the radical sign: $\sqrt{2 - \frac{7}{4}} = \sqrt{2 - 1\frac{3}{4}} = \sqrt{\frac{1}{4}} = \frac{1}{2}; \frac{1}{2} = \frac{1}{2}$. So $x = \frac{1}{2}$ is a solution. Check the solution $x = -4$ in the original equation: $\sqrt{2 - \frac{7}{2}(-4)} = (-4)$. Simplify the expression: $\sqrt{16} = (-4)$ or $4 = -4$. This is not true. There is one solution for the original equation, $x = \frac{1}{2}$.

456. Square both sides of the equation: $x^2 = \frac{3}{2}x + 10$. Add $(\frac{-3}{2x} - 10)$ to both sides of the equation: $x^2 - \frac{3}{2}x - 10 = 0$. Multiply the equation by 2 to eliminate the fraction: $2x^2 - 3x - 20 = 0$. Factor using the trinomial factor form: $2x^2 - 3x - 20 = (x-4)(2x+5) = 0$. Letting each factor of the trinomial factors equal zero results in two possible solutions for the original equation, $x = 4$ and/or $x = -2\frac{1}{2}$. Check the first possible solution in the original equation: $4 = \sqrt{\frac{3}{2}(4) + 10}$. Simplify the radical expression: $4 = \sqrt{16}$ or $4 = 4$. The solution $x = 4$ checks out as a solution for the original equation. Check the second possible solution in the original equation: $-2\frac{1}{2} = \sqrt{\frac{3}{2}(-2\frac{1}{2}) + 10}$. A negative number cannot be equal to a positive square root, which the radical sign in the original expression calls for. Therefore, $x = -2\frac{1}{2}$ is *not* a solution to the original equation. The only solution for this equation is $x = 4$.

457. **d.** Let $x =$ the number. The sentence "The square of a positive number is 49" translates to the equation $x^2 = 49$. Take the square root of each side to get $\sqrt{x^2} = \sqrt{49}$, so $x = 7$ or -7. Since you are looking for a positive number, the final solution is 7.

458. **d.** Let $x =$ the number. The statement "The square of a number added to 25 equals 10 times the number" translates to the equation $x^2 + 25 = 10x$. Put the equation in standard form $(ax^2 + bx + c = 0)$, and set it equal to zero: $x^2 - 10x + 25 = 0$. Factor the left side of the equation: $(x-5)(x-5) = 0$. Set each factor equal to zero and solve: $x - 5 = 0$ or $x - 5 = 0$; $x = 5$ or $x = 5$. The number is 5.

459. **b.** Let $x =$ the number. The statement "The sum of the square of a number and 12 times the number is -27" translates to the equation $x^2 + 12x = -27$. Put the equation in standard form and set it equal to zero: $x^2 + 12x + 27 = 0$. Factor the left side of the equation: $(x + 3)(x + 9) = 0$. Set each factor equal to zero and solve: $x + 3 = 0$ or $x + 9 = 0$; $x = -3$ or $x = -9$. The possible values of this number are -3 or -9, the smaller of which is -9.

7 ▶ Coordinate Geometry

COORDINATE GEOMETRY IS where algebra and geometry come together. Equations that we looked at back in Chapter 6 can be graphed on the coordinate plane. The **coordinate plane** is a set of axes, *x* and *y*, on and from which we can graph points and lines. The **x-axis** is the horizontal axis and the **y-axis** is the vertical axis. We describe where a point lies on the coordinate plane relative to these axes. Look at the following coordinate plane.

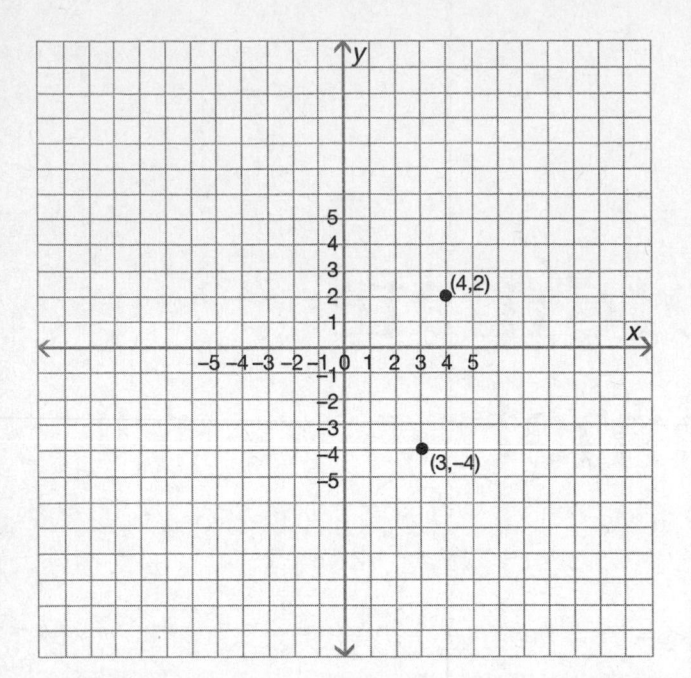

The point at which the y-axis and x-axis intersect is called the **origin**. The coordinates of this point are the ordered pair (0,0). An **ordered pair** is the x-value followed by the y-value of a point. Each value represents a distance from an axis. The x-value is the distance from the y-axis to the point, and the y-value is the distance from the x-axis to the point. The origin has the coordinates (0,0) because at that point we are at both axes, and we are therefore zero units from them.

The point (4,2) is labeled on the coordinate plane, or graph. That is because the point is found four units to the right of the y-axis and two units above the x-axis. Units to the right of the y-axis are positive, and units to the left of the y-axis are negative. Units above the x-axis are positive and units below the x-axis are negative. The point (3,–4) is found three units to the right of the y-axis and four units below the x-axis. That is why its x-value is positive and its y-value is negative.

The points (4,2) and (3,–4) are **coplanar**, because they exist on the same plane. If the points were on different planes, we would say that they are **non-**

coplanar. The two points do not have a common x-coordinate or a common y-coordinate. Two or more points are **collinear** if they could be connected to form a straight line, and two or more points are **noncollinear** if they could not be connected to form a straight line.

The coordinate plane is divided into four areas, or quadrants, by the two axes. The upper right quadrant is called quadrant I. Points in this quadrant have a positive x-value and a positive y-value. The upper left quadrant is called quadrant II. Points in this quadrant have a negative x-value and a positive y-value. Quadrant III is located in the lower left area, and quadrant IV is located in the lower right area. The following chart summarizes the values of x and y in each quadrant:

QUADRANT	X-VALUE	Y-VALUE
I	positive	positive
II	negative	positive
III	negative	negative
IV	positive	negative

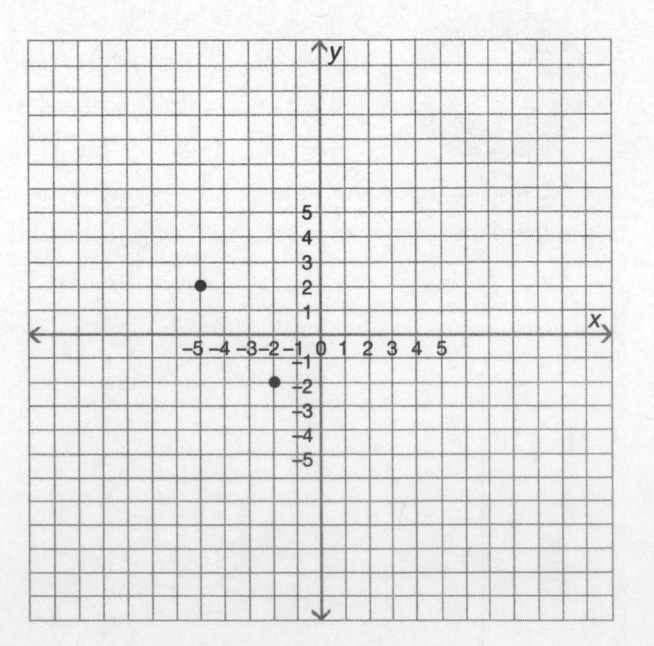

Example

What are the coordinates of each point on the preceding graph?

The point in quadrant II is located five units to the left of the y-axis, so its x-value is –5. The point is also located two units above the x-axis, so its y-value is 2. The coordinates of the point are (–5,2). The point in quadrant III is located two units to the left of the x-axis and two units below the y-axis. The coordinates of this point are (–2,–2).

▶ Distance

What if we want to find the distance between two points on a graph? We have a formula for that: $D = \sqrt{(x_2 - y_1)^2 + (y_2 - y_1)^2}$. In other words, square the difference between the x-values of the two points and add that to the square of the difference between the y-values of the two points. Then, take the square root of that sum. This sounds complicated, so let's look at an example.

Example

What is the distance between (–5,2) and (–2,–2)?

We just saw these points on the graph. First, let's use the formula, and then let's look at the graph. The second x-value, or x_2, is –2, and the first x-value, or x_1, is –5. The second y-value, or y_2, is –2, and the first y-value, or y_1, is 2. Plug these values into the formula: $D = \sqrt{(x_2 - y_1)^2 + (y_2 - y_1)^2}$
$= \sqrt{[(-2)-(-5)]^2 + [(-2)-2]^2}$
$= \sqrt{(3)^2 + (-4)^2} = \sqrt{9 + 16} = \sqrt{25} = 5$
units. The distance from (–5,2) to (–2,–2) is 5 units. Now, let's look at the graph, and

connect those points with a third point, point (–5,–2):

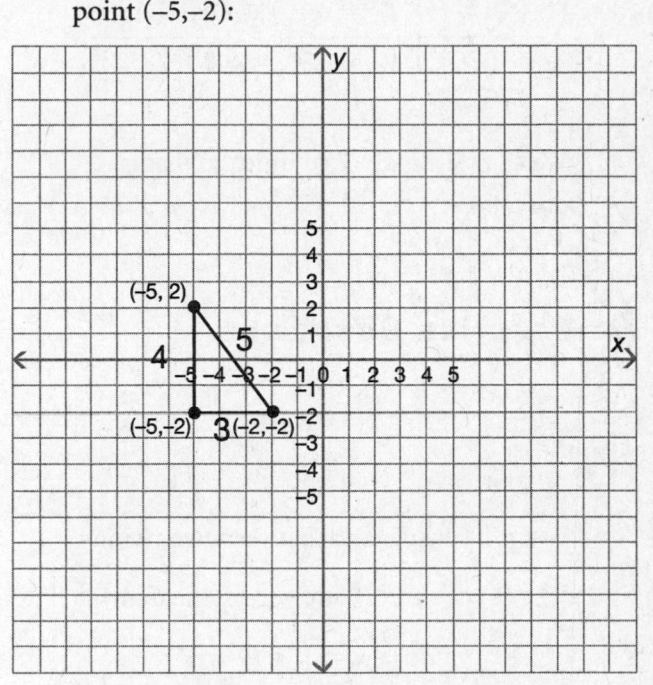

Those points form a common right triangle, a 3-4-5 right triangle. The distance formula might have felt a lot like the Pythagorean theorem, which we'll review in Chapter 12. The length of the short leg, from (–5,–2) to (–2,–2), squared, plus the length of the long leg, from (–5,2) to (–5,–2), squared, is equal to the length of the hypotenuse, from (–5,2) to (–2,–2), squared. The distance formula comes directly from the Pythagorean theorem. The change in x-values represents the length of one leg of a triangle, and the change in y-values represents the length of the other leg of the triangle.

The distance between two points doesn't always work out to a whole number.

Example

Find the distance between points (–1,5) and (2,–3).

Plug these values into the formula

$$D = \sqrt{(x_2 - x_1)^2 + (y_2 - y_1)^2}:$$

$$\sqrt{[2-(-1)]^2 + [(-3)-5]^2} = \sqrt{(3)^2 + (-8)^2}$$

$$= \sqrt{9 + 64} = \sqrt{73}$$

We can't simplify $\sqrt{73}$ units, so that is our answer.

▶ Practice Questions

$$a = x - x$$

$$b = y - y$$

$$c = d \text{ (the distance between two points)}$$

$$c^2 = a^2 + b^2 \text{ (Pythagorean theorem)}$$

$$d^2 = (x - x)^2 + (y - y)^2$$

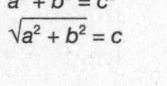

Pythagorean theorem
$$a^2 + b^2 = c^2$$
$$\sqrt{a^2 + b^2} = c$$

Distance = $\sqrt{\Delta x^2 + \Delta y^2}$
$D = \sqrt{(-2 - 4)^2 + (2 - -2)^2}$
$D = \sqrt{(-6)^2 + (4)^2}$
$D = \sqrt{36 + 16}$
$D = \sqrt{52} = 2\sqrt{13}$

Choose the best answer.

460. The origin is
  a. where the x-axis begins.
  b. where the y-axis begins.
  c. where the x-axis intersects the y-axis.
  d. not a location.

461. The point $(-3,-2)$ lies in quadrant
  a. I.
  b. II.
  c. III.
  d. IV.

462. The point $(-109,0.3)$ lies in quadrant
  a. I.
  b. II.
  c. III.
  d. IV.

463. The point $(0.01,100)$ lies in quadrant
  a. I.
  b. II.
  c. III.
  d. IV.

464. A point is three spaces right and one space above the point $(-1,-2)$. It lies in quadrant
  a. I.
  b. II.
  c. III.
  d. IV.

465. A point is 40 spaces left and 0.02 spaces above the point $(20,0.18)$. It lies in quadrant
  a. I.
  b. II.
  c. III.
  d. IV.

466. A point is 15 spaces right and 15 spaces below the point $(-15,0)$. It lies on the
  a. x-axis.
  b. y-axis.
  c. z-axis.
  d. origin.

467. On a coordinate plane, $y = 0$ is
   a. the x-axis.
   b. the y-axis.
   c. a solid line.
   d. finitely long.

468. A baseball field is divided into quadrants. The pitcher is the point of origin. The second baseman and the hitter lie on the y-axis; the first baseman and the third baseman lie on the x-axis. If the hitter bats a ball into the far left field, the ball lies in quadrant
   a. I.
   b. II.
   c. III.
   d. IV.

469. Points (12,3), (0,3), and (−12,3) are
   a. noncoplanar.
   b. collinear.
   c. noncollinear.
   d. a line.

470. Points (14,−2), (−1,15), and (3,0)
   a. determine a plane.
   b. are collinear.
   c. are noncoplanar.
   d. are a line.

471. The distance between the point (4,−5) and the point (−2,0) is
   a. $\sqrt{11}$.
   b. $\sqrt{29}$.
   c. $\sqrt{61}$.
   d. $\sqrt{22}$.

State the ordered pair for each point.

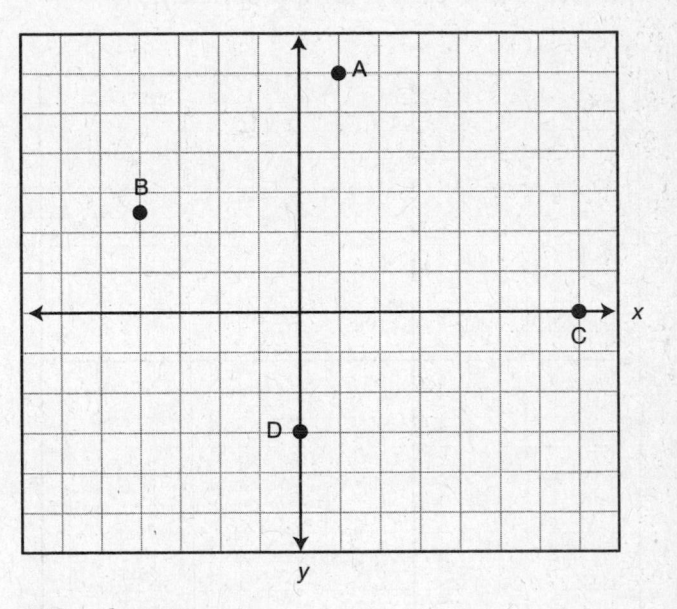

472. the point A

473. the point B

474. the point C

475. the point D

Plot each point on the same coordinate plane. Remember to label each point appropriately.

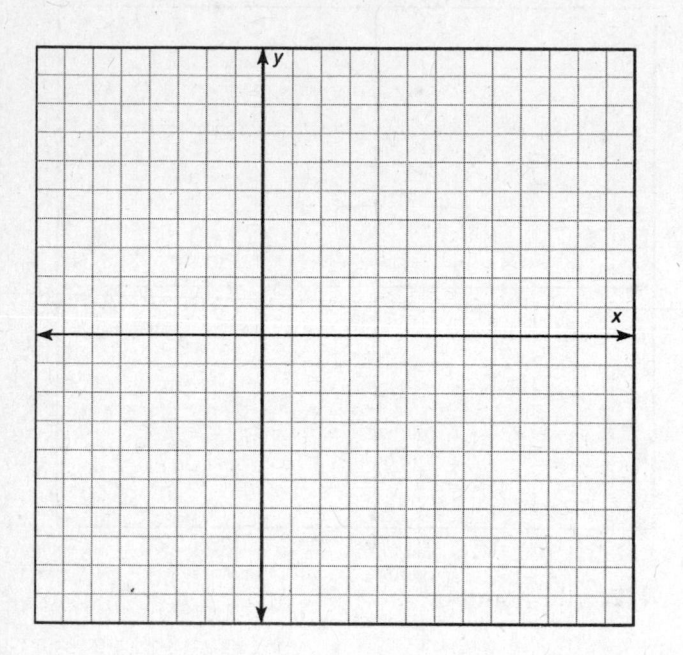

476. From the origin, plot the point M (4,5).

477. From the origin, plot the point N (12,–1).

478. From the origin, plot the point O (–3,–6).

479. From the point M, plot the point P (0,1).

480. From the point N, plot the point Q (–4,0).

481. From the point O, plot the point R (–7,–3).

Find the distance between each given pair of points.

482. points (0,4) and (0,32)

483. points (–1,–2) and (4,–1)

484. points (–3,3) and (7,3)

485. points (17,0) and (–3,0)

▶ Equation of a Line

Now that we've seen how to connect two points to form a line, we can find the equation of a line. The equation of a line has two parts: the slope and the y-intercept. The **slope** of a line is the change in its y-values divided by the change in its x-values, written formally as $\frac{y_2 - y_1}{x_2 - x_1}$ or $\frac{\Delta y}{\Delta x}$. The Δ symbol is called a **delta**, and it represents change. The **y-intercept** is the y-value of the point where the line crosses the y-axis.

When a line is in slope-intercept form, with y alone on the left side of the equation, the y-intercept is added to or subtracted from the x-value multiplied by the slope. We say that $y = mx + b$, where m is the slope of the line and b is the y-intercept of the line. The line $y = 3x + 4$ has a slope of 3 and a y-intercept of 4.

Example

What is the slope of the line $y = -4x + 6$?
What is the y-intercept?

The slope of the line is –4, and the y-intercept is 6.

PITFALL

The slope of a line is the coefficient of the x term only when the equation is in slope-intercept form. Be sure an equation is in slope-intercept form before finding the slope and y-intercept. If the equation is not in that form, convert the equation to slope-intercept form.

Example

What is the slope of the line $4y = 2x - 40$? What is the y-intercept?

This equation is not in slope-intercept form, so we must put it in slope-intercept form. To get y alone on the left side of the equation, divide both sides of the equation by 4: $y = \frac{1}{2}x - 10$. The slope of the line is $\frac{1}{2}$ and the y-intercept is -10.

Example

What is the slope of the line shown here? What is the y-intercept?

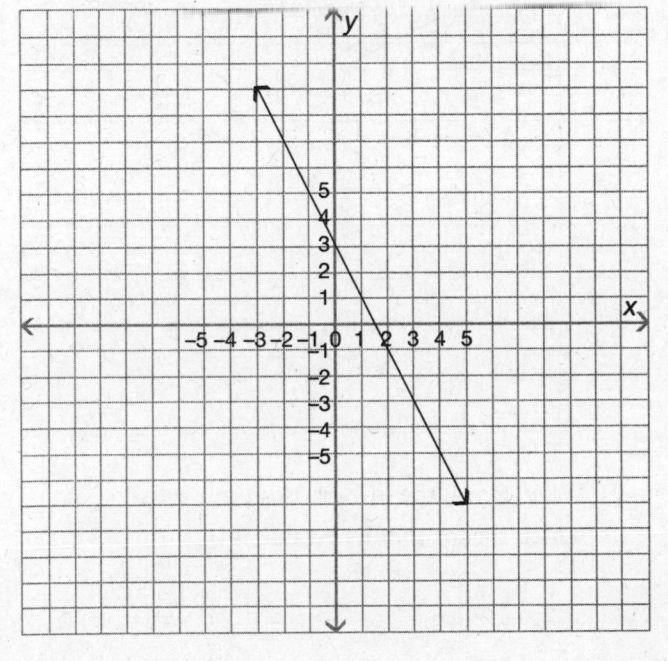

Choose any two points on the line and divide the difference between their y-values by the difference between their x-values. Let's use the points (0,3) and (1,1). Plug these values into the formula: slope $= \frac{y_2 - y_1}{x_2 - x_1} = \frac{1-3}{1-0} = -2$. The slope of this line is -2. The y-intercept is the y-value of the point where the line crosses the y-axis. The line crosses the y-axis at (0,3), which means that the y-intercept is 3. Now that we have the slope and the y-intercept of the line, we can write the equation of the line. Remember, the y-intercept is added to the product of the slope and the variable x: $y = -2x + 3$.

The slope of a line is said to be increasing when it is positive and decreasing when it is negative. The preceding graph showed a slope that was decreasing, and sure enough, the slope was negative (-2). The line $y = 2x + 3$ has a positive slope, so that slope is increasing, as shown on the following graph.

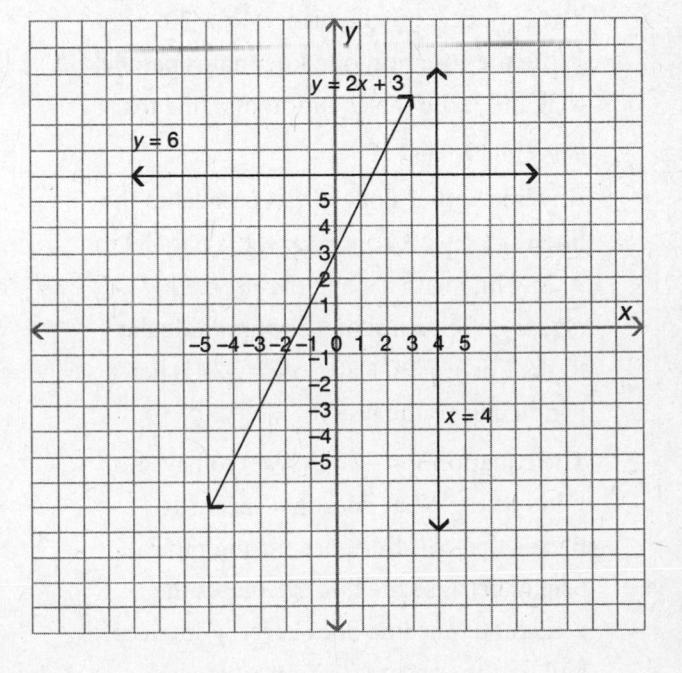

The slope of a line is zero if the line is horizontal. A line such as $y = 6$, shown here, is horizontal. The slope is zero, so you could also think of the equation as $y = 0x + 6$.

If the line is vertical, it is said to have no slope, which is not the same as zero slope. A line with zero slope has no change in y-value, so the slope, written as a fraction, is 0 over a number other than 0. A line with no slope has no change in x-value, so that slope, written has a fraction, would have a 0 in the denominator, which would make the fraction undefined. A line such as $x = 4$ has no slope.

If two lines have the same slope, then they are parallel. The lines $y = -3x + 2$ and $y = -3x - 13$ are parallel. If the slope of a line is the negative reciprocal of the slope of another line, then those lines are perpendicular. Remember, to find the reciprocal of a number, switch the numerator and the denominator of the number. The line $y = 9x + 2$ is perpendicular to the line $y = -\frac{1}{9}x + 7$, since 9 and $-\frac{1}{9}$ are negative reciprocals of each other.

Example

The equation of a line is $y = \frac{1}{2}x + 20$. Write the equation of a line that is parallel to it and write the equation of a line that is perpendicular to it.

Any line with a slope of $\frac{1}{2}$ is parallel to the line $y = \frac{1}{2}x + 20$, so $y = \frac{1}{2}x + 1$ is parallel to it. Any line with a slope that is -2, the negative reciprocal of $\frac{1}{2}$, is perpendicular to the line $y = \frac{1}{2}x + 20$, so $y = -2x$ is perpendicular to the line $y = \frac{1}{2}x + 20$.

The equation $y = -2x$ appears to have no y-intercept. What does that mean? It means that this line goes through the origin, because the line intercepts the y-axis at 0. The line does have a y-intercept, and that intercept is 0. You could also think of the equation as $y = -2x + 0$.

We can use the equation of a line to find the x-value of a point given the y-value, or we can find the y-value of a point given the x-value.

Example

What is the y-coordinate of a point that has an x-value of 3 on the line $y = 4x - 8$?

Plug the x-value into the equation and solve for y: $y = 4(3) - 8 = 12 - 8 = 4$.

We can even use graphs to help us solve equations. Given the x-value of a point on a line, we can use that point to locate the y-value on the graph.

Example

The following graph shows the equation $y = -\frac{1}{2}x + 2$. Use the graph to find the value of y when $x = 4$.

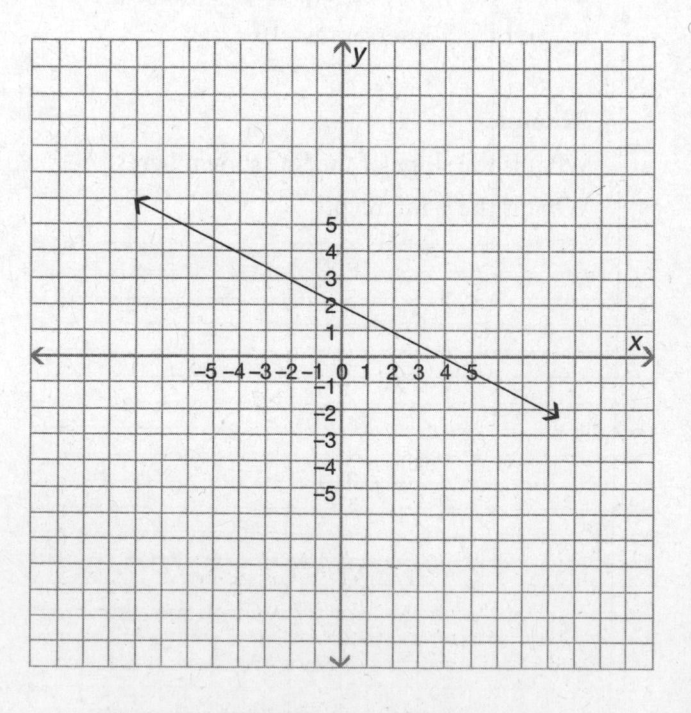

When $x = 4$, the line crosses the x-axis, which means that its y value is 0.

▶ Practice Questions

Choose the best answer.

486. Pam and Sam are climbing different hills with the same incline. If each hill were graphed, they would have the same
   a. equation.
   b. slope.
   c. length.
   d. coordinates.

487. In American homes, a standard stair rises 7″ for every 9″. The slope of a standard staircase is
   a. $\frac{7}{9}$.
   b. $\frac{2}{7}$.
   c. $\frac{16}{9}$.
   d. $\frac{9}{7}$.

488. Which equation is a line perpendicular to $y = -\frac{1}{2}x + 4$?
   a. $y = \frac{1}{2}x + 4$
   b. $y = 2x + 8$
   c. $y = -?x + 8$
   d. $y = \frac{1}{2}x + 8$

489. Bethany's ramp to her office lobby rises 3 feet for every 36 feet. The incline is
   a. $\frac{36 \text{ feet}}{1 \text{ foot}}$.
   b. $\frac{12 \text{ feet}}{1 \text{ foot}}$.
   c. $\frac{1 \text{ foot}}{12 \text{ feet}}$.
   d. $\frac{36 \text{ feet}}{3 \text{ feet}}$.

490. Which equation is a line parallel to $y = -\frac{14}{15}x + 7$?
   a. $y = \frac{14}{15}x + 12$
   b. $y = \frac{15}{14}x + 7$
   c. $y = -\frac{14}{15}x + 12$
   d. $y = \frac{15}{14}x + 12$

491. The y-axis has
   a. zero slope.
   b. undefined slope.
   c. positive slope.
   d. negative slope.

State the slope for each of the following diagrams.

492.

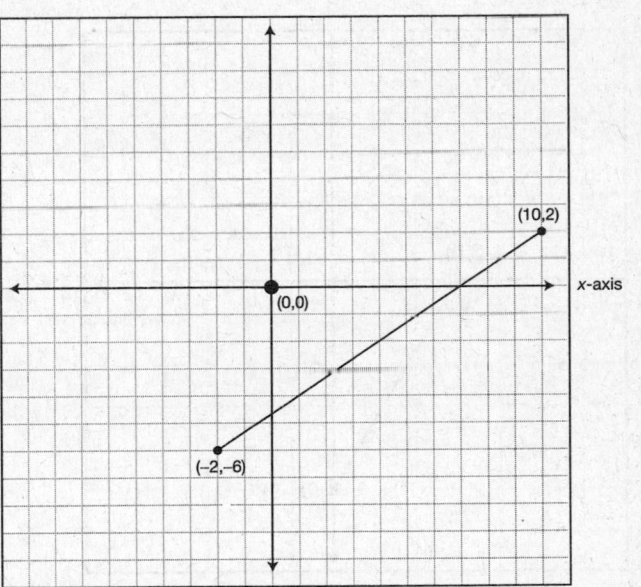

493.

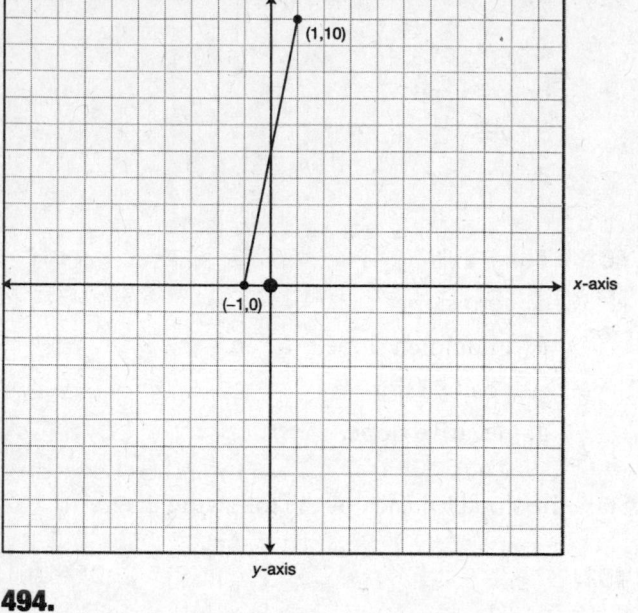

494.

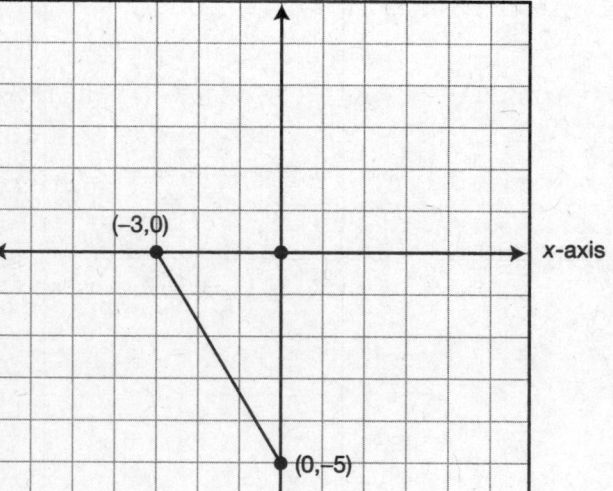

▶ The Slope-Intercept Equation

A special arrangement of the linear equation looks like $y = mx + b$. m represents the line's slope. b represents the y coordinate where the line crosses the y-axis.

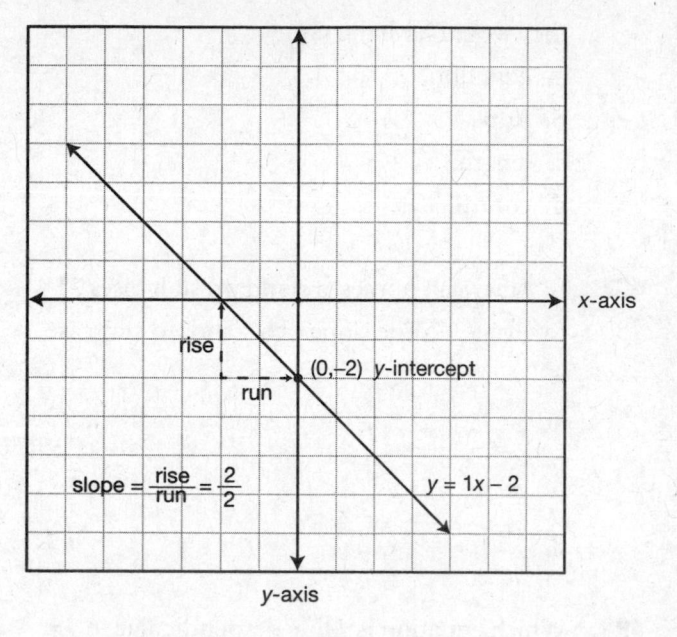

▶ Practice Questions

Choose the best answer.

495. In the linear equation $y = -4x + 5$, what is the y-intercept?
　　a. (5,0)
　　b. (−4,0)
　　c. (0,−4)
　　d. (0,5)

496. What is the slope of the linear equation $y = \frac{2}{3}x - 1$?
　　a. 2
　　b. $\frac{2}{3}$
　　c. $\frac{3}{2}$
　　d. 3

497. What is the value of b if $(-2,3)$ satisfies the equation $y = \frac{1}{2}x + b$?

a. -2

b. -1

c. 3

d. 4

498. What is the value of y if $(1,y)$ satisfies the equation $y = -\frac{12}{5}x + \frac{2}{5}$?

a. 1

b. -2

c. -3

d. -1

499. Convert the linear equation $4x - 2y = 4$ into a slope-intercept equation.

a. $y = 2x - 2$

b. $y = -2x + 2$

c. $x = \frac{1}{2}y - 2$

d. $x = -\frac{1}{2}y + 2$

500. The points $(-4,0)$, $(0,3)$, and $(8,9)$ satisfy which equation?

a. $y = \frac{4}{3}x + 3$

b. $y = \frac{3}{4}x + 0$

c. $y = \frac{3}{4}x + 3$

d. $y = \frac{6}{8}x + 9$

501. Find the missing y value if the points $(-3,-1)$, $(0,y)$, and $(3,-9)$ are collinear.

a. 1

b. -1

c. -3

d. -5

502. Which line perpendicularly meets line $1x + 2y = 4$ on the y-axis?

a. $y = -\frac{1}{2}x + 2$

b. $y = 2x + 2$

c. $y = -2x - 2$

d. $y = \frac{1}{2}x - 2$

503. A $(0,-2)$ satisfies which equation that parallels $\frac{1}{2}x + \frac{1}{4}y = \frac{1}{8}$?

a. $y = 2x + \frac{1}{2}$

b. $y = \frac{1}{2}x + \frac{1}{2}$

c. $y = -2x - 2$

d. $y = -2x + \frac{1}{2}$

▶ Graphing a Line

To graph a line, begin by putting its equation into slope-intercept form. Then, plot the y-intercept. This is the easiest point to plot, because you can simply place a point along the y-axis according to the value of the y-intercept. Next, use the slope to find a few more points on the graph. Finally, connect your points and label your graph.

Example

Graph the line $y = 3x - 3$.

Begin with the y-intercept. Place the point $(0,-3)$. Next, count three units up and one unit to the right to place the next point, since the slope of the line is 3. Remember, the slope is the change in y over the change in x, so in this example, the y-value must increase by three units for every unit by which the x-value increases. In the same way, plot a point that is three units down and one unit to the left, since $\frac{-3}{-1}$ is also equal to 3.

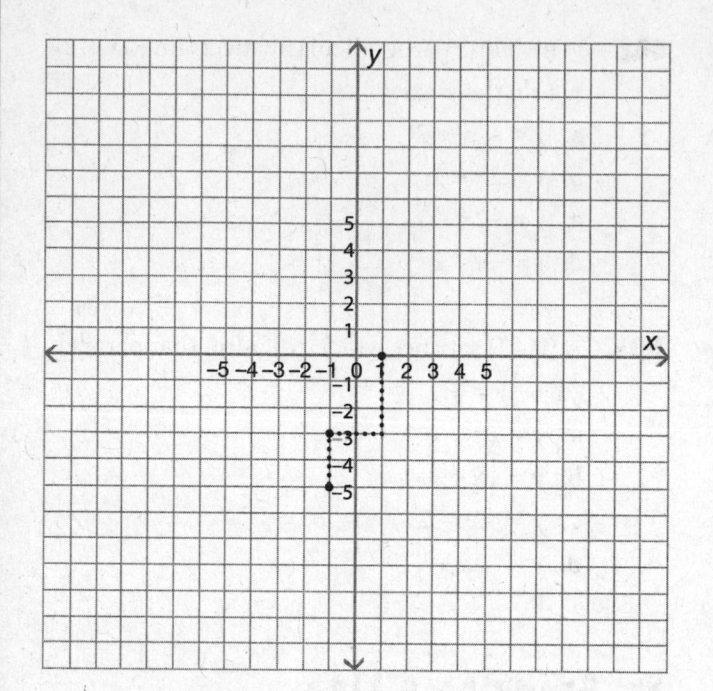

Plot a few more points and connect them.

Label your line $y = 3x - 3$.

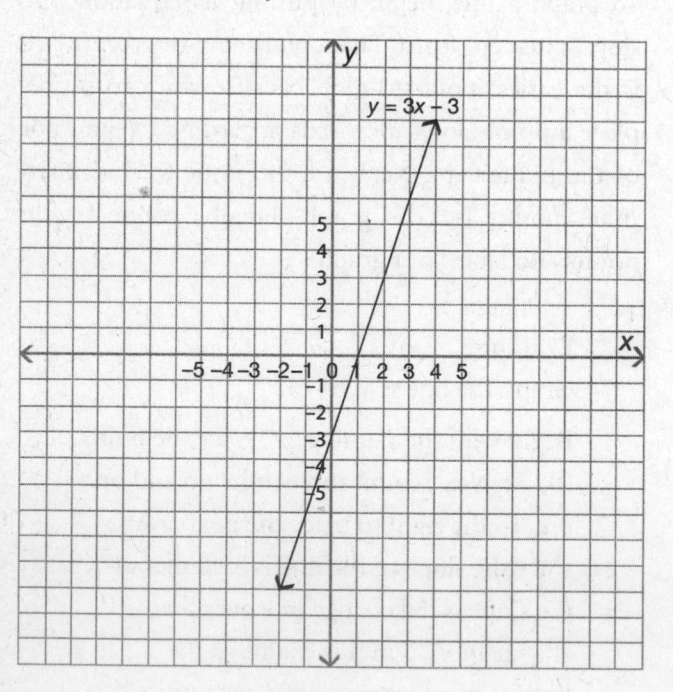

▶ **Practice Questions**

Graph the following equations using the slope and y-intercept method.

504. $y = 2x + 3$

505. $y = 5x - 2$

506. $y = -2x + 9$

507. $y = \frac{3}{4}x - 1$

508. $y = \frac{5}{2}x - 3$

509. $y - 2x = 4$

510. $y + 3x = -2$

511. $y - \frac{1}{2}x = 3\frac{1}{2}$

512. $2x + 5y = 30$

513. $2y + 4x = 10$

514. $x - 3y = 12$

515. $3x + 9y = -27$

516. $-5x - y = -\frac{7}{2}$

517. $x = 7y - 14$

518. $0 = 3x + 2y$

519. $3x + 12y = -18$

520. $y - 0.6x = -2$

521. $\frac{2}{3}y - \frac{1}{2}x = 0$

522. $\frac{5}{6}x - \frac{1}{3}y = 2$

523. $7x = 4y + 8$

524. $20x - 15 = 5y$

525. $6y + 13x = 12$

526. $0.1x = 0.7y + 1.4$

527. $-34x + 85 = 17y$

528. $6y + 27x = -42$

For the following problems, use the slope/y-intercept method to write an equation that would enable you to draw a graphic solution for each problem.

529. A glider has a 25:1 descent ratio when there are no updrafts to raise its altitude. That is, for every 25 feet it moves parallel to the ground, it will lose 1 foot of altitude. Write an equation to represent the glider's descent from an altitude of 250 feet.

530. An Internet service provider charges $15 plus $0.25 per hour of usage per month. Write an equation that would represent the monthly bill of a user.

531. A scooter rental agency charges $20 per day plus $0.05 per mile for the rental of a motor scooter. Write an equation to represent the cost of one day's rental.

532. A dive resort rents scuba equipment at a weekly rate of $150 per week and charges $8 per tank of compressed air used during the week of diving. Write an equation to represent a diver's cost for one week of diving at the resort.

533. A recent backyard bird count showed that one out of every seven birds that visited backyard feeders was a chickadee. Write an equation to represent this ratio.

TERMS TO REVIEW

coordinate plane

x-axis

y-axis

origin

ordered pair

coplanar

noncoplanar

collinear

noncollinear

slope

delta

y-intercept

FORMULAS TO REVIEW

distance $= \sqrt{(x_2 - x_1)^2 + (y_2 - y_1)^2}$

slope $= \frac{y_2 - y_1}{x_2 - x_1}$ or $\frac{\Delta y}{\Delta x}$

▶ Answers

460. c. The origin, whose coordinate pair is (0,0), is in fact a location. It is where the *x*-axis meets the *y*-axis. It is not the beginning of either axis because both axes extend infinitely in opposite directions, which means they have no beginning and no end.

461. c. Both coordinates are negative: count three spaces left of the origin; then count two spaces down from the *x*-axis. The point (−3,−2) is in quadrant III.

462. b. You do not need to actually count 109 spaces left of the origin to know that the point (−109,0.3) lies left of the *y*-axis. Nor do you need to count three-tenths of a space to know that it lies above the *x*-axis. Points left of the *y*-axis and above the *x*-axis are in quadrant II.

463. a. Again, you do not need to count one-hundredth of a space right of the origin or a hundred spaces up from the *x*-axis to find in which quadrant the point (0.01,100) lies. You only need to know that the point is right of the *y*-axis and above the *x*-axis. Points right of the *y*-axis and above the *x*-axis lie in quadrant I.

464. d. To find a new coordinate pair, add like coordinates: $3 + (−1) = 2$, and $1 + (−2) = −1$. This new coordinate pair is (2,−1), which lies in quadrant IV.

465. b. To find a new coordinate pair, add like coordinates: $(−40) + 20 = −20$, and $0.02 + 0.18 = 0.20$. This new coordinate pair is (−20,0.20), which lies in quadrant II.

466. b. To find a new coordinate pair, add like coordinates: $15 + (−15) = 0$, and $(−15) + 0 = −15$. This new coordinate pair is (0,−15); any point whose *x*-coordinate is zero lies on the *y*-axis.

467. a. The *y*-coordinate of every point on the *x*-axis is zero.

468. b. Draw a baseball field—its exact shape is irrelevant; only the alignment of the players matters. They form the axes of the coordinate plane. The ball passes the pitcher and veers left of the second baseman; it is in the second quadrant.

469. b. The three points are collinear; they could be connected to make a horizontal line, but they are not a line. Choice **a** is incorrect because all points on a coordinate plane are coplanar.

470. a. Three noncollinear points determine a plane. Choices **b** and **d** are incorrect because the points do not lie on a common line, nor can they be connected to form a straight line. Caution: Do not assume points are noncollinear because they do not share a common *x*- or *y*-coordinate. To be certain, plot the points on a coordinate plane and try to connect them with one straight line.

471. c. First, find the difference between like coordinates: $x − x$ and $y − y$: $4 − (−2) = 6$, and $−5 − 0 = −5$. Square both differences: $6^2 = 36$, and $(−5)^2 = 25$. Remember a negative number multiplied by a negative number is a positive number. Add the squared differences together, and take the square root of their sum: $36 + 25 = 61$, and $d = \sqrt{61}$. If you chose choice **a**, then your mistake began after you squared −5; the square of a negative number is positive. If you chose choice **b**, then your mistake began when subtracting the *x*-coordinates; two negatives make a positive. If you chose **d**, then you didn't square your differences; you doubled your differences.

472. To locate the point A from the origin, count one space right of the origin and six spaces up. The coordinate pair is (1,6).

473. To locate the point B from the origin, count four spaces left of the origin and two and a half spaces up. The coordinate pair is (−4, 2.5).

474. To locate the point C from the origin, count seven spaces right of the origin and no spaces up or down. This point lies on the *x*-axis. The coordinate pair is (7,0).

475. To locate the point D from the origin, count no spaces left or right, but count three spaces down from the origin. This point lies on the *y*-axis, and *x* equals zero. The coordinate pair is (0,−3).

For answers to questions 476–481, see the graph shown here.

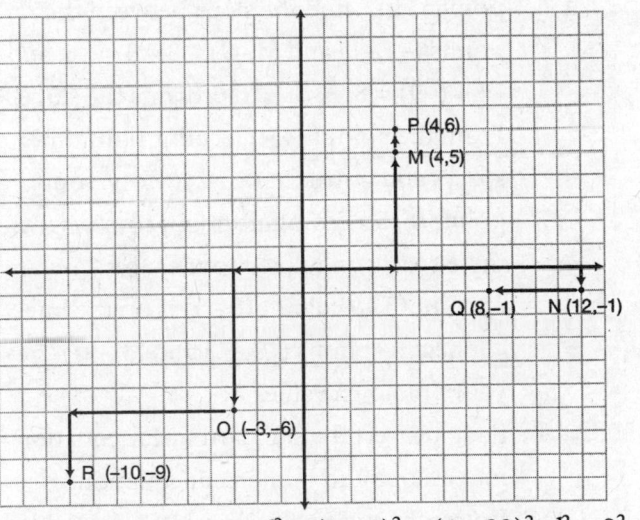

482. Distance = 28. $d^2 = (0 − 0)^2 + (4 − 32)^2$; $d^2 = 0^2 + (−28)^2$; $d^2 = 784$; $d = 28$. Because these two points form a vertical line, you could just count the number of spaces along the line's length to find the distance between the points.

483. Distance = $\sqrt{26}$. $d^2 = (−1 − 4)^2 + [−2 − (−1)]^2$; $d^2 = (−5)^2 + (−1)^2$; $d^2 = 25 + 1$; $d = \sqrt{26}$.

484. Distance = 10. $d^2 = (−3 − 7)^2 + (3 − 3)^2$; $d^2 = (−10)^2 + 0^2$; $d^2 = 100$; $d = 10$. Again, because these two points form a horizontal line, you could just count the number of spaces along the line's length to find the distance between the points.

485. Distance = 20. $d^2 = [17 − (−3)]^2 + (0 − 0)^2$; $d^2 = (20)^2 + 0^2$; $d^2 = 400$; $d = 20$. Because these two points also form a horizontal line, you could just count the spaces along the line's length to find the distance between the points.

486. **b.** If two lines have the same incline, they rise the same amount over the same distance; the relationship of rise over distance is slope.

487. **a.** If every step rises 7″ for every 9″, then the relationship of rise over distance is $\frac{7}{9}$.

488. **b.** In the slope-intercept formula, the constant preceding the variable *x* is the line's slope. Because perpendicular lines have slopes that are negative reciprocals, a line perpendicular to $y = −\frac{1}{2}x + 4$ must have a $\frac{2}{1}$ slope.

489. **c.** If the ramp rises 3 feet for every 36 feet, then the relationship of rise over distance is $\frac{3 \text{ feet}}{36 \text{ feet}}$. The simplified ratio is $\frac{1 \text{ foot}}{12 \text{ feet}}$.

490. **c.** Parallel lines have the same rise-over-distance ratio, or slope. That means in slope-intercept equations, the constant before the *x*-variable will be the same. In this case, $−\frac{14}{15}$ must precede *x* in both equations. Choices **b** and **d** are perpendicular line equations because their slopes are negative reciprocals of the given slope. Choice **a** is an entirely different line.

491. **b.** The y-axis is a vertical line; its slope is $\frac{1}{0}$ or undefined (sometimes referred to as "no slope"). The x-axis is an example of a horizontal line; horizontal lines have zero slope. Positive slopes are nonvertical lines that rise from left to right; negative slopes are nonvertical lines that descend from left to right.

492. Subtract like coordinates: $-2 - 10 = -12$, and $-6 - 2 = -8$. Place the vertical change in distance over the horizontal change in distance: $\frac{-8}{-12}$. Then reduce the top and bottom of the fraction by -4. The final slope is $\frac{2}{3}$.

493. Subtract like coordinates: $-1 - 1 = -2$, and $0 - 10 = -10$. Place the vertical change in distance over the horizontal change in distance: $\frac{-10}{-2}$. Then reduce the top and bottom of the fraction by -2. The final slope is 5.

494. Subtract like coordinates: $-3 - 0 = -3$, and $0 - (-5) = 5$. Place the vertical change in distance over the horizontal change in distance: $\frac{5}{-3}$. The slope is $-\frac{5}{3}$.

495. **d.** When a line intercepts the y-axis, its x-value is always zero. Immediately, choices **a** and **b** are eliminated. In the slope and y-intercept equation, the number without a variable beside it is the y-value of the y-intercept coordinate pair. Choice **c** is eliminated because -4 is actually the line's slope value.

496. **b.** In the slope and y-intercept equation, the number preceding the x variable is the line's slope. In this case that number is the entire fraction $\frac{2}{3}$.

497. **d.** Plug the values of x and y into the equation and solve: $3 = \frac{1}{2}(-2) + b$; $3 = (-1) + b$; $4 = b$.

498. **b.** Plug the value of x into the equation and solve: $y = -\frac{12}{5}(1) + \frac{2}{5}$; $y = -\frac{12}{5} + \frac{2}{5}$; $y = -\frac{10}{5}$; $y = -2$.

499. **a.** To convert a standard linear equation into a slope-intercept equation, single out the y variable. Subtract $4x$ from both sides: $-2y = -4x + 4$. Divide both sides by -2: $y = 2x - 2$. Choices **c** and **d** are incorrect because they single out the x variable. Choice **b** is incorrect because after both sides of the equation are divided by -2, the signs were not reversed on the right-hand side.

500. **c.** Find the slope between any two of the given points: $\frac{(0-3)}{(-4-0)} = \frac{-3}{-4}$, or $\frac{3}{4}$. The point $(0,3)$ is the y-intercept. Plug the slope and y-value of the point $(0,3)$ into the formula $y = mx + b$: $y = \frac{3}{4}x + 3$.

501. **d.** The unknown y value is also the intercept value of a line that connects all three points. First, find the slope between the points $(-3,-1)$ and $(3,-9)$: $-3 - 3 = -6$, and $-1 - (-9) = 8$; $\frac{8}{-6}$ or $\frac{-4}{3}$ represents the slope. From the point $(-3,-1)$, count right three spaces and down four spaces. You are at point $(0,-5)$. From this point, count right three spaces and down four spaces. You are at point $(3,-9)$. Point $(0,-5)$ is on the line connecting points $(-3,-1)$ and $(3,-9)$; -5 is your unknown value.

502. **b.** First, convert the standard linear equation into a slope and y-intercept equation. Isolate the y variable: $2y = -1x + 4$. Divide both sides by 2: $y = -\frac{1}{2}x + 2$. A line that perpendicularly intercepts this line on the y-axis has a negative reciprocal slope but has the same y intercept value: $y = 2x + 2$.

503. **c.** First, convert the standard linear equation into a slope-intercept equation. Isolate the y variable: $\frac{1}{4}y = -\frac{1}{2}x + \frac{1}{8}$. Multiply both sides by 4: $y = -2x + \frac{1}{2}$. A parallel line will have the same slope as the given equation; however, the y intercept will be different: $y = -2x - 2$.

504. The equation is in the proper slope and y-intercept form: $m = 2 = \frac{2}{1} = \frac{\text{change in } y}{\text{change in } x}$. $b = 3$. The y-intercept is at the point $(0,3)$. A change in y of 2 and in x of 1 gives the point $(0 + 1, 3 + 2)$ or $(1,5)$.

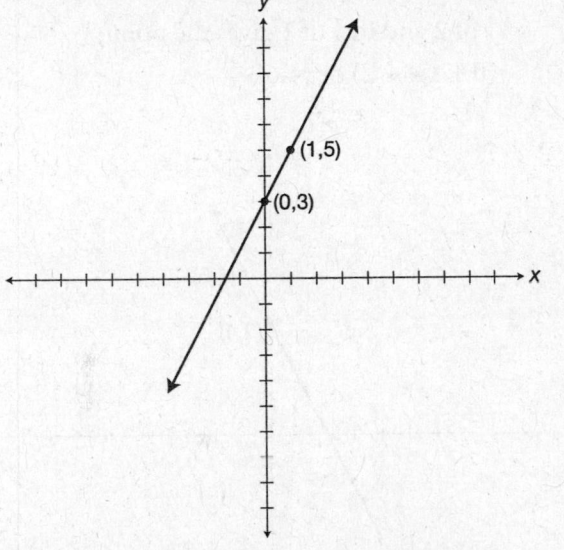

505. The equation is in the proper slope/ y-intercept form: $m = 5 = \frac{5}{1} = \frac{\text{change in } y}{\text{change in } x}$. $b = -2$. The y-intercept is at the point $(0,-2)$. A change in y of 5 and in x of 1 gives the point $(0 + 1, -2 + 5)$ or $(1,3)$.

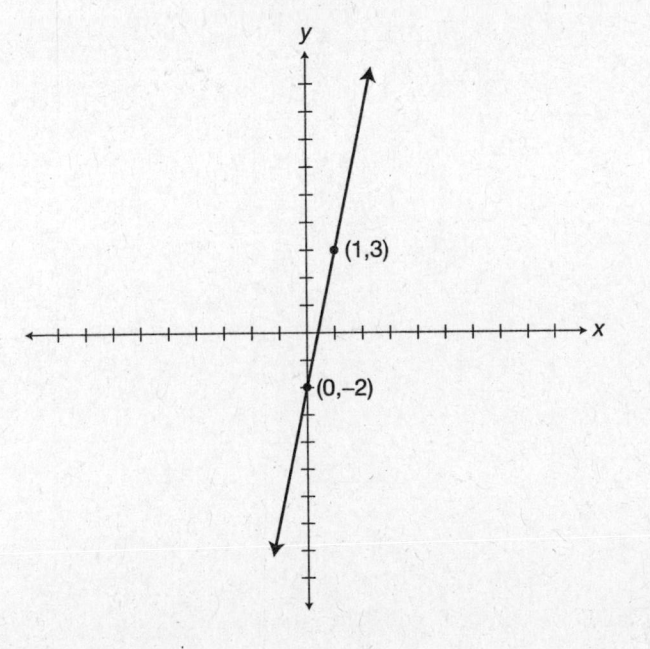

506. The equation is in the proper slope/ y-intercept form: $m = -2 = -\frac{2}{1} = \frac{\text{change in } y}{\text{change in } x}$. $b = 9$. The y-intercept is at the point $(0,9)$. A change in y of -2 and in x of 1 gives the point $(0 + 1, 9 - 2)$ or $(1,7)$.

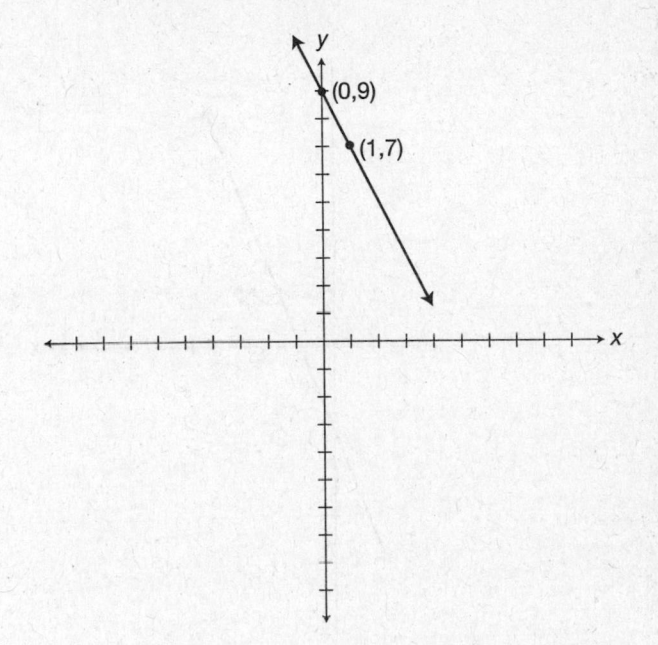

507. The equation is in the proper slope/ y-intercept form: $m = \frac{3}{4} = \frac{\text{change in } y}{\text{change in } x}$. $b = -1$. The y-intercept is at the point $(0,-1)$. A change in y of 3 and in x of 4 gives the point $(0 + 4, -1 + 3)$ or $(4,2)$.

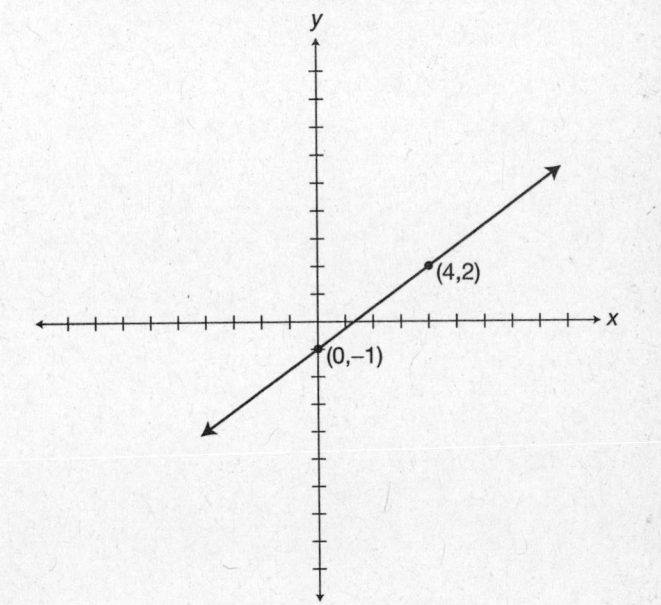

508. The equation is in the proper slope/ y-intercept form: $m = \frac{5}{2} = \frac{\text{change in } y}{\text{change in } x}$. $b = -3$. The y-intercept is at the point $(0,-3)$.

A change in y of 5 and in x of 2 gives the point $(0+2,-3+5)$ or $(2,2)$.

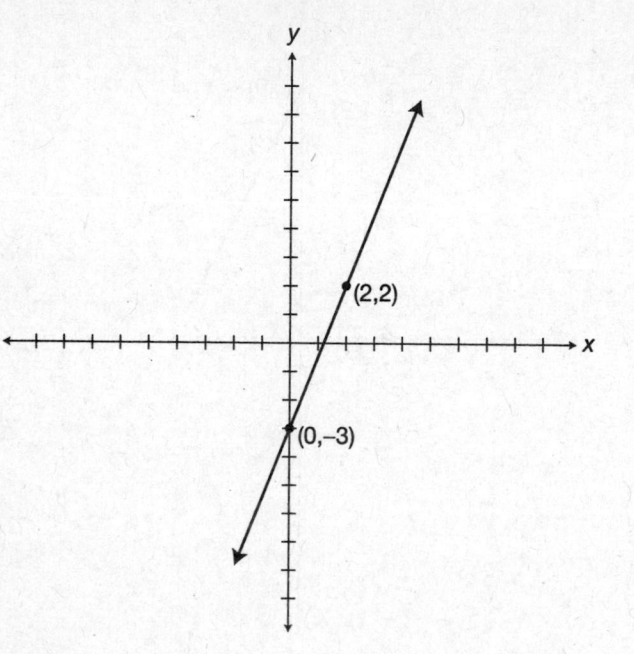

509. Put the equation in the proper form. Add $2x$ to both sides of the equation: $y + 2x - 2x = 2x + 4$. Simplify the equation: $y = 2x + 4$. The equation is in the proper slope/y-intercept form: $m = 2 = \frac{2}{1} = \frac{\text{change in } y}{\text{change in } x}$. $b = 4$. The y-intercept is at the point $(0,4)$. A change in y of 2 and in x of 1 gives the point $(0+1,4+2)$ or $(1,6)$.

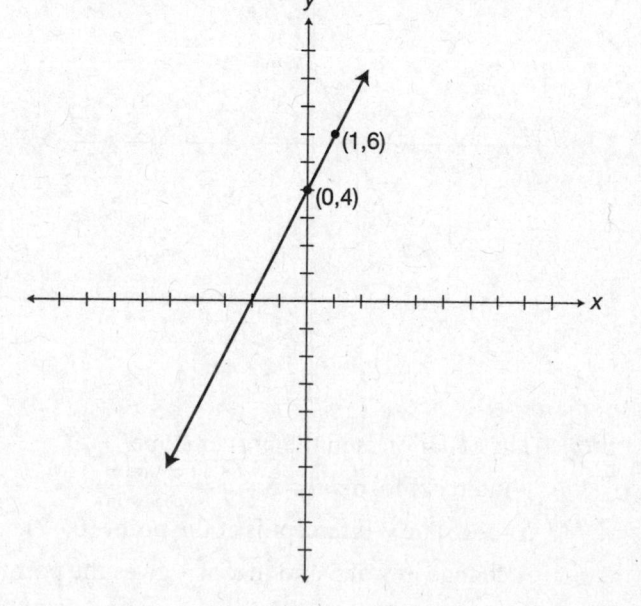

510. Put the equation in the proper form. Subtract $3x$ from both sides of the equation: $y + 3x - 3x = -3x - 2$. Simplify the equation: $y = -3x - 2$. The equation is in the proper slope/ y-intercept form: $m = -3 = -\frac{3}{1} = \frac{\text{change in } y}{\text{change in } x}$. $b = -2$. The y-intercept is at the point $(0, -2)$. A change in y of -3 and in x of 1 gives the point $(0 + 1, -2 - 3)$ or $(1, -5)$.

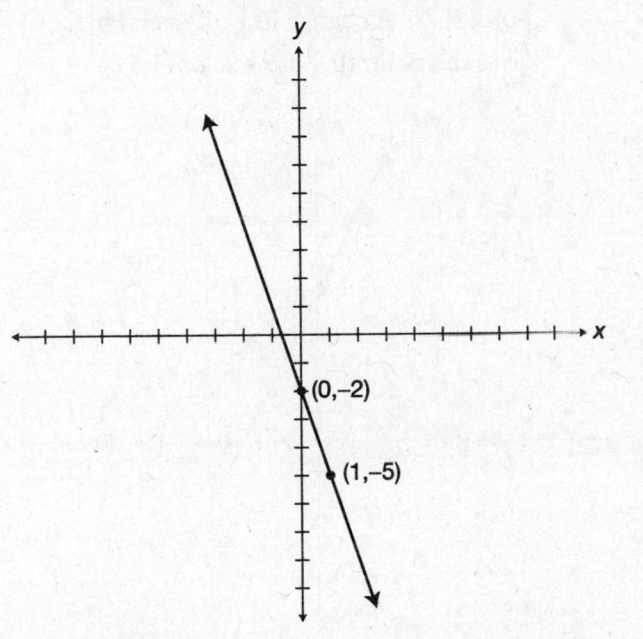

511. Put the equation in the proper form. Add $\frac{1}{2}x$ to both sides of the equation: $y + \frac{1}{2}x - \frac{1}{2}x = \frac{1}{2}x + 3\frac{1}{2}$. Simplify the equation: $y = \frac{1}{2}x + 3\frac{1}{2}$. The equation is in the proper slope/ y-intercept form: $m = \frac{1}{2} = \frac{\text{change in } y}{\text{change in } x}$. $b = 3\frac{1}{2}$. The y-intercept is at the point $(0, 3\frac{1}{2})$. A change in y of 1 and in x of 2 gives the point $(0 + 2, 3\frac{1}{2} + 1)$ or $(2, 4\frac{1}{2})$.

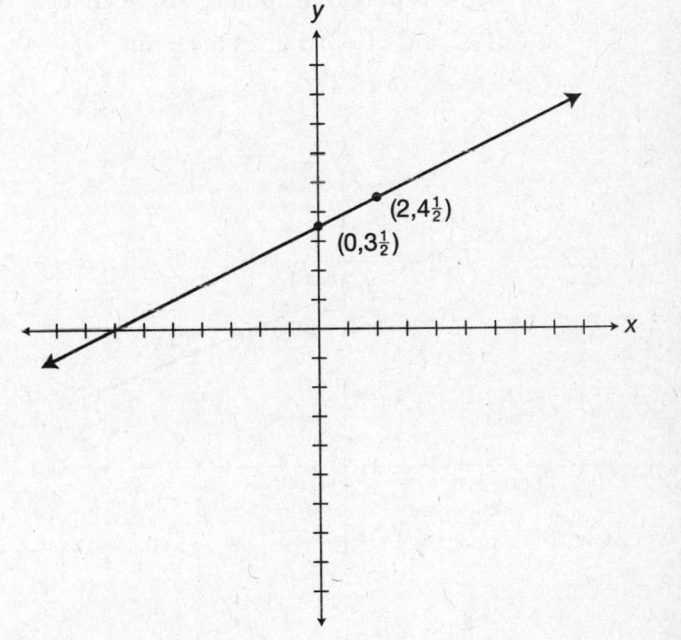

512. Put the equation in the proper form. Subtract $2x$ from both sides of the equation: $2x - 2x + 5y = -2x + 30$. Simplify the equation: $5y = -2x + 30$. Divide both sides of the equation by 5: $\frac{5y}{5} = \frac{(-2x + 30)}{5}$. Simplify the equation: $y = \frac{-2x}{5} + \frac{30}{5}$; $y = \frac{-2}{5}x + 6$. The equation is in the proper slope/ y-intercept form: $m = \frac{-2}{5} = \frac{\text{change in } y}{\text{change in } x}$. $b = 6$. The y-intercept is at the point $(0,6)$. A change in y of -2 and in x of 5 gives the point $(0 + 5, 6 - 2)$ or $(5,4)$.

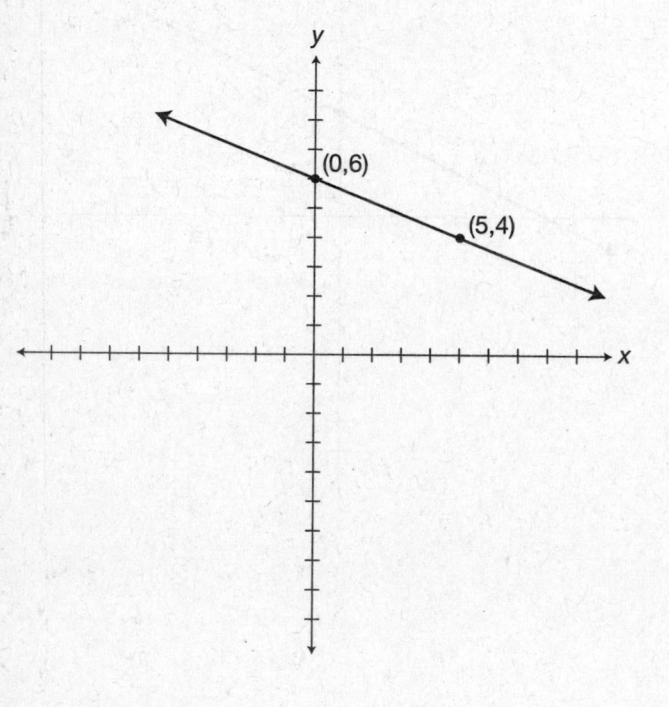

513. Put the equation in the proper form. Subtract $4x$ from both sides of the equation: $2y + 4x - 4x = -4x + 10$. Simplify the equation: $2y = -4x + 10$. Divide both sides of the equation by 2: $\frac{2y}{2} = \frac{(-4x + 10)}{2}$. Simplify the equation: $y = \frac{-4x}{2} + \frac{10}{2}$; $y = -2x + 5$. The equation is in the proper slope/y-intercept form: $m = -2 = -\frac{2}{1} = \frac{\text{change in } y}{\text{change in } x}$. $b = 5$. The y-intercept is at the point $(0,5)$. A change in y of -2 and in x of 1 gives the point $(0 + 1, 5 - 2)$ or $(1,3)$.

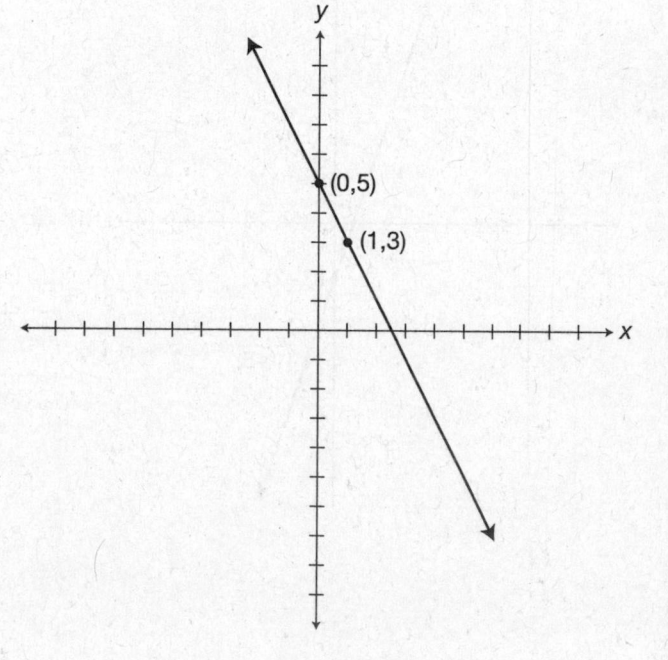

514. Put the equation in the proper form. Add $3y$ to both sides of the equation: $x - 3y + 3y = 12 + 3y$. Simplify the equation: $x = 12 + 3y$. Subtract 12 from both sides of the equation: $x - 12 = 12 - 12 + 3y$. Simplify the equation: $x - 12 = 3y$. Divide both sides of the equation by 3: $\frac{(x-12)}{3} = y$. Simplify the equation: $\frac{x}{3} - \frac{12}{3} = y$; $\frac{x}{3} - 4 = y$; $\frac{x}{3} = \frac{(1)(x)}{(3)(1)} = \frac{1}{3} \times \frac{x}{1} = \frac{1}{3}x$; $\frac{1}{3}x - 4 = y$. The equation is equivalent to the proper form: $y = \frac{1}{3}x - 4$. The equation is in the proper slope/y-intercept form: $m = \frac{1}{3} = \frac{\text{change in } y}{\text{change in } x}$. $b = -4$. The y-intercept is at the point $(0,-4)$. A change in y of 1 and in x of 3 gives the point $(0 + 3,-4 + 1)$ or $(3,-3)$.

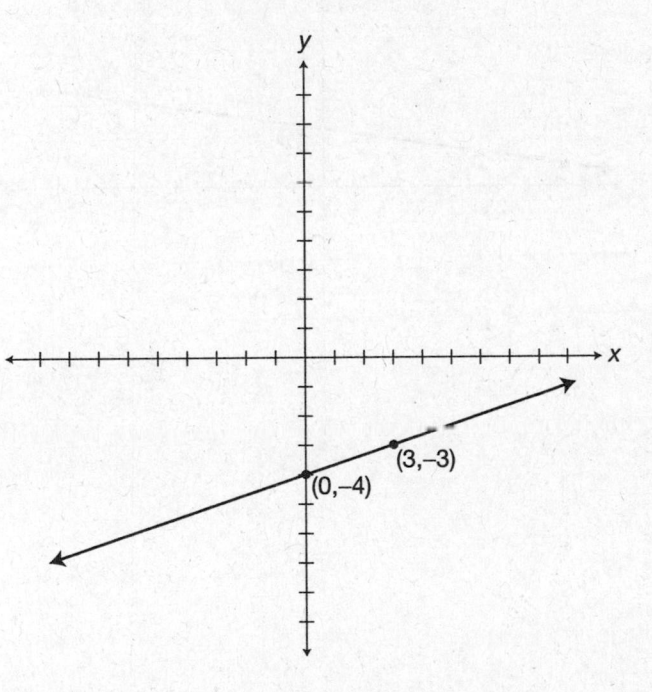

515. Put the equation in the proper form. Divide both sides of the equation by 3: $\frac{(3x + 9y)}{3} = \frac{-27}{3}$. Simplify the equation: $\frac{3x}{3} + \frac{9y}{3} = -9$; $x + 3y = -9$. Subtract x from both sides of the equation: $x - x + 3y = -x - 9$. Simplify the equation: $3y = -x - 9$. Divide both sides of the equation by 3: $\frac{3y}{3} = \frac{(-x-9)}{3}$. Simplify the equation: $y = \frac{-x}{3} - \frac{9}{3}$; $y = \frac{1}{3}x - 3$. The equation is in the proper slope/y-intercept form: $m = -\frac{1}{3} = \frac{\text{change in } y}{\text{change in } x}$. $b = -3$. The y-intercept is at the point $(0,-3)$. A change in y of -1 and in x of 3 gives the point $(0 + 3,-3 - 1)$ or $(3,-4)$.

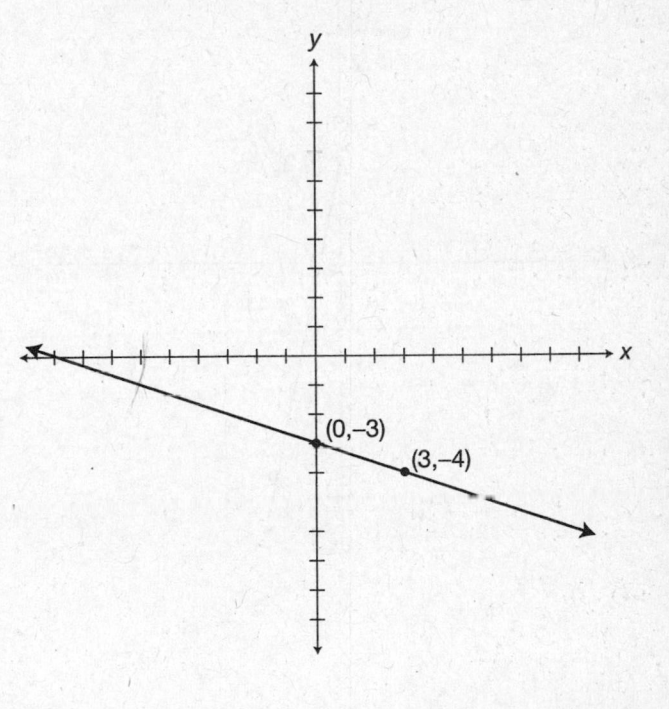

516. Put the equation in the proper form. Add $5x$ to both sides of the equation: $5x - 5x - y = 5x - \frac{7}{2}$. Simplify the equation: $-y = 5x - \frac{7}{2}$. Multiply both sides of the equation by -1: $-1(-y) = -1(5x - \frac{7}{2})$. Simplify the equation: $y = -5x + \frac{7}{2}$. The equation is in the proper slope/y-intercept form: $m = -5 = -\frac{5}{1} = \frac{\text{change in } y}{\text{change in } x}$. $b = \frac{7}{2} = 3\frac{1}{2}$. The y-intercept is at the point $(0, 3\frac{1}{2})$. A change in y of -5 and in x of 1 gives the point $(0 + 1, 3\frac{1}{2} - 5)$ or $(1, -1\frac{1}{2})$.

517. Put the equation in the proper form. Add 14 to both sides of the equation: $x + 14 = 7y + 14 - 14$. Simplify the equation: $x + 14 = 7y$. Divide both sides of the equation by 7: $\frac{(x + 14)}{7} = \frac{7y}{7}$. Simplify the equation: $\frac{x}{7} + 2 = y$; $\frac{1}{7}x + 2 = y$; $y = \frac{1}{7}x + 2$. The equation is in the proper slope/y-intercept form: $m = \frac{1}{7} = \frac{\text{change in } y}{\text{change in } x}$. $b = 2$. The y-intercept is at the point $(0, 2)$. A change in y of 1 and in x of 7 gives the point $(0 + 7, 2 + 1)$ or $(7, 3)$.

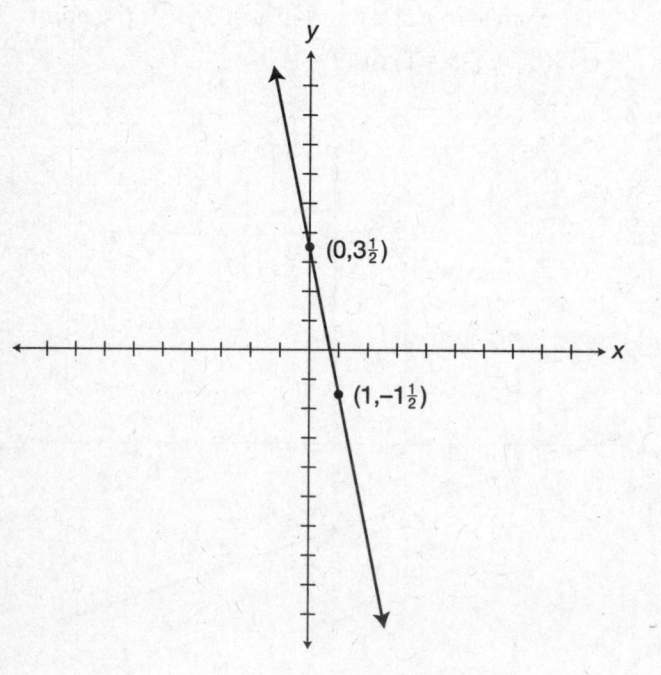

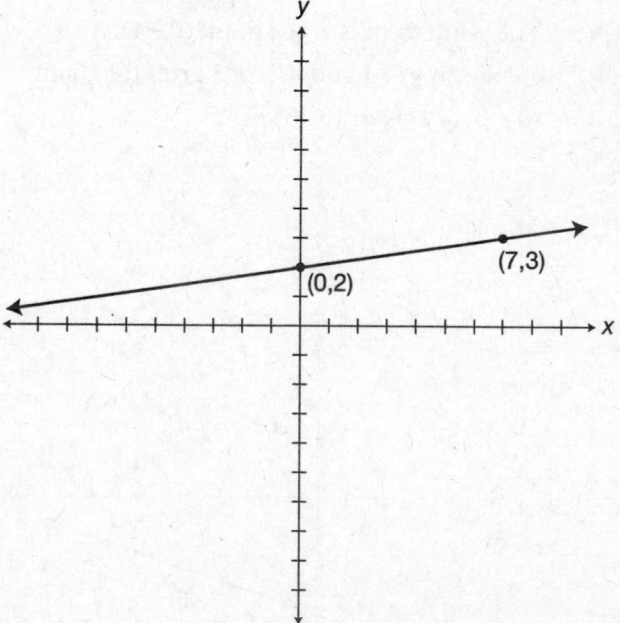

518. Put the equation in the proper form. Subtract $2y$ from both sides of the equation: $0 - 2y = 3x + 2y - 2y$. Simplify the equation: $-2y = 3x$. Divide both sides of the equation by -2: $\frac{-2y}{-2} = \frac{3x}{-2}$. Simplify the equation: $y = \frac{-3}{2}x$. The equation is in the proper slope/y-intercept form: $m = \frac{-3}{2} = \frac{\text{change in } y}{\text{change in } x}$. There is no b showing in the equation, so $b = 0$. The y-intercept is at the point $(0,0)$. A change in y of -3 and in x of 2 gives the point $(0 + 2, 0 - 3)$ or $(2, -3)$.

519. Put the equation in the proper form. Divide both sides of the equation by 3: $\frac{(3x + 12y)}{3} = \frac{-18}{3}$. Simplify the equation: $\frac{3x}{3} + \frac{12y}{3} = -6$; $x + 4y = -6$. Subtract x from both sides of the equation: $x - x + 4y = -x - 6$. Simplify the equation: $4y = -x - 6$. Divide both sides of the equation by 4: $\frac{4y}{4} = \frac{(-x - 6)}{4}$. Simplify the equation: $y = \frac{-x}{4} - \frac{6}{4}$: $y = -\frac{1}{4}x - 1\frac{1}{2}$. The equation is in the proper slope/y-intercept form: $m = \frac{-1}{4} = \frac{\text{change in } y}{\text{change in } x}$. $b = -\frac{3}{2}$. The y-intercept is at the point $(0, -1\frac{1}{2})$. A change in y of -1 and in x of 4 gives the point $(0 + 4, -1\frac{1}{2} - 1)$ or $(4, -2\frac{1}{2})$.

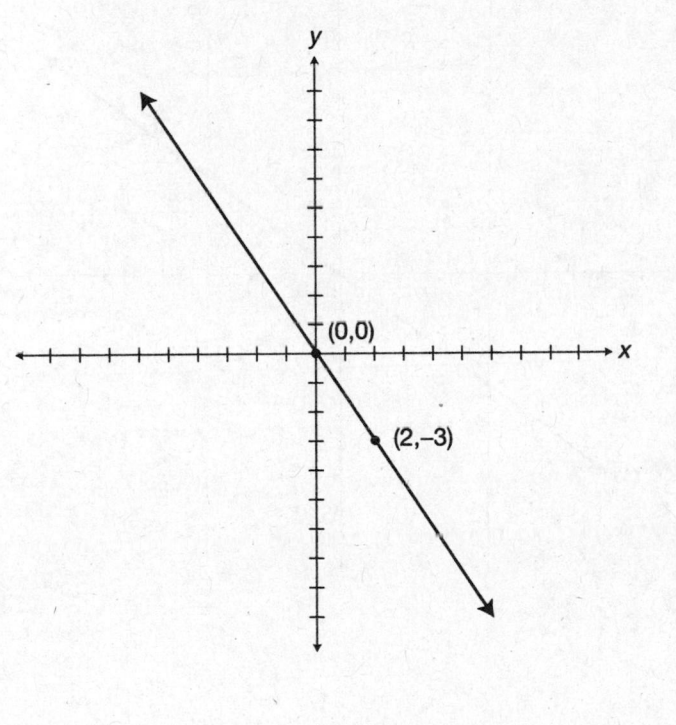

520. Put the equation in the proper form. Add $0.6x$ to both sides of the equation: $y + 0.6x - 0.6x = 0.6x - 2$. Simplify the equation: $y = 0.6x - 2$. The equation is in the proper slope/ y-intercept form: $m = 0.6 = \frac{6}{10} = \frac{3}{5} = \frac{\text{change in } y}{\text{change in } x}$. $b = -2$. The y-intercept is at the point $(0,-2)$. A change in y of 3 and in x of 5 gives the point $(0 + 5, -2 + 3)$ or $(5,1)$.

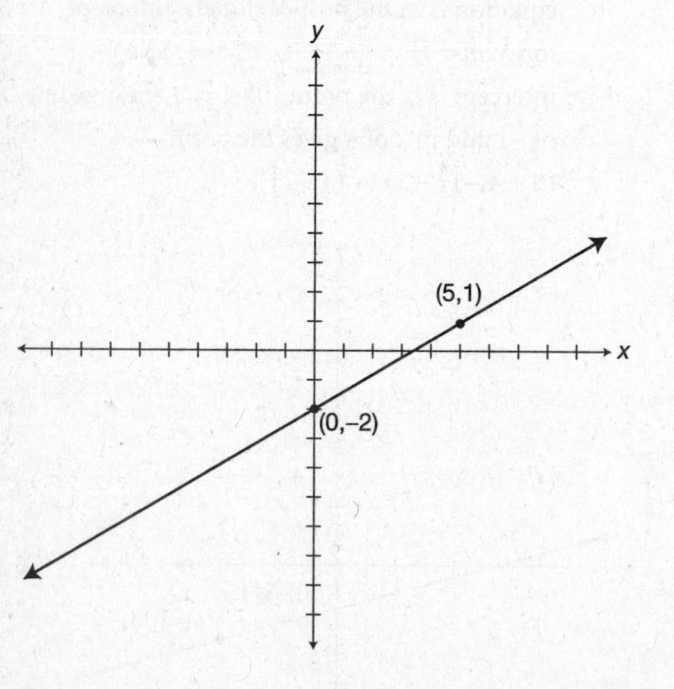

521. Put the equation in the proper form. Add $\frac{1}{2}x$ to both sides of the equation: $\frac{2}{3}y - \frac{1}{2}x + \frac{1}{2}x = 0 + \frac{1}{2}x$. Simplify the equation: $\frac{2}{3}y = \frac{1}{2}x$. Multiply both sides of the equation by $\frac{3}{2}$: $\frac{3}{2}(\frac{2}{3})y = \frac{3}{2}(\frac{1}{2}x)$. Simplify the equation: $1y = \frac{3}{4}x$; $y = \frac{3}{4}x$. The equation is in the proper slope/y-intercept form: $m = \frac{3}{4} = \frac{\text{change in } y}{\text{change in } x}$. $b = 0$. The y-intercept is at the point $(0,0)$. A change in y of 3 and in x of 4 gives the point $(0 + 4, 0 + 3)$ or $(4,3)$.

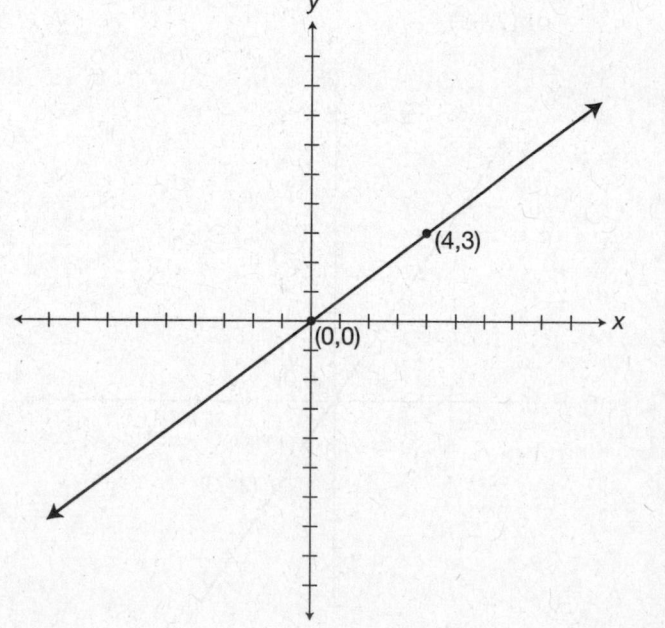

522. Simplify the equation. It would be easier to operate with an equation that doesn't have fractional coefficients. So, if you multiply the whole equation by the lowest common multiple of the denominators, you will have whole numbers with coefficients. Multiply both sides of the equation by 6: $6(\frac{5}{6}x - \frac{1}{3}y) = 6(2)$. Use the distributive property of multiplication: $6(\frac{5}{6}x) - 6(\frac{1}{3}y) = 6(2)$. Simplify the equation: $5x - 2y = 12$. Subtract $5x$ from both sides of the equation: $5x - 5x - 2y = -5x + 12$. Simplify the equation: $-2y = -5x + 12$. Divide both sides of the equation by -2: $\frac{-2y}{-2} = \frac{-5x}{-2} + \frac{12}{-2}$. Simplify the equation: $y = \frac{5}{2}x - 6$. The equation is in the proper slope/y-intercept form: $m = \frac{5}{2} = \frac{\text{change in } y}{\text{change in } x}$. $b = -6$. The y-intercept is at the point $(0,-6)$. A change in y of 5 and in x of 2 gives the point $(0 + 2, -6 + 5)$ or $(2,-1)$.

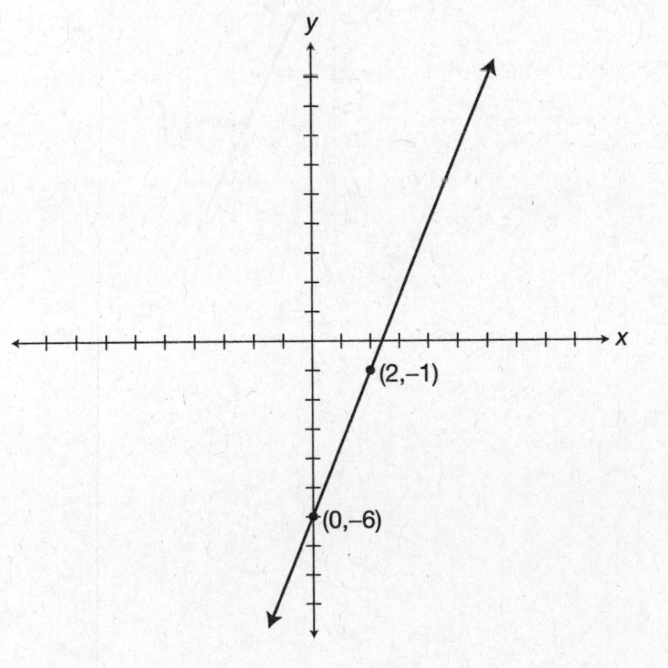

523. Put the equation in the proper form. Subtract 8 from both sides of the equation: $7x - 8 = 4y + 8 - 8$. Simplify the equation: $7x - 8 = 4y$. Exchange the terms on each side of the equal sign: $4y = 7x - 8$. Divide both sides of the equation by 4: $\frac{4y}{4} = \frac{7x}{4} - \frac{8}{4}$. Simplify the equation: $y = \frac{7}{4}x - 2$. The equation is in the proper slope/y-intercept form: $m = \frac{7}{4} = \frac{\text{change in } y}{\text{change in } x}$. $b = -2$. The y-intercept is at the point $(0,-2)$. A change in y of 7 and in x of 4 gives the point: $(0 + 4, -2 + 7)$ or $(4,5)$.

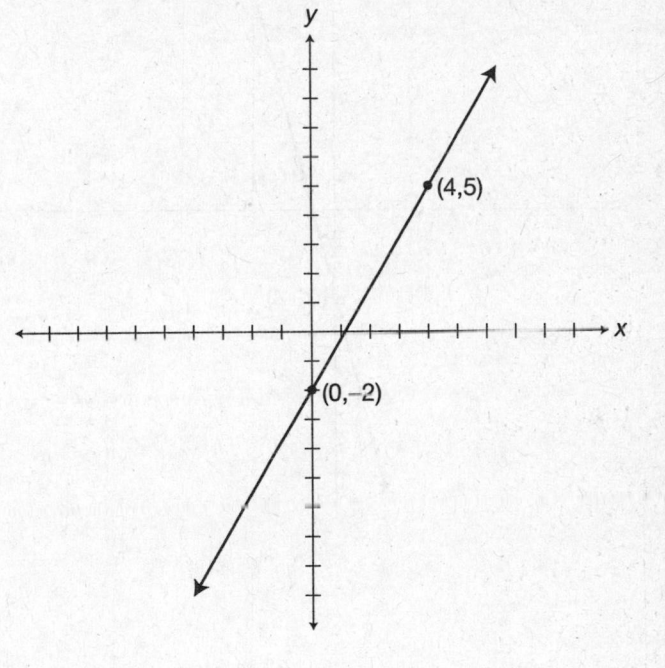

524. Exchange the terms on each side of the equal sign: $5y = 20x - 15$. Divide both sides of the equation by 5: $\frac{5y}{5} = \frac{20x}{5} - \frac{15}{5}$. Simplify the equation: $y = 4x - 3$. The equation is in the proper slope/y-intercept form: $m = 4 = \frac{4}{1} = \frac{\text{change in } y}{\text{change in } x}$. $b = -3$. The y-intercept is at the point $(0,-3)$. A change in y of 4 and in x of 1 gives the point $(0 + 1, -3 + 4)$ or $(1,1)$.

525. Put the equation in the proper form. Subtract $13x$ from both sides of the equation: $6y + 13x - 13x = -13x + 12$. Simplify the equation: $6y = -13x + 12$. Divide both sides of the equation by 6: $\frac{6y}{6} = \frac{-13x}{6} + \frac{12}{6}$. Simplify the equation: $y = \frac{-13}{6}x + 2$. The equation is in the proper slope/y-intercept form: $m = \frac{-13}{6} = \frac{\text{change in } y}{\text{change in } x}$. $b = 2$. The y-intercept is at the point $(0,2)$. A change in y of -13 and in x of 6 gives the point $(0 + 6, 2 - 13)$ or $(6,-11)$.

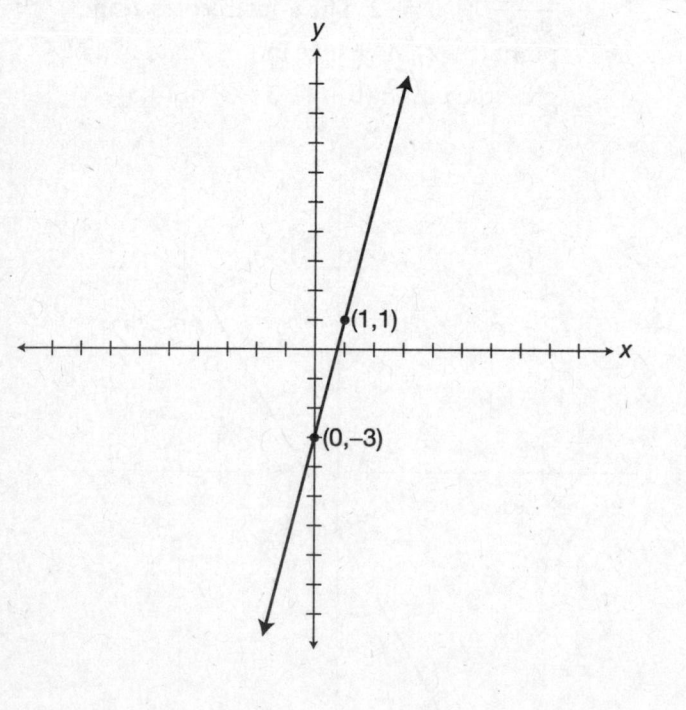

526. Once again, if it would be easier for you to operate with whole number coefficients instead of decimals to start, you could multiply the whole equation by 10. Multiply both sides of the equation by 10: $10(0.1x) = 10(0.7y + 1.4)$. Simplify the expression: $x = 7y + 14$. Subtract 14 from both sides of the equation: $x - 14 = 7y + 14 - 14$. Simplify the equation: $x - 14 = 7y$. If $a = b$, then $b = a$: $7y = x - 14$. Divide both sides of the equation by 7: $\frac{7y}{7} = \frac{x}{7} - \frac{14}{7}$. Simplify the equation: $y = \frac{1}{7}x - 2$. The equation is in the proper slope/y-intercept form: $m = \frac{1}{7} = \frac{\text{change in } y}{\text{change in } x}$. $b = -2$. The y-intercept is at the point $(0, -2)$. A change in y of 1 and in x of 7 gives the point $(0 + 7, -2 + 1)$ or $(7, -1)$.

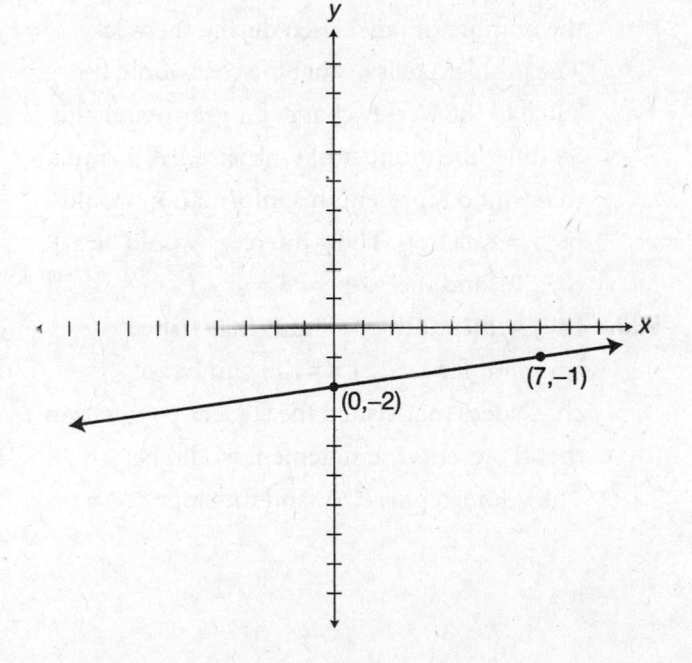

527. Exchange the terms on each side of the equal sign: $17y = -34x + 85$. Divide both sides of the equation by 17: $\frac{17y}{17} = \frac{-34x}{17} + \frac{85}{17}$. Simplify the equation: $y = -2x + 5$. The equation is in the proper slope/y-intercept form: $m = -2 = \frac{-2}{1} = \frac{\text{change in } y}{\text{change in } x}$. $b = 5$. The y-intercept is at the point $(0, 5)$. A change in y of -2 and in x of 1 gives the point $(0 + 1, 5 - 2)$ or $(1, 3)$.

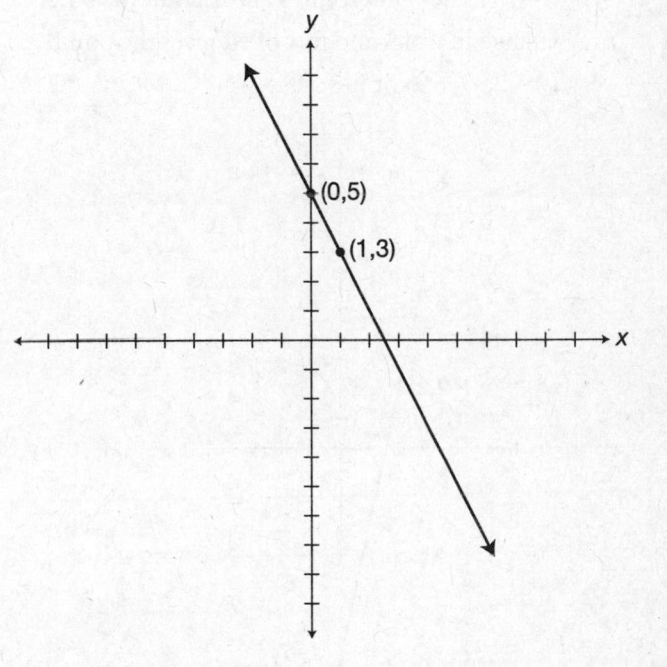

528. Put the equation in the proper form. Add $-27x$ to both sides of the equation: $6y + 27x - 27x = -27x - 42$. Simplify the equation: $6y = -27x - 42$. Divide both sides of the equation by 6: $\frac{6y}{6} = \frac{-27x}{6} - \frac{42}{6}$. Simplify the equation: $y = \frac{-27}{6}x - 7$. Simplify the coefficient of x by a common factor of 3: $y = \frac{-9}{2}x - 7$. The equation is in the proper slope/y-intercept form: $m = \frac{-9}{2} = \frac{9}{-2} = \frac{\text{change in } y}{\text{change in } x}$. $b = -7$. The y-intercept is at the point $(0,-7)$. A change in y of 9 and in x of -2 gives the point $(0 - 2, -7 + 9)$ or $(-2,2)$.

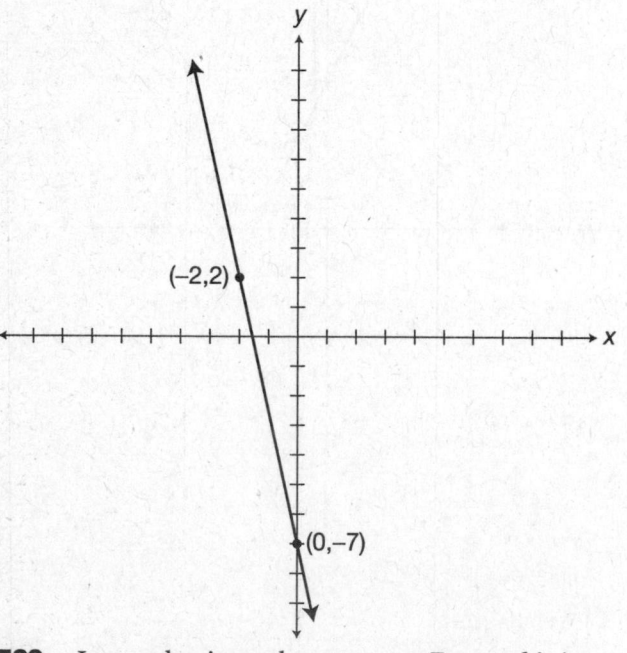

529. Let x = horizontal movement. Forward is in the positive direction. Let y = vertical movement. Ascending is in the positive direction. Descending is in the negative. The change in position of the glider is described by the slope. The change in y is -1 for every change in x of $+25$: slope $= \frac{\text{change in } y}{\text{change in } x} = \frac{-1}{25} = m$. The starting position for the purposes of this graphic solution is at an altitude of 250 feet. So: $b = 250$. Using the standard form $y = mx + b$, you substitute the given values into the formula: $y = \frac{-1}{25}x + 250$. A graph of this equation would have a slope of $-\frac{1}{25}$. The y-intercept would be at the point: $(0,250)$.

530. Let y = the amount of a monthly bill. Let x = the hours of Internet use for the month. The costs for the month will equal \$15 plus \$0.25 times the number of hours of use. Written as an equation, this information would be: $y = 0.25x + 15$. A graph of this equation would have a slope of 0.25 or $\frac{25}{100} = \frac{1}{4}$. The y-intercept would be at $(0,15)$.

531. Let y = the cost of a scooter rental for one day. Let x = the number of miles driven in one day. The problem tells us that the cost would be equal to the daily charges plus \$0.05 times the number of miles driven. Written as an equation, this would be: $y = \$0.05x + 20$. The graph would have a y-intercept at $(0,20)$ and the slope would be $\frac{5}{100} = \frac{1}{20}$.

532. Let y = the total cost for equipment. Let x = the number of tanks used during the week. The problem tells us that the cost would be equal to the weekly charge for gear rental plus \$8 times the number of tanks used. A formula that would represent this information would be: $y = 8x + 150$. The y-intercept would be at $(0,150)$, and the slope $= 8 = \frac{8}{1}$.

533. Let y = the number of birds that visited a backyard feeder. Let x = the number of chickadees that visited the feeder. An equation that represents the statement would be: $y = 7x$. The y-intercept is $(0,0)$ and the slope $= 7 = \frac{7}{1}$.

8 ▶ Inequalities

AN EQUATION IS A NUMBER SENTENCE with an equal sign. The quantity on the left side of the equation is equal to the quantity on the right side of the equation. An **inequality** is a number sentence in which the quantity on the left side is greater than, greater than or equal to, less than, or less than or equal to the quantity on the right side of the equation.

There are four symbols used in inequalities:

SYMBOL	NAME
less than	<
greater than	>
less than or equal to	≤
greater than or equal to	≥

Example

$4 < 5$

This is read as "four is less than five." The open end of the inequality symbol points toward the larger value, and the pointed end of the symbol points toward the smaller value. The less than sign looks a little like a bent L, which might help you remember that "<" stands for "less than."

We can make another statement about the values 4 and 5:

Example

$5 > 4$

This is read as "five is greater than four." Again, the wider end of the inequality symbol points toward the larger value, and the smaller end of the symbol points toward the smaller value.

We can also say $x \geq y$, meaning x is greater than or equal to y. We can use the $\geq$ symbol when the left side of the inequality is either greater than the right side or equal to the right side. In the same way, we can also write $y \leq x$.

▶ Solving Inequalities

A linear equation with a single variable has one solution. The equation $5x + 1 = 11$ can be solved by subtracting 1 from both sides, and then dividing both sides by 5. An inequality has a set of solutions, but we go about solving inequalities in the same way we go about solving equations.

Example

$5x + 1 < 11$

Solve for x just as you would if this were an equation:

$5x + 1 < 11$

$5x + 1 - 1 < 11 - 1$

$5x < 10$

$\frac{5x}{5} < \frac{10}{5}$

$x < 2$

The answer to this inequality is $x < 2$. All values of x that are less than 2 will make this inequality true. We can test our answer by substituting a value that is less than 2 into the equation. Zero is a good choice for

Pitfall

Unlike checking an answer to an equation, no check can guarantee that an answer to an inequality is correct. There are many correct solutions (and many incorrect ones), so it is impossible to test every possibility. In this example, if we were to test a value that is less than 2 and found that it made the inequality untrue, we would know that our answer was incorrect. However, the fact that a check of a single value shows the inequality to be true does not prove that our answer is correct. It is helpful to check a few values, but there is no surefire check.

that value, since it makes the computations easy: $5(0) + 1 < 11, 0 + 1 < 11, 1 < 11$. It's true that 1 is less than 11, which proves that 0 is definitely in the solution set.

Besides the absence of a surefire check, there is one other difference between solving an equation and solving an inequality. If you multiply or divide both sides of an inequality by a negative number, you must change the direction of the inequality symbol.

Example

$-3x - 7 \geq 14$

Begin by adding 7 to both sides of the inequality:

$-3x - 7 + 7 \geq 14 + 7$

$-3x \geq 21$

Now divide by -3 and change the direction of the inequality symbol, since we are dividing by a negative number:

$-3x \geq 21$

$\frac{-3x}{-3} \leq \frac{21}{-3}$

$x \leq -7$

The solution set is $x \leq -7$. Why do we change the direction of the inequality symbol? Before dividing, the greater side of the inequality was multiplied by a negative number. After dividing, the greater side must still be multiplied by a negative number in order for the inequality to remain true; think of -7 as $(-1)(7)$. A check of the answer shows that flipping the symbol was the correct action to take. Let's use $x = -8$: $-3(-8) - 7 \geq 14$, $24 - 7 \geq 14$, $17 \geq 14$.

▶ **Practice Questions**

Solve the following inequalities.

534. $3x + 2 < 11$

535. $4x - 6 > 30$

536. $\frac{2}{5}x \leq 18$

537. $4x + 26 \geq 90$

538. $8 - 6x < 50$

539. $5x - 9 \leq -2$

540. $2x + 0.29 > 0.79$

541. $-6(x + 1) \geq 60$

542. $3(5 - 4x) < x - 63$

543. $4(x + 1) < 5(x + 2)$

544. $2(7x - 3) \geq -2(5 + 3x)$

545. $16x - 1 < 4(6 - x)$

546. $\frac{-x}{0.3} \leq 20$

547. $\frac{4}{3}x - 5 > x - 2$

548. $3x + 5 \geq -2(x + 10)$

549. $-4x + 3(x + 5) \geq 3(x + 2)$

550. $x - \frac{3}{4} < -\frac{3}{4}(x + 2)$

551. $\frac{3}{2}x + 0.1 \geq 0.9 + x$

552. $x - 4\frac{1}{3} < 9 + \frac{2}{3}x$

553. $-7(x + 3) < -4x$

554. $\frac{5}{4}(x + 4) > \frac{1}{2}(x + 8) - 8$

555. $3(1 - 3x) \geq -3(x + 27)$

556. $-5[9 + (x - 4)] \geq 2(13 - x)$

557. $11(1 - x) \geq 3(3 - x) - 1$

558. $3(x - 16) - 2 < 9(x - 2) - 7x$

Solve the following word problems.

559. Nine minus five times a number is no less than 39. Which of the following expressions represents all the possible values of the number?
  a. $x \leq 6$
  b. $x \geq -6$
  c. $x \leq -6$
  d. $x \geq 6$

560. Will has a bag of gumdrops. If he eats two of his gumdrops, he will have between two and six of them left. Which of the following represents how many gumdrops were originally in his bag?
  a. $4 < x < 8$
  b. $0 < x < 4$
  c. $0 > x > 4$
  d. $4 > x > 8$

561. The value of y is between negative 3 and positive 8 inclusive. Which of the following represents y?
  a. $-3 \leq y \leq 8$
  b. $-3 < y \leq 8$
  c. $-3 \leq y < 8$
  d. $-3 \geq y \geq 8$

562. Five more than the quotient of a number divided by 2 is at least that same number. What is the greatest value of the number?
  a. 7
  b. 10
  c. 5
  d. 2

563. Carl worked three more than twice as many hours as Cindy did. What is the maximum number of hours Cindy worked if together they worked 48 hours at most?
  a. 17
  b. 33
  c. 37
  d. 15

▶ Graphing Inequalities

We begin to graph an inequality in the same way we begin to graph the equation of a line. Put the inequality in slope-intercept form (which will look like $y < mx + b$, $y > mx + b$, $y \leq mx + b$, or $y \geq mx + b$) and then plot the y-intercept. Use the slope to find a few more points on the graph. If the inequality has the equal sign in it ($\leq$ or $\geq$), connect the points with a solid line, just as you would if you were graphing the equation of a line. If the inequality doesn't have the equal sign in it ($<$ or $>$), connect the points with a dashed line. Then, shade the side of the line where the solution set lies.

Why are some inequalities graphed with a solid line and others graphed with a dashed line? A solid line indicates that the points on the lines are part of the solution set, and a dashed line indicates that the points on the line are not part of the solution set. The points on the line are the values that make the left side and the right side of the inequality equal, so we need to include them in the solutions to less-than-or-equal-to and greater-than-or-equal-to inequalities, but keep them out of the solutions to less-than and greater-than inequalities.

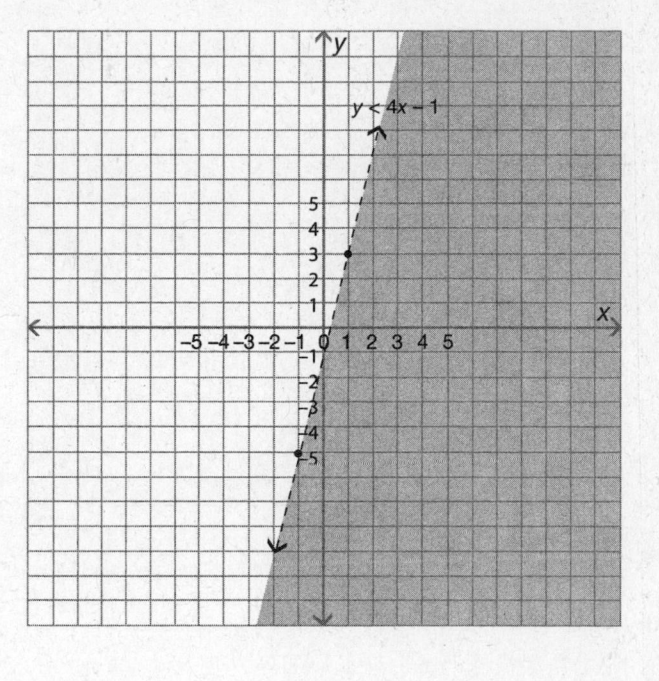

Example

Graph $y < 4x - 1$.

First, plot the y-intercept, $(0,-1)$. Then, use the slope to plot a few more points, such as $(1,3)$ and $(-1,-5)$. Next, connect them with a dashed line, since the points on this line are not part of the solution set. These points would make y equal to $4x - 1$ instead of making y less than $4x - 1$. Finally, we need to shade one side of the graph. Test the point $(0,0)$. If the inequality holds true, shade the side of the line where $(0,0)$ lies. If the inequality does not hold true, shade the other side of the line. $0 < 4(0) - 1$, and $0 < -1$. The inequality is not true for the point $(0,0)$, so we must shade below and to the right of the dashed line. The graph of $y < 4x - 1$ is shown here.

Example

Graph $-2y \leq 14$.

First, divide both sides of the inequality by -2 to put the inequality in slope-intercept form. Because we're dividing by a negative number, switch the direction of the inequality symbol. As shown on the following graph, $y \geq -7$ is a horizontal line, because it has no slope. Plot a few points along the line $y = -7$ and connect them with a solid line, since these points are part of the solution set. Shade the area above the line, since points above the line $y = -7$ have y values that are greater than -7.

$-2y \le 14$

▶ Practice Questions

Graph the following inequalities on a coordinate plane. (Use graph paper.)

564. $y < x + 1$

565. $y \ge -x + 2$

566. $y < 4x - 5$

567. $\frac{1}{2}x + y \le 3$

568. $2y - 3x < 8$

569. $y + 2 \le 3x + 5$

570. $3x - 4 \ge 2y$

571. $\frac{3}{4}y + 6 \ge 3x$

572. $0.5y - x + 3 > 0$

573. $x - y \le 7$

574. $\frac{y}{3} < \frac{2}{3} - x$

575. $-3y + 9x \le -6$

576. $0.5x > 0.3y - 0.9$

577. $3x - y \le 7x + y - 8$

578. $3y + 4x < 9 - 2x$

579. $-12 \le -3(x + y)$

580. $9y + 7 \ge 2(x + 8)$

581. $\frac{x}{3} + y \le 3x - 5$

582. $2(y + 3) - x \ge 6(1 - x)$

583. $-28y \ge 2x - 14(y + 10)$

TERM TO REVIEW

inequality

▶ Answers

534. Subtract 2 from both sides of the inequality: $3x + 2 - 2 < 11 - 2$. Simplify the inequality: $3x < 9$. Divide both sides of the inequality by 3: $\frac{3x}{3} < \frac{9}{3}$. Simplify: $x < 3$.

535. Add 6 to both sides of the inequality: $4x - 6 + 6 > 30 + 6$. Simplify the inequality: $4x > 36$. Divide both sides of the inequality by 4: $\frac{4x}{4} > \frac{36}{4}$. Simplify: $x > 9$.

536. Multiply both sides of the inequality by $\frac{5}{2}$: $\frac{5}{2}(\frac{2}{5})x \le \frac{5}{2}(18)$. Simplify: $(\frac{10}{10})x \le \frac{5}{2}(\frac{18}{1})$; $(1)x \le 45; x \le 45$.

537. Subtract 26 from both sides of the inequality: $4x + 26 - 26 \geq 90 - 26$. Simplify: $4x \geq 64$. Divide both sides of the inequality by 4: $\frac{4x}{4} \geq \frac{64}{4}$. Simplify the inequality: $x \geq 16$.

538. Subtract 8 from both sides of the inequality: $8 - 8 - 6x < 50 - 8$. Simplify the inequality: $-6x < 42$. Divide both sides of the inequality by -6 and change the direction of the inequality sign: $\frac{-6x}{-6} > \frac{42}{-6}$. Simplify: $x > -7$.

539. Add 9 to both sides of the inequality: $5x - 9 + 9 \leq -2 + 9$. Combine like terms on each side of the inequality: $5x \leq 7$. Divide both sides of the inequality by 5: $\frac{5x}{5} \leq \frac{7}{5}$. Simplify: $x \leq \frac{7}{5}; x \leq 1\frac{2}{5}$.

540. Subtract 0.29 from both sides of the inequality: $2x + 0.29 - 0.29 > 0.79 - 0.29$. Combine like terms on each side of the inequality: $2x > 0.50$. Divide both sides of the inequality by 2: $\frac{2x}{2} > \frac{0.50}{2}$. Simplify the inequality: $x > 0.25$.

541. Divide both sides of the inequality by -6 and change the direction of the inequality sign: $\frac{-6(x+1)}{-6} \leq \frac{60}{-6}$. Simplify the expressions: $x + 1 \leq -10$. Subtract 1 from both sides of the inequality: $x + 1 - 1 \leq -10 - 1$. Simplify: $x \leq -11$.

542. Use the distributive property of multiplication: $3(5) - 3(4x) < x - 63$. Simplify: $15 - 12x < x - 63$. Add $12x$ to both sides of the inequality: $15 - 12x + 12x < x + 12x - 63$. Combine like terms on each side of the inequality: $15 < 13x - 63$. Add 63 to both sides of the inequality: $15 + 63 < 13x - 63 + 63$. Simplify the inequality: $78 < 13x$. Divide both sides of the inequality by 13: $\frac{78}{13} < \frac{13x}{13}$. Simplify: $6 < x$.

543. Use the distributive property of multiplication: $4(x) + 4(1) < 5(x) + 5(2)$. Simplify: $4x + 4 < 5x + 10$. Subtract 4 from both sides of the inequality: $4x + 4 - 4 < 5x + 10 - 4$. Combine like terms on each side of the inequality: $4x < 5x + 6$. Subtract $5x$ from both sides of the inequality: $4x - 5x < 5x - 5x + 6$. Simplify the inequality: $-x < 6$. Multiply both sides of the equation by -1 and change the direction of the inequality sign: $-1(-x) > -1(6)$. Simplify: $x > -6$.

544. Use the distributive property of multiplication: $2(7x) - 2(3) \geq -2(5) - 2(3x)$. Simplify the expressions: $14x - 6 \geq -10 - 6x$. Add $6x$ to both sides of the inequality: $14x + 6x - 6 \geq -10 - 6x + 6x$. Combine like terms: $20x - 6 \geq -10$. Add 6 to both sides of the inequality: $20x - 6 + 6 \geq -10 + 6$. Simplify: $20x \geq -4$. Divide both sides of the inequality by 20: $\frac{20x}{20} \geq \frac{-4}{20}$. Simplify: $x \geq \frac{-4}{20}$. Reduce the fraction to lowest terms: $x \geq \frac{-1}{5}$.

545. Use the distributive property of multiplication: $16x - 1 < 4(6) - 4(x)$. Simplify: $16x - 1 < 24 - 4x$. Add 1 to both sides of the inequality: $16x - 1 + 1 < 1 + 24 - 4x$. Combine like terms: $16x < 25 - 4x$. Add $4x$ to both sides of the inequality: $16x + 4x < 25 - 4x + 4x$. Combine like terms on each side of the inequality: $20x < 25$. Divide both sides of the inequality by 20: $\frac{20x}{20} < \frac{25}{20}$. Simplify and express the fraction in simplest terms: $x < 1\frac{1}{4}$.

546. Multiply both sides of the inequality by 0.3: $0.3(\frac{-x}{0.3}) \leq 0.3(20)$. Simplify the expressions on both sides: $-x \leq 6$. Multiply both sides of the inequality by -1 and change the direction of the inequality sign: $-1(-x) \geq -1(6)$. Simplify the expressions: $x \geq -6$.

547. Add 5 to both sides of the inequality: $\frac{4}{3}x - 5 + 5 > x - 2 + 5$. Simplify: $\frac{4}{3}x > x + 3$. Subtract $1x$ from both sides of the inequality: $\frac{4}{3}x - x > x - x + 3$; $\frac{4}{3}x - \frac{3}{3}x > x - x + 3$. Simplify the expressions: $\frac{1}{3}x > 3$. Multiply both sides of the inequality by 3: $3(\frac{1}{3}x) > 3(3)$; $\frac{3}{1}(\frac{1}{3}x) > 3(3)$. Simplify the expressions: $x > 9$.

548. Use the distributive property of multiplication: $3x + 5 \geq -2(x) - 2(10)$. Simplify: $3x + 5 \geq -2x - 20$. Subtract 5 from both sides of the inequality: $3x + 5 - 5 \geq -2x - 20 - 5$. Combine like terms on each side of the inequality: $3x \geq -2x - 25$. Add $2x$ to both sides of the inequality: $3x + 2x \geq 2x - 2x - 25$. Combine like terms: $5x \geq -25$. Divide both sides of the inequality by 5: $\frac{5x}{5} \geq \frac{-25}{5}$. Simplify: $x \geq -5$.

549. Use the distributive property of multiplication: $-4x + 3(x) + 3(5) \geq 3(x) + 3(2)$. Simplify: $-4x + 3x + 15 \geq 3x + 6$. Combine like terms: $(-4x + 3x) + 15 \geq 3x + 6$. Simplify: $-x + 15 \geq 3x + 6$. Add x to both sides of the inequality: $x - x + 15 \geq x + 3x + 6$. Combine like terms: $15 \geq 4x + 6$. Subtract 6 from both sides of the inequality: $15 - 6 \geq 4x + 6 - 6$. Simplify: $9 \geq 4x$. Divide both sides of the inequality by 4: $\frac{9}{4} \geq \frac{4x}{4}$. Simplify: $\frac{9}{4} \geq x$. Express the fraction in its simplest form: $2\frac{1}{4} \geq x$.

550. You can simplify equations (and inequalities) with fractions by multiplying them by a common multiple of the denominators. Multiply both sides of the inequality by 4: $4(x - \frac{3}{4}) < 4[\frac{-3}{4}(x + 2)]$. Use the distributive property of multiplication: $4(x) - 4(\frac{3}{4}) < 4(\frac{-3}{4})(x + 2)$. Simplify the expressions: $4x - 3 < -3(x + 2)$. Use the distributive property of multiplication: $4x - 3 < -3(x) - 3(2)$. Simplify: $4x - 3 < -3x - 6$. Add 3 to both sides of the

inequality: $4x - 3 + 3 < -3x - 6 + 3$. Combine like terms: $4x < -3x - 3$. Add $3x$ to both sides of the equation: $3x + 4x < 3x - 3x - 3$. Combine like terms: $7x < -3$. Divide both sides of the inequality by 7: $\frac{7x}{7} < \frac{-3}{7}$. Even though you have a fraction for an answer, it has been easier to operate with whole numbers until the last step. Simplify the expressions: $x < \frac{-3}{7}$.

551. Subtract 0.1 from both sides of the inequality: $\frac{3}{2}x + 0.1 - 0.1 \geq 0.9 - 0.1 + x$. Combine like terms on each side of the inequality: $\frac{3}{2}x \geq 0.8 + x$. Subtract x from both sides of the inequality: $\frac{3}{2}x - x \geq 0.8 + x - x$. Simplify: $\frac{1}{2}x \geq 0.8$. Multiply both sides of the inequality by 2: $2(\frac{1}{2}x) \geq 2(0.8)$. Simplify the expressions: $x \geq 1.6$.

552. Change the term to an improper fraction: $x - \frac{13}{3} < 9 + \frac{2}{3}x$. Multiply both sides of the inequality by 3: $3(x - \frac{13}{3}) < 3(9 + \frac{2}{3}x)$. Use the distributive property of multiplication: $3(x) - 3(\frac{13}{3}) < 3(9) + 3(\frac{2}{3}x)$. Simplify the terms: $3x - 13 < 27 + 2x$. Add 13 to both sides of the inequality: $3x - 13 + 13 < 13 + 27 + 2x$. Combine like terms and simplify: $3x < 40 + 2x$. Subtract $2x$ from both sides of the inequality: $3x - 2x < 40 + 2x - 2x$. Simplify: $x < 40$.

553. Use the distributive property of multiplication: $-7(x) - 7(3) < -4x$. Simplify the terms: $-7x - 21 < -4x$. Add 21 to both sides of the inequality: $-7x - 21 + 21 < -4x + 21$. Simplify by combining like terms: $-7x < -4x + 21$. Add $4x$ to both sides of the inequality: $-7x + 4x < -4x + 4x + 21$. Simplify the terms: $-3x < 21$. Divide both sides of the inequality by -3 and change the direction of the inequality sign: $\frac{-3x}{-3} > \frac{21}{-3}$. Simplify the expressions: $x > -7$.

554. Use the distributive property of multiplication: $\frac{5}{4}(x) + \frac{5}{4}(4) > \frac{1}{2}(x) + \frac{1}{2}(8) - 8$. Simplify the terms: $\frac{5}{4}x + 5 > \frac{1}{2}x + 4 - 8$. Combine like terms: $\frac{5}{4}x + 5 > \frac{1}{2}x - 4$. Subtract $\frac{1}{2}x$ from both sides of the inequality: $\frac{5}{4}x - \frac{1}{2}x + 5 > \frac{1}{2}x - \frac{1}{2}x - 4$. Combine like terms: $\frac{3}{4}x + 5 > -4$. Subtract 5 from both sides of the inequality: $\frac{3}{4}x + 5 - 5 > 4 - 5$. Simplify: $\frac{3}{4}x > -9$. Multiply both sides by $\frac{4}{3}$ (the reciprocal of $\frac{3}{4}$): $\frac{4}{3}(\frac{3}{4}x) > \frac{4}{3}(-9)$. Simplify the expressions: $x > -12$.

555. Use the distributive property of multiplication: $3(1) - 3(3x) \geq -3(x) - 3(27)$. Simplify terms: $3 - 9x \geq -3x - 81$. Add $9x$ to both sides: $3 - 9x + 9x \geq 9x - 3x - 81$. Combine like terms: $3 \geq 6x - 81$. Add 81 to both sides of the inequality: $3 + 81 \geq 6x - 81 + 81$. Combine like terms: $84 \geq 6x$. Divide both sides of the inequality by 6: $\frac{84}{6} \geq \frac{6x}{6}$. Simplify: $14 \geq x$.

556. Remove the inner parentheses, change the brackets to parentheses, use the commutative property of addition, and combine like terms: $-5(9 + x - 4) \geq 2(13 - x); -5(x + 9 - 4) \geq 2(13 - x); -5(x + 5) \geq 2(13 - x)$. Use the distributive property of multiplication: $-5(x) - 5(5) \geq 2(13) - 2(x)$. Simplify terms: $-5x - 25 \geq 26 - 2x$. Add $5x$ to both sides of the inequality: $-5x + 5x - 25 \geq 26 - 2x + 5x$. Combine like terms: $-25 \geq 26 + 3x$. Subtract 26 from both sides of the inequality: $-25 - 26 \geq 26 - 26 + 3x$. Combine like terms: $-51 \geq 3x$. Divide both sides of the inequality by 3: $\frac{-51}{3} \geq \frac{3x}{3}$. Simplify terms: $-17 \geq x$.

557. Use the distributive property of multiplication: $11(1) - 11(x) \geq 3(3) - 3(x) - 1$. Simplify terms: $11 - 11x \geq 9 - 3x - 1$. Use the commutative property: $11 - 11x \geq 9 - 1 - 3x$. Combine like terms: $11 - 11x \geq 8 - 3x$. Subtract 8 from both sides of the inequality: $11 - 8 - 11x \geq 8 - 8 - 3x$. Combine like terms: $3 - 11x \geq -3x$. Add $11x$ to both sides: $3 - 11x + 11x \geq -3x + 11x$. Combine like terms: $3 \geq 8x$. Divide both sides by 8: $\frac{3}{8} \geq \frac{8x}{8}$. Simplify: $\frac{3}{8} \geq x$.

558. Use the distributive property of multiplication: $3(x) - 3(16) - 2 < 9(x) - 9(2) - 7x$. Simplify terms: $3x - 48 - 2 < 9x - 18 - 7x$. Use the commutative property to associate like terms: $3x - 48 - 2 < 9x - 7x - 18$. Simplify terms: $3x - 50 < 2x - 18$. Add 50 to both sides of the inequality: $3x - 50 + 50 < 2x - 18 + 50$. Combine like terms: $3x < 2x + 32$. Subtract $2x$ from both sides of the inequality: $3x - 2x < 2x - 2x + 32$. Combine like terms: $x < 32$.

559. **c.** Translate the sentence "Nine minus five times a number is no less than 39" into symbols: $9 - 5x \geq 39$. Subtract 9 from both sides of the inequality: $9 - 9 - 5x \geq 39 - 9$. Simplify: $-5x \geq 30$; divide both sides of the inequality by -5. Remember that when dividing or multiplying each side of an inequality by a negative number, the inequality symbol changes direction: $\frac{-5x}{-5} \leq \frac{30}{-5}$. The variable is now alone: $x \leq -6$.

560. **a.** This problem is an example of a compound inequality, where there is more than one inequality in the question. In order to solve it, let x = the total number of gumdrops Will has. Set up the compound inequality, and then solve it as two separate inequalities. Therefore, the second sentence in the problem can be written as: $2 < x - 2 < 6$. The two inequalities are: $2 < x - 2$ and $x - 2 < 6$. Add 2 to both sides of both inequalities: $2 + 2 < x - 2 + 2$ and $x - 2 + 2 < 6 + 2$; simplify: $4 < x$ and $x < 8$. If x is greater than four gumdrops and less than eight, it means that the solution is between 4 and 8. This can be shortened to: $4 < x < 8$.

561. **a.** This inequality shows a solution set where y is greater than or equal to 3 and less than or equal to 8. Both −3 and 8 are in the solution set because of the word *inclusive*, which indicates that the set includes them. The only choice that shows values between −3 and 8 and also includes them is choice **a.**

562. **b.** Let x = the number. Remember that *quotient* is a key word for division, and *at least* means greater than or equal to. From the question, the sentence would translate to: $\frac{x}{2} + 5 \geq x$. Subtract 5 from both sides of the inequality: $\frac{x}{2} + 5 - 5 \geq x - 5$; simplify: $\frac{x}{2} \geq x - 5$. Multiply both sides of the inequality by 2: $\frac{x}{2} \times 2 \geq (x - 5) \times 2$; simplify: $x \geq (x - 5)2$. Use the distributive property on the right side of the inequality: $x \geq 2x - 10$. Add 10 to both sides of the inequality: $x + 10 \geq 2x - 10 + 10$; simplify: $x + 10 \geq 2x$. Subtract x from both sides of the inequality: $x - x + 10 \geq 2x - x$. The variable is now alone: $10 \geq x$. The number is at most 10.

563. **d.** Let x = the number of hours Cindy worked. Let $2x + 3$ = the number of hours Carl worked. Because the total hours added together were at most 48, the inequality would be $(x) + (2x + 3) \leq 48$. Combine like terms on the left side of the inequality: $3x + 3 \leq 48$. Subtract 3 from both sides of the inequality: $3x + 3 - 3 \leq 48 - 3$; simplify: $3x \leq 45$. Divide both sides of the inequality by 3: $\frac{3x}{3} \leq \frac{45}{3}$; the variable is now alone: $x \leq 15$. The maximum number of hours Cindy worked was 15.

564. The inequality is in the proper slope/y-intercept form: $m = 1 = \frac{1}{1} = \frac{\text{change in } y}{\text{change in } x}$. $b = 1$. The y-intercept is at the point $(0,1)$. A change in y of 1 and in x of 1 gives the point $(0 + 1, 1 + 1)$ or $(1,2)$. Draw a dashed boundary line and shade below it.

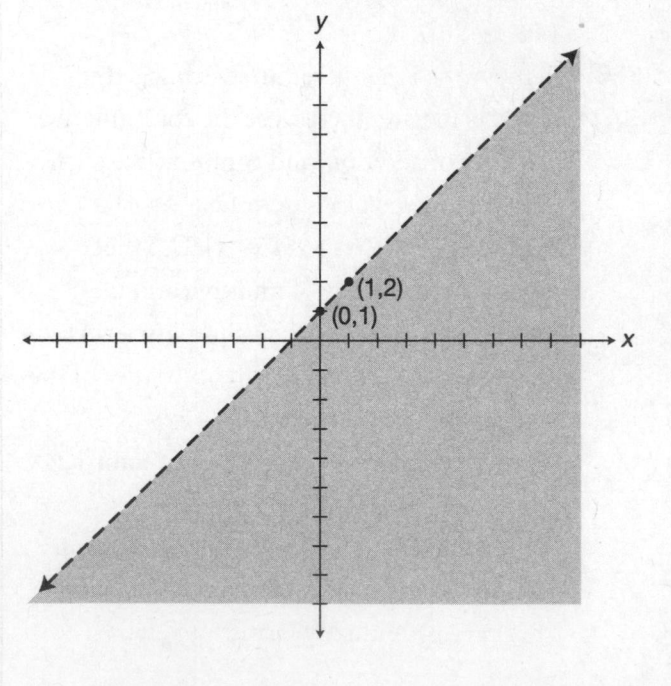

565. The inequality is in the proper slope/ y-intercept form: $m = -1 = \frac{-1}{1} = \frac{\text{change in } y}{\text{change in } x}$. $b = 2$. The y-intercept is at the point (0,2). A change in y of -1 and in x of 1 gives the point $(0 + 1, 2 - 1)$ or (1,1). Draw a solid boundary line and shade above it.

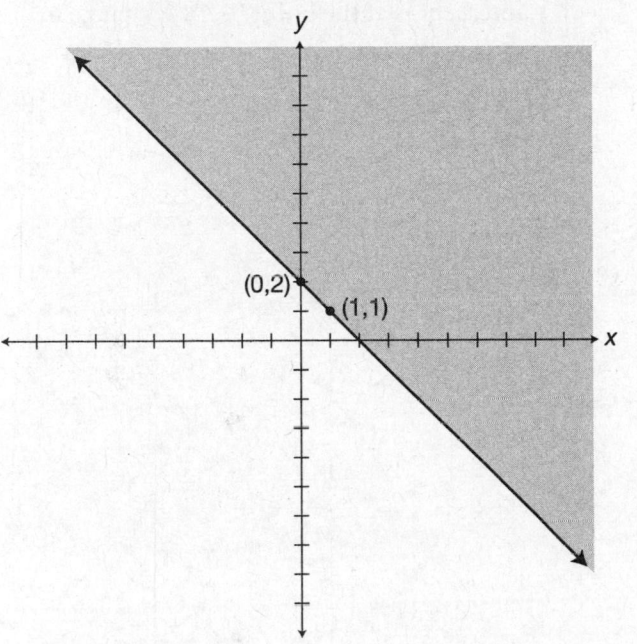

566. The inequality is in the proper slope/ y-intercept form: $m = 4 = \frac{4}{1} = \frac{\text{change in } y}{\text{change in } x}$. $b = -5$. The y-intercept is at the point (0,-5). A change in y of 4 and in x of 1 gives the point $(0 + 1, -5 + 4)$ or (1,-1). Draw a dashed boundary line and shade below it.

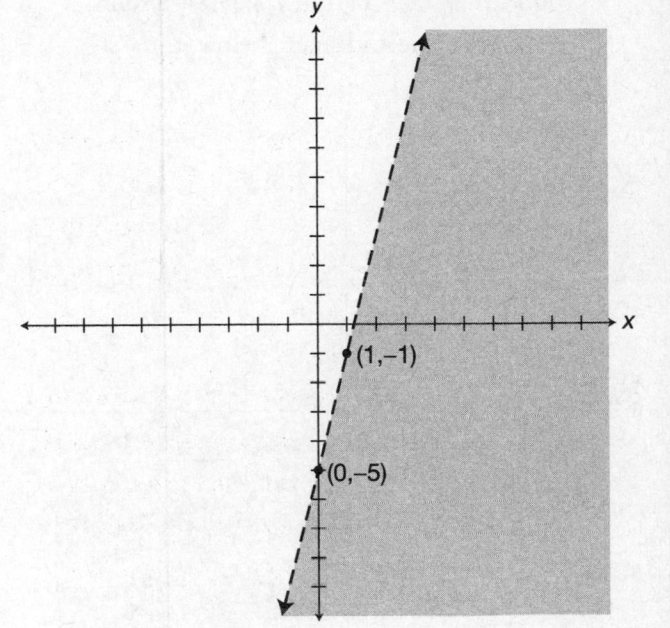

567. Subtract $\frac{1}{2}x$ from both sides of the inequality: $\frac{1}{2}x - \frac{1}{2}x + y \le -\frac{1}{2}x + 3$. Combine like terms: $y \le -\frac{1}{2}x + 3$. The inequality is in the proper slope/y-intercept form: $m = \frac{-1}{2} = \frac{\text{change in } y}{\text{change in } x}$. $b = 3$. The y-intercept is at the point $(0,3)$. A change in y of -1 and in x of 2 gives the point $(0 + 2, 3 - 1)$ or $(2,2)$. Draw a solid boundary line and shade below it.

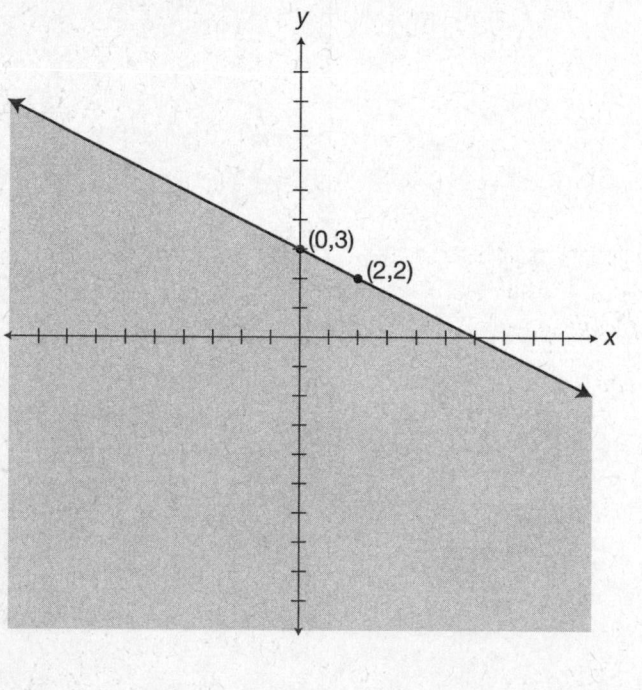

568. Add $3x$ to both sides of the inequality: $2y - 3x + 3x < 3x + 8$. Combine like terms: $2y < 3x + 8$. Divide both sides of the inequality by 2: $\frac{2y}{2} < \frac{3x}{2} + \frac{8}{2}$. Simplify terms: $y < \frac{3}{2}x + 4$. The inequality is in the proper slope/y-intercept form: $m = \frac{3}{2} = \frac{\text{change in } y}{\text{change in } x}$. $b = 4$. The y-intercept is at the point $(0,4)$. A change in y of 3 and in x of 2 gives the point $(0 + 2, 4 + 3)$ or $(2,7)$. Draw a dashed boundary line and shade below it.

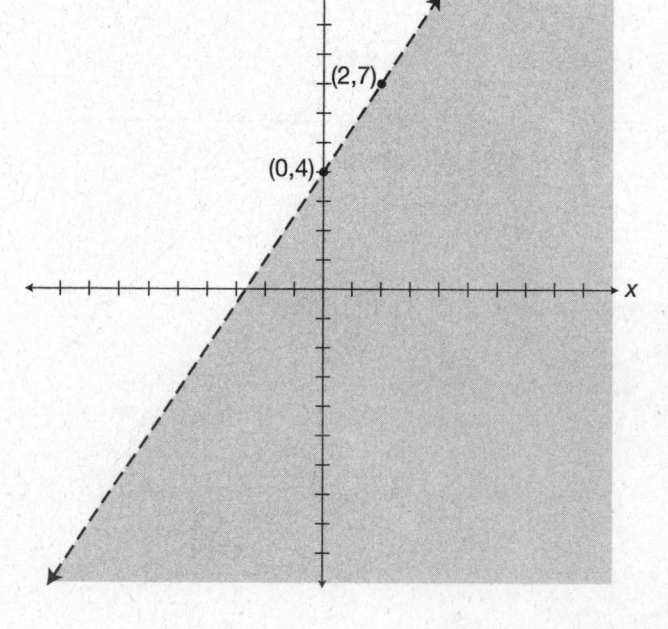

569. Subtract 2 from both sides of the inequality: $y + 2 - 2 \le 3x + 5 - 2$. Combine like terms: $y \le 3x + 3$. The inequality is in the proper slope/y-intercept form: $m = 3 = \frac{3}{1} = \frac{\text{change in } y}{\text{change in } x}$. $b = 3$. The y-intercept is at the point $(0,3)$. A change in y of 3 and in x of 1 gives the point $(0 + 1, 3 + 3)$ or $(1,6)$. Draw a solid boundary line and shade below it.

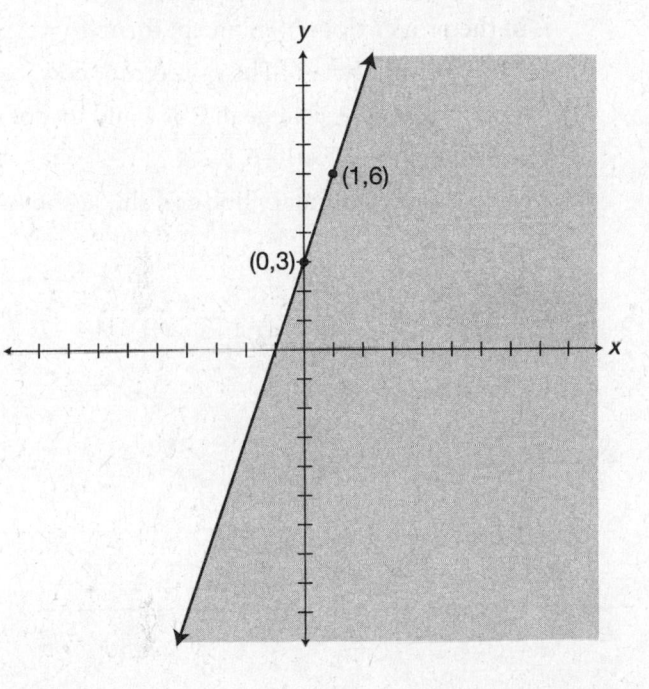

570. In an equation, if $c = d$, then $d = c$. But for an inequality, the direction of the inequality symbol must change when you change sides of the statement. If $c \ge d$, then $d \le c$. Rewrite the inequality with sides exchanged and the symbol reversed: $2y \le 3x - 4$. Divide both sides of the inequality by 2: $\frac{2y}{2} \le \frac{3x}{2} - \frac{4}{2}$. Simplify terms: $y \le \frac{3}{2}x - 2$. The inequality is in the proper slope/y-intercept form: $m = \frac{3}{2} = \frac{\text{change in } y}{\text{change in } x}$. $b = -2$. The y-intercept is at the point $(0,-2)$. A change in y of 3 and in x of 2 gives the point $(0 + 2, -2 + 3)$ or $(2,1)$. Draw a solid boundary line and shade below it.

571. Subtract 6 from both sides of the inequality: $\frac{3}{4}y + 6 - 6 \geq 3x - 6$. Combine like terms: $\frac{3}{4}y \geq 3x - 6$. Multiply both sides of the inequality by the reciprocal $\frac{4}{3}$: $\frac{4}{3}(\frac{3}{4}y) \geq \frac{4}{3}(3x - 6)$. Use the distributive property of multiplication: $\frac{4}{3}(\frac{3}{4}y) \geq \frac{4}{3}(3x) - \frac{4}{3}(6)$. Simplify terms: $y \geq 4x - 8$. The inequality is in the proper slope/y-intercept form: $m = \frac{4}{1} = \frac{\text{change in } y}{\text{change in } x}$. $b = -8$. The y-intercept is at the point $(0, -8)$. A change in y of 4 and in x of 1 gives the point $(0 + 1, -8 + 4)$ or $(1, -4)$. Draw a solid boundary line and shade above it.

572. Subtract 3 from both sides of the inequality: $0.5y - x + 3 - 3 > 0 - 3$. Combine like terms on each side of the inequality: $0.5y - x > -3$. Add $1x$ to both sides of the inequality: $0.5y + x - x > x - 3$. Combine like terms: $0.5y > x - 3$. Divide both sides of the inequality by 0.5: $\frac{0.5y}{0.5} > \frac{x}{0.5} - \frac{3}{0.5}$. Simplify the expressions: $y > \frac{x}{0.5} - \frac{3}{0.5}$. Simplify terms: $y > 2x - 6$. The inequality is in the proper slope/y-intercept form: $m = 2 = \frac{2}{1} = \frac{\text{change in } y}{\text{change in } x}$. $b = -6$. The y-intercept is at the point $(0, -6)$. A change in y of 2 and in x of 1 gives the point $(0 + 1, -6 + 2)$ or $(1, -4)$. Draw a dashed boundary line and shade above it.

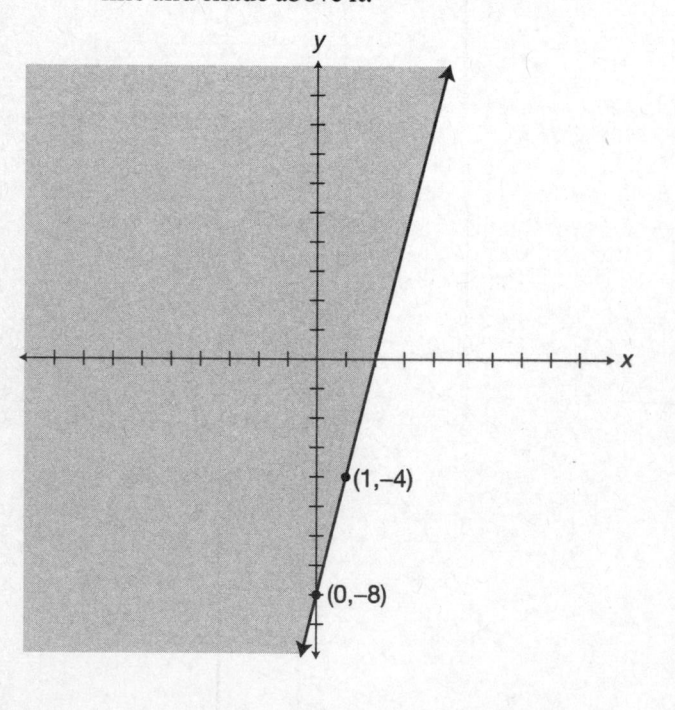

573. Subtract x from both sides of the inequality: $x - y - x \leq 7 - x$. Use the commutative property of addition to associate like terms: $x - x - y \leq 7 - x$. Simplify the expression: $-y \leq 7 - x$. Multiply both sides of the inequality by -1 and change the direction of the inequality symbol: $(-1)(-y) \geq (-1)(7 - x)$. Use the distributive property of multiplication: $(-1)(-y) \geq (-1)(7) - (-1)(x)$. Simplify terms: $y \geq -7 + x$. Use the commutative property of addition: $y \geq x - 7$. The inequality is in the proper slope/y-intercept form: $m = 1 = \frac{1}{1} = \frac{\text{change in } y}{\text{change in } x}$. $b = -7$. The y-intercept is at the point $(0,-7)$. A change in y of 1 and in x of 1 gives the point $(0 + 1, -7 + 1)$ or $(1,-6)$. Draw a solid boundary line and shade above it.

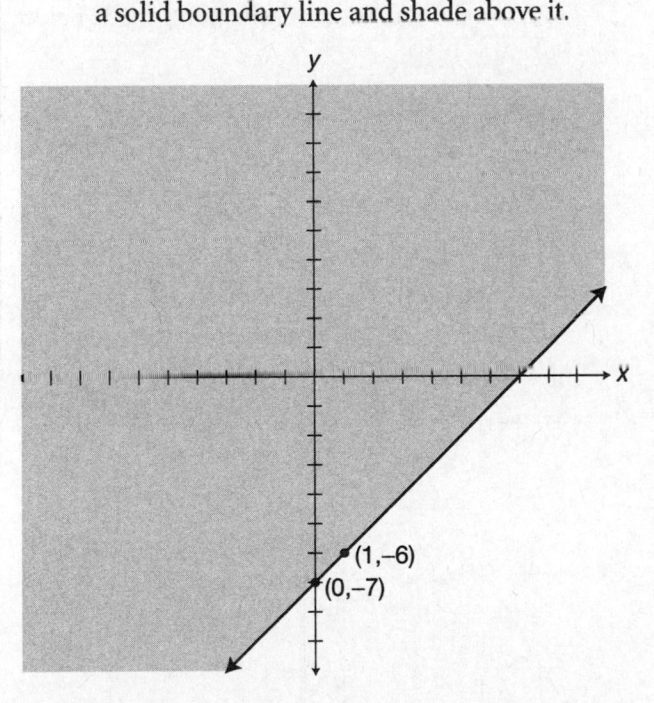

574. Multiply both sides of the inequality by 3: $3(\frac{y}{3}) < 3(\frac{2}{3} - x)$. Use the distributive property of multiplication: $3(\frac{y}{3}) < 3(\frac{2}{3}) - 3(x)$. Simplify terms: $y < 2 - 3x$. Use the commutative property of addition: $y < -3x + 2$. The inequality is in the proper slope/y-intercept form: $m = -3 = \frac{-3}{1} = \frac{\text{change in } y}{\text{change in } x}$. $b = 2$. The y-intercept is at the point $(0,2)$. A change in y of -3 and in x of 1 gives the point: $(0 + 1, 2 - 3)$ or $(1,-1)$. Draw a dashed boundary line and shade below it.

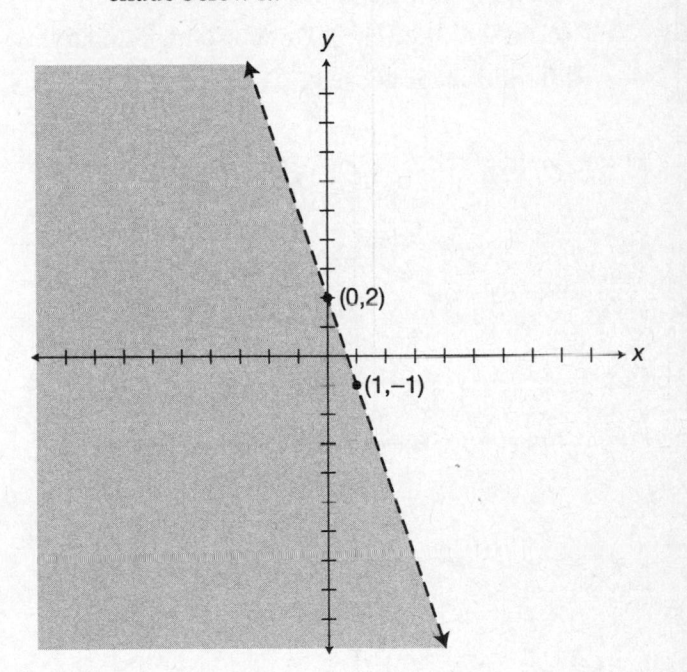

575. Subtract $9x$ from both sides of the inequality: $-3y + 9x - 9x \leq -6 - 9x$. Combine like terms: $^-3y \leq -6 - 9x$. Divide both sides of the inequality by -3 and change the direction of the inequality symbol: $\frac{-3y}{-3} \geq \frac{-6}{-3} - \frac{9x}{-3}$. Simplify the terms: $y \geq \frac{-6}{-3} - (\frac{9x}{-3})$; $y \geq 2 - (-3x)$; $y \geq 2 + 3x$. Use the commutative property of addition: $y \geq 3x + 2$. The inequality is in the proper slope/y-intercept form: $m = 3 = \frac{3}{1} = \frac{\text{change in } y}{\text{change in } x}$. $b = 2$. The y-intercept is at the point $(0,2)$. A change in y of 3 and in x of 1 gives the point $(0 + 1, 2 + 3)$ or $(1,5)$. Draw a solid boundary line and shade above it.

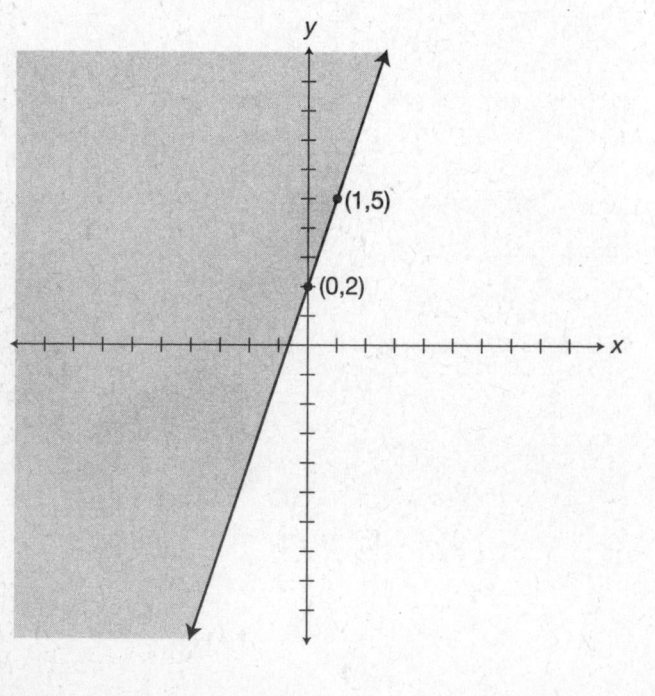

576. Subtract $0.3y$ from both sides of the inequality: $0.5x - 0.3y > 0.3y - 0.3y - 0.9$. Combine like terms: $0.5x - 0.3y > -0.9$. Subtract $0.5x$ from both sides of the inequality: $0.5x - 0.5x - 0.3y > -0.5x - 0.9$. Combine like terms: $-0.3y > -0.5x - 0.9$. Divide both sides of the inequality by -0.3 and change the direction of the inequality symbol: $\frac{-0.3y}{-0.3} < \frac{-0.5x}{-0.3} - \frac{0.9}{-0.3}$. Simplify the terms: $y < \frac{5}{3}x - (-3)$. Subtracting a negative number is the same as adding a positive: $y < \frac{5}{3}x + 3$. The inequality is in the proper slope/y-intercept form: $m = \frac{5}{3} = \frac{\text{change in } y}{\text{change in } x}$. $b = 3$. The y-intercept is at the point $(0,3)$. A change in y of 3 and in x of 1 gives the point $(0 + 3, 3 + 5)$ or $(3,8)$. Draw a dashed boundary line and shade below it.

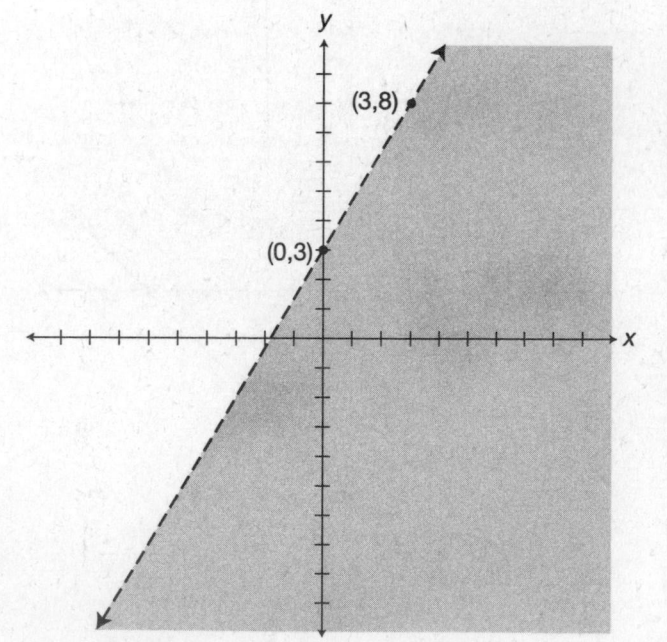

577. Subtract $3x$ from both sides of the inequality: $3x - 3x - y \leq 7x - 3x + y - 8$. Combine like terms: $-y \leq 4x + y - 8$. Subtract y from both sides of the inequality: $-y - y \leq 4x + y - y - 8$. Combine like terms: $-2y \leq 4x - 8$. Divide both sides of the inequality by -2 and change the direction of the inequality symbol: $\frac{-2y}{-2} \geq \frac{4x-8}{-2}$; $y \geq \frac{4x}{-2} - (\frac{8}{-2})$. Simplify terms: $y \geq -2x - (-4)$. Simplify: $y \geq -2x + 4$. The inequality is in the proper slope/y-intercept form: $m = -2 = \frac{-2}{1} = \frac{\text{change in } y}{\text{change in } x}$. $b = 4$. The y-intercept is at the point $(0,4)$. A change in y of -2 and in x of 1 gives the point $(0 + 1, 4 - 2)$ or $(1,2)$. Draw a solid boundary line and shade above it.

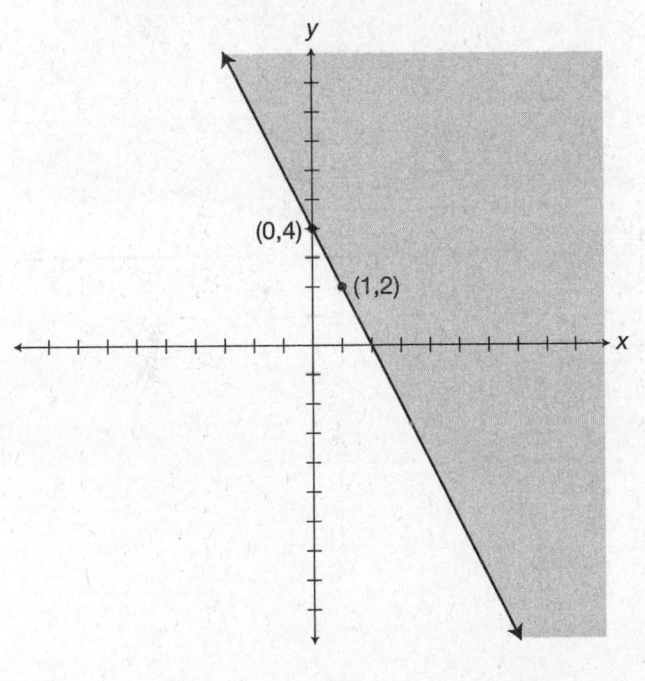

578. Subtract $4x$ from both sides of the inequality: $3y + 4x - 4x < 9 - 2x - 4x$. Combine like terms: $3y < 9 - 6x$. Divide both sides of the inequality by 3: $\frac{3y}{3} < \frac{9}{3} - \frac{6x}{3}$. Simplify the expressions: $y < (\frac{9}{3}) - (\frac{6x}{3})$. Simplify the terms: $y < 3 - 2x$. Use the commutative property: $y < -2x + 3$. The inequality is in the proper slope/y-intercept form: $m = \frac{-2}{1} = \frac{\text{change in } y}{\text{change in } x}$. $b = 3$. The y-intercept is at the point $(0,3)$. A change in y of -2 and in x of 1 gives the point $(0 + 1, 3 - 2)$ or $(1,1)$. Draw a dashed boundary line and shade below it.

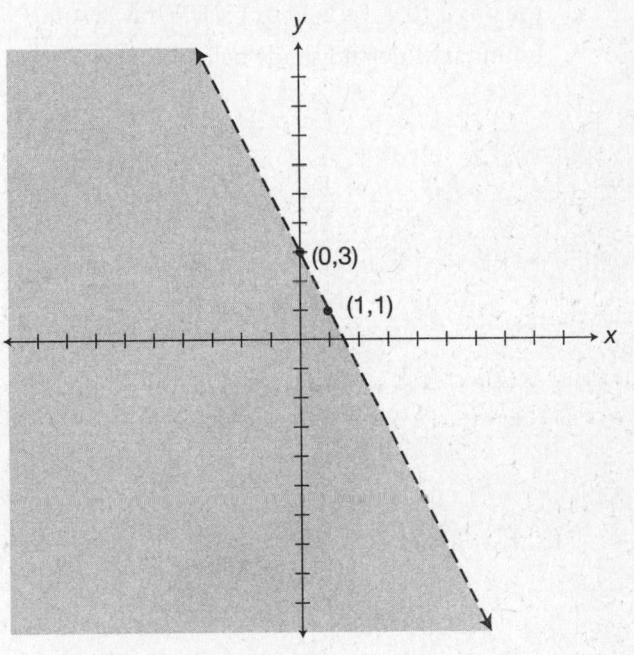

579. Use the distributive property of multiplication: $-12 \leq -3x - 3y$. Add $3y$ to both sides of the inequality: $3y - 12 \leq -3x - 3y + 3y$. Combine like terms: $3y - 12 \leq -3x$. Add 12 to both sides of the inequality: $3y - 12 + 12 \leq -3x + 12$. Combine like terms: $3y \leq -3x + 12$. Divide both sides of the inequality by 3: $\frac{3y}{3} \leq \frac{-3x}{3} + \frac{12}{3}$. Simplify the expressions: $y \leq -x + 4$. The inequality is in the proper slope/y-intercept form: $m = -1 = \frac{-1}{1} = \frac{\text{change in } y}{\text{change in } x}$. $b = 4$. The y-intercept is at the point $(0,4)$. A change in y of -1 and in x of 1 gives the point $(0 + 1, 4 - 1)$ or $(1,3)$. Draw a solid boundary line and shade below it.

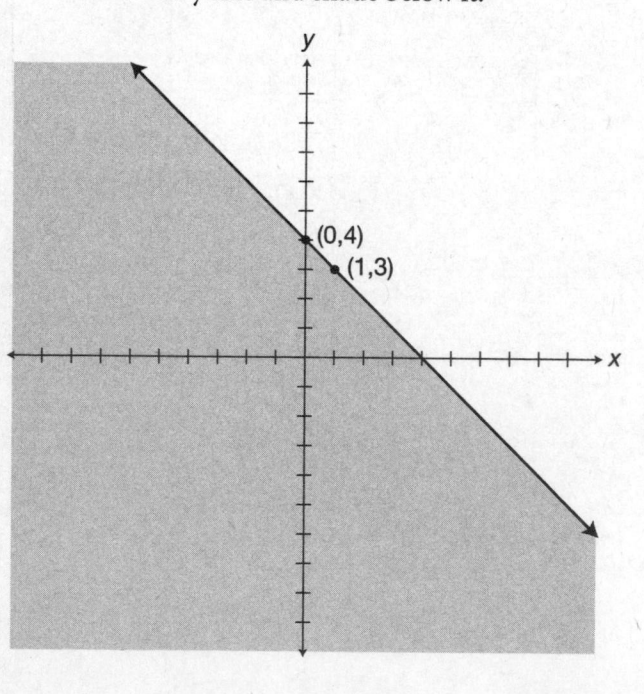

580. Use the distributive property of multiplication: $9y + 7 \geq 2x + 16$. Subtract 7 from both sides of the inequality: $9y + 7 - 7 \geq 2x + 16 - 7$. Combine like terms: $9y \geq 2x + 9$. Divide both sides of the inequality by 9: $\frac{9y}{9} \geq \frac{2x}{9} + \frac{9}{9}$. Simplify the expressions: $y \geq \frac{2}{9}x + 1$. The inequality is in the proper slope/y-intercept form: $m = \frac{2}{9} = \frac{\text{change in } y}{\text{change in } x}$. $b = 1$. The y-intercept is at the point $(0,1)$. A change in y of 2 and in x of 9 gives the point $(0 + 9, 1 + 2)$ or $(9,3)$. Draw a solid boundary line and shade above it.

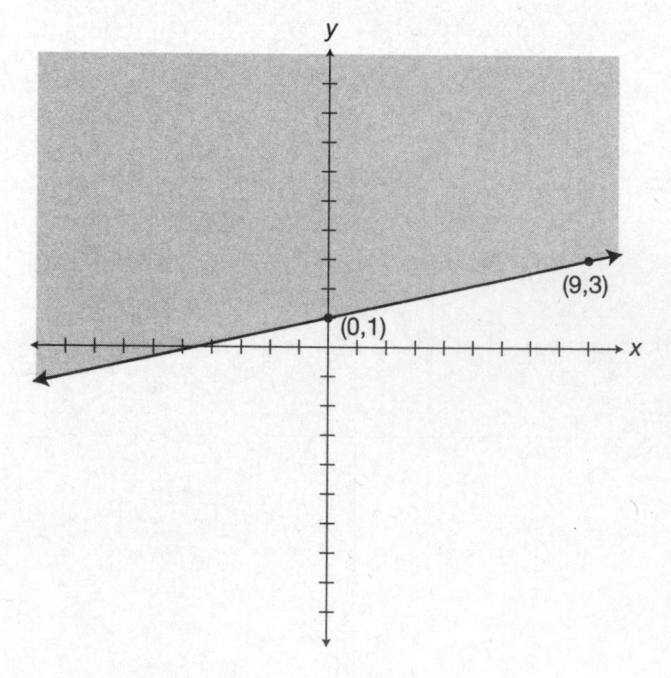

581. Multiply both sides of the inequality by 3: $3(\frac{x}{3} + y) \le 3(3x - 5)$. Use the distributive property of multiplication: $3(\frac{x}{3}) + 3(y) \le 3(3x) - 3(5)$. Simplify terms: $x + 3y \le 9x - 15$. Subtract x from both sides of the inequality: $x - x + 3y \le 9x - x - 15$. Combine like terms: $3y \le 8x - 15$. Divide both sides of the inequality by 3: $\frac{3y}{3} \le \frac{8x}{3} - \frac{15}{3}$. Simplify the expressions: $y \le \frac{8}{3}x - 5$. The inequality is in the proper slope/y-intercept form: $m = \frac{8}{3} = \frac{\text{change in } y}{\text{change in } x}$. $b = -5$. The y-intercept is at the point $(0,-5)$. A change in y of 8 and in x of 3 gives the point $(0 + 3, -5 + 8)$ or $(3,3)$. Draw a solid boundary line and shade below it.

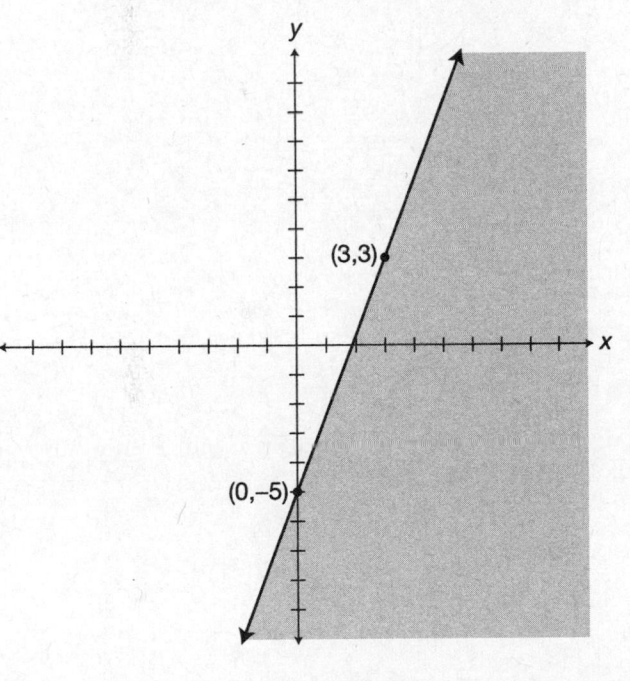

582. Use the distributive property of multiplication: $2(y) + 2(3) - x \ge 6(1) - 6(x)$. Simplify terms: $2y + 6 - x \ge 6 - 6x$. Add x to both sides of the inequality: $2y + 6 - x + x \ge 6 - 6x + x$. Combine like terms: $2y + 6 \ge 6 - 5x$. Subtract 6 from both sides of the inequality: $2y + 6 - 6 \ge 6 - 5x - 6$. Use the commutative property with like terms: $2y + 6 - 6 \ge 6 - 6 - 5x$. Combine like terms: $2y \ge -5x$. Divide both sides of the inequality by 2: $\frac{2y}{2} \ge \frac{-5x}{2}$. Simplify the terms: $y \ge \frac{-5}{2}x$. The inequality is in the proper slope/y-intercept form: $m = \frac{-5}{2} = \frac{\text{change in } y}{\text{change in } x}$. $b = 0$. The y-intercept is at the point $(0,0)$. A change in y of -5 and in x of 2 gives the point $(0 + 2, 0 - 5)$ or $(2,-5)$. Draw a solid boundary line and shade above it.

583. Use the distributive property of multiplication: $-28y \geq 2x - 14(y) - 14(10)$. Simplify terms: $-28y \geq 2x - 14y - 140$. Add $14y$ to both sides of the inequality: $28y + 14y \geq 2x - 14y + 14y - 140$. Combine like terms on each side of the inequality: $-14y \geq 2x - 140$. Divide both sides of the inequality by -14 and change the direction of the inequality symbol: $\frac{-14y}{-14} \leq \frac{2x}{-14} - \frac{140}{-14}$. Simplify the terms: $y \leq \frac{-1}{7}x + 10$. The inequality is in the proper slope/y-intercept form: $m = \frac{-1}{7} = \frac{\text{change in } y}{\text{change in } x}$. $b = 10$. The y-intercept is at the point $(0,10)$. A change in y of -1 and in x of 7 gives the point $(0 + 7, 10 - 1)$ or $(7,9)$. Draw a solid boundary line and shade below it.

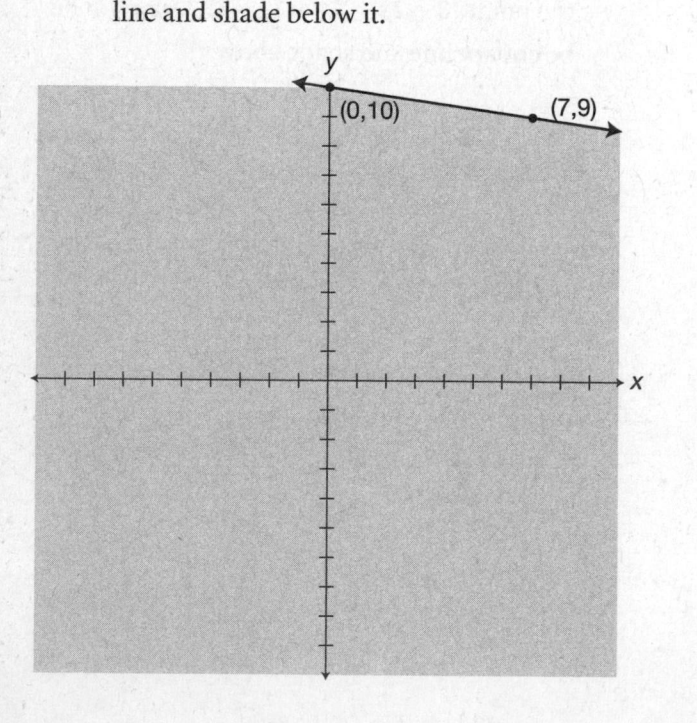

9 ▶ Systems of Equations

WE'VE SEEN HOW TO SOLVE a single equation with one variable. To find the values of two variables, we need two equations. A pair or set of related equations is called a **system of equations**. The values of each unique variable in a system of equations are the same in each equation.

Example

$3x + 2y = 6$

$-x + 5y = -19$

These two equations together are a system of equations. In both equations, $x = 4$ and $y = -3$. How do we know that? There are three ways to solve a system of equations: the elimination method, the substitution method, and by using a graph to find a point of intersection.

▶ Elimination Method

In this method, we add the two equations together to eliminate one variable. Usually, before we can add, we must multiply or divide one equation by a constant, so that when we add the equations together, one variable drops out. This will leave us with one variable and one equation—and we know how to solve those. After solving for the value of that variable, substitute the value into either of the two equations to find the value of the other variable.

Example

$2x - 4y = 4$

$x + y = 5$

If we add these two equations together, we will still have two variables, x and y. We need to change one of them so that when they are added, either x or y disappears. If we multiply the second equation by 4, it becomes $4x + 4y = 20$. Because we have multiplied both sides of the equation by 4, we have not changed the value of the equation. Now we are ready to add the equations:

$$
\begin{array}{r}
2x - 4y = 4 \\
+\ 4x + 4y = 20 \\
\hline
6x \phantom{{}-4y} = 24
\end{array}
$$

To solve $6x = 24$, we divide both sides of the equation by 6 and find that $x = 4$. Now that we have the value of one variable, we can substitute it into either equation to find the value of the other variable. Let's use the second equation: $4 + y = 5$, so $y = 1$. The solution to this system of equations is $(4,1)$. Check the answer by substituting the solution into both equations:

$2x - 4y = 4$; $2(4) - 4(1) = 4$; $8 - 4 = 4$; $4 = 4$

$x + y = 5$; $4 + 1 = 5$; $5 = 5$

Both equations are true, so our answer is correct.

PITFALL

When multiplying or dividing an equation in a system, be sure to multiply and divide both sides of the equation. It is easy to forget to multiply or divide the side of the equation that does not contain variables, but this will lead to an incorrect answer.

▶ Substitution Method

Instead of adding the equations together, we can solve for one variable in terms of the other. For example, we could write one equation in the form $y =$, and then take the expression that is equal to y and use it in place of y in the other equation. That will give us an equation with only one variable. After solving for the value of x, we can use the value of x to find the value of y in either equation.

Example

$-7x - 5y = 4$

$y - x = 4$

We can rewrite either equation so that it is in the form $y =$ or the form $x =$. Let's solve the second equation for y in terms of x. Why? Because it can be done easily— just add x to both sides of the equation, and $y = x + 4$. Now that we have the value of y in terms of x, substitute that expression for y in the first equation and solve for x:

$-7x - 5(x + 4) = 4$

$-7x - 5x - 20 = 4$

$-12x - 20 = 4$

$-12x = 24$

$x = -2$

Now that we have the value of x, we can use it to find the value of y. Substitute -2 for x in either equation. Let's use the second equation, because it will be easier to solve:

$y - (-2) = 4$

$y + 2 = 4$

$y = 2$

The solution to this system of equations is $(-2, 2)$. Check the solution:

$-7(-2) - 5(2) = 4; 14 - 10 = 4; 4 = 4$

$(2) - (-2) = 4; 2 + 2 = 4; 4 = 4$

TIP

Before beginning a system of equations problem, look to see whether combining the equations is easier or solving for one variable in terms of the other is easier. Which method to use is your choice, so don't make the problem difficult for yourself! If you choose substitution, choose the equation and variable that are easier to manipulate.

▶ Practice Questions

Use the elimination method to solve the following systems of equations.

584. $x + y = 4$
$2x - y = -1$

585. $3x + 4y = 17$
$-x + 2y = 1$

586. $7x + 3y = 11$
$2x + y = 3$

587. $0.5x + 5y = 28$
$3x - y = 13$

588. $3(x + y) = 18$
$5x + y = -2$

589. $\frac{1}{2}x + 2y = 11$
$2x - y = 17$

590. $5x + 8y = 25$
$3x - 15 = y$

591. $6y + 3x = 30$
$2y + 6x = 0$

592. $3x = 5 - 7y$
$2y = x - 6$

593. $3x + y = 20$
$\frac{x}{3} + 10 = y$

594. $2x + 7y = 45$
$3x + 4y = 22$

595. $3x - 5y = -21$
$2(2y - x) = 16$

596. $\frac{1}{4}x + y = 12$

$2x - \frac{1}{3}y = 21$

Use the substitution method to solve the following systems of equations.

597. $y = 5x$

$4x + 5y = 87$

598. $x + y = 3$

$3x + 101 = 7y$

599. $5x + y = 3.6$

$y + 21x = 8.4$

600. $8x - y = 0$

$10x + y = 27$

601. $\frac{x}{3} = y + 2$

$2x - 4y = 32$

602. $y + 3x = 0$

$y - 3x = 24$

603. $5x + y = 20$

$3x = \frac{1}{2}y + 1$

604. $2x + y = 2 - 5y$

$x - y = 5$

605. $\frac{2x}{10} + \frac{y}{5} = 1$

$3x + 2y = 12$

606. $x + 6y = 11$

$x - 3 = 2y$

607. $4y + 31 = 3x$

$y + 10 = 3x$

608. $2(2 - x) = 3y - 2$

$3x + 9 = 4(5 - y)$

► Graphing a System of Equations

We can also graph the two equations in a system. The point at which they intersect is the solution to the system. However, if the two lines don't intersect, the system has no solution, and if the two lines are identical, then the solution is every point on the line. We can spot these two cases before graphing, though. If the two equations have the same slope but different y-intercepts, then they are parallel and will never cross—there is no solution to that system. If the two equations are of the same line, then both equations will reduce to the same equation.

Example

$y = 5x + 2$

$x + y = -4$

First, write the second equation in slope-intercept form. Divide both sides by $-x$, and $x + y = -4$ becomes $y = -x - 4$. Now graph both lines:

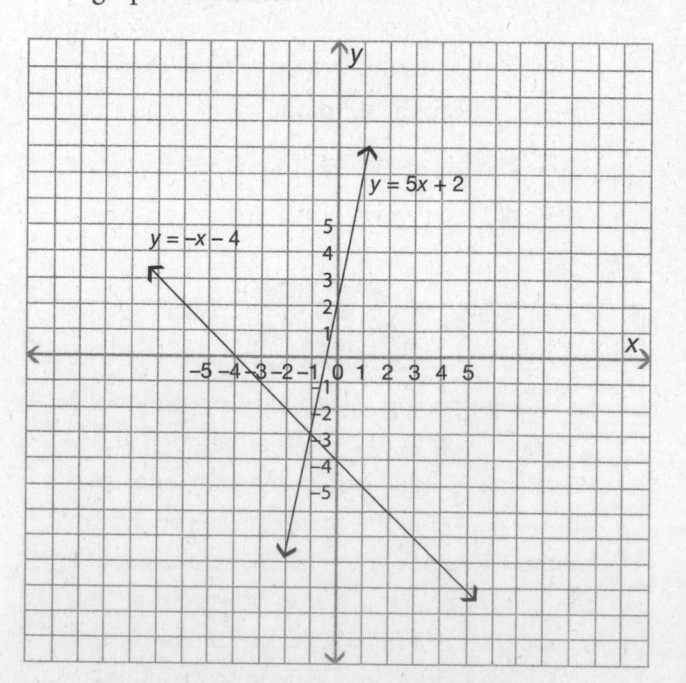

The lines intersect at the point $(-1, -3)$, which means that this is the solution to the system of equations. Substitute those values into the equations to check:

$-3 = 5(-1) + 2; -3 = -5 + 2; -3 = -3$

$(-1) + (-3) = -4; -4 = -4$

Example

$6x = 4 + 12y$

$y - \frac{1}{2}x = 8$

Begin by writing each equation in slope-intercept form: $6x = 4 + 12y$ becomes $y = \frac{1}{2}x - \frac{1}{3}$, and $y - \frac{1}{2}x = 8$ becomes $y = \frac{1}{2}x + 8$. These two lines have the same slope, but different y-intercepts. They will never intersect, so there is no solution to this system of equations.

Example

$3y = 9x + 12$

$-3x + y = 4$

Again, begin by writing each equation in slope-intercept form: $3y = 9x + 12$ becomes $y = 3x + 4$, and $-3x + y = 4$ becomes $y = 3x + 4$, too. These two lines are the same, which means that the solution is all points on that line.

▶ **Practice Questions**

Find the solutions for the following systems of equations by graphing on graph paper.

609. $y = x + 4$
$y = -x + 2$

610. $2y - x = 2$
$3x + y = 8$

611. $4y = -7(x + 4)$
$4y = x + 4$

612. $y - x = 5 - x$
$-4y = 8 - 7x$

613. $2y = 6x + 14$
$4y = x - 16$

614. $2x + y = 4$
$3(y + 9) = 7x$

615. $y = x + 9$
$4y = 16 - x$

616. $4x - 5y = 5$
$5y = 20 - x$

617. $6y = 9(x - 6)$
$3(2y + 5x) = -6$

618. $15y = 6(3x + 15)$
$y = 6(1 - x)$

619. $3y = 6x + 6$
$5y = 10(x - 5)$

620. $3(2x + 3y) = 63$

$27y = 9(x - 6)$

621. $x - 20 = 5y$

$10y = 8x + 20$

622. $3x + 4y = 12$

$y = 3 - \frac{6}{8}x$

623. $16y = 10(x - 8)$

$8y - 17x = 56$

▶ Graphing Systems of Inequalities

It would be very difficult to describe the solution of a system of inequalities without using a graph. If we used the elimination or substitution methods to find the value of one variable, we would have $x < c, y < c, x > c$, or $y > c$ (where c is some number). We couldn't substitute that for x or y into either equation, because

TIP

If no shaded area overlaps the other, then there is no solution to the system of inequalities. This happens when the plotted lines (dashed or solid) are parallel to each other. If two lines do not have the same slope, they will intersect at some point, even if that point doesn't appear to be on the coordinate plane that you have drawn. If the lines intersect, then the inequalities that they represent will have overlapping areas somewhere, and that is the solution set. If you plot inequalities with different slopes and you do not see the solution set, expand the size of your coordinate plane.

x or y would represent a whole set of values, not a single value. To show the solution set for a system of inequalities, we use a graph. Plot and shade for the first inequality, and then plot and shade for the second inequality. The area that is shaded by both inequalities is the solution to the system of inequalities.

In addition to no overlap, discussed in the "Tip" box, there are a few other special cases:

- If the two inequalities are the same in slope-intercept form, but the inequality symbols point in opposite directions (such as $y > x + 4$ and $y < x + 4$), then the system has no solution, as the two shaded areas will be opposite sides of the line.
- If the two inequalities are the same in slope-intercept form, but the inequality symbols point in opposite directions and only one symbol contains the equal sign (such as $y \geq x + 4$ and $y < x + 4$, or, $y > x + 4$ and $y \leq x + 4$), then the system has no solution, as the two shaded areas will be opposite sides of the line and the points on the line will not be part of the solution to the inequality that does not contain the equal sign.
- If the two inequalities are the same in slope-intercept form, but the inequality symbols point in opposite directions and contain the equal sign (such as $y \leq x + 4$ and $y \geq x + 4$), then the solution to the system is all points on the graphed line.

Example

$4y + 28 > x$

$1 \leq 2x - y$

Write each inequality in slope-intercept form: $4y + 28 > x$ is the same as $y > \frac{1}{4}x - 7$, and $1 \leq 2x - y$ is the same as $y \leq 2x - 1$. Graph $y > \frac{1}{4}x - 7$. The graph will have a dashed line, since y is greater than, not greater than or equal to, $\frac{1}{4}x - 7$. Test the inequality with point (0,0): $0 > \frac{1}{4}(0) - 7$, so $0 > -7$. The inequality holds true, so shade the area that includes the point (0,0), which is the area above the line. Now graph $y \leq 2x - 1$. Be sure to make the line solid, since this is a less-than-or-equal-to inequality. Test this inequality with (0,0): $0 \leq 2(0) - 1$, so $0 \leq -1$, which is not true. Shade the area that does not include the point (0,0), which is the area to the right of the line. The area that is shaded by both inequalities, which is mostly in the first and fourth quadrants, is the solution to the system of inequalities.

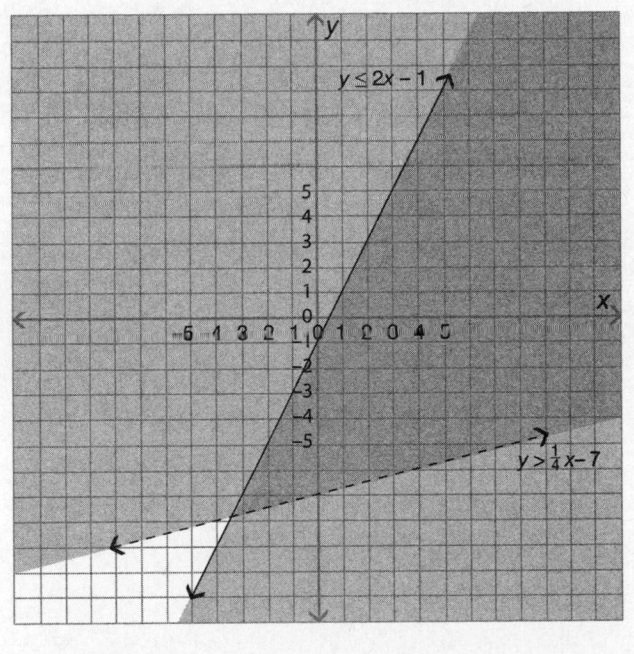

Example

$-6y < x - 18$

$2x + 12y < 36$

Write each inequality in slope-intercept form: $-6y < x - 18$ is the same as $y > -\frac{1}{6}x + 3$ (remember to switch the direction of the inequality symbol when dividing both

sides by a negative number), and $2x + 12y < 36$ is the same as $y < -\frac{1}{6}x + 3$. These two inequalities would be plotted along the same dashed line, but shaded to opposite sides, since the inequality symbols point in different directions. Since the points along the dashed line are not part of the solution set for either inequality, this system of inequalities has no solution.

▶ Practice Questions

Find the solution for each of the following systems of inequalities by graphing and shading.

624. $2y - 3x \geq -6$

$y \geq 5 - \frac{5}{2}x$

625. $6y < 5x - 30$

$2y < -x + 4$

626. $y - x \geq 6$

$11y \geq -2(x + 11)$

627. $5y \leq 8(x + 5)$

$5y \leq 12(5 - x)$

628. $2(x + 5y) > 5(x + 6)$

$4x + y < 4x + 5$

629. $3y \geq -2(x + 3)$

$3y \leq 2(6 - x)$

630. $9(y - 4) < 4x$

$-9y < 2(x + 9)$

631. $7(y-5) < -5x$

$-3 < \frac{1}{4}(2x - 3y)$

632. $y > \frac{7}{4}(4 - x)$

$3(y + 5) > 7x$

633. $5x - 2(y + 10) \leq 0$

$2x + y \leq -3$

TERM TO REVIEW

system of equations

▶ **Answers**

584. Add the equations: $x + y = 4$ and $2x - y = -1$. Total: $3x + 0 = 3$. Additive identity: $3x = 3$. Now solve for x. Divide both sides by 3: $\frac{3x}{3} = \frac{3}{3}$. Simplify terms: $x = 1$. Substitute the value of x into one of the equations in the system and solve for y: $(1) + y = 4$. Subtract 1 from both sides: $1 - 1 + y = 4 - 1$. Simplify: $y = 3$. The solution for the system of equations is $(1,3)$.

585. We could add the equations together if we had a $-3x$ in the second equation, so multiply the second equation by 3: $3(-x + 2y = 1)$. Simplify: $-3x + 6y = 3$. Add the first equation to the transformed second: $3x + 4y = 17$. Total: $0 + 10y = 20$. Identity element of addition: $10y = 20$. Divide both sides by 10: $y = 2$. Substitute the value of y into one of the equations in the system and solve for x: $-x + 2(2) = 1$. Simplify: $-x + 4 = 1$. Subtract 4 from both sides: $-x + 4 - 4 = 1 - 4$. Simplify: $-x = -3$. Multiply both sides by -1: $x = 3$. The solution for the system of equations is $(3,2)$.

586. Transform the second equation so you can add it to the first and eliminate y. Multiply the equation by -3: $-3(2x + y = 3)$. Simplify: $-6x - 3y = -9$. Add the first equation to the transformed second: $7x + 3y = 11$. Total: $x + 0 = 2$; $x = 2$. Substitute the value of x into one of the equations in the system and solve for y: $2(2) + y = 3$; $4 + y = 3$. Subtract 4 from both sides: $4 - 4 + y = 3 - 4$. Simplify: $y = -1$. The solution for the system of equations is $(2,-1)$.

587. If you multiply the second equation by 5 and add the equations together, you can eliminate the y: $5(3x - y) = 5(13)$. Use the distributive property of multiplication: $5(3x) - 5(y) = 5(13)$. Simplify terms: $15x - 5y = 65$. Add the first equation: $0.5x + 5y = 28$. Total: $15.5x + 0 = 93$. Additive identity: $15.5x = 93$. Divide both sides by 15.5: $x = 6$. Substitute the value of x into one of the equations in the system and solve for y: $3(6) - y = 13$. Simplify: $18 - y = 13$. Subtract 18 from both sides: $18 - 18 - y = 13 - 18$. Simplify: $-y = -5$. Multiply both sides by -1: $(-1)(-y) = (-1)(-5)$. Simplify: $y = 5$. The solution for the system of equations is $(6,5)$.

588. See what the first equation looks like after distributing the multiplication on the left. Use the distributive property of multiplication: $3x + 3y = 18$. Multiply the second equation by -3 and add the two equations to eliminate y: $-3(5x + y) = -3(-2)$. Use the distributive property of multiplication: $-3(5x) - 3(y) = -3(-2)$. Simplify terms: $-15x - 3y = 6$. Add the transformed first equation: $3x + 3y = 18$. Total: $-12x + 0 = 24$. Additive identity: $-12x = 24$. Divide both sides by -12: $x = -2$. Substitute the value of x into one of the equations in the system and solve for y: $5(-2) + y = -2; -10 + y = -2$. Add 10 to both sides: $10 - 10 + y = 10 - 2$. Simplify: $y = 8$. The solution for the system of equations is $(-2,8)$.

589. Multiply the second equation by 2 and add to the first to eliminate y: $2(2x - y) = 2(17)$. Use the distributive property of multiplication: $2(2x) - 2(y) = 2(17)$. Simplify: $4x - 2y = 34$. Add the first equation to the second: $\frac{1}{2}x + 2y = 11$. Total: $4\frac{1}{2}x + 0 = 45$. Additive identity: $4\frac{1}{2}x = 45$. Multiply the equation by 2 to simplify the fraction: $2(4\frac{1}{2}x = 45)$. Simplify: $9x = 90$. Divide both sides by 9: $x = 10$. Substitute the value of x into one of the equations in the system and solve for y: $2(10) - y = 17$. Subtract 20 from both sides: $20 - 20 - y = 17 - 20$. Combine like terms on each side: $-y = -3$. Multiply the equation by -1: $y = 3$. The solution for the system of equations is $(10,3)$.

590. Transform the second equation into a similar format to the first equation, then line up like terms. Add 15 to both sides: $3x - 15 + 15 = y + 15$. Simplify: $3x = y + 15$. Subtract y from both sides: $3x - y = y - y + 15$. Simplify: $3x - y = 15$. Multiply the second equation by 8 and add the first equation to the second: $8(3x - y) = 15$. Use the distributive property of multiplication: $24x - 8y = 120$. Add the first equation to the second: $5x + 8y = 25$. Total: $29x + 0 = 145$. Additive identity: $29x = 145$. Divide both sides by 29: $x = 5$. Substitute the value of x into one of the equations in the system and solve for y: $3(5) - 15 = y$. Simplify: $15 - 15 = y; 0 = y$. The solution for the system of equations is $(5,0)$.

591. Multiply the second equation by -3 and add it to the first equation to eliminate y: $-3(2y + 6x = 0)$ becomes $-6y - 18x = 0$; add to $6y + 3x = 30$. Total: $0 - 15x = 30$. Additive identity. Divide both sides by -15: $x = -2$. Substitute the value of x into one of the equations in the system and solve for y: $2y + 6(-2) = 0$; $2y - 12 = 0$. Add 12 to both sides: $2y - 12 + 12 = 0 + 12$. Simplify: $2y = 12$. Divide both sides by 2: $y = 6$. The solution for the system of equations is $(-2,6)$.

592. Transform the first equation into familiar form ($ax + by = c$): $3x = 5 - 7y$. Add $7y$ to both sides: $3x + 7y = 5 - 7y + 7y$. Simplify: $3x + 7y = 5$. Transform the second equation into familiar form ($ax + by = c$): $2y = x - 6$. Subtract x from both sides: $-x + 2y = x - x - 6$. Simplify: $-x + 2y = -6$. Multiply equation by 3: $3(-x + 2y = -6)$; $3(-x) + 3(2y) = 3(-6)$. Simplify terms: $-3x + 6y = -18$. Add the transformed first equation: $3x + 7y = 5$. Total: $0 + 13y = -13$. Additive identity. Divide both sides by 13: $y = -1$. Substitute the value of y into one of the equations in the system and solve for x: $3x = 5 - 7(-1)$; $3x = 5 + 7$; $3x = 12$. Divide both sides by 3: $x = 4$. The solution for the system of equations is $(4, -1)$.

593. Transform the second equation into familiar form ($ax + by = c$). Multiply the equation by 3: $3(\frac{x}{3} + 10 = y)$. Use the distributive property: $3(\frac{x}{3}) + 3(10) = 3(y)$. Simplify terms: $x + 30 = 3y$. Subtract 30 from each side: $x + 30 - 30 = 3y - 30$. Simplify: $x = 3y - 30$. Subtract $3y$ from both sides: $x - 3y = 3y - 3y - 30$. Simplify: $x - 3y = -30$. Multiply the first equation by 3 and add to the second equation to eliminate y: $3(3x + y = 20)$. Use the distributive property: $3(3x) + 3(y) = 3(20)$ Simplify terms: $9x + 3y = 60$. Add the transformed second equation to the first: $x - 3y = -30$. Total: $10x + 0 = 30$. Additive identity. Divide both sides by 10: $x = 3$. Substitute the value of x into one of the equations in the system and solve for y: $\frac{(3)}{3} + 10 = y$. Simplify terms: $1 + 10 = y$; $11 = y$. The solution for the system of equations is $(3, 11)$.

594. Transform the first equation by multiplying by 3, the second by multiplying by -2, and eliminate the x variable by adding the equations together: $3(2x + 7y = 45)$. Use the distributive property: $3(2x) + 3(7y) = 3(45)$. Simplify terms: $6x + 21y = 135$. Second equation: $-2(3x + 4y = 22)$. Use the distributive property: $-2(3x) - 2(4y) = -2(22)$. Simplify terms: $-6x - 8y = -44$. Add the transformed first equation to the second: $6x + 21y = 135$. Total: $0 + 13y = 91$. Additive identity. Divide both sides by 13: $y = 7$. Substitute the value of y into one of the equations in the system and solve for x: $3x + 4(7) = 22$. Simplify terms: $3x + 28 = 22$. Subtract 28 from both sides: $3x + 28 - 28 = 22 - 28$. Simplify: $3x = -6$. Divide both sides by 3: $x = -2$. The solution for the system of equations is $(-2, 7)$.

595. Transform the second equation into a similar form to the first equation. Use the distributive property of multiplication: $2(2y) - 2(x) = 16$. Simplify terms: $4y - 2x = 16$. Commutative property of addition: $-2x + 4y = 16$. Multiply the first equation by 2 and the second equation by 3, and add the transformed equations to eliminate the variable x: $2(3x - 5y = -21)$. Use the distributive property. Simplify terms: $6x - 10y = -42$; $3(-2x + 4y = 16)$. Use the distributive property. Simplify terms: $-6x + 12y = 48$. Add the first equation to the second: $6x - 10y = -42$. Total: $2y = 6$. Divide both sides by 2: $y = 3$. Substitute the value of y into one of the equations in the system and solve for x: $3x - 5(3) = -21$. Simplify terms and add 15 to each side: $3x - 15 + 15 = -21 + 15$. Combine like terms on each side: $3x = -6$. Divide both sides by 3: $x = -2$. The solution for the system of equations is $(-2, 3)$.

596. Transform the second equation by multiplying it by 3. Then, add the equations together to eliminate y: $3(2x - \frac{1}{3}y = 21)$. Use the distributive property of multiplication: $3(2x) - 3(\frac{1}{3}y) = 3(21)$. Simplify terms: $6x - y = 63$. Add the first equation to the second: $\frac{1}{4}x + y = 12$. Total: $6\frac{1}{4}x + 0 = 75$. Additive identity: $6\frac{1}{4}x = 75$. Divide both sides by $6\frac{1}{4}$: $\frac{6\frac{1}{4}x}{6\frac{1}{4}} = \frac{75}{6\frac{1}{4}}$. Simplify: $x = 12$. Substitute the value of x into one of the equations in the system and solve for y: $\frac{1}{4}(12) + y = 12$. Simplify the first term and subtract from both sides: $3 - 3 + y = 12 - 3$. Simplify: $y = 9$. The solution for the system of equations is $(12,9)$.

597. The first equation tells you that $y = 5x$. Substitute $5x$ for y in the second equation and then solve for x: $4x + 5(5x) = 87$. Simplify term and add like terms: $4x + 25x = 87$; $29x = 87$. Divide both sides by 29: $x = \frac{87}{29} = 3$. Substitute 3 for x in one of the equations: $y = 5 \times (3) = 15$. The solution for the system of equations is $(3,15)$.

598. Transform the first equation so that the value of x is expressed in terms of y. Subtract y from both sides of the equation: $x + y - y = 3 - y$. Simplify: $x = 3 - y$. Substitute $3 - y$ for x in the second equation and solve for y: $3(3 - y) + 101 = 7y$. Use the distributive property of multiplication: $9 - 3y + 101 = 7y$. Use the commutative property of addition: $9 + 101 - 3y = 7y$. Add like terms. Add $3y$ to both sides: $110 - 3y + 3y = 7y + 3y$. Combine like terms: $110 = 10y$. Divide both sides by 10: $11 = y$. Substitute the value of y into one of the equations in the system and solve for x: $x + (11) = 3$. Subtract 11 from both sides: $x + 11 - 11 = 3 - 11$. Combine like terms on each side: $x = -8$. The solution for the system of equations is $(-8,11)$.

599. Transform the first equation so that y is expressed in terms of x: $5x + y = 3.6$. Subtract $5x$ from both sides of the equation: $5x - 5x + y = 3.6 - 5x$. Combine like terms on each side: $y = 3.6 - 5x$. Substitute the value of y into the second equation: $(3.6 - 5x) + 21x = 8.4$. Combine like terms: $3.6 + 16x = 8.4$. Subtract 3.6 from both sides: $3.6 - 3.6 + 16x = 8.4 - 3.6$. Combine like terms on each side: $16x = 4.8$. Divide both sides by 16: $x = 0.3$. Substitute the value of x into one of the equations in the system and solve for y: $5(0.3) + y = 3.6$. Simplify terms: $1.5 + y = 3.6$. Subtract 1.5 from both sides: $1.5 - 1.5 + y = 3.6 - 1.5$. Combine like terms on each side: $y = 2.1$. The solution for the system of equations is $(0.3,2.1)$.

600. Transform the first equation so that y is expressed in terms of x: $8x - y = 0$. Add y to both sides of the equation: $8x + y - y = y + 0$. Combine like terms on each side and simplify: $8x = y$. Substitute the value of y into the second equation: $10x + 8x = 27$. Combine like terms: $18x = 27$. Divide both sides by 18: $x = \frac{3}{2}$. Substitute the value of x into one of the equations in the system and solve for y: $8(\frac{3}{2}) - y = 0$. Simplify terms: $12 - y = 0$. Add y to both sides of the equation: $12 - y + y = 0 + y$. Simplify: $12 = y$. The solution for the system of equations is $(\frac{3}{2}, 12)$.

601. Transform the first equation so that the value of x is expressed in terms of y. Multiply the equation by 3: $3[(\frac{x}{3}) = y + 2]$. Use the distributive property: $3(\frac{x}{3}) = 3y + 6$. Simplify: $x = 3y + 6$. Substitute the value of x into the second equation in the system and solve for y: $2(3y + 6) - 4y = 32$. Use the distributive property of multiplication: $6y + 12 - 4y = 32$. Use the commutative property of addition: $6y - 4y + 12 = 32$. Combine like terms. Subtract 12 from both sides: $2y + 12 - 12 = 32 - 12$. Combine like terms on each side: $2y = 20$. Divide both sides by 2: $y = 10$. Substitute the value of y into one of the equations in the system and solve for x: $2x - 4(10) = 32$. Simplify and add 40 to both sides: $2x - 40 + 40 = 32 + 40$. Combine like terms: $2x = 72$. Divide both sides by 2: $x = 36$. The solution for the system of equations is $(36, 10)$.

602. Express y in terms of x in the first equation: $y + 3x = 0$. Subtract $3x$ from both sides: $y + 3x - 3x = 0 - 3x$. Combine like terms and simplify: $y = -3x$. Substitute the value of y into the second equation in the system and solve for x: $(-3x) - 3x = 24$. Combine like terms: $-6x = 24$. Divide both sides by -6: $x = -4$. Substitute the found value for x into one of the equations and solve for y: $y + 3(-4) = 0$. Simplify: $y - 12 = 0$. Add 12 to both sides: $y = 12$. The solution for the system of equations is $(-4, 12)$.

603. Transform the first equation so that the value of y is expressed in terms of x. Subtract $5x$ from both sides of the equation: $5x - 5x + y = 20 - 5x$. Combine like terms: $y = 20 - 5x$. Substitute the value of y into the second equation in the system and solve for x: $3x = \frac{1}{2}(20 - 5x) + 1$. Use the distributive property of multiplication: $3x = 10 - \frac{5}{2}x + 1$. Combine like terms: $3x = 11 - \frac{5}{2}x$. Add $\frac{5}{2}x$ to both sides: $3x + \frac{5}{2}x = 11 + \frac{5}{2}x - \frac{5}{2}x$. Combine like terms: $5\frac{1}{2}x = 11$. Divide both sides by $5\frac{1}{2}$: $\frac{5\frac{1}{2}x}{5\frac{1}{2}} = \frac{11}{5\frac{1}{2}}$. Simplify terms: $x = 2$. Substitute the found value for x into one of the equations and solve for y: $5(2) + y = 20$. Simplify: $10 + y = 20$. Subtract 10 from both sides: $y = 10$. The solution for the system of equations is $(2, 10)$.

604. Transform the second equation so that the value of x is expressed in terms of y. Add y to both sides of the equation: $x - y + y = 5 + y$. Combine like terms on each side: $x = 5 + y$. Substitute the value of x into the second equation in the system and solve for y: $2(5 + y) + y = 2 - 5y$. Use the distributive property of multiplication: $10 + 2y + y = 2 - 5y$. Add $5y$ to both sides of the equation: $10 + 2y + y + 5y = 2 - 5y + 5y$. Combine like terms on each side: $10 + 8y = 2$. Subtract 10 from both sides: $10 - 10 + 8y = 2 - 10$. Combine like terms on each side: $8y = -8$. Divide both sides by 8: $y = -1$. Substitute the found value for y into one of the equations and solve for x: $x - (-1) = 5$. Simplify: $x + 1 = 5$; $x = 4$. The solution for the system of equations is $(4,-1)$.

605. Transform the first equation by eliminating the denominators. Multiply both sides of the equation by 10: $10(\frac{2x}{10} + \frac{y}{5}) = 10(1)$. Use the distributive property of multiplication: $10(\frac{2x}{10}) + 10(\frac{y}{5}) = 10$. Simplify terms: $2x + 2y = 10$. Divide both sides by 2: $\frac{2x + 2y}{2} = \frac{10}{2}$. Simplify terms: $x + y = 5$. Now express x in terms of y. Subtract y from both sides of the equation: $x + y - y = 5 - y$. Simplify: $x = 5 - y$. Substitute the value of x into the second equation and solve for y: $3(5 - y) + 2y = 12$. Use the distributive property of multiplication: $3(5) - 3y + 2y = 12$. Combine like terms on each side: $15 - y = 12$. Add y to both sides: $15 - y + y = 12 + y$. Combine like terms: $15 = 12 + y$. Subtract 12 from both sides: $15 - 12 = 12 - 12 + y$. Simplify: $3 = y$. Substitute the value of y into one of the equations in the system and solve for x: $3x + 2(3) = 12$. Simplify the term and subtract 6 from both sides: $3x + 6 - 6 = 12 - 6$. Combine like terms on each side: $3x = 6$. Divide both sides by 3: $x = 2$. The solution for the system of equations is $(2,3)$.

606. Transform the second equation so that the value of x is expressed in terms of y. Add 3 to both sides: $x - 3 + 3 = 2y + 3$. Combine like terms on each side: $x = 2y + 3$. Substitute the value of x into the first equation in the system and solve for y: $(2y + 3) + 6y = 11$. Use the commutative property of addition: $2y + 6y + 3 = 11$. Combine like terms: $8y + 3 = 11$. Subtract 3 from both sides: $8y + 3 - 3 = 11 - 3$. Combine like terms on each side: $8y = 8$. Divide both sides by 8: $y = 1$. Substitute the found value for y into the first equation and solve for x: $x + 6(1) = 11$. Subtract 6 from both sides: $x + 6 - 6 = 11 - 6$. Simplify: $x = 5$. The solution for the system of equations is $(5,1)$.

607. Transform the second equation so that the value of y is expressed in terms of x. Subtract 10 from both sides of the equation: $y + 10 - 10 = 3x - 10$. Combine like terms on each side: $y = 3x - 10$. Substitute the value of y into the first equation in the system and solve for x: $4(3x - 10) + 31 = 3x$. Use the distributive property of multiplication: $4(3x) - 4(10) + 31 = 3x$. Simplify terms: $12x - 40 + 31 = 3x$. Combine like terms on each side: $12x - 9 = 3x$. Add 9 to both sides of the equation: $12x - 9 + 9 = 3x + 9$. Combine like terms on each side: $12x = 3x + 9$. Subtract $3x$ from both sides: $12x - 3x = 3x - 3x + 9$. Combine like terms on each side: $9x = 9$. Divide both sides by 9: $x = 1$. Substitute the value of x into the second equation and solve for y: $y + 10 = 3(1)$. Subtract 10 from both sides: $y + 10 - 10 = 3 - 10$. Combine like terms on each side: $y = -7$. The solution for the system of equations is $(1,-7)$.

608. Begin with the second equation and express x in terms of y. Use the distributive property of multiplication: $3x + 9 = 20 - 4y$. Subtract 9 from both sides: $3x + 9 - 9 = 20 - 9 - 4y$. Combine like terms on each side: $3x = 11 - 4y$. Divide both sides by 3: $x = \frac{11 - 4y}{3}$. Substitute the value of x into the second equation and solve for y. First, use the distributive property to simplify the equation: $4 - 2x = 3y - 2$; $4 - 2(\frac{11 - 4y}{3}) = 3y - 2$. Multiply the numerator by the factor 2: $4 - (\frac{22 - 8y}{3}) = 3y - 2$. Multiply both sides of the equation by 3 to eliminate the denominator: $3[4 - (\frac{22 - 8y}{3})] = 3(3y - 2)$. Use the distributive property of multiplication: $3(4) - 3(\frac{22 - 8y}{3}) = 3(3y) - 3(2)$. Simplify each term: $12 - (22 - 8y) = 9y - 6$. Simplify the second term and the − sign: $12 - 22 + 8y = 9y - 6$. Combine like terms: $-10 + 8y = 9y - 6$. Add 6 to both sides: $6 - 10 + 8y = 9y + 6 - 6$. Combine like terms on each side: $-4 + 8y = 9y$. Subtract $8y$ from both sides: $-4 + 8y - 8y = 9y - 8y$. Combine like terms on each side: $-4 = y$. Substitute the value of y into the first equation in the system and solve for x: $2(2 - x) = 3(-4) - 2$. Use the distributive property of multiplication: $4 - 2x = -12 - 2$. Subtract 4 from both sides: $4 - 4 - 2x = -12 - 2 - 4$. Simplify: $-2x = -18$. Divide both sides by −2: $x = 9$. The solution for the system of equations is $(9, -4)$.

609. Transform equations into slope/y-intercept form. $y = x + 4$. The equation is in the proper slope/y-intercept form: $m = 1 = \frac{1}{1} = \frac{\text{change in } y}{\text{change in } x}$. $b = 4$: The y-intercept is at the point $(0,4)$. The slope tells you to go up 1 space and right 1 for $(1,5)$. $y = -x + 2$. The equation is in the proper slope/y-intercept form: $m = -1 = \frac{-1}{1} = \frac{\text{change in } y}{\text{change in } x}$. $b = 2$: The y-intercept is at the point $(0,2)$. The slope tells you to go down 1 space and right 1 for $(1,1)$. The solution is $(-1,3)$.

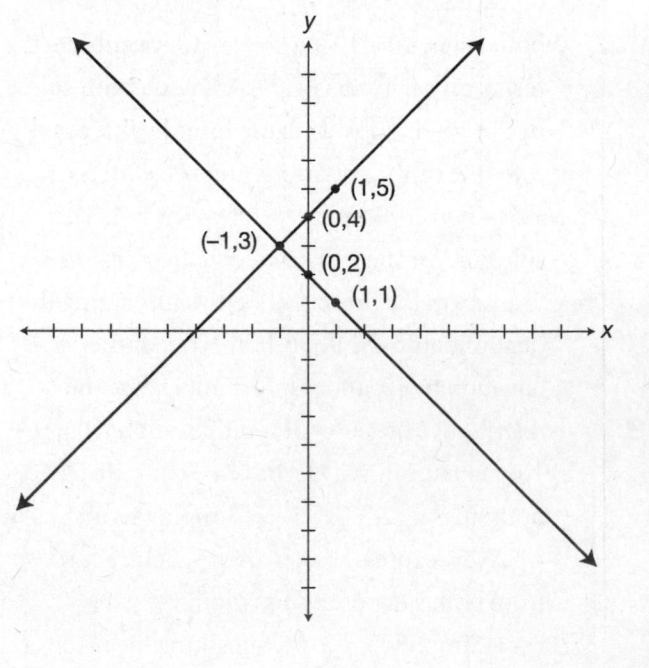

610. Transform equations into slope/y-intercept form. $2y - x = 2$. Add x to both sides: $2y - x + x = x + 2$. Combine like terms: $2y = x + 2$. Divide both sides by 2: $y = \frac{1}{2}x + 1$. The equation is in the proper slope/y-intercept form: $m = \frac{1}{2} = \frac{\text{change in } y}{\text{change in } x}$. $b = 1$: The y-intercept is at the point $(0,1)$. The slope tells you to go up 1 space and right 2 for $(2,2)$. $3x + y = 8$. Subtract $3x$ from both sides: $3x - 3x + y = -3x + 8$. Simplify: $y = -3x + 8$. The equation is in the proper slope/y-intercept form: $m = -3 = \frac{-3}{1} = \frac{\text{change in } y}{\text{change in } x}$. $b = 8$: The y-intercept is at the point $(0,8)$. The slope tells you to go down 3 spaces and right 1 for $(1,5)$. The solution is $(2,2)$.

611. Transform equations into slope/y-intercept form. $4y = -7(x + 4)$. Use the distributive property of multiplication: $4y = -7x - 28$. Divide both sides by 4: $y = \frac{-7}{4}x - 7$. The equation is in the proper slope/y-intercept form: $m = \frac{-7}{4} = \frac{\text{change in } y}{\text{change in } x}$. $b = -7$: The y-intercept is at the point $(0,-7)$. The slope tells you to go down 7 spaces and right 4 for $(4,-14)$. $4y = x + 4$. Divide both sides by 4: $y = \frac{1}{4}x + 1$. The equation is in the proper slope/y-intercept form: $m = \frac{1}{4} = \frac{\text{change in } y}{\text{change in } x}$. $b = 1$: The y-intercept is at the point $(0,1)$. The slope tells you to go up 1 space and right 4 for $(4,2)$. The solution is $(-4,0)$.

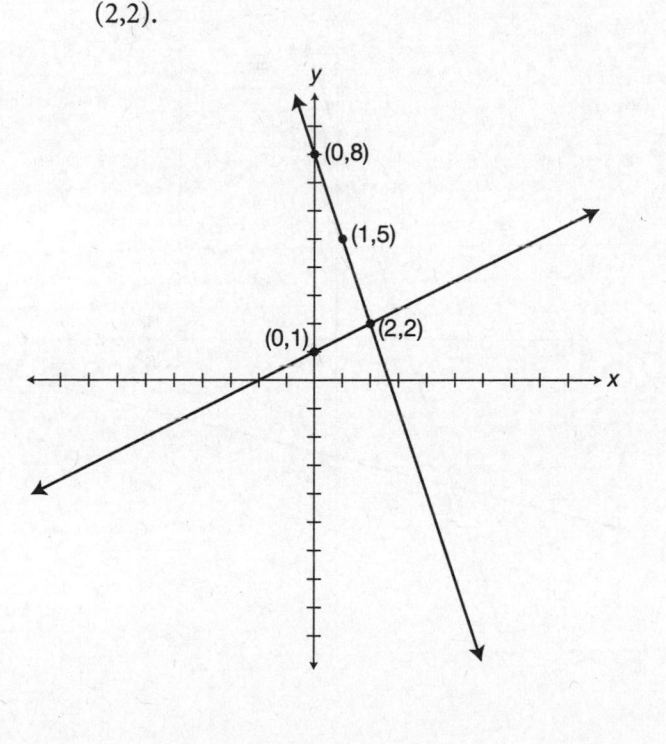

612. Transform equations into slope/y-intercept form. $y - x = 5 - x$. Add x to both sides: $y - x + x = 5 - x + x$. Combine like terms on each side: $y = 5$. The graph is a line parallel to the x-axis through $(0,5)$. $-4y = 8 - 7x$. Divide both sides by -4: $\frac{-4y}{-4} = \frac{8}{-4} - \frac{7x}{-4}$. Simplify terms: $y = -2 + \frac{7}{4}x$. Use the commutative property: $y = \frac{7}{4}x - 2$. The equation is in the proper slope/y-intercept form: $m = \frac{7}{4} = \frac{\text{change in } y}{\text{change in } x}$. $b = -2$: The y-intercept is at the point $(0,-2)$. The slope tells you to go up 7 spaces and right 4 for $(4,5)$. The solution is $(4,5)$.

613. Transform equations into slope/y-intercept form. $2y = 6x + 14$. Divide both sides by 2: $y = \frac{6}{2}x + 7$. The equation is in the proper slope/y-intercept form. Use the negatives to keep the coordinates near the origin: $m = \frac{6}{2} = \frac{-6}{-2} = \frac{\text{change in } y}{\text{change in } x}$. $b = 7$: The y-intercept is at the point $(0,7)$. The slope tells you to go down 6 spaces and left 2 for $(-2,1)$. $4y = x - 16$. Divide both sides by 4: $y = \frac{1}{4}x - 4$. The equation is in the proper slope/y-intercept form: $m = \frac{1}{4} = \frac{\text{change in } y}{\text{change in } x}$. $b = -4$: The y-intercept is at the point $(0,-4)$. The slope tells you to go up 1 space and right 4 for $(4,-3)$. The solution for the system of equations is $(-4,-5)$.

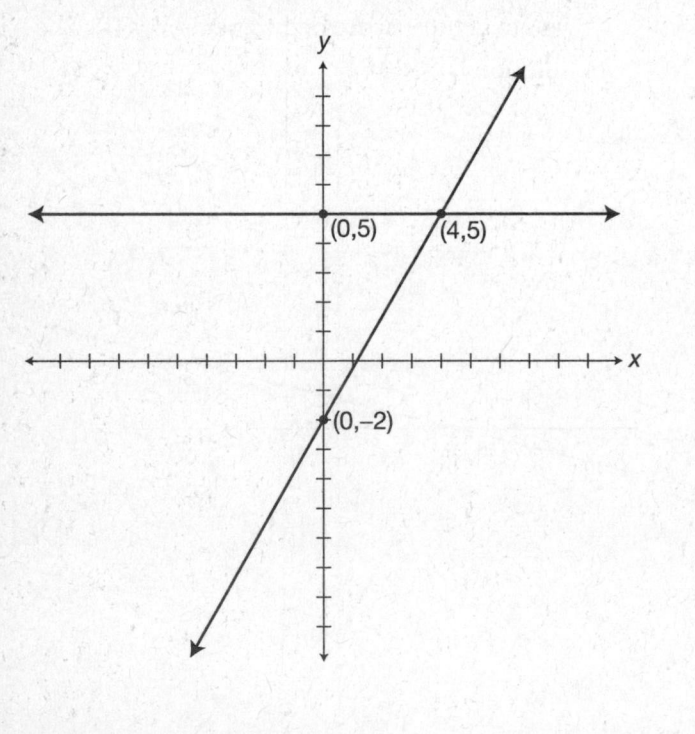

614. Transform equations into slope/y-intercept form. $2x + y = 4$. Subtract $2x$ from both sides: $2x - 2x + y = 4 - 2x$. Combine like terms on each side: $y = 4 - 2x$. Use the commutative property: $y = -2x + 4$. The equation is in the proper slope/y-intercept form: $m = -2 = \frac{-2}{1} = \frac{\text{change in } y}{\text{change in } x}$. $b = 4$: The y-intercept is at the point $(0,4)$. The slope tells you to go down 2 spaces and right 1 for $(1,2)$. $3(y + 9) = 7x$. Use the distributive property of multiplication: $3y + 27 = 7x$. Subtract 27 from both sides: $3y + 27 - 27 = 7x - 27$. Simplify: $3y = 7x - 27$. Divide both sides by 3: $y = \frac{7}{3}x - 9$. The equation is in the proper slope/y-intercept form: $m = \frac{7}{3} = \frac{\text{change in } y}{\text{change in } x}$. $b = -9$: The y-intercept is at the point $(0,-9)$. The slope tells you to go up 7 spaces and right 3 for $(3,-2)$. The solution for the system of equations is $(3,-2)$.

615. Transform equations into slope/y-intercept form. $y = x + 9$. The equation is in the proper slope/y-intercept form: $m = 1 = \frac{1}{1} = \frac{\text{change in } y}{\text{change in } x}$. $b = 9$: The y-intercept is at the point $(0,9)$. The slope tells you to go up 1 space and right 1 for $(1,10)$. $4y = 16 - x$. Use the commutative property: $4y = -x + 16$. Divide both sides by 4: $y = -\frac{1}{4}x + 4$. The equation is in the proper slope/y-intercept form: $m = \frac{-1}{4} = \frac{\text{change in } y}{\text{change in } x}$. $b = 4$: The y-intercept is at the point $(0,4)$. The slope tells you to go down 1 space and right 4 for $(4,3)$. The solution for the system of equations is $(-4,5)$.

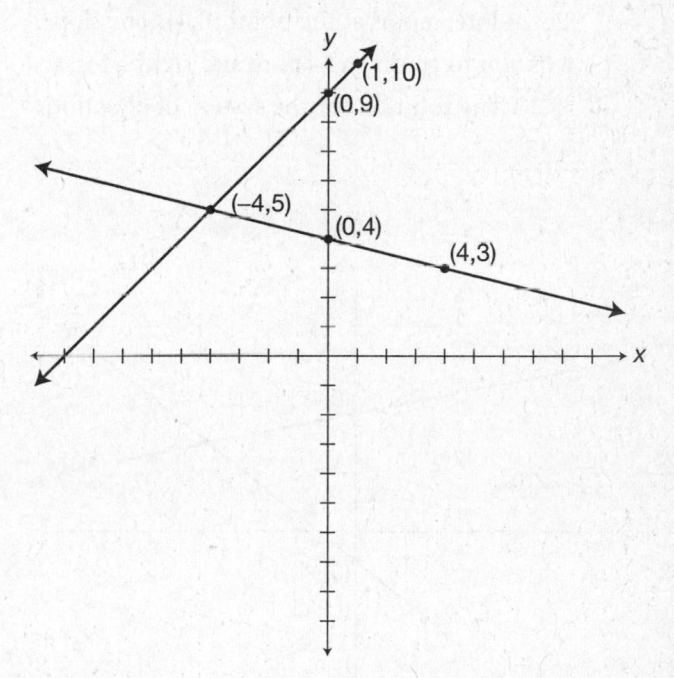

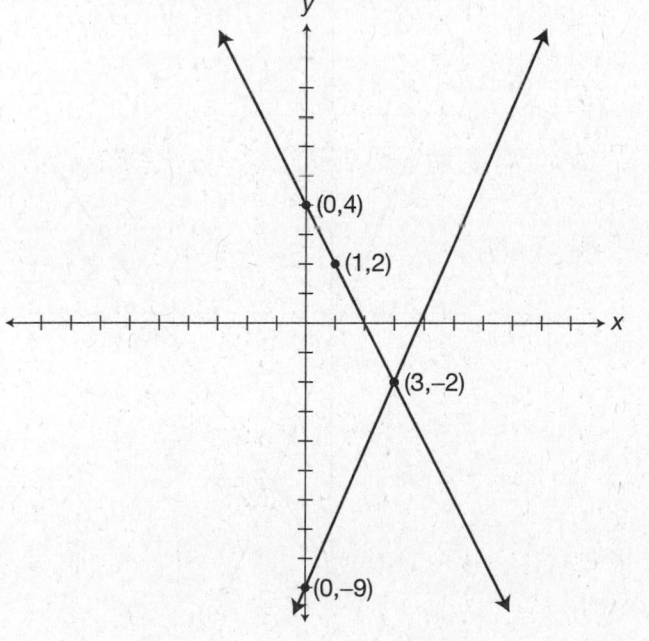

616. Transform equations into slope/y-intercept form. $4x - 5y = 5$. Subtract $4x$ from both sides: $4x - 4x - 5y = 5 - 4x$. Simplify: $-5y = 5 - 4x$. Use the commutative property: $-5y = -4x + 5$. Divide both sides by -5: $y = \frac{4}{5}x - 1$. The equation is in the proper slope/y-intercept form: $m = \frac{4}{5} = \frac{\text{change in } y}{\text{change in } x}$. $b = -1$: The y-intercept is at the point $(0,-1)$. The slope tells you to go up 4 spaces and right 5 for $(5,3)$. $5y = 20 - x$. Divide both sides by 5: $y = 4 - \frac{1}{5}x$. Use the commutative property: $y = \frac{-1}{5}x + 4$. The equation is in the proper slope/y-intercept form: $m = \frac{-1}{5} = \frac{\text{change in } y}{\text{change in } x}$. $b = 4$: The y-intercept is at the point $(0,4)$. The slope tells you to go down 1 space and right 5 for $(5,3)$. The solution for the system of equations is $(5,3)$.

617. Transform equations into slope/y-intercept form. $6y = 9(x - 6)$. Use the distributive property of multiplication: $6y = 9x - 54$. Divide both sides by 6: $y = \frac{9}{6}x - 9$. The equation is in the proper slope/y-intercept form: $m = \frac{9}{6} = \frac{\text{change in } y}{\text{change in } x}$. $b = -9$: The y-intercept is at the point $(0,-9)$. The slope tells you to go up 9 spaces and right 6 for $(6,0)$. $3(2y + 5x) = -6$. Use the distributive property of multiplication: $6y + 15x = -6$. Subtract $15x$ from both sides: $6y + 15x - 15x = -15x - 6$. Simplify: $6y = -15x - 6$. Divide both sides by 6: $y = \frac{-15}{6}x - 1$. The equation is in the proper slope/y-intercept form: $m = \frac{-15}{6} = \frac{-5}{2} = \frac{\text{change in } y}{\text{change in } x}$. $b = -1$: The y-intercept is at the point $(0,-1)$. The slope tells you to go down 5 spaces and right 2 for $(2,-6)$. The solution for the system of equations is $(2,-6)$.

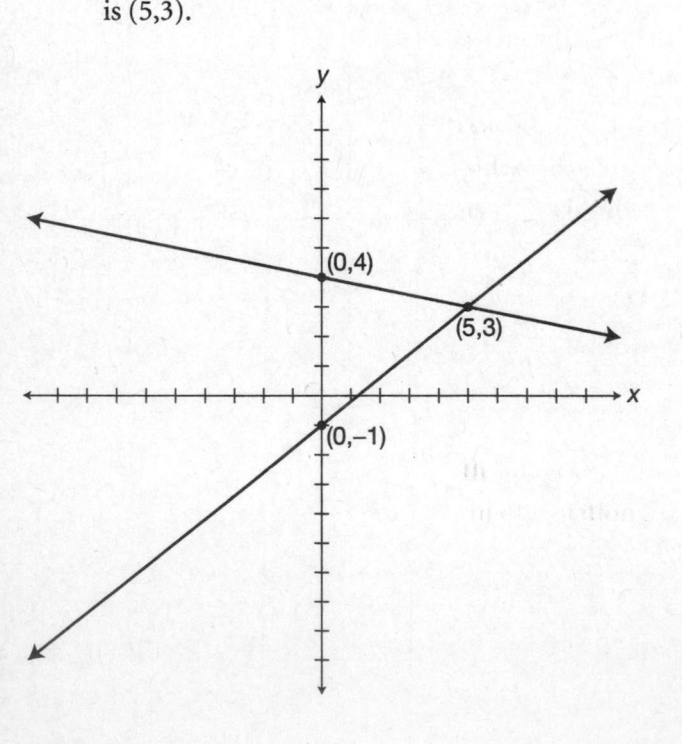

618. Transform equations into slope/y-intercept form. $15y = 6(3x + 15)$. Use the distributive property of multiplication: $15y = 18x + 90$. Divide both sides by 15: $y = \frac{18}{15}x + 6$. The equation is in the proper slope/y-intercept form: $m = \frac{18}{15} = \frac{6}{5} = \frac{-6}{-5} = \frac{\text{change in } y}{\text{change in } x}$. $b = 6$: The y-intercept is at the point (0,6). The slope tells you to go down 6 spaces and left 5 for (−5,0). $y = 6(1 - x)$. Use the distributive property of multiplication: $y = 6 - 6x$. Use the commutative property: $y = -6x + 6$. The equation is in the proper slope/y-intercept form: $m = -6 = \frac{-6}{1} = \frac{\text{change in } y}{\text{change in } x}$. $b = 6$. The y-intercept is at the point (0,6). The slope tells you to go down 6 spaces and right 1 for (1,0). The solution for the system of equations is (0,6).

619. Transform equations into slope/y-intercept form. $3y = 6x + 6$. Divide both sides by 3: $y = 2x + 2$. The equation is in the proper slope/y-intercept form: $m = 2 = \frac{2}{1} = \frac{\text{change in } y}{\text{change in } x}$. $b = 2$: The y-intercept is at the point (0,2). The slope tells you to go up 2 spaces and right 1 for (1,4). $5y = 10(x - 5)$. Use the distributive property: $5y = 10x - 50$. Divide both sides by 5: $y = 2x - 10$. The equation is in the proper slope/y-intercept form: $m = 2 = \frac{2}{1} = \frac{\text{change in } y}{\text{change in } x}$. $b = -10$: The y-intercept is at the point (0,−10). The slope tells you to go up 2 spaces and right 1 for (1,−8). The slopes are the same, so the line graphs are parallel and do not intersect. There is no solution.

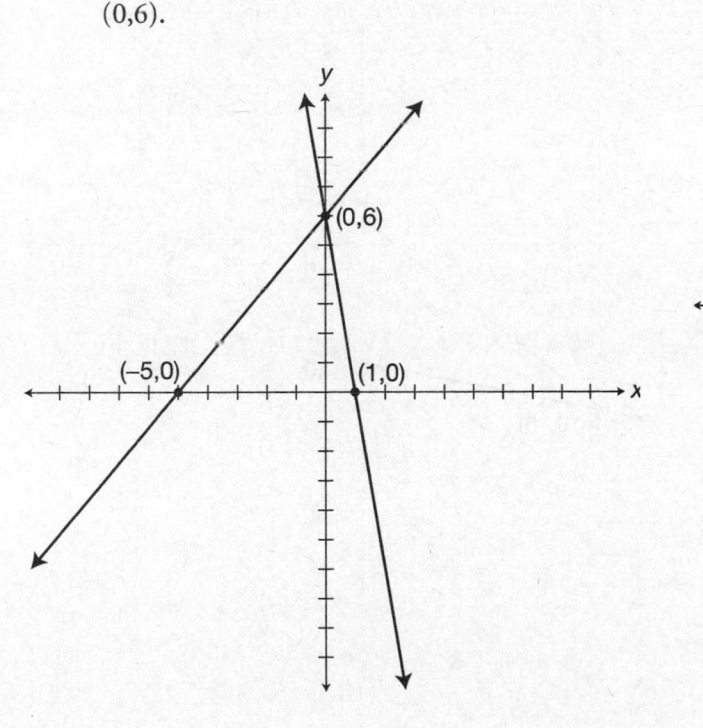

620. Transform equations into slope/y-intercept form. $3(2x + 3y) = 63$. Use the distributive property of multiplication: $6x + 9y = 63$. Subtract $6x$ from both sides: $6x - 6x + 9y = 63 - 6x$. Simplify: $9y = 63 - 6x$. Use the commutative property: $9y = -6x + 63$. Divide both sides by 9: $y = \frac{-6}{9}x + 7$. The equation is in the proper slope/y-intercept form: $m = \frac{-6}{9} = \frac{-2}{3} = \frac{\text{change in } y}{\text{change in } x}$. $b = 7$: The y-intercept is at the point $(0,7)$. The slope tells you to go down 2 spaces and right 3 for $(3,5)$. $27y = 9(x - 6)$. Use the distributive property of multiplication: $27y = 9x - 54$. Divide both sides by 27: $y = \frac{9}{27}x - 2$. The equation is in the proper slope/y-intercept form: $m = \frac{9}{27} = \frac{1}{3} = \frac{\text{change in } y}{\text{change in } x}$. $b = -2$: The y-intercept is at the point $(0,-2)$. The slope tells you to go up 1 space and right 3 for $(3,-1)$. The solution for the system of equations is $(9,1)$.

621. Transform equations into slope/y-intercept form. $x - 20 = 5y$. If $a = b$, then $b = a$: $5y = x - 20$. Divide both sides by 5: $y = \frac{1}{5}x - 4$. The equation is in the proper slope/y-intercept form: $m = \frac{1}{5} = \frac{\text{change in } y}{\text{change in } x}$. $b = -4$. The y-intercept is at the point $(0,-4)$. The slope tells you to go up 1 space and right 5 for $(5,-3)$. $10y = 8x + 20$. Divide both sides by 10: $y = \frac{8}{10}x + 2$. The equation is in the proper slope/y-intercept form: $m = \frac{8}{10} = \frac{4}{5} = \frac{\text{change in } y}{\text{change in } x}$. $b = 2$: The y-intercept is at the point $(0,2)$. The slope tells you to go up 4 spaces and right 5 for $(5,6)$. The solution for the system of equations is $(-10,-6)$.

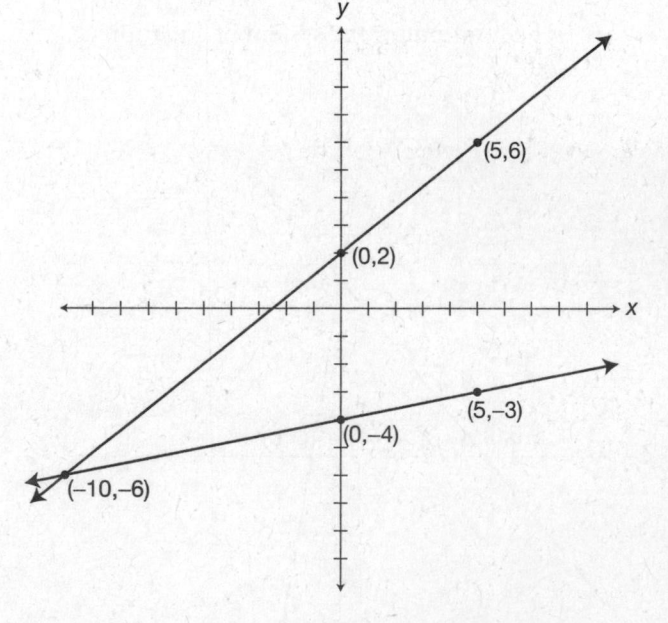

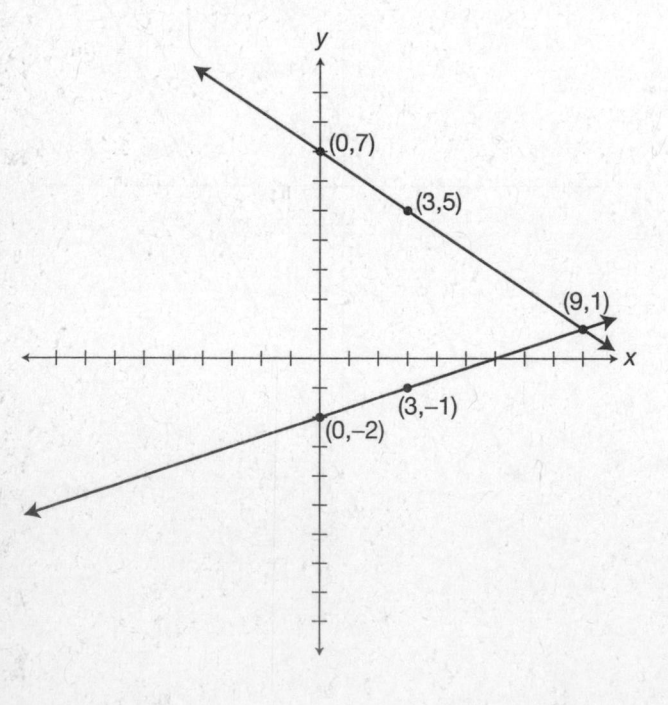

622. Transform equations into slope/y-intercept form. $3x + 4y = 12$. Subtract $3x$ from both sides: $3x - 3x + 4y = -3x + 12$. Simplify: $4y = -3x + 12$. Divide both sides by 4: $y = \frac{-3}{4}x + 3$. The equation is in the proper slope/y-intercept form: $m = \frac{-3}{4} = \frac{\text{change in } y}{\text{change in } x}$. $b = 3$: The y-intercept is at the point $(0,3)$. The slope tells you to go down 3 spaces and right 4 for $(4,0)$. $y = 3 - \frac{6}{8}x$. Use the commutative property: $y = \frac{-6}{8}x + 3$. The equation is in the proper slope/y-intercept form: $m = \frac{-6}{8} = \frac{\text{change in } y}{\text{change in } x}$. $b = 3$: The y-intercept is at the point $(0,3)$. The slope tells you to go down 6 spaces and right 8 for $(8,-3)$. The solution for the system of equations is the entire line because the graphs coincide.

623. Transform equations into slope/y-intercept form. $16y = 10(x - 8)$. Use the distributive property of multiplication: $16y = 10x - 80$. Divide both sides by 16: $y = \frac{10}{16}x - 5$. The equation is in the proper slope/y-intercept form: $m = \frac{10}{16} = \frac{5}{8} = \frac{-5}{-8} = \frac{\text{change in } y}{\text{change in } x}$. Use the negatives to keep the coordinates near the origin. $b = -5$: The y-intercept is at the point $(0,-5)$. The slope tells you to go down 5 spaces and left 8 for $(-8,-10)$. $8y - 17x = 56$. Add $17x$ to both sides: $8y - 17x + 17x = 17x + 56$. Simplify: $8y = 17x + 56$. Divide both sides by 8: $y = \frac{17}{8}x + 7$. The equation is in the proper slope/y-intercept form: $m = \frac{17}{8} = \frac{-17}{-8} = \frac{\text{change in } y}{\text{change in } x}$. Use the negatives to keep the coordinates near the origin. $b = 7$: The y-intercept is at the point $(0,7)$. The slope tells you to go down 17 spaces and left 8 for $(-8,-10)$. The solution for the system of equations is $(-8,-10)$.

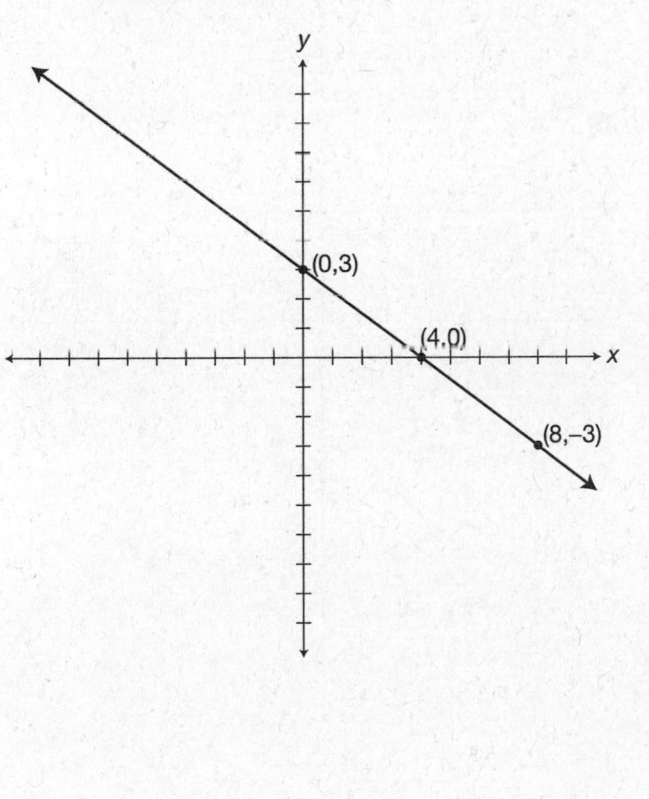

624. Transform the inequalities into slope/y-intercept form. $2y - 3x \geq -6$. Add $3x$ to both sides: $2y - 3x + 3x \geq {}^+3x - 6$. Simplify: $2y \geq 3x - 6$. Divide both sides by 2: $y \geq \frac{3}{2}x - 3$; $m = \frac{3}{2} = \frac{\text{change in } y}{\text{change in } x}$. $b = -3$: The y-intercept is at the point $(0,-3)$. The slope tells you to go up 3 spaces and right 2 for $(2,0)$. Use a **solid** line for the border and shade **above** the line because the symbol is $\geq$. $y \geq 5 - \frac{5}{2}x$. Use the commutative property: $y \geq -\frac{5}{2}x + 5$; $m = \frac{-5}{2} = \frac{\text{change in } y}{\text{change in } x}$. Use the negatives to keep the coordinates near the origin. $b = 5$: The y-intercept is at the point $(0,5)$. The slope tells you to go down 5 spaces and right 2 for $(2,0)$. Use a **solid** line for the border and shade **above** the line because the symbol is $\geq$. The solution for the system of inequalities is where the shaded areas overlap.

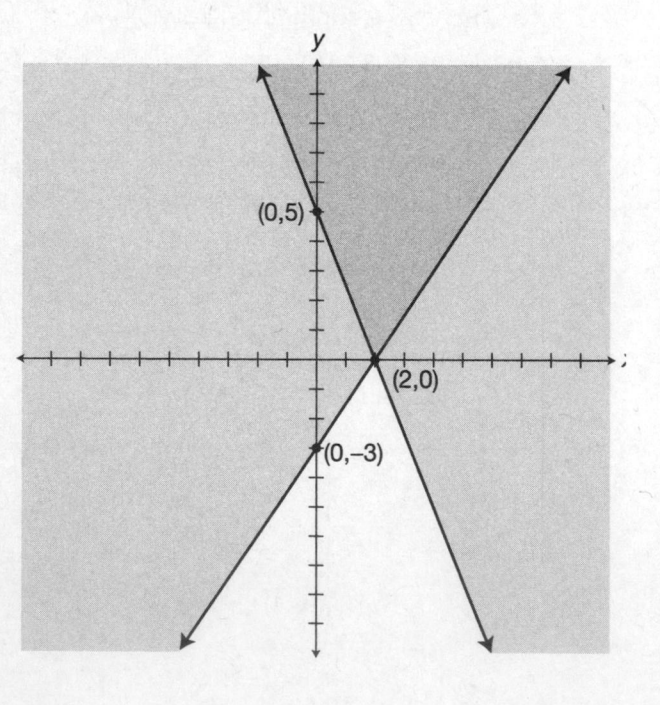

625. Transform equations into slope/y-intercept form. $6y < 5x - 30$. Divide both sides by 6: $y < \frac{5}{6}x - 5$; $m = \frac{5}{6} = \frac{\text{change in } y}{\text{change in } x}$. $b = -5$: The y-intercept is at the point $(0,-5)$. The slope tells you to go up 5 spaces and right 6 for $(6,0)$. Use a **dotted** line for the border and shade **below** it because the symbol is $<$. $2y < -x + 4$. Divide both sides by 2: $y < \frac{-1}{2}x + 2$; $m = \frac{-1}{2} = \frac{\text{change in } y}{\text{change in } x}$. $b = 2$: The y-intercept is at the point $(0,2)$. The slope tells you to go down 1 space and right 2 for $(2,1)$. Use a **dotted** line for the border and shade **below** it because the symbol is $<$. The solution for the system of inequalities is the double-shaded area on the graph.

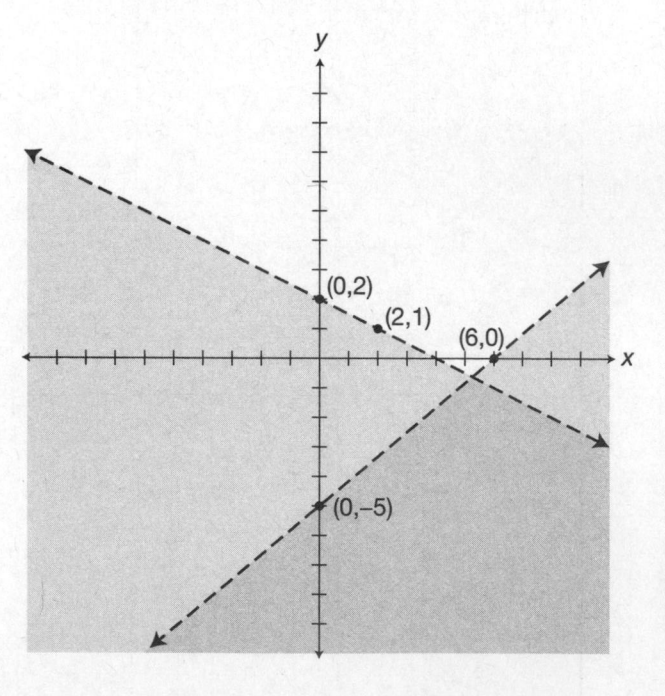

626. Transform equations into slope/y-intercept form. $y - x \geq 6$. Add x to both sides: $y - x + x \geq x + 6$. Simplify: $y \geq x + 6$; $m = 1 = \frac{-1}{-1} = \frac{\text{change in } y}{\text{change in } x}$. $b = 6$: The y-intercept is at the point $(0,6)$. The slope tells you to go down 1 space and left 1 for $(-1,5)$. Use a **solid** line for the border and shade **above** it because the symbol is $\geq$. $11y \geq -2(x + 11)$. Use the distributive property of multiplication: $11y \geq -2x - 22$. Divide both sides by 11: $y \geq \frac{-2}{11}x - 2$; $m = \frac{-2}{11} = \frac{\text{change in } y}{\text{change in } x}$. $b = -2$: The y-intercept is at the point $(0,-2)$. The slope tells you to go down 2 spaces and right 11 for $(11,-4)$. Use a **solid** line for the border and shade **above** the line because the symbol is $\geq$. The solution for the system of inequalities is the double-shaded area on the graph.

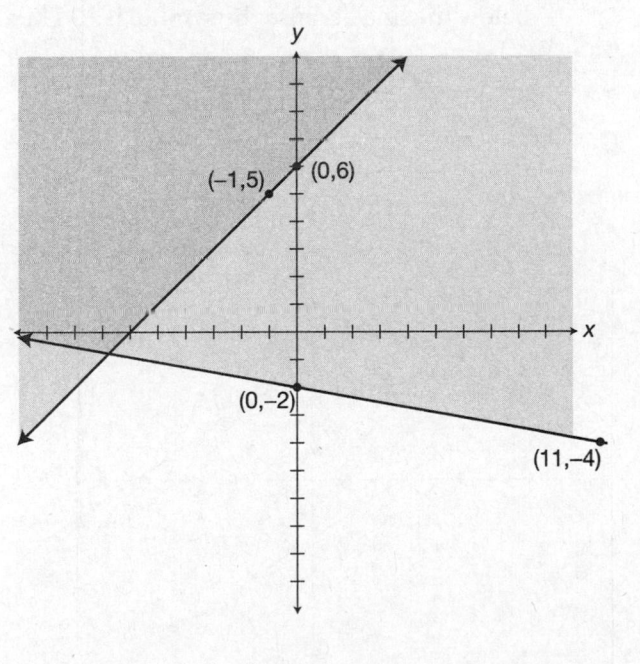

627. Transform equations into slope/y-intercept form. $5y \leq 8(x + 5)$. Use the distributive property of multiplication: $5y \leq 8x + 40$. Divide both sides by 5: $y \leq \frac{8}{5}x + 8$; $m = \frac{8}{5} = \frac{-8}{-5} = \frac{\text{change in } y}{\text{change in } x}$. Use the negatives to keep the coordinates near the origin. $b = 8$: The y-intercept is at the point $(0,8)$. The slope tells you to go down 8 spaces and left 5 for $(-5,0)$. Use a **solid** line for the border and shade **below** it because the symbol is $\leq$. $5y \leq 12(5 - x)$. Use the distributive property of multiplication: $5y \leq 60 - 12x$. Use the commutative property of addition: $5y \leq -12x + 60$. Divide both sides by 5: $y \leq \frac{-12}{5}x + 12$; $m = \frac{-12}{5} = \frac{\text{change in } y}{\text{change in } x}$. $b = 12$: The y-intercept is at the point $(0,12)$. The slope tells you to go down 12 spaces and right 5 for $(5,0)$. Use a **solid** line for the border and shade **below** the line because the symbol is $\leq$. The solution for the system of inequalities is the double-shaded area on the graph.

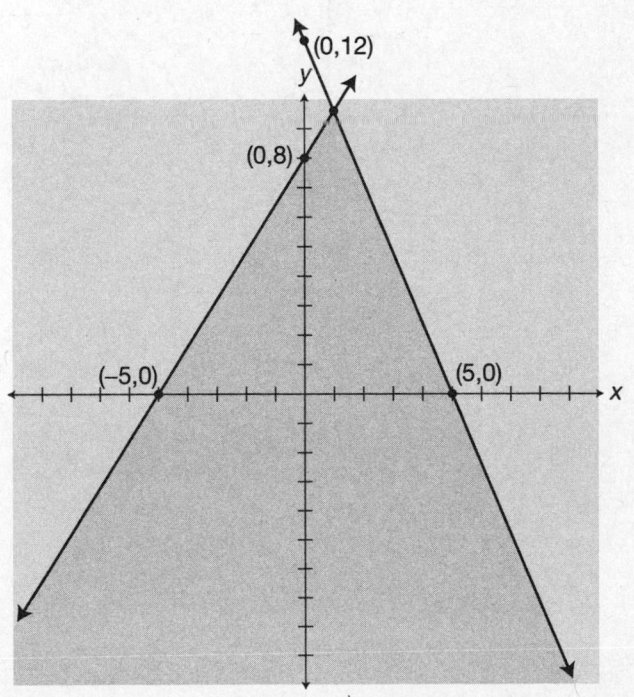

628. Transform equations into slope/y-intercept form. $2(x + 5y) > 5(x + 6)$. Use the distributive property of multiplication: $2x + 10y > 5x + 30$. Subtract $2x$ from both sides: $2x - 2x + 10y > 5x - 2x + 30$. Simplify the inequality: $10y > 3x + 30$. Divide both sides by 10: $y > \frac{3}{10}x + 3$; $m = \frac{3}{10} = \frac{\text{change in } y}{\text{change in } x}$. $b = 3$: The y-intercept is at the point $(0,3)$. The slope tells you to go up 3 spaces and right 10 for $(10,6)$. Use a **dotted** line for the border and shade **above** it because the symbol is $>$. $4x + y < 4x + 5$. Subtract $4x$ from both sides: $4x - 4x + y < 4x - 4x + 5$. Simplify: $y < 0x + 5$. With a slope of zero, the line is parallel to the x-axis. The y-intercept is at the point $(0,5)$. Use a **dotted** line for the border and shade **below** it because the symbol is $<$. The solution for the system of inequalities is the double-shaded area on the graph.

629. Transform equations into slope/y-intercept form. $3y \geq -2(x + 3)$. Use the distributive property of multiplication: $3y \geq -2x - 6$. Divide both sides by 3: $y \geq \frac{-2}{3}x - 2$; $m = \frac{-2}{3} = \frac{\text{change in } y}{\text{change in } x}$. $b = -2$: The y-intercept is at the point $(0,-2)$. The slope tells you to go down 2 spaces and right 3 for $(3,-4)$. Use a **solid** line for the border and shade **above** it because the symbol is $\geq$. $3y \leq 2(6 - x)$. Use the distributive property of multiplication: $3y \leq 12 - 2x$. Use the commutative property: $3y \leq -2x + 12$. Divide both sides by 3: $y \leq \frac{-2}{3}x + 4$; $m = \frac{-2}{3} = \frac{\text{change in } y}{\text{change in } x}$. Slopes that are the same will result in parallel lines. $b = 4$: The y-intercept is at the point $(0,4)$. The slope tells you to go down 2 spaces and right 3 for $(3,2)$. Use a **solid** line for the border and shade **below** the line because the symbol is $\leq$. The solution for the system of inequalities is the double-shaded area on the graph.

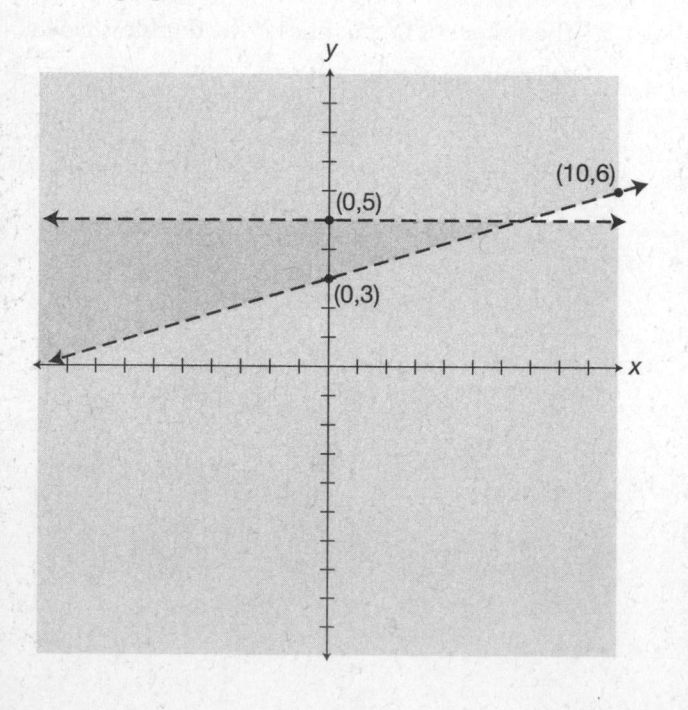

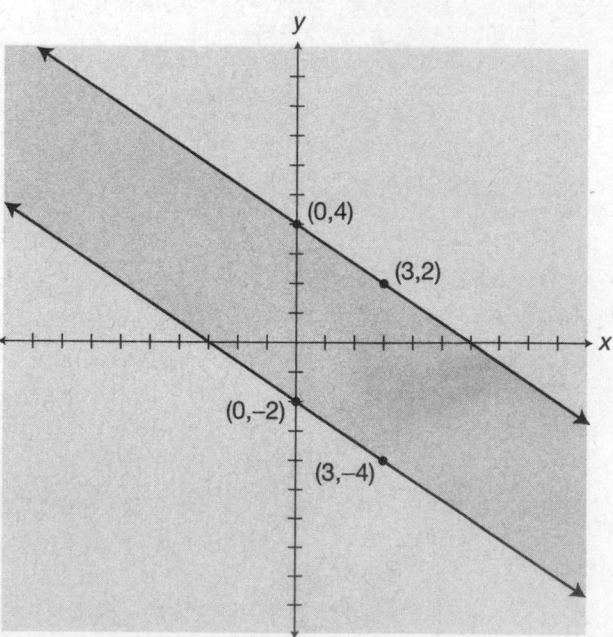

630. Transform equations into slope/y-intercept form. $9(y-4) < 4x$. Use the distributive property of multiplication: $9y - 36 < 4x$. Add 36 to both sides: $9y - 36 + 36 < 4x + 36$. Divide both sides by 9: $y < \frac{4}{9}x + 4$; $m = \frac{4}{9} = \frac{\text{change in } y}{\text{change in } x}$. $b = 4$: The y-intercept is at the point $(0,4)$. The slope tells you to go up 4 spaces and right 9 for $(9,8)$. Use a **dotted** line for the border and shade **below** it because the symbol is $<$. $-9y < 2(x+9)$. Use the distributive property of multiplication: $-9y < 2x + 18$. Divide both sides by -9. Change the direction of the symbol when dividing by a negative: $y > \frac{-2}{9}x - 2$; $m = \frac{-2}{9} = \frac{\text{change in } y}{\text{change in } x}$. $b = -2$: The y-intercept is at the point $(0,-2)$. The slope tells you to go down 2 spaces and right 9 for $(9,-4)$. Use a **dotted** line for the border and shade **above** the line because the symbol is $>$. The solution for the system of inequalities is the double-shaded area on the graph.

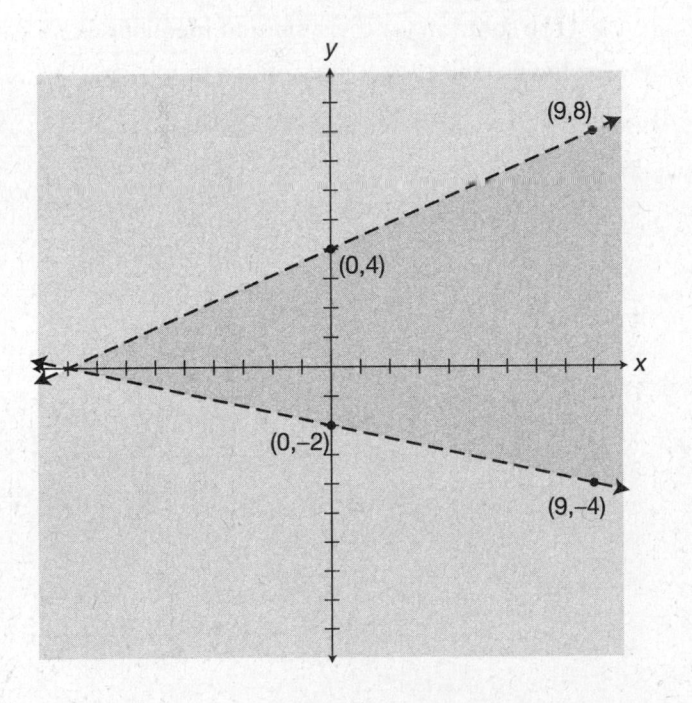

631. Transform equations into slope/y-intercept form. $7(y-5) < -5x$. Use the distributive property of multiplication: $7y - 35 < -5x$. Add 35 to both sides: $7y - 35 + 35 < -5x + 35$. Simplify: $7y < -5x + 35$. Divide both sides by 7: $y < \frac{-5}{7}x + 5$; $m = \frac{-5}{7} = \frac{\text{change in } y}{\text{change in } x}$. $b = 5$: The y-intercept is at the point $(0,5)$. The slope tells you to go down 5 spaces and right 7 for $(7,0)$. Use a **dotted** line for the border and shade **below** it because the symbol is $<$. $-3 < \frac{1}{4}(2x - 3y)$. Multiply both sides of the inequality by 4: $4(-3) < 4(\frac{1}{4})(2x - 3y)$. Simplify the inequality: $-12 < 1(2x - 3y)$; $-12 < 2x - 3y$. Add $3y$ to both sides: $-12 + 3y < 2x - 3y + 3y$. Simplify the inequality: $-12 + 3y < 2x$. Add 12 to both sides: $-12 + 12 + 3y < 2x + 12$. Simplify the inequality: $3y < 2x + 12$. Divide both sides by 3: $y < \frac{2}{3}x + 4$; $m = \frac{2}{3} = \frac{\text{change in } y}{\text{change in } x}$. $b = 4$: The y-intercept is at the point $(0,4)$. The slope tells you to go up 2 spaces and right 3 for $(3,6)$. Use a **dotted** line for the border and shade **below** the line because the symbol is $<$. The solution for the system of inequalities is the double-shaded area on the graph.

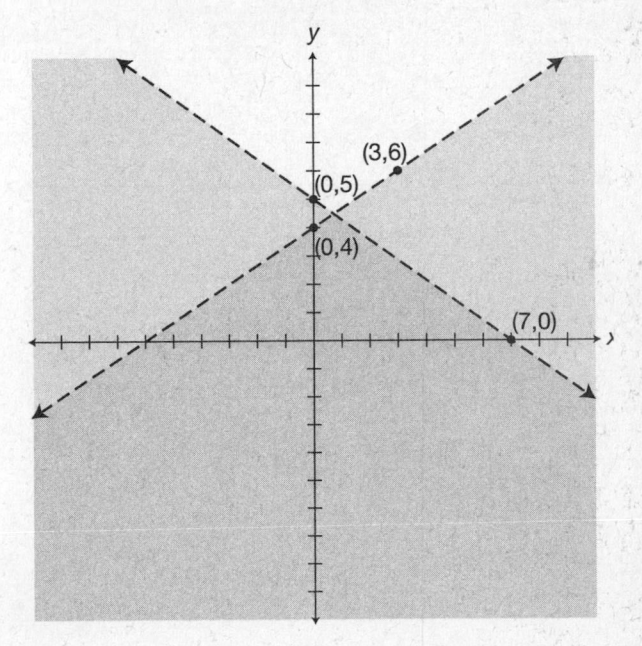

632. Transform equations into slope/y-intercept form. $y > \frac{7}{4}(4 - x)$. Use the distributive property of multiplication: $y > 7 - \frac{7}{4}x$. Use the commutative property: $y > \frac{-7}{4}x + 7$; $m = \frac{-7}{4} = \frac{\text{change in } y}{\text{change in } x}$. $b = 7$: The y-intercept is at the point $(0,7)$. The slope tells you to go down 7 spaces and right 4 for $(4,0)$. Use a **dotted** line for the border and shade **above** it because the symbol is $>$. $3(y + 5) > 7x$. Use the distributive property of multiplication: $3y + 15 > 7x$. Subtract 15 from both sides: $3y + 15 - 15 > 7x - 15$. Simplify the inequality: $3y > 7x - 15$. Divide both sides by 3: $y > \frac{7}{3}x - 5$; $m = \frac{7}{3} = \frac{\text{change in } y}{\text{change in } x}$. $b = -5$: The y-intercept is at the point $(0,-5)$. The slope tells you to go up 7 spaces and right 3 for $(3,2)$. Use a **dotted** line for the border and shade **above** the line because the symbol is $>$. The solution for the system of inequalities is the double-shaded area on the graph.

633. Transform equations into slope/y-intercept form. $5x - 2(y + 10) \leq 0$. Use the distributive property of multiplication: $5x - 2y - 20 \leq 0$. Subtract $5x$ from both sides: $5x - 5x - 2y - 20 \leq -5x$. Add 20 to both sides: $-2y - 20 + 20 \leq -5x + 20$. Simplify the inequality: $-2y \leq -5x + 20$. Divide both sides by -2. Change the direction of the symbol when dividing by a negative: $y \geq \frac{5}{2}x - 10$; $m = \frac{5}{2} = \frac{\text{change in } y}{\text{change in } x}$. $b = -10$: The y-intercept is at the point $(0,-10)$. The slope tells you to go up 5 spaces and right 2 for $(2,-5)$. Use a **solid** line for the border and shade **above** it because the symbol is $\geq$. $2x + y \leq -3$. Subtract $2x$ from both sides: $2x - 2x + y \leq -2x - 3$. Simplify the inequality: $y \leq -2x - 3$; $m = \frac{-2}{1} = \frac{\text{change in } y}{\text{change in } x}$. $b = -3$: The y-intercept is at the point $(0,-3)$. The slope tells you to go down 2 spaces and right 1 for $(1,-5)$. Use a **solid** line for the border and shade **below** the line because the symbol is $\leq$. The solution for the system of inequalities is the double-shaded area on the graph.

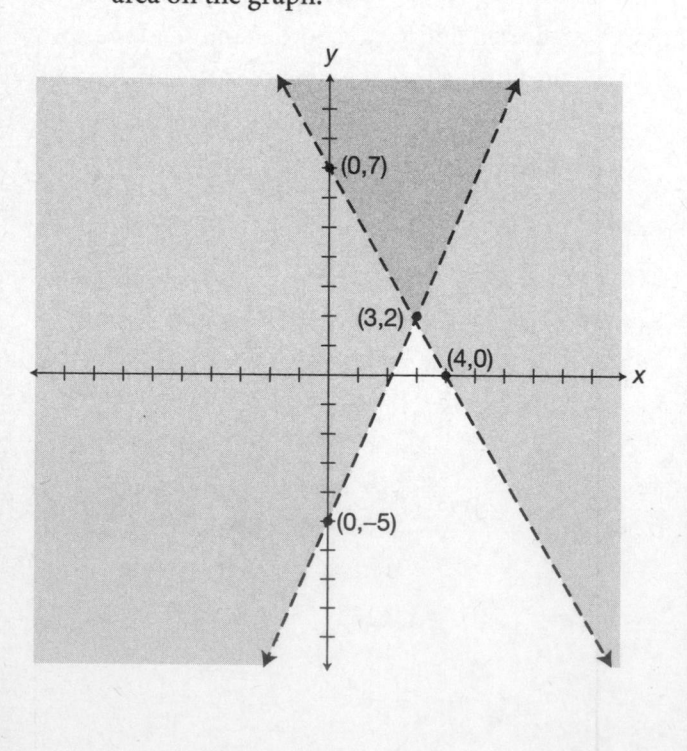

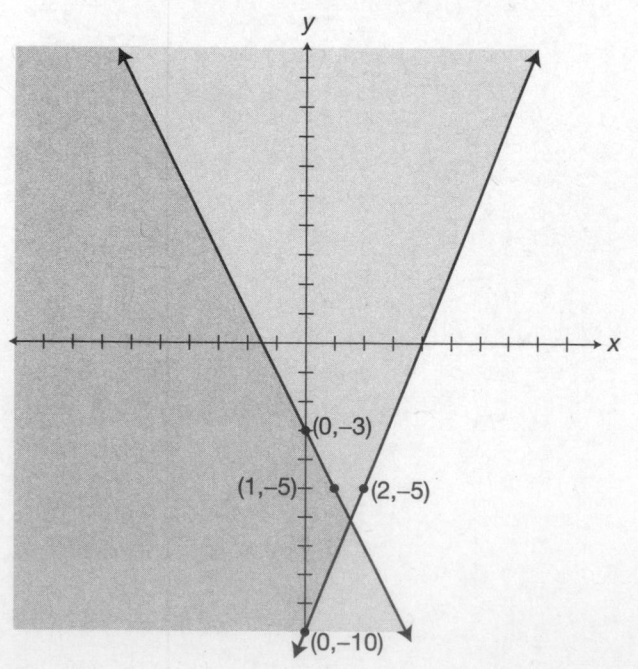

CHAPTER

10 ▶ Multiplying and Factoring Polynomials

BACK IN CHAPTER 4, we looked at multiplying algebraic terms. We learned that you could multiply $5x^2y^6$ and $4xy^3$ by multiplying their coefficients and adding the exponents of the variables they had in common: $(5x^2y^6)(4xy^3) = 20x^3y^9$. Each of these terms is called a monomial. A **monomial** is a single term made up of a variable or a constant. The coefficient, if there is one, and the variable are multiplied, not added or subtracted. For instance, $4x$, y^4, and $21x^2y^3z^5$ are all monomials. In this chapter, we'll learn how to multiply **binomials** (expressions with two terms), **trinomials** (three terms), and other **polynomials** (more than one term).

▶ Multiplying a Monomial by a Polynomial

To multiply a monomial by a polynomial, multiply the monomial by each term in the polynomial, and then combine any like terms.

Example

$2x(3x + 4)$

In this example, a monomial ($2x$) is multiplied by a binomial ($3x + 4$). Multiply $2x$ by $3x$, and then multiply $2x$ by 4: $(2x)(3x)$ $= 6x^2$, and $(2x)(4) = 8x$. The product of $2x$ and $3x + 4$ is $6x^2 + 8x$. We cannot combine these terms, because they are unlike; so our answer is a binomial.

Example

$5a^3(b^2 + 3a + 4)$

In this example, a monomial is multiplied by a trinomial. The same rule applies: Multiply the monomial $5a^3$ by each term in the polynomial: $(5a^3)(b^2) = 5a^3b^2$; $(5a^3)(3a) = 15a^4$; and $(5a^3)(4) = 20a^3$. Our answer is $5a^3b^2 + 15a^4 + 20a^3$. These terms are unlike and cannot be combined.

The general rule for multiplying a monomial by a polynomial is $a(b + c + d) = ab + ac + ad$.

▶ Multiplying a Binomial by a Binomial

To multiply a binomial by a binomial, we must multiply each term in the first binomial by each term in the second binomial. The acronym FOIL helps us re-member how to do this. FOIL stands for first, outside, inside, last.

- ■ When multiplying binomials, begin by finding the product of the **first** terms in each expression.
- ■ Next, find the product of the **outside** terms— the first term in the first binomial and the last term in the second binomial.
- ■ Then, find the product of the **inside** terms—the second term in the first binomial and the first term in the second binomial.
- ■ Finally, find the product of the **last** terms in each expression.

Add these four products together and combine any like terms. The general rule for multiplying binomials is $(a + b)(c + d) = ac + ad + bc + bd$ (the same as FOIL). By multiplying in this way, we are sure to multiply each term in the first binomial by each term in the second binomial.

Example

$(x + 5)(x - 3)$

Begin by multiplying the first terms: $(x)(x) = x^2$.

Next, multiply the outside terms: $(x)(-3)$ $= -3x$.

Then, multiply the inside terms: $(5)(x) = 5x$.

Finally, multiply the last terms: $(5)(-3) = -15$.

Our product is $x^2 - 3x + 5x - 15$, or $x^2 + 2x - 15$ after the like terms are combined.

Example

$(4a + 1)(b + 7)$

Multiply the first terms: $(4a)(b) = 4ab$.

Multiply the outside terms: $(4a)(7) = 28a$.

Multiply the inside terms: $(1)(b) = b$.

Multiply the last terms: $(1)(7) = 7$.

Our product is $4ab + 28a + b + 7$.

▶ Multiplying Two Polynomials

To multiply two polynomials, multiply every term in the first polynomial by every term in the second polynomial. Does that sound familiar? That's the same way we multiplied two binomials. In fact, that's the same way we multiplied a monomial by a binomial; it's just that a monomial only has *one* term. So, to sum up: When multiplying a nomial by another nomial, multiply every term in the first nomial by every term in the second nomial, no matter how many terms there are in either nomial.

The general rule for multiplying polynomials is $(a + b + c)(d + e + f) = ad + ae + af + bd + be + bf + cd + ce + cf$.

Example
$(x - 1)(x^2 + 5x + 6)$

Multiply the first term in the first polynomial, x, by each term in the second polynomial: $(x)(x^2 + 5x + 6) = x^3 + 5x^2 + 6x$.

Now, multiply the second term in the first polynomial, -1, by each term in the second polynomial: $(-1)(x^2 + 5x + 6) = -x^2 - 5x - 6$.

Finally, add these products and combine like terms: $x^3 + 5x^2 + 6x + -x^2 - 5x - 6 = x^3 + 4x^2 + x - 6$.

▶ Practice Questions

Multiply the following polynomials.

634. $x(5x + 3y - 7)$

635. $2a(5a^2 - 7a + 9)$

636. $4bc(3b^2c + 7b - 9c + 2bc^2 - 8)$

637. $3mn(-4m + 6n + 7mn^2 - 3m^2n)$

638. $4x(9x^3 + \frac{3}{x^2} - x^4 + \frac{6x - 1}{x^2})$

639. $(x + 3)(x + 6)$

640. $(x - 4)(x - 9)$

641. $(2x + 1)(3x - 7)$

642. $(x + 2)(x - 3y)$

643. $(7x + 2y)(2x - 4y)$

644. $(5x + 7)(5x - 7)$

645. $(28x + 7)(\frac{x}{7} - 11)$

646. $(3x^2 + y^2)(x^2 - 2y^2)$

647. $(4 + 2x^2)(9 - 3x)$

648. $(2x^2 + y^2)(x^2 - y^2)$

649. $(x + 2)(3x^2 - 5x + 2)$

650. $(2x - 3)(x^3 + 3x^2 - 4x)$

651. $(4a + b)(5a^2 + 2ab - b^2)$

652. $(3y - 7)(6y^2 - 3y + 7)$

653. $(3x + 2)(3x^2 - 2x - 5)$

654. $(x + 2)(2x + 1)(x - 1)$

655. $(3a - 4)(5a + 2)(a + 3)$

656. $(2n - 3)(2n + 3)(n + 4)$

657. $(5r - 7)(3r^4 + 2r^2 + 6)$

658. $(3x^2 + 4)(x - 3)(3x^2 - 4)$

▶ Factoring Polynomials

Now that we know how to multiply polynomials, let's look at how to break them down by factoring them. When we factor a polynomial, we split it up into the smallest parts that multiply to the original polynomial. The three methods for factoring polynomial expressions are: greatest common factor method, difference between two perfect squares method, and trinomial method.

First, try to find the greatest common factor of all the terms in the polynomial. The greatest common factor could be a whole number, a variable, or a variable with a coefficient.

Example

$2x + 4$

The greatest common factor of $2x$ and 4 is 2, since both $2x$ and 4 can be divided evenly by 2. Pull a 2 out of the polynomial by dividing both terms by 2: $\frac{2x}{2} = x$, and $\frac{4}{2} = 2$. Therefore, $2x + 4$ factors to $2(x + 2)$.

Example

$5x^2 + 6x$

The greatest common factor of $5x^2$ and $6x$ is x. Divide each term by x: $\frac{5x^2}{x} = 5x$, and $\frac{6x}{x} = 6$. Therefore, $5x^2 + 6x = x(5x + 6)$.

Example

$12x^3 - 6x^2$

The greatest common factor of these terms is $6x^2$: $12x^3 - 6x^2$ factors to $6x^2(2x - 1)$.

Sometimes a binomial is the product of two binomials. After the original binomials were multiplied, the outside and inside terms canceled each other out, leaving just the products of the first terms and the last terms. When this happens, both the first term and the last term of the product binomial are either perfect squares or multiples of perfect squares. If one (or both) is a multiple of a perfect square, we can begin by factoring out that multiple. Then we must factor the difference of the perfect squares. One factor of the difference between perfect squares will be the sum of the square root of the first term and the square root of the second term; the second factor will be the difference between those two square roots. It's a lot easier to see in practice, so let's look at an example.

Example

$x^2 - 36$

Both the first term and the second term are perfect squares. There is no common multiple we can factor out of this binomial, so we are ready to split it into two factors. The square root of x^2 is x, and the square root of 36 is 6. The first factor is the sum of these two square roots, $(x + 6)$, and the second factor is the difference between them, $(x - 6)$. Therefore, $x^2 - 36$ factors into $(x + 6)(x - 6)$.

PITFALL

Be sure to list the terms in each factor in the correct order. For example, $x + 6$ is the same as $6 + x$, but $x - 6$ is not the same as $6 - x$. When writing the difference between perfect squares, it is always the square root of the first term minus the square root of the second term. To be sure your factoring is correct, multiply the factors to be sure they equal the original polynomial. In the preceding example, we factored $x^2 - 36$ into $(x + 6)$ $(x - 6)$. Use FOIL to check the answer: $(x)(x) + (x)(-6) + (6)(x) + (6)(-6) = x^2 - 6x + 6x - 36 = x^2 - 36$. Our answer is correct.

Example
$4y^2 - 36$

First, factor 4 out of this binomial, since it is the greatest common factor of the two terms: $4y^2 - 36 = 4(y^2 - 9)$. Now, factor $(y^2 - 9)$. It is the difference between two perfect squares. The square root of y^2 is y, and the square root of 9 is 3. $(y^2 - 9) = (y - 3)(y + 3)$, so $4y^2 - 36 = 4(y - 3)(y + 3)$.

If those two methods don't work, we will need to factor the outside terms of the polynomial and use them to help us figure out what two binomials were multiplied to form the polynomial.

Example
$x^2 + 10x + 21$

Factor x^2 and 21. The only possible factors of x^2 are x and x. The factors of 21 are 1, 3, 7, and 21. We need to find a pair of numbers that multiply to 21 and add to 10, because each of these numbers will be multiplied and then added, and we know from the polynomial that the sum of these products is $10x$. The numbers 3 and 7 multiply to 21 and add to 10, so these are the inside terms of our factors: $(x + 3)$ $(x + 7)$. Checking our answer, we find that $(x + 3)(x + 7) = x^2 + 7x + 3x + 21 = x^2 + 10x + 21$.

Example
$6x^2 + 19x + 15$

This problem is a little tougher. Two pairs of expressions that multiply to $6x^2$ are x and $6x$ and $2x$ and $3x$. Two pairs of numbers that multiply to 15 are 1 and 15 and 3 and 5. Let's test out some combinations:

$(6x + 1)(x + 15) = 6x^2 + 91x + 15$. The middle term is way too large, so these are not the right factors.

$(6x + 15)(x + 1) = 6x^2 + 21x + 15$. The middle term is still too large.

$(6x + 5)(x + 3) = 6x^2 + 23x + 15$. The middle term is still too large.

$(6x + 3)(x + 5) = 6x^2 + 33x + 15$. The middle term is still too large—let's try $3x$ and $2x$ as the first factors.

$(3x + 15)(2x + 1) = 6x^2 + 33x + 15$. The middle term is still too large.

$(3x + 1)(2x + 15) = 6x^2 + 47x + 15$. The middle term is way too large.

$(3x + 3)(2x + 5) = 6x^2 + 21x + 15$. The middle term is still too large.

$(3x + 5)(2x + 3) = 6x^2 + 19x + 15$. Got it!

TIP

You can see that some polynomials take a lot of trial and error to factor. Make a list of all factors of the first term and all factors of the last term before you begin factoring. This will help you keep track of which combinations you've tested, so that you don't test the same combinations more than once. That preceding problem took eight tries. Some of these practice problems may take even more tries, but keep track of your combinations and stick with it!

▶ **Practice Questions**

Factor the following polynomials.

659. $9a + 15$

660. $3a^2x + 9ax$

661. $x^2 - 16$

662. $4a^2 - 25$

663. $7n^2 - 28n$

664. $7x^4y^2 - 35x^2y^2 + 14x^2y^4$

665. $x^2 + 3x + 2$

666. $9r^2 - 49$

667. $x^2 - 2x - 8$

668. $x^2 + 5x + 6$

669. $x^2 + x - 6$

670. $b^2 - 100$

671. $x^2 + 7x + 12$

672. $x^2 - 3x - 18$

673. $b^2 - 6b + 8$

674. $b^2 - 4b - 21$

675. $a^2 + 11a - 12$

676. $x^2 + 10x + 25$

677. $36y^4 - 4z^2$

678. $x^2 + 20x + 99$

679. $c^2 - 12c + 32$

680. $h^2 - 12h + 11$

681. $m^2 - 11m + 18$

682. $v^4 - 13v^2 - 48$

683. $x^2 - 20x + 36$

684. $2x^2 + 7x + 6$

685. $3x^2 + 13x + 12$

686. $5x^2 - 14x - 3$

687. $9x^2 + 15x + 4$

688. $9x^2 + 34x - 8$

689. $3x^2 - 3x - 18$

690. $4a^2 - 16a - 9$

691. $6a^2 - 13a - 15$

692. $6a^2 - 5a - 6$

693. $16y^2 - 100$

694. $6x^2 + 15x - 36$

695. $4bc^2 + 22bc - 42b$

696. $2a^6 + a^3 - 21$

697. $6a^2x - 39ax - 72x$

698. $8x^2 - 6x - 9$

699. $5c^2 - 9c - 2$

700. $9x^3 - 4x$

701. $8r^2 + 46r + 63$

702. $4x^4 - 37x^2 + 9$

703. $12d^2 + 7d - 12$

704. $4xy^3 + 6xy^2 - 10xy$

705. $4ax^2 - 38ax - 66a$

706. $3c^2 + 19c - 40$

707. $2a^2 + 17a - 84$

708. $4x^4 + 2x^2 - 30$

TERMS TO REVIEW

monomial

binomials

trinomials

polynomials

FORMULAS TO REVIEW

multiplying a monomial by a polynomial: $a(b + c + d)$
$= ab + ac + ad$

multiplying a binomial by a binomial: $(a + b)(c + d) =$
$ac + ad + bc + bd$

multiplying a polynomial by a polynomial: $(a + b + c)$
$(d + e + f) = ad + ae + af + bd + be + bf + cd + ce + cf$

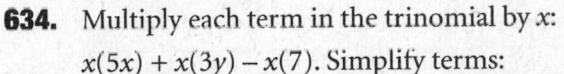 **Answers**

634. Multiply each term in the trinomial by x:
$x(5x) + x(3y) - x(7)$. Simplify terms:
$5x^2 + 3xy - 7x$.

635. Multiply each term in the trinomial by $2a$:
$2a(5a^2) - 2a(7a) + 2a(9)$. Simplify terms:
$10a^3 - 14a^2 + 18a$.

636. Multiply each term in the polynomial by $4bc$:
$4bc(3b^2c) + 4bc(7b) - 4bc(9c) + 4bc(2bc^2) -$
$4bc(8)$. Simplify terms: $12b^3c^2 + 28b^2c - 36bc^2$
$+ 8b^2c^3 - 32bc$.

637. Multiply each term in the polynomial by $3mn$:
$3mn(-4m) + 3mn(6n) + 3mn(7mn^2) -$
$3mn(3m^2n)$. Simplify terms: $-12m^2n + 18mn^2$
$+ 21m^2n^3 - 9m^3n^2$.

638. Multiply each term in the polynomial by $4x$: $4x(9x^3) + 4x(\frac{3}{x^2}) - 4x(x^4) + 4x(\frac{6x-1}{x^2})$. Simplify terms: $36x^4 + \frac{12x}{x^2} - 4x^5 + \frac{4x(6x-1)}{x^2}$. Use the distributive property in the numerator of the fourth term: $36x^4 + \frac{12x}{x^2} - 4x^5 + \frac{24x^2 - 4x}{x^2}$. When similar factors, or bases, are being divided, subtract the exponent in the denominator from the exponent in the numerator: $36x^4 + 12x^{1-2} - 4x^5 + 24x^{2-2} - 4x^{1-2}$. Simplify operations in the exponents: $36x^4 + 12x^{-1} - 4x^5 + 24x^0 - 4x^{-1}$. Use the associative property of addition: $36x^4 + 12x^{-1} - 4x^{-1} - 4x^5 + 24x^0$. Combine like terms: $36x^4 + 8x^{-1} - 4x^5 + 24x^0$. A base with a negative exponent in the numerator is equivalent to the same variable or base in the denominator with the inverse sign for the exponent: $36x^4 + \frac{8}{x} - 4x^5 + 24x^0$. A variable to the power of zero equals 1: $36x^4 + \frac{8}{x} - 4x^5 + 24(1)$. Simplify and put in order: $-4x^5 + 36x^4 + \frac{8}{x} + 24$.

639. Use FOIL to multiply binomials: $(x + 3)(x + 6)$. Multiply the first terms in each binomial: $x \times x = x^2$. Multiply the outer terms in each binomial: $x \times 6 = 6x$. Multiply the inner terms in each binomial: $3 \times x = 3x$. Multiply the last terms in each binomial: $3 \times 6 = 18$. Add the products of FOIL together: $x^2 + 6x + 3x + 18$. Combine like terms: $x^2 + 9x + 18$.

640. Use FOIL to multiply binomials: $(x - 4)(x - 9)$. Multiply the first terms in each binomial: $x \times x = x^2$. Multiply the outer terms in each binomial: $x \times -9 = -9x$. Multiply the inner terms in each binomial: $-4 \times x = -4x$. Multiply the last terms in each binomial: $-4 \times -9 = 36$. Add the products of FOIL together: $x^2 - 9x - 4x + 36$. Combine like terms: $x^2 - 13x + 36$.

641. Use FOIL to multiply binomials: $(2x + 1)(3x - 7)$. Multiply the first terms in each binomial: $2x \times 3x = 6x^2$. Multiply the outer terms in each binomial: $2x \times -7 = -14x$. Multiply the inner terms in each binomial: $1 \times 3x = 3x$. Multiply the last terms in each binomial: $1 \times -7 = -7$. Add the products of FOIL together: $6x^2 - 14x + 3x - 7$. Combine like terms: $6x^2 - 11x - 7$.

642. Use FOIL to multiply binomials: $(x + 2)(x - 3y)$. Multiply the first terms in each binomial: $x \times x = x^2$. Multiply the outer terms in each binomial: $x \times -3y = -3xy$. Multiply the inner terms in each binomial: $2 \times x = 2x$. Multiply the last terms in each binomial: $2 \times -3y = -6y$. Add the products of FOIL together: $x^2 - 3xy + 2x - 6y$.

643. Use FOIL to multiply binomials: $(7x + 2y)(2x - 4y)$. Multiply the first terms in each binomial: $7x \times 2x = 14x^2$. Multiply the outer terms in each binomial: $7x \times 4y = -28xy$. Multiply the inner terms in each binomial: $2y \times 2x = 4xy$. Multiply the last terms in each binomial: $2y \times -4y = -8y^2$. Add the products of FOIL together: $14x^2 - 28xy + 4xy - 8y^2$. Combine like terms: $14x^2 - 24xy - 8y^2$.

644. Use FOIL to multiply binomials: $(5x + 7)(5x - 7)$. Multiply the first terms in each binomial: $5x \times 5x = 25x^2$. Multiply the outer terms in each binomial: $5x \times -7 = -35x$. Multiply the inner terms in each binomial: $7 \times 5x = 35x$. Multiply the last terms in each binomial: $7 \times -7 = -49$. Add the products of FOIL together: $25x^2 - 35x + 35x - 49$. Combine like terms: $25x^2 - 49$.

645. Use FOIL to multiply binomials: $(28x + 7)$ $(\frac{x}{7} - 11)$. Multiply the first terms in each binomial: $28x \times \frac{x}{7} = 4x^2$. Multiply the outer terms in each binomial: $28x \times 11 = -308x$. Multiply the inner terms in each binomial: $7 \times \frac{x}{7} = x$. Multiply the last terms in each binomial: $7 \times -11 = -77$. Add the products of FOIL together: $4x^2 - 308x + x - 77$. Combine like terms: $4x^2 - 307x - 77$.

646. Use FOIL to multiply binomials: $(3x^2 + y^2)$ $(x^2 - 2y^2)$. Multiply the first terms in each binomial: $3x^2 \times x^2 = 3x^4$. Multiply the outer terms in each binomial: $3x^2 \times 2y^2 = -6x^2y^2$. Multiply the inner terms in each binomial: $y^2 \times x^2 = x^2y^2$. Multiply the last terms in each binomial: $y^2 \times -2y^2 = -2y^4$. Add the products of FOIL together: $3x^4 - 6x^2y^2 + x^2y^2 - 2y^4$. Combine like terms: $3x^4 - 5x^2y^2 - 2y^4$.

647. Use FOIL to multiply binomials: $(4 + 2x^2)$ $(9 - 3x)$. Multiply the first terms in each binomial: $4 \times 9 = 36$. Multiply the outer terms in each binomial: $4 \times -3x = -12x$. Multiply the inner terms in each binomial: $2x^2 \times 9 = 18x^2$. Multiply the last terms in each binomial: $2x^2 \times -3x = -6x^3$. Add the products of FOIL together: $36 - 12x + 18x^2 - 6x^3$. Simplify and put them in order from the highest power: $-6x^3 + 18x^2 - 12x + 36$.

648. Use FOIL to multiply binomials: $(2x^2 + y^2)$ $(x^2 - y^2)$. Multiply the first terms in each binomial: $2x^2 \times x^2 = 2x^4$. Multiply the outer terms in each binomial: $2x^2 \times -y^2 = -2x^2y^2$. Multiply the inner terms in each binomial: $y^2 \times x^2 = x^2y^2$. Multiply the last terms in each binomial: $y^2 \times -y^2 = -y^4$. Add the products of FOIL together: $2x^4 - 2x^2y^2 + x^2y^2 - y^4$. Combine like terms: $2x^4 - x^2y^2 - y^4$.

649. Multiply the trinomial by the first term in the binomial, x: $x(3x^2 - 5x + 2)$; $x(3x^2) - x(5x)$ $+ x(2)$. Simplify terms: $3x^3 - 5x^2 + 2x$. Multiply the trinomial by the second term in the binomial, 2: $2(3x^2 - 5x + 2)$; $2(3x^2) - 2(5x)$ $+ 2(2)$. Simplify terms: $6x^2 - 10x + 4$. Add together the results of multiplying by the terms in the binomial: $3x^3 - 5x^2 + 2x + 6x^2$ $- 10x + 4$. Use the commutative property of addition: $3x^3 - 5x^2 + 6x^2 + 2x - 10x + 4$. Combine like terms: $3x^3 + x^2 - 8x + 4$.

650. Multiply the trinomial by the first term in the binomial, $2x$: $2x(x^3 + 3x^2 - 4x)$. Use the distributive property of multiplication: $2x(x^3) + 2x(3x^2) - 2x(4x)$. Simplify terms: $2x^4 + 6x^3 - 8x^2$. Multiply the trinomial by the second term in the binomial, -3: $-3(x^3 + 3x^2 - 4x)$. Use the distributive property of multiplication: $-3(x^3) - 3(3x^2)$ $- 3(-4x)$. Simplify terms: $-3x^3 - 9x^2 + 12x$. Add together the results of multiplying by the terms in the binomial: $2x^4 + 6x^3 - 8x^2 + -3x^3$ $- 9x^2 + 12x)$. Use the commutative property of addition: $2x^4 + 6x^3 - 3x^3 - 8x^2 - 9x^2 + 12x$. Combine like terms: $2x^4 + 3x^3 - 17x^2 + 12x$.

651. Multiply the trinomial by the first term in the binomial, $4a$: $4a(5a^2 + 2ab - b^2)$. Use the distributive property of multiplication: $4a(5a^2) + 4a(2ab) - 4a(b^2)$. Simplify terms: $20a^3 + 8a^2b - 4ab^2$. Multiply the trinomial by the second term in the binomial, b: $b(5a^2 + 2ab - b^2)$. Use the distributive property of multiplication: $b(5a^2) + b(2ab)$ $- b(b^2)$. Simplify terms: $5a^2b + 2ab^2 - b^3$. Add together the results of multiplying by the terms in the binomial: $20a^3 + 8a^2b - 4ab^2$ $+ 5a^2b + 2ab^2 - b^3)$. Use the commutative property of addition: $20a^3 + 8a^2b + 5a^2b$ $- 4ab^2 + 2ab^2 - b^3$. Combine like terms: $20a^3 + 13a^2b - 2ab^2 - b^3$.

652. Multiply the trinomial by the first term in the binomial, $3y$: $3y(6y^2 - 3y + 7)$. Use the distributive property of multiplication: $3y(6y^2) - 3y(3y) + 3y(7)$. Simplify terms: $18y^3 - 9y^2 + 21y$. Multiply the trinomial by the second term in the binomial, -7: $-7(6y^2 - 3y + 7)$. Use the distributive property of multiplication: $-7(6y^2) - 7(-3y) - 7(7)$. Simplify terms: $-42y^2 + 21y - 49$. Add together the results of multiplying by the terms in the binomial: $18y^3 - 9y^2 + 21y + -42y^2 + 21y - 49$. Use the commutative property of addition: $18y^3 - 9y^2 - 42y^2 + 21y + 21y - 49$. Combine like terms: $18y^3 - 51y^2 + 42y - 49$.

653. Multiply the trinomial by the first term in the binomial, $3x$: $3x(3x^2 - 2x - 5)$. Use the distributive property of multiplication: $3x(3x^2) - 3x(2x) - 3x(5)$. Simplify terms: $9x^3 - 6x^2 - 15x$. Multiply the trinomial by the second term in the binomial, 2: $2(3x^2 - 2x - 5)$. Use the distributive property of multiplication: $2(3x^2) - 2(2x) - 2(5)$. Simplify terms: $6x^2 - 4x - 10$. Add together the results of multiplying by the terms in the binomial: $9x^3 - 6x^2 - 15x + 6x^2 - 4x - 10$. Use the commutative property of addition: $9x^3 - 6x^2 + 6x^2 - 15x - 4x - 10$. Combine like terms: $9x^3 - 19x - 10$.

654. Multiply the first two parenthetical terms in the expression using FOIL: $(x + 2)(2x + 1)$. Multiply the first terms in each binomial: $x \times 2x = 2x^2$. Multiply the outer terms in each binomial: $x \times 1 = x$. Multiply the inner terms in each binomial: $2 \times 2x = 4x$. Multiply the last terms in each binomial: $2 \times 1 = 2$. Add the products of FOIL together: $2x^2 + x + 4x + 2$. Combine like terms: $2x^2 + 5x + 2$. Multiply the resulting trinomial by the last binomial in the original expression: $(x - 1)(2x^2 + 5x + 2)$. Multiply the trinomial by the first term in the binomial, x: $x(2x^2 + 5x + 2)$. Use the distributive property of multiplication: $x(2x^2) + x(5x) + x(2)$. Simplify terms: $2x^3 + 5x^2 + 2x$. Multiply the trinomial by the second term in the binomial, -1: $-1(2x^2 + 5x + 2)$. Use the distributive property of multiplication: $-2x^2 - 5x - 2$. Add together the results of multiplying by the terms in the binomial: $2x^3 + 5x^2 + 2x + -2x^2 - 5x - 2$. Use the commutative property of addition: $2x^3 + 5x^2 - 2x^2 + 2x - 5x - 2$. Combine like terms: $2x^3 + 3x^2 - 3x - 2$.

655. Multiply the first two parenthetical terms in the expression using FOIL: $(3a - 4)(5a + 2)$. Multiply the first terms in each binomial: $3a \times 5a = 15a^2$. Multiply the outer terms in each binomial: $3a \times 2 = 6a$. Multiply the inner terms in each binomial: $-4 \times 5a = -20a$. Multiply the last terms in each binomial: $-4 \times 2 = -8$. Add the products of FOIL together: $15a^2 + 6a - 20a - 8$. Combine like terms: $15a^2 - 14a - 8$. Multiply the resulting trinomial by the last binomial in the original expression: $(a + 3)(15a^2 - 14a - 8)$. Multiply the trinomial by the first term in the binomial, a: $a(15a^2 - 14a - 8)$. Use the distributive property of multiplication: $a(15a^2) - a(14a) - a(8)$. Simplify terms: $15a^3 - 14a^2 - 8a$. Multiply the trinomial by the second term in the binomial, 3: $3(15a^2) - 3(14a) - 3(8)$. Use the distributive property of multiplication: $45a^2 - 42a - 24$. Add together the results of multiplying by the terms in the binomial: $15a^3 - 14a^2 - 8a + 45a^2 - 42a - 24$. Use the commutative property of addition: $15a^3 - 14a^2 + 45a^2 - 8a - 42a - 24$. Combine like terms: $15a^3 + 31a^2 - 50a - 24$.

656. Multiply the first two parenthetical terms in the expression using FOIL: $(2n - 3)(2n + 3)$. Multiply the first terms in each binomial: $2n \times 2n = 4n^2$. Multiply the outer terms in each binomial: $2n \times 3 = 6n$. Multiply the inner terms in each binomial: $-3 \times 2n = -6n$. Multiply the last terms in each binomial: $-3 \times 3 = -9$. Add the products of FOIL together: $4n^2 + 6n - 6n - 9$. Combine like terms: $4n^2 - 9$. Now we again have two binomials. Use FOIL to find the solution: $(n + 4)(4n^2 - 9)$. Multiply the first terms in each binomial: $n \times 4n^2 = 4n^3$. Multiply the outer terms in each binomial: $n \times -9 = -9n$. Multiply the inner terms in each binomial: $4 \times 4n^2 = 16n^2$. Multiply the last terms in each binomial: $4 \times -9 = -36$. Add the products of FOIL together: $4n^3 - 9n + 16n^2 - 36$. Order terms from the highest to lowest power: $4n^3 + 16n^2 - 9n - 36$.

657. Multiply the trinomial by the first term in the binomial, $5r$: $5r(3r^4 + 2r^2 + 6)$. Use the distributive property of multiplication: $5r(3r^4) + 5r(2r^2) + 5r(6)$. Simplify terms: $15r^5 + 10r^3 + 30r$. Multiply the trinomial by the second term in the binomial, -7: $-7(3r^4 + 2r^2 + 6)$. Use the distributive property of multiplication: $-7(3r^4) - 7(2r^2) - 7(6)$. Simplify terms: $-21r^4 - 14r^2 - 42$. Add together the results of multiplying by the terms in the binomial: $15r^5 + 10r^3 + 30r + -21r^4 - 14r^2 - 42$. Use the commutative property of addition: $15r^5 - 21r^4 + 10r^3 - 14r^2 + 30r - 42$.

658. Multiply the first two parenthetical terms in the expression using FOIL: $(3x^2 + 4)(x - 3)$. Multiply the first terms in each binomial: $3x^2 \times x = 3x^3$. Multiply the outer terms in each binomial: $3x^2 \times -3 = -9x^2$. Multiply the inner terms in each binomial: $4 \times x = 4x$. Multiply the last terms in each binomial: $4 \times -3 = -12$. Add the products of FOIL together: $3x^3 - 9x^2 + 4x - 12$. Multiply the resulting polynomial by the last binomial in the original expression: $(3x^2 - 4)(3x^3 - 9x^2 + 4x - 12)$. Multiply the polynomial by the first term in the binomial, $3x^2$: $3x^2(3x^3 - 9x^2 + 4x - 12)$. Use the distributive property of multiplication: $3x^2(3x^3) - 3x^2(9x^2) + 3x^2(4x) - 3x^2(12)$. Simplify terms: $9x^5 - 27x^4 + 12x^3 - 36x^2$. Multiply the polynomial by the second term in the binomial, -4: $-4(3x^3 - 9x^2 + 4x - 12)$. Use the distributive property of multiplication: $-12x^3 + 36x^2 - 16x + 48$. Add together the results of multiplying by the terms in the binomial: $9x^5 - 27x^4 + 12x^3 - 36x^2 + -12x^3 + 36x^2 - 16x + 48$. Use the commutative property of addition: $9x^5 - 27x^4 + 12x^3 - 12x^3 - 36x^2 + 36x^2 - 16x + 48$. Combine like terms: $9x^5 - 27x^4 - 16x + 48$.

659. The terms have a common factor of 3. Factor 3 out of each term and write the expression in factored form: $9a + 15 = 3(3a + 5)$.

660. The terms have a common factor of $3ax$. Factor $3ax$ out of each term and write the expression in factored form: $3a^2x + 9ax = 3ax(a + 3)$.

661. Both terms in the polynomial are perfect squares. Use the form for factoring the difference of two perfect squares and put the roots of each factor in the proper place: $x^2 - 16 = (x + 4)(x - 4)$. Check using FOIL: $(x + 4)(x - 4) = x^2 - 4x + 4x - 16 = x^2 - 16$.

662. Both terms in the polynomial are perfect squares. $4a^2 = (2a)^2$, and $25 = 5^2$. Use the form for factoring the difference of two perfect squares and put the roots of each factor in the proper place: $4a^2 - 25 = (2a + 5)(2a - 5)$. Check using FOIL: $(2a + 5)(2a - 5) = 4a^2 - 10a + 10a - 25 = 4a^2 - 25$.

663. The terms have a common factor of $7n$. Factor $7n$ out of each term and write the expression in factored form: $7n^2 - 28n = 7n(n - 4)$.

664. The terms have a common factor of $7x^2y^2$. Factor $7x^2y^2$ out of each term in the expression and write it in factored form: $7x^2y^2(x^2 - 5 + 2y^2)$.

665. This expression can be factored using the trinomial method. The factors of x^2 are x and x, and the factors of 2 are 1 and 2. Place the factors into the trinomial factor form and check using FOIL: $(x + 2)(x + 1) = x^2 + x + 2x + 2 = x^2 + 3x + 2$. The factors are correct.

666. Both terms in the polynomial are perfect squares: $9r^2 = (3r)^2$ and $49 = 7^2$. Use the form for factoring the difference between two perfect squares and put the roots of each factor in the proper place: $9r^2 - 49 = (3r + 7)(3r - 7)$.

667. This expression can be factored using the trinomial method. The factors of x^2 are x and x, and the factors of 8 are $(1)(8)$ and $(2)(4)$. You want the result of the outer and inner products of the FOIL method for multiplying factors to add up to $-2x$. Only terms with opposite signs will result in a negative numerical term, which is what you need, since the third term is -8. Place the factors $(2)(4)$ into the trinomial factor form and check using FOIL: $(x + 4)(x - 2) = x^2 - 2x + 4x - 8 = x^2 + 2x - 8$. Almost correct! Change the position of the factors of the numerical term and check using FOIL: $(x + 2)(x - 4) = x^2 - 4x + 2x - 8 = x^2 - 2x - 8$. The factors of the trinomial are now correct.

668. This expression can be factored using the trinomial method. The factors of x^2 are x and x, and the factors of 6 are $(1)(6)$ and $(2)(3)$. Since the numerical term of the polynomial is positive, the signs in the factor form for trinomials will be the same, because only two like signs multiplied together will result in a positive. Now consider the second term in the trinomial. In order to add up to $5x$, the result of multiplying the inner and outer terms of the trinomial factors will have to be positive. Try using two positive signs and the factors 2 and 3, which add up to 5. Check using FOIL: $(x + 2)(x + 3) = x^2 + 2x + 3x + 6 = x^2 + 5x + 6$.

669. This expression can be factored using the trinomial method. The factors of x^2 are x and x, and the factors of 6 are $(1)(6)$ and $(2)(3)$. You want the result of the outer and inner products of the FOIL method for multiplying factors to add up to $1x$. Only terms with opposite signs will result in a negative numerical term that you need with the third term being -6. Place the factors $(2)(3)$ into the trinomial factor form and check using FOIL: $(x + 3)(x - 2) = x^2 - 2x + 3x - 6 = x^2 + x - 6$. The factors of the trinomial are now correct.

670. Both terms in the polynomial are perfect squares: $b^2 = (b)^2$ and $100 = 10^2$. Use the form for factoring the difference between two perfect squares and put the roots of each factor in the proper place: $b^2 - 100 = (b + 10)(b - 10)$. Check using FOIL: $(b + 10)(b - 10) = b^2 - 10b + 10b - 100 = b^2 - 100$.

671. This expression can be factored using the trinomial method. The factors of x^2 are x and x, and the factors of 12 are $(1)(12)$ or $(2)(6)$ or $(3)(4)$. You want the result of the outside and inside terms of the FOIL method for multiplying factors to add up to $7x$. The factors $(3)(4)$ would give terms that add up to 7. Since all signs are positive, use positive signs in the factored form for the trinomial: $(x + 3)(x + 4) = x^2 + 3x + 4x + 12 = x^2 + 7x + 12$. The result is correct. This is not just luck. You can use logical guesses to find the correct combination of factors and signs.

672. This expression can be factored using the trinomial method. The factors of x^2 are x and x, and the factors of 18 are $(1)(18)$ or $(2)(9)$ or $(3)(6)$. Only the product of a positive and a negative numerical term will result in -18. The sum of the results of multiplying the outer and inner terms of the trinomial factors needs to add up to $-3x$. So use $(3)(6)$ in the trinomial factors form and check using FOIL: $(x+3)(x-6) = x^2 - 6x + 3x - 18 = x^2 - 3x - 18$.

673. This expression can be factored using the trinomial method. The factors of b^2 are b and b, and the factors of 8 are $(1)(8)$ or $(2)(4)$. You want the result of the outer and inner products of the FOIL method for multiplying factors to add up to $-6b$. The signs within the parentheses of the factorization of the trinomial must be the same to result in a positive numerical term in the trinomial. The middle term has a negative sign, so let's try two negative signs. How can you get 6 from adding two of the factors of 8? Right! Use the $(2)(4)$. Check the answer using FOIL: $(b-2)(b-4) = b^2 - 4b - 2b + 8 = b^2 - 6b + 8$.

674. This expression can be factored using the trinomial method. The factors of b^2 are b and b, and the factors of 21 are $(1)(21)$ or $(3)(7)$. You want the result of the outer and inner products of the FOIL method for multiplying factors to add up to $-4b$. Only the product of a positive and a negative numerical term will result in -21. So let's use (3) and (7) in the trinomial factors form, because the difference between 3 and 7 is 4. Check using FOIL: $(b+3)(b-7) = b^2 - 7b + 3b - 21 = b^2 - 4b - 21$.

675. This expression can be factored using the trinomial method. The factors of a^2 are a and a, and the factors of 12 are $(1)(12)$ or $(2)(6)$ or $(3)(4)$. You want the result of the outer and inner products of the FOIL method for multiplying factors to add up to $11a$. Only the product of a positive and a negative numerical term will result in -12. Since the signs in the factors must be one positive and one negative, use the factors 12 and 1 in the trinomial factors form. Use FOIL to check the answer: $(a+12)(a-1) = a^2 - 1a + 12a - 12 = a^2 + 11a - 12$.

676. This expression can be factored using the trinomial method. The factors of x^2 are x and x, and the factors of 25 are $(1)(25)$ or $(5)(5)$. To get a positive 25 after multiplying the factors of the trinomial expression, the signs in the two factors must both be positive or both be negative. The sum of the results of multiplying the outer and inner terms of the trinomial factors needs to add up to $10x$. So let's use $(5)(5)$ in the trinomial factors form and check using FOIL: $(x+5)(x+5) = x^2 + 5x + 5x + 25 = x^2 + 10x + 25$.

677. Both terms in the polynomial are perfect squares: $36y^4 = (6y^2)^2$ and $4z^2 = (2z)^2$. Use the form for factoring the difference between two perfect squares and put the roots of each factor in the proper place: $(6y^2 + 2z)(6y^2 - 2z)$. Check using FOIL: $(6y^2 + 2z)(6y^2 - 2z) = 36y^4 - 12y^2z + 12y^2z - 4z^2 = 36y^4 - 4z^2$.

678. This expression can be factored using the trinomial method. The factors of x^2 are x and x, and the factors of 99 are $(1)(99)$ or $(3)(33)$ or $(9)(11)$. To get a positive 99 after multiplying the factors of the trinomial expression, the signs in the two factors must both be positive or both be negative. The sum of the results of multiplying the outer and inner terms of the trinomial factors needs to add up to $20x$. So let's use $(9)(11)$ in the trinomial factors form, because $9 + 11 = 20$. Check using FOIL: $(x + 9)(x + 11) = x^2 + 9x + 11x + 99 = x^2 + 20x + 99$.

679. This expression can be factored using the trinomial method. The factors of c^2 are c and c, and the factors of 32 are $(1)(32)$ or $(2)(16)$ or $(4)(8)$. The sign of the numerical term is positive, so the signs in the factors of our trinomial factorization must be the same (both positve or both negative). The sign of the first-degree term (the variable to the power of 1) is negative. This leads one to believe that the signs in the trinomial factors will both be negative. The only factors of 32 that add up to 12 are 4 and 8. Check using FOIL: $(c - 4)(c - 8) = c^2 - 8c - 4c + 32 = c^2 - 12c + 32$.

680. The factors of h^2 are h and h, and the factors of 11 are $(1)(11)$. The sign of the numerical term is positive, so the signs in the factors of our trinomial factorization must be the same (both positive or both negative). The sign of the first-degree term (the variable to the power of 1) is negative. So use negative signs in the trinomial factors. Check your answer: $(h - 1)(h - 11) = h^2 - 11h - 1h + 11 = h^2 - 12h + 11$.

681. This expression can be factored using the trinomial method. The factors of m^2 are m and m, and the factors of 18 are $(1)(18)$ or $(2)(9)$ or $(3)(6)$. The sign of the numerical term is positive, so the signs in the factors of our trinomial factorization must be the same (both positive or negative). The sum of the results of multiplying the outer and inner terms of the trinomial factors needs to add up to $-11m$. The only factors of 18 that can be added or subtracted in any way to equal 11 are 2 and 9. Use them and two subtraction signs in the trinomial factor terms. Check your answer using FOIL. $(m - 2)(m - 9) = m^2 - 9m - 2m + 18 = m^2 - 11m + 18$.

682. This expression can be factored using the trinomial method. The factors of v^4 are $(v^2)(v^2)$, and the factors of 48 are $(1)(48)$ or $(2)(24)$ or $(3)(16)$ or $(4)(12)$ or $(6)(8)$. Only the product of a positive and a negative numerical term will result in -48. The only factors of 48 that can be added or subtracted in any way to equal 13 are 3 and 16. Use 3 and 16 and a positive and negative sign in the terms of the trinomial factors. Check your answer using FOIL: $(v^2 + 3)(v^2 - 16) = v^4 - 16v^2 + 3v^2 - 48 = v^4 - 13v - 48$. You may notice that one of the two factors of the trinomial expression can itself be factored. The second term is the difference between two perfect squares. Factor $(v^2 - 16)$ using the form for factoring the difference between two perfect squares: $(v + 4)(v - 4) = v^2 - 4v + 4v - 16 = v^2 - 16$. This now makes the factorization complete: $v^4 - 13v^2 - 48 = (v^2 + 3)(v + 4)(v - 4)$.

683. The factors of x^2 are x and x, and the factors of 36 are $(1)(36)$ or $(2)(18)$ or $(4)(9)$ or $(6)(6)$. The sign of the numerical term is positive, so the signs in the factors of our trinomial factorization must be the same (both positive or both negative). The sign of the first-degree term (the variable to the power of 1) is negative. This leads one to believe that the signs in the trinomial factors will both be negative. The only factors of 36 that add up to 20 are 2 and 18. Use them and two negative signs in the trinomial factor form. Check your answer using FOIL: $(x - 2)(x - 18) = x^2 - 18x - 2x + 36 = x^2 - 20x + 36$.

684. Both signs in the trinomial are positive, so use positive signs in the trinomial factor form: $(ax + \ )(bx + \ )$. The factors of the second-degree term $2x^2$ are $(x)(2x)$. The factors of the numerical term 6 are $(1)(6)$ or $(2)(3)$. You want to get $7x$ from adding the result of the outer and inner multiplications when using FOIL. You could make the following guesses for the factors of the original expression: $(2x + 1)(x + 6)$; $(2x + 6)(x + 1)$; $(2x + 2)(x + 3)$; $(2x + 3)(x + 2)$. Now just consider the results of the outer and inner products of the terms for each guess. The one that results in a first-degree term of $7x$ is the factorization you want to fully check. $(2x + 1)(x + 6)$ will result in outer product plus inner product: $2x(6) + (1)x = 12x + x = 13x$. $(2x + 6)(x + 1)$ will result in outer product plus inner product: $2x(1) + (6)x = 2x + 6x = 8x$. $(2x + 2)(x + 3)$ will result in outer product plus inner product: $2x(3) + (2)x = 6x + 2x = 8x$. $(2x + 3)(x + 2)$ will result in outer product plus inner product: $2x(2) + (3)x = 4x + 3x = 7x$. Place the factors in the trinomial factor form so that the product of the outer terms $(2x)(2) = 4x$ and the product of the inner terms $(3)(x) = 3x$. That way, $4x + 3x = 7x$, the middle term of the trinomial. $(2x + 3)(x + 2)$. Check using FOIL: first—$(2x)(x) = 2x^2$; outer—$(2x)(2) = 4x$; inner—$(3)(x) = 3x$; last—$(3)(2) = 6$. Add the products of multiplication using FOIL: $2x^2 + 4x + 3x + 6 = 2x^2 + 7x + 6$. The factors check out. $(2x + 3)(x + 2) = 2x^2 + 7x + 6$.

685. Both signs in the trinomial are positive, so use positive signs in the trinomial factor form: $(ax + \quad)(bx + \quad)$. The factors of the second-degree term $3x^2$ are $(x)(3x)$. The factors of the numerical term 12 are $(1)(12)$ or $(2)(6)$ or $(3)(4)$. You want to get $13x$ from adding the result of the outer and inner multiplications when using FOIL. Place the factors in the trinomial factor form so that the product of the outer terms $(3x)(3) = 9x$ and the product of the inner terms $(4)(x) = 4x$. Then $9x + 4x = 13x$, the middle term of the trinomial. $(3x + 4)(x + 3)$. Check using FOIL: first—$(3x)(x) = 3x^2$; outer—$(3x)(3) = 9x$; inner—$(4)(x) = 4x$; last—$(4)(3) = 12$. Add the products of multiplication using FOIL: $3x^2 + 9x + 4x + 12 = 3x^2 + 13x + 12$. The factors check out: $(3x + 4)(x + 3) = 3x^2 + 13x + 12$.

686. Both signs in the trinomial are negative. To get a negative sign for the numerical term, the signs in the factors must be + and –: $(ax + \quad)(bx - \quad)$. The factors of the second-degree term $5x^2$ are $(x)(5x)$. The factors of the numerical term 3 are $(1)(3)$. When you multiply the outer and inner terms of the trinomial factors, the results must add up to be $-14x$. Multiplying, $5x(-3) = -15x$, and $1x(1) = 1x$. Adding $(-15x) + (1x) = -14x$. Place those terms into the trinomial factor form. $(5x + 1)(x - 3)$. Check using FOIL: first—$(5x)(x) = 5x^2$; outer—$(5x)(-3) = -15x$; inner—$(1)(x) = x$; last—$(1)(-3) = -3$. Add the products of multiplication using FOIL: $5x^2 - 15x + 1x - 3 = 5x^2 - 14x - 3$. The factors check out: $(5x + 1)(x - 3) = 5x^2 - 14x - 3$.

687. Both signs in the trinomial are positive, so use positive signs in the trinomial factor form: $(ax + \quad)(bx + \quad)$. The factors of the second-degree term $9x^2$ are $(x)(9x)$ or $(3x)(3x)$. The factors of the numerical term 4 are $(1)(4)$ or $(2)(2)$. To get $15x$ from adding the result of the outer and inner multiplications when using FOIL, place the factors in the trinomial factor form so that the product of the outer terms $(3x)(1) = 3x$ and the product of the inner terms $(4)(3x) = 12x$. Then $3x + 12x = 15x$, the middle term of the trinomial. $(3x + 4)(3x + 1)$. Check using FOIL: first—$(3x)(3x) = 9x^2$; outer—$(3x)(1) = 3x$; inner—$(4)(3x) = 12x$; last—$(4)(1) = 4$. The factors check out: $(3x + 4)(3x + 1) = 9x^2 + 3x + 12x + 4 = 9x^2 + 15x + 4$.

688. The sign of the numerical term is negative, so the signs in the trinomial factor form will have to be + and –. That is the only way to get a negative sign by multiplying the last terms when checking with FOIL: $(ax + \quad)(bx - \quad)$. The factors of the second-degree term $9x^2$ are $(x)(9x)$ or $(3x)(3x)$. The factors of the numerical term 8 are $(1)(8)$ or $(2)(4)$. Let's try putting in factors in the trinomial factor form and see what we get: $(9x + 1)(x - 8)$. Using FOIL to check, we get $9x^2 - 72x + 1x - 8 = 9x^2 - 71x - 8$. No, that doesn't work. You are looking for a positive x term in the middle of the expression. Changing the position of the signs would help, but not with these factors, because the term would be $71x$. Try different factors of 8: $(9x - 2)(x + 4)$. Check using FOIL: $9x^2 + 36x - 2x - 8 = 9x^2 + 34x - 8$. There it is! And on only the second try. Be persistent and learn from your mistakes.

689. The three terms have a common factor of 3. You can factor out 3 and represent the trinomial as $3(x^2 - x - 6)$. Now factor the trinomial in the parentheses, and don't forget to include the factor 3 when you are done. The sign of the numerical term is negative, so the signs in the trinomial factor form will have to be + and −, because that is the only way to get a negative sign when multiplying the last terms when checking with FOIL: $(ax + \quad)(bx - \quad)$. The factors of the second-degree term $3x^2$ are $(x)(3x)$. The sign of the second term is negative. That tells you that the result of adding the products of the outer and inner terms of the trinomial factors must result in a negative sum for the x term. The factors of the numerical term 6 are $(1)(6)$ or $(2)(3)$. Put the + with the 2 and the − with the 3: $(x + 2)(x - 3)$. Check using FOIL: first—$(x)(x) = x^2$; outer—$(x)(-3) = -3x$; inner—$(2)(x) = 2x$; last—$(2)(-3) = -6$. The factors check out: $(x + 2)(x - 3) = x^2 - 3x + 2x - 6 = x^2 - x - 6$. Include the common factor of 3: $3(x + 2)(x - 3)$ $= 3(x^2 - x - 6) = 3x^2 - 3x - 18$.

690. Both signs in the trinomial expression are negative. To get a negative sign for the numerical term, the signs within the trinomial factors must be + and −: $(ax + \quad)(bx - \quad)$. The factors of the second-degree term $4a^2$ are $(a)(4a)$ or $(2a)(2a)$. The factors of the numerical term 9 are $(1)(9)$ or $(3)(3)$. The coefficient of the first-degree term is 2 less than 18. You can multiply $2a$ and (9) to get $18a$, leaving the factors $(2a)$ and (1) to get a $2a$. Use this information to place factors within the trinomial factor form: $(2a + 1)$ $(2a - 9)$. Check using FOIL: first—$(2a)(2a) = 4a^2$; outer—$(2a)(-9) = -18a$; inner—$(1)(2a) = 2a$; last—$(1)(-9) = -9$. The result of multiplying the factors is $(2a + 1)(2a - 9) = 4a^2 - 18a + 2a - 9 = 4a^2 - 16a - 9$.

691. Both signs in the trinomial expression are negative. To get a negative sign for the numerical term, the signs within the trinomial factors must be + and −: $(ax + \quad)(bx - \quad)$. The factors of the second-degree term $6a^2$ are $(a)(6a)$ or $(2a)(3a)$. The factors of the numerical term 15 are $(1)(15)$ or $(3)(5)$. We can predict that $13 = 18 - 5$. The factors $(6a)(3) = 18a$. The remaining factors $(a)(5) = 5a$. But we need the 13 to be negative, so arrange the $6a$ and the (3) so their product is $-18a$: $(6a + 5)(a - 3)$. Check using FOIL: first—$(6a)(a) = 6a^2$; outer—$(6a)(-3) = -18a$; inner—$(5)(a) = 5a$; last—$(5)(-3) = -15$. Combining the results of multiplying using FOIL results in $(6a + 5)(a - 3) = 6a^2 - 18a + 5a - 15 = 6a^2 - 13a - 15$. The factors check out.

692. Both signs in the trinomial expression are negative. To get a negative sign for the numerical term, the signs within the trinomial factors must be + and −: $(ax + \quad)(bx − \quad)$. The factors of the second-degree term $6a^2$ are $(a)(6a)$ or $(2a)(3a)$. The factors of the numerical term 6 are $(1)(6)$ or $(2)(3)$. The trinomial looks balanced with a 6 on each end and a 5 in the middle. Try a balanced factor arrangement: $(3a + 2)(2a − 3)$. Check using FOIL: first—$(3a)(2a) = 6a^2$; outer—$(3a)(−3) = −9a$; inner—$(2)(2a) = 4a$; last—$(2)(−3) = −6$. Combining the results of multiplying using FOIL results in $(3a + 2)(2a − 3) = 6a^2 − 9a + 4a − 6 = 6a^2 − 5a − 6$. Didn't that work out nicely? A sense of balance can be useful.

693. This expression is the difference between two perfect squares. Using the form for the difference between two perfect squares gives you the factors $(4y + 10)(4y − 10)$. However, there is a greatest common factor that could be factored out first to leave $4(4y^2 − 25)$. Now you need only factor the difference between two simpler perfect squares: $4(4y^2 − 25) = 4(2y + 5)(2y − 5)$. The first factorization is equivalent to the second because you can factor out 2 from each of the factors. This will result in $(2)(2)(2y + 5)(2y − 5)$ or $4(2y + 5)(2y − 5) = 16y^2 − 100$. When factoring polynomials, watch for the greatest common factors first.

694. The terms of the trinomial have a greatest common factor of 3. So the term $3(2x^2 + 5x − 12)$ will simplify the trinomial factoring. You need only factor the trinomial within the parentheses. The sign of the numerical term is negative, so the signs in the trinomial factor form will have to be + and −: $(ax + \quad)(bx − \quad)$. The factors of the second-degree term $2x^2$ are $(x)(2x)$. The factors of the numerical term 12 are $(1)(12)$ or $(2)(6)$ or $(3)(4)$. The factors $(2x)(4) = 8x$ and the remaining factors $(x)(3) = 3x$. It's clear that $8x − 3x = 5x$. Use those factors in the trinomial factor form: $(2x − 3)(x + 4)$. Check using FOIL: first—$(2x)(x) = 2x^2$; outer—$(2x)(4) = 8x$; inner—$(−3)(x) = −3x$; last—$(−3)(4) = −12$. The result of multiplying factors is $2x^2 + 8x − 3x − 12 = 2x^2 + 5x − 12$. Now, include the greatest common factor of 3 for the final solution: $3(2x − 3)(x + 4) = 6x^2 + 15x − 36$.

695. Each term in the polynomial has a common factor of $2b$. The resulting expression looks like this: $2b(2c^2 + 11c - 21)$. The sign of the numerical term is negative, so the signs in the trinomial factor form will have to be $+$ and $-$, because that is the only way to get a negative sign when multiplying the last terms when checking with FOIL: $(ax + \quad)(bx - \quad)$. The factors of the second-degree term $2c^2$ are $(c)(2c)$. The factors of the numerical term 21 are $(1)(21)$ or $(3)(7)$. The factors $(2c)(7) = 14c$ and the associated factors $(c)(3) = 3c$. Place these factors in the trinomial factor form so that the result of the outer and inner products when using FOIL to multiply are $14c$ and $-3c$: $(2c - 3)(c + 7)$. Check using FOIL: first—$(2c)(c) = 2c^2$; outer—$(2c)(7) = 14c$; inner—$(-3)(c) = -3c$; last—$(-3)(7) = -21$. The product of the factors is $(2c - 3)(c + 7) = 2c^2 + 14c - 3c - 21 = 2c^2 + 11c - 21$. Now, include the greatest common factor term: $2b(2c - 3)(c + 7) = 2b(2c^2 + 11c - 21) = 4bc^2 + 22bc - 42b$.

696. This expression appears to be in the familiar trinomial form, but what's with those exponents? Think of $a^6 = (a^3)^2$. Then the expression becomes $2(a^3)^2 + (a^3) - 21$. Now, you factor as though it were a trinomial expression. The sign of the numerical term is negative, so the signs in the trinomial factor form will have to be $+$ and $-$: $(ax + \quad)$ $(bx - \quad)$. The factors of the second-degree term $2(a^3)^2$ are $2(a^3)(a^3)$. The factors of the numerical term 21 are $(1)(21)$ or $(3)(7)$. The factors $(a^3)(7) = 7(a^3)$ and the factors $[2(a^3)](3) = 6(a^3)$. The difference between 7 and 6 is 1. Place these factors in the trinomial factor form so that the first-degree term is $1(a^3)$: $(a^3 - 3)(2a^3 + 7)$. Check using FOIL: first—$(a^3)(2a^3) = 2a^6$; outer—$(a^3)(7) = 7a^3$; inner—$(-3)(2a^3) = -6a^3$; last—$(-3)(7) = -21$. The product of the factors is $(a^3 - 3)$ $(2a^3 + 7) = 2a^6 + 7a^3 - 6a^3 - 21 = 2a^6 + a^3 - 21$. The factors of the trinomial are correct.

697. The greatest common factor of the terms in the trinomial expression is $3x$. Factoring $3x$ out results in the expression $3x(2a^2 - 13a - 24)$. Factor the trinomial expression inside the parentheses. The sign of the numerical term is negative, so the signs in the trinomial factor form will have to be $+$ and $-$: $(ax + \quad)$ $(bx - \quad)$. The factors of the term $2a^2$ are $(a)(2a)$. The factors of the numerical term 24 are $(1)(24)$ or $(2)(12)$ or $(3)(8)$ or $(4)(6)$. The factors $(2a)(8) = 16a$, and the related factors $(a)(3) = 3a$. The difference between 16 and 3 is 13. Place these numbers in the trinomial factor form: $(2a + 3)(a - 8)$. Check the expression using FOIL: first—$(2a)(a) = 2a^2$; outer—$(2a)(-8) = -16a$; inner—$(3)(a) = 3a$; last—$(3)(-8) = -24$. The result is $(2a + 3)$ $(a - 8) = 2a^2 - 16a + 3a - 24 = 2a^2 - 13a - 24$. Now, include the greatest common factor of $3x$: $3x(2a + 3)(a - 8) = 3x(2a^2 - 13a - 24) = 6a^2x - 39ax - 72x$.

698. The numerical term of the trinomial has a negative sign, so the signs within the factors of the trinomial will be $+$ and $-$: $(ax + \quad)$ $(bx - \quad)$. The factors of the second-degree term $8x^2$ are $(x)(8x)$ or $(2x)(4x)$. The numerical term 9 of the trinomial has factors of $(1)(9)$ or $(3)(3)$. What combination will result in $-6x$ when the outer and inner products of the multiplication of the trinomial factors are added together? Consider just the coefficients of x and the numerical term factors. The numbers $2(3) = 6$, and the corresponding $4(3) = 12$. The difference between 12 and 6 is 6. So use the second-degree term factors $(2x)(4x)$ and the numerical factors $(3)(3)$: $(2x - 3)(4x + 3)$. Check using FOIL: first—$(2x)(4x) = 8x^2$; outer—$(2x)(3) = 6x$; inner—$(-3)(4x) = -12x$; last—$(-3)(3) = -9$. The product of the factors is $(2x - 3)(4x + 3) = 8x^2 + 6x - 12x - 9 = 8x^2 - 6x - 9$.

699. The numerical term of the trinomial has a negative sign, so the signs within the factors of the trinomial will be $+$ and $-$: $(ax + \quad)$ $(bx - \quad)$. The only factors of the second-degree term are $(c)(5c)$. The numerical term 2 of the trinomial has factors of $(1)(2)$. What combination will result in a $-9c$ when the outer and inner products of the trinomial factors are added together? Our choices are $(5)(1)c + (1)(-2)c$, $(5)(-1)c + (1)(2)$, $(5)(2)c + (1)(-1)c$, or $(5)(-2)c + (1)(1)c$. The last of these is equal to the desired $-9c$, which gives the factoring $(5c + 1)(c - 2)$. Check using FOIL to multiply terms: $(5c + 1)(c - 2)$. Check using FOIL: first—$(5c)(c) = 5c^2$; outer—$(5c)(-2) = -10c$; inner—$(1)(c) = c$; last—$(1)(-2) = -2$. The product of the factors is $(5c + 1)(c - 2) = 5c^2 - 10c + c - 2 = 5c^2 - 9c - 2$.

700. The terms of the expression have a greatest common factor of x. Factoring x out of the expression results in $x(9x^2 - 4)$. The expression inside the parentheses is the difference between two perfect squares. Factor that expression using the form for the difference between two perfect squares. Include the greatest common factor to complete the factorization of the original expression: $9x^2 = (3x)^2$; $4 = 2^2$. Using the form, the factorization of the difference between two perfect squares is $(3x - 2)(3x + 2)$. Check using FOIL: first—$(3x)(3x) = 9x^2$; outer—$(3x)(2) = 6x$; inner—$(-2)(3x) = -6x$; last—$(-2)(2) = -4$. Include the greatest common factor x in the complete factorization: $x(3x - 2)(3x + 2) = x(9x^2 - 4) = 9x^3 - 4x$.

701. The terms in the trinomial expression are all positive, so the signs in the trinomial factor form will be positive: $(ax + \quad)(bx + \quad)$. The factors of the second-degree term $8r^2$ are $(r)(8r)$ or $(2r)(4r)$. The numerical term 63 has the factors $(1)(63)$ or $(3)(21)$ or $(7)(9)$. You need two sets of factors that when multiplied and added will result in 46. Let's look at the possibilities using the $2r$ and $4r$. $2r(1) + 4r(63) = 254r$ is too much. $2r(3) + 4r(21) = 87r$ is still too much. Try $2r(21)$ and $4r(3) = 54r$. We're getting closer. $2r(7) + 4r(9) = 50r$. We're still not there. Now try $2r(9) + 4r(7) = 46r$. Bingo! $(2r + 7)(4r + 9)$. Check using FOIL: first—$(2r)(4r) = 8r^2$; outer—$(2r)(9) = 18r$; inner—$(4r)(7) = 28r$; last—$(7)(9) = 63$. Add the result of the multiplication. $8r^2 + 18r + 28r + 63 = 8r^2 + 46r + 63$. The factors check out: $(2r + 7)(4r + 9) = 8r^2 + 46r + 63$.

702. When you think of $x^4 = (x^2)^2$, you can see that the expression is a trinomial that is easy to factor. The numerical term is positive, so the signs in the trinomial factor form will be the same (both positive or negative). The sign of the first-degree term is negative, so you will use two $-$ signs: $(ax - \quad)(bx - \quad)$. The factors of the second-degree term $4x^4$ are $(x^2)(4x^2)$ or $(2x^2)(2x^2)$. The numerical term 9 has $(1)(9)$ or $(3)(3)$ as factors. What combination will result in a total of 37 when the outer and inner products are determined? $4x^2(9) = 36x^2$, $1x^2(1) = 1x^2$, and $36x^2 + 1x^2 = 37x^2$. Use these factors in the trinomial factor form. $(4x^2 - 1)(x^2 - 9)$. Check using FOIL and you will find: $(4x^2 - 1)(x^2 - 9) = 4x^4 - 36x^2 - x^2 + 9 = 4x^4 - 37x^2 + 9$. Now, notice that the factors of the original trinomial expression are both factorable. Why? Because they are both the difference between two perfect squares. Use the factor form for the difference between two perfect squares for each factor of the trinomial: $(4x^2 - 1) = (2x + 1)(2x - 1)$; $(x^2 - 9) = (x + 3)(x - 3)$. Put the factors together to complete the factorization of the original expression: $(4x^2 - 1)(x^2 - 9) = (2x + 1)(2x - 1)(x + 3)(x - 3) = 4x^4 - 37x^2 + 9$.

703. The negative sign in front of the numerical term tells you that the signs of the trinomial factors will be + and −: $(ax + \quad)(bx − \quad)$. This expression has a nice balance to it with 12 at the extremities and a modest 7 in the middle. Let's guess at some middle-of-the-road factors to plug in. Use FOIL to check: $(4d + 3)(3d − 4)$. Using FOIL, you find: $(4d + 3)(3d − 4) = 12d^2 − 16d + 9d − 12 = 12d^2 − 7d − 12$. Those are the right terms but the wrong signs. Try changing the signs around: $(4d − 3)(3d + 4)$. Multiply the factors using FOIL: $(4d − 3)(3d + 4) = 12d^2 + 16d − 9d − 12 = 12d^2 + 7d − 12$. This is the correct factorization of the original expression.

704. Each term in the expression has a common factor of $2xy$. When factored out, the expression becomes $2xy(2y^2 + 3y − 5)$. Now factor the trinomial in the parentheses. The last sign is negative, so the signs within the factor form will be + and −: $(ax + \quad)(bx − \quad)$. The factors of the second-degree term $2y^2$ are $(y)(2y)$. The numerical term 5 has factors $(1)(5)$. Place the factors of the second-degree term and the numerical term so that the result of the outer and inner multiplication of terms within the factor form of a trinomial expression results in $3x$: $(2y + 5)(y − 1)$. Multiply using FOIL: $(2y + 5)(y − 1) = 2y^2 − 2y + 5y − 5 = 2y^2 + 3y − 5$. The factors of the trinomial expression are correct. Now include the greatest common factor to complete the factorization of the original expression: $2xy(2y + 5)(y − 1) = 2xy(2y^2 − 2y + 5y − 5) = 2xy(2y^2 + 3y − 5)$.

705. The terms of the trinomial have a greatest common factor of $2a$. When factored out, the resulting expression is $2a(2x^2 − 19x − 33)$. The expression within the parentheses is a trinomial and can be factored. The signs within the terms of the factor form will be + and −, because the numerical term has a negative sign. Only $(+)(−) = (−)$: $(ax + \quad)(bx − \quad)$. The factors of the second-degree term $2x^2$ are $(x)(2x)$. The numerical term 33 has $(1)(33)$ or $(3)(11)$ as factors. Since $2x(11) = 22x$, and $x(3) = 3x$, and $22x − 3x = 19x$, use those factors in the trinomial factor form so that multiplication of the outer and inner terms results in $−19x$. $(2x + 3)(x − 11)$. Check using FOIL: $(2x + 3)(x − 11) = 2x^2 − 22x + 3x − 33 = 2x^2 − 19x − 33$. The factorization of the trinomial factor is correct. Now include the greatest common factor of the original expression to get the complete factorization of the original expression: $2a(2x + 3)(x − 11) = 2a(2x^2 − 22x + 3x − 33) = 2a(2x^2 − 19x − 33)$.

706. The signs within the terms of the factor form will be + and −, because the numerical term has a negative sign: $(ax + \quad)(bx − \quad)$. The factors of the second-degree term $3c^2$ are $(c)(3c)$. The numerical term 40 has $(1)(40)$ or $(2)(20)$ or $(4)(10)$ or $(5)(8)$ as factors. You want the result of multiplying and then adding the outer and inner terms of the trinomial factor form to be $19c$ when the like terms are combined. Using trial and error, you can determine that $3c(8) = 24c$, and $c(5) = 5c$, and $24c − 5c = 19c$. Use those factors in the factor form in such a way that you get the result you seek: $(3c − 5)(c + 8)$. The complete factorization of the original expression is: $(3c − 5)(c + 8) = 3c^2 + 24c − 5c − 40 = 3c^2 + 19c − 40$.

707. The signs within the terms of the factor form will be + and − because the numerical term has a negative sign: $(ax + \quad)(bx − \quad)$. The factors of the second-degree term $2a^2$ are $(a)(2a)$. The numerical term 84 has $(1)(84)$ or $(2)(42)$ or $(3)(28)$ or $(4)(21)$ or $(6)(14)$ or $(7)(12)$ as factors. You want the result of multiplying and then adding the outer and inner terms of the trinomial factor form to result in $17a$ when the like terms are combined. $2a(12) = 24a$, and $a(7) = 7a$, and $24a − 7a = 17a$. Use the factors $(2a)$ and (a) as the first terms in the factor form and use (7) and (12) as the numerical terms. Place them in position so you get the result that you want: $(2a − 7)(a + 12)$. The complete factorization of the original expression is $(2a − 7)(a + 12) = 2a^2 + 24a − 7a − 84 = 2a^2 + 17a − 84$.

708. If you think of the variable as x^2, you can see that the expression is in the trinomial form. Use x^2 where you usually put a first-degree variable. The trinomial you will be factoring looks like this: $4(x^2)^2 + 2(x^2) − 30$. The signs within the terms of the factor form will be + and −, because the numerical term has a negative sign: $(ax + \quad)(bx − \quad)$. The term $4(x^2)^2$ can be factored as $4(x^2)^2 = (x^2)4x^2$ or $(2x^2)(2x^2)$. The numerical term 30 can be factored as $(1)(30)$ or $(2)(15)$ or $(3)(10)$ or $(5)(6)$. The factors $(4x^2)(3) = 12x^2$ and $x^2(10) = 10x^2$ will give you $12x^2 − 10x^2 = 2x^2$ when you perform the inner and outer multiplications and combine like terms using FOIL with the terms in the trinomial factor form. The factors of the expression will be $(4x^2 − 10)(x^2 + 3)$. Check using FOIL: $(4x^2 − 10)(x^2 + 3) = 4x^4 + 12x^2 − 10x^2 − 30 = 4x^4 + 2x^2 − 30$. The expression $(4x^2 − 10)(x^2 + 3)$ is the correct factorization of the original expression. However, the first factor has a greatest common factor of 2. So a complete factorization would be: $2(2x^2 − 5)(x^2 + 3) = 4x^4 + 2x^2 − 30$. Did you notice that you could have used the greatest common factor method to factor out a 2 from each term in the original polynomial? If you did, you would have had to factor the trinomial expression $2x^4 + x^2 − 15$ and multiply the result by the factor 2 to equal the original expression. Let's see: $2(2x^4 + x^2 − 15) = 2(2x^2 − 5)(x^2 + 3) = (4x^2 − 10)(x^2 + 3) = 4x^4 + 2x^2 − 30$. It all comes out the same, but if you left the factor of 2 in the term $(4x^2 − 10)$, you wouldn't have done a complete factorization of the original trinomial expression.

CHAPTER 11 ▶ Quadratic Equations and the Quadratic Formula

WE'VE SEEN HOW TO MULTIPLY and factor different kinds of polynomials, and we've seen how to solve algebraic equations. In this chapter, we'll look at a specific kind of equation: a quadratic equation.

A **quadratic equation** is an equation that is written in the form $ax^2 + bx + c = 0$. The equation has one variable that is raised to the second power, and could have one variable that is raised to the first power (if b is not zero) and one constant (if c is not zero). A **constant** is a real number (not a variable). So although a quadratic equation may not have an x term or a constant, it will always have an x^2 term, and the right side will always be equal to zero.

The following are all quadratic equations:

$$3x^2 + 17x + 10 = 0$$
$$x^2 - 16 = 0$$
$$6x^2 = 0$$

Not all quadratic equations are given to you in the forms shown, though. The first step to solving a quadratic equation is to move the terms onto the left side of the equation, so that the right side of the equation is equal to zero. Once the equation is in the form $ax^2 + bx + c = 0$, factor the polynomial using the skills we reviewed in the preceding chapter. Then, set each factor equal to zero and solve for the value of x. Most quadratic equations break down into two factors, and when we set those factors equal to zero, we find the roots of the equation. The roots of an equation are the values that, when substituted for x, make the equation equal to zero.

Let's look at an example.

Example

What are the roots of $x^2 + 4x = -3$?

Step 1: Put the equation in the form $ax^2 + bx + c = 0$. To do that, add 3 to both sides of the equation:

$x^2 + 4x + 3 = -3 + 3$

$x^2 + 4x + 3 = 0$

Step 2: Factor the equation; x^2 can only be found by multiplying x by itself, so one part of each factor must be x:

$(x + \quad)(x + \quad)$

Two pairs of integers multiply to 3 (1 and 3, and −1 and −3), but only one pair multiplies to 3 and adds to 4, and that pair is 1 and 3. We now have our factors:

$(x + 1)(x + 3)$

We can check ourselves by multiplying those two binomials: $(x + 1)(x + 3) = x^2 + 3x + x + 3 = x^2 + 4x + 3$, so we know we've factored correctly.

Step 3: Set each factor equal to zero and solve for x:

$x + 1 = 0; x = -1$

$x + 3 = 0; x = -3$

The values −1 and −3 are the roots of the equation $x^2 + 4x = -3$, because these are the values that make the equation true (or the values that make $x^2 + 4x + 3$ equal to zero). Let's check these roots by substituting them into the original equation:

$x^2 + 4x = -3; (-1)^2 + 4(-1) = -3; 1 - 4 = -3; -3 = -3$

$x^2 + 4x = -3; (-3)^2 + 4(-3) = -3; 9 - 12 = -3; -3 = -3$

Both equations hold true, so our answers must be correct.

Sometimes a quadratic equation doesn't have an x term. Let's look at an example.

Example

$3x^2 = 48$

Even though there is no x term, we follow the same steps:

Step 1: Put the equation in the form $ax^2 + bx + c = 0$. To do that, subtract 48 from both sides of the equation:

$3x^2 - 48 = 48 - 48$

$3x^2 - 48 = 0$

Step 2: Factor the equation; $3x^2$ can only be found by multiplying $3x$ by x, so our factors must begin like this:

$(3x + \quad)(x + \quad)$

We're looking for two numbers that multiply to -48, but three times one of those numbers must be equal to the other number, because there is no x term in this quadratic equation. That means that if we were to multiply these factors, the product of the outside terms and the product of the inside terms will add to zero: 12 times 4 is 48, and 12 is three times 4.

$(3x + 12)(x - 4)$

or

$(3x - 12)(x + 4)$

How do we know which set of factors to use? It actually doesn't matter—both give us the same roots. Let's test both sets of factors.

Step 3: Set each factor equal to zero and solve for x:

$3x + 12 = 0; 3x = -12; x = -4$

$x - 4 = 0; x = 4$

The first set of factors gives us the roots -4 and 4. Now, look at the second set:

$3x - 12 = 0; 3x = 12; x = 4$

$x + 4 = 0; x = -4$

This set also gives us the roots -4 and 4.

Finally, let's check our answers. Substitute both into the original equation:

$3x^2 = 48; 3(-4)^2 = 48; 3(16) = 48; 48 = 48$

$3x^2 = 48; 3(4)^2 = 48; 3(16) = 48; 48 = 48$

Some quadratic solutions appear to have two roots, but it's really the same root twice. This can happen if a binomial, such as $2x + 3$, is squared.

Example

Find the roots of $x^2 + 12x + 36 = 0$.

Step 1: This time, the equation is already in the form $ax^2 + bx + c = 0$.

Step 2: Factor the equation; x^2 can only be found by multiplying x by itself, so one part of each factor must be x:

$(x + \quad)(x + \quad)$

The numbers 6 and 6 multiply to 36 and add to 12. We can't use -6 and -6 because although these numbers multiply to 36, they add to -12, not 12.

$(x + 6)(x + 6)$

Step 3: Set each factor equal to zero and solve for x. Because both factors are the same, both factors will yield the same root:

$x + 6 = 0; x = -6$

$x + 6 = 0; x = -6$

This quadratic equation has only one root, -6.

▶ **Practice Questions**

Find the solutions to the following quadratic equations.

709. $x^2 - 25 = 0$

710. $n^2 - 169 = 0$

711. $a^2 + 12a + 32 = 0$

712. $y^2 - 15y + 56 = 0$

713. $b^2 + b - 90 = 0$

714. $4x^2 = 49$

715. $25r^2 = 144$

716. $2n^2 + 20n + 42 = 0$

717. $3c^2 - 33c - 78 = 0$

718. $100r^2 = 144$

719. $3x^2 - 36x + 108 = 0$

720. $7a^2 - 21a - 28 = 0$

721. $8y^2 + 56y + 96 = 0$

722. $2x^2 + 9x = -10$

723. $4x^2 + 4x = 15$

724. $9x^2 + 12x = -4$

725. $3x^2 = 19x - 20$

726. $8b^2 + 10b = 42$

727. $14n^2 = 7n + 21$

728. $6b^2 + 20b = -9b - 20$

729. $15x^2 - 70x - 120 = 0$

730. $7x^2 = 52x - 21$

731. $36z^2 + 78z = -36$

732. $12r^2 = 192 - 40r$

733. $24x^2 = 3(43x - 15)$

Some quadratic equations cannot be solved by factoring. If you find that you are unable to factor a quadratic equation, use the **quadratic formula**. The formula looks complex, but with a little practice, you'll become comfortable with it:

$$x = \frac{-b \pm \sqrt{b^2 - 4ac}}{2a}$$

Before you can use the quadratic formula to solve an equation, you must put the quadratic equation in the form $ax^2 + bx + c = 0$. Remember, a is the coefficient of the squared term, b is the coefficient of the x term, and c is the constant. If there is no x term in the quadratic equation, substitute 0 for b in the formula. If there is no constant in the quadratic equation, substitute 0 for c in the formula. The $\pm$ symbol represents a plus sign and a minus sign. A quadratic equation has two roots; one of them is found by using the plus sign:

$$x = \frac{-b + \sqrt{b^2 - 4ac}}{2a}$$

and the other is found by using the minus sign:

$$x = \frac{-b - \sqrt{b^2 - 4ac}}{2a}$$

Example

Find the solutions to $2x^2 - 5x = -1$.

This quadratic equation cannot be factored, so we must use the quadratic formula. Add 1 to both sides of the equation so that the equation is in the form $ax^2 + bx + c = 0$.

$2x^2 - 5x + 1 = -1 + 1$

$2x^2 - 5x + 1 = 0$

The coefficient of the x^2 term is 2, so $a = 2$. The coefficient of the x term is -5, so $b = -5$, and the constant is 1, so $c = 1$. Substitute these values into the quadratic equation and simplify:

$$x = \frac{-(-5) \pm \sqrt{(-5)^2 - 4(2)(1)}}{2(2)}$$

$$= \frac{5 + \sqrt{25 - 8}}{4}$$

$$= \frac{5 + \sqrt{17}}{4} \text{ and } \frac{5 - \sqrt{17}}{4}$$

The two solutions are $\frac{5+\sqrt{17}}{4}$ and $\frac{5-\sqrt{17}}{4}$, as these cannot be simplified any further. To find exact solutions, use a calculator to find an approximation for the square root of 17. Then add that value to 5 and divide by 4 for one solution, and subtract that value from 5 and divide by 4 for the other solution.

TIP

Don't worry if you're not sure when to factor and when to use the quadratic formula. The quadratic formula *always* works, even on quadratic equations that can be factored. Let's try it on the first example in this chapter, $x^2 + 4x = -3$.

$x^2 + 4x + 3 = 0$

$a = 1, b = 4, c = 3$

$$x = \frac{-4 \pm \sqrt{(4)^2 - 4(1)(3)}}{2(1)}$$

$$= \frac{-4 \pm \sqrt{16 - 12}}{2}$$

$$= \frac{-4 \pm \sqrt{4}}{2}$$

$$= \frac{-4 \pm 2}{2}$$

$$= \frac{-4 + 2}{2} \text{ and } \frac{-4 - 2}{2}$$

$$= -1 \text{ and } -3$$

The roots -1 and -3 are the same roots we found by factoring. Try using the quadratic formula on $3x^2 = 48$ and $x^2 + 12x + 36 = 0$.

▶ **Practice Questions**

Solve the following equations using the quadratic formula. Reduce answers to their simplest form or to the simplest radical form. Use of a calculator is recommended.

734. $x^2 + 2x - 8 = 0$

735. $2x^2 - 7x - 30 = 0$

736. $6x^2 + 13x - 28 = 0$

737. $18x^2 + 9x + 1 = 0$

738. $6x^2 + 17x = 28$

739. $14x^2 = 12x + 32$

740. $4x^2 + 5x = 0$

741. $5x^2 = 27$

742. $5x^2 = 18x - 17$

743. $3x^2 + 11x - 7 = 0$

744. $5x^2 + 52x + 20 = 0$

745. $x^2 = \frac{-5x - 2}{2}$

746. $x^2 + 8x = 5$

747. $x^2 = 20x - 19$

748. $23x^2 = 2(8x - 1)$

749. $x^2 + 10x + 11 = 0$

750. $24x^2 + 18x - 6 = 0$

751. $7x^2 = 4(3x + 1)$

752. $\frac{1}{3}x^2 + \frac{3}{4}x - 3 = 0$

753. $5x^2 - 12x + 1 = 0$

Find the solutions to the following equations to the nearest hundredth.

754. $11r^2 - 4r - 7 = 0$

755. $3m^2 + 21m - 8 = 0$

756. $4y^2 = 16y - 5$

757. $5s^2 + 12s - 1 = 0$

758. $4c^2 - 11c + 2 = 0$

759. $11k^2 - 32k + 10 = 0$

TERMS TO REVIEW

quadratic equation

constant

quadratic formula

FORMULA TO REVIEW

quadratic formula: $x = \dfrac{-b \pm \sqrt{b^2 - 4ac}}{2a}$

▶ **Answers**

709. The expression is the difference between two perfect squares. Factor the equation: $(x + 5)(x - 5) = 0$. Applying the zero product property [if $(a)(b) = 0$, then $a = 0$ or $b = 0$ or both $= 0$], the first factor or the second factor or both must equal zero: $(x + 5) = 0$. Subtract 5 from both sides of the equation: $x + 5 - 5 = 0 - 5$. Combine like terms on each side: $x = -5$. Let the second factor equal zero: $x - 5 = 0$. Add 5 to both sides of the equation: $x - 5 + 5 = 0 + 5$. Combine like terms on each side: $x = 5$. The solutions for the equation are $x = 5$ and $x = -5$.

710. The expression is the difference between two perfect squares. Factor the equation: $(n + 13)(n - 13) = 0$. Applying the zero product property [if $(a)(b) = 0$, then $a = 0$ or $b = 0$ or both $= 0$], the first factor or the second factor or both must equal zero: $(n + 13) = 0$. Subtract 13 from both sides of the equation: $n + 13 - 13 = 0 - 13$. Combine like terms on each side: $n = -13$. Let the second factor equal zero: $n - 13 = 0$. Add 13 to both sides of the equation: $n - 13 + 13 = 0 + 13$. Combine like terms on each side: $n = 13$. The solutions for the equation are $n = 13$ and $n = -13$.

711. Factor the trinomial expression using the trinomial factor form: $(a + 4)(a + 8) = 0$. Use the zero product property with the first factor: $(a + 4) = 0$. Subtract 4 from both sides: $a + 4 - 4 = 0 - 4$. Combine like terms on each side: $a = -4$. Let the second factor equal zero: $(a + 8) = 0$. Subtract 8 from both sides: $a + 8 - 8 = 0 - 8$. Combine like terms on each side: $a = -8$. The solutions for the quadratic equation $a^2 + 12a + 32 = 0$ are $a = -4$. and $a = -8$.

712. Factor the trinomial expression using the trinomial factor form: $(y - 8)(y - 7) = 0$. Use the zero product property: $(y - 8) = 0$. Add 8 to both sides: $y - 8 + 8 = 0 + 8$. Combine like terms on each side: $y = 8$. Let the second factor equal zero: $(y - 7) = 0$. Add 7 to both sides: $y - 7 + 7 = 0 + 7$. Combine like terms on each side: $y = 7$. The solutions for the equation $y^2 - 15y + 56 = 0$ are $y = 8$ and $y = 7$.

713. Factor the trinomial expression using the trinomial factor form: $(b + 10)(b - 9) = 0$. Use the zero product property: $(b + 10) = 0$. Subtract 10 from both sides: $b + 10 - 10 = 0 - 10$. Combine like terms on each side: $b = -10$. Let the second factor equal zero: $(b - 9) = 0$. Add 9 to both sides: $b - 9 + 9 = 0 + 9$. Combine like terms on each side: $b = 9$. The solutions for the quadratic equation $b^2 + b - 90 = 0$ are $b = -10$ and $b = 9$.

714. To transform the equation so that all terms are on one side and are equal to zero, first subtract 49 from both sides: $4x^2 - 49 = 49 - 49$. Combine like terms on each side: $4x^2 - 49 = 0$. The expression is the difference between two perfect squares. Factor the equation: $(2x + 7)(2x - 7) = 0$. Applying the zero product property [if $(a)(b) = 0$, then $a = 0$ or $b = 0$ or both $= 0$], the first factor or the second factor or both must equal zero: $(2x + 7) = 0$. Subtract 7 from both sides of the equation: $2x + 7 - 7 = 0 - 7$. Combine like terms on each side: $2x = -7$. Divide both sides by 2: $\frac{2x}{2} = \frac{-7}{2}$. Simplify: $x = -3\frac{1}{2}$. Let the second factor equal zero: $(2x - 7) = 0$. Add 7 to both sides of the equation: $2x - 7 + 7 = 0 + 7$. Combine like terms on both sides: $2x = 7$. Divide both sides by 2: $x = 3\frac{1}{2}$. The solutions for the quadratic equation $4x^2 = 49$ are $x = -3\frac{1}{2}$ and $x = 3\frac{1}{2}$.

715. To transform the equation so that all terms are on one side and are equal to zero, first subtract 144 from both sides: $25r^2 - 144 = 144 - 144$. Combine like terms on each side: $25r^2 - 144 = 0$. The expression is the difference between two perfect squares. Factor the equation: $(5r + 12)(5r - 12) = 0$. Applying the zero product property [if $(a)(b) = 0$, then $a = 0$ or $b = 0$ or both $= 0$], the first factor or the second factor or both must equal zero: $(5r + 12) = 0$. Subtract 12 from both sides of the equation: $5r + 12 - 12 = 0 - 12$. Combine like terms on each side: $5r = -12$. Divide both sides by 5: $\frac{5r}{5} = \frac{-12}{5}$. Simplify: $r = -2\frac{2}{5}$. Let the second factor equal zero: $(5r - 12) = 0$. Add 12 to both sides of the equation: $5r - 12 + 12 = 0 + 12$. Combine like terms on each side: $5r = 12$. Divide both sides by 5: $\frac{5r}{5} = \frac{12}{5}$. Simplify: $r = 2\frac{2}{5}$. The solution for the quadratic equation $25r^2 = 144$ is $r = \pm 2\frac{2}{5}$.

716. Factor the trinomial expression using the trinomial factor form: $(2n + 6)(n + 7) = 0$. Use the zero product property: $(2n + 6) = 0$. Subtract 6 from both sides: $2n + 6 - 6 = 0 - 6$. Combine like terms on each side: $2n = -6$. Divide both sides by 2: $\frac{2n}{2} = \frac{-6}{2}$. Simplify terms: $n = -3$. Let the second factor equal zero: $(n + 7) = 0$. Subtract 7 from both sides: $n + 7 - 7 = 0 - 7$. Combine like terms on each side: $n = -7$. The solutions for the quadratic equation $2n^2 + 20n + 42 = 0$ are $n = -3$ and $n = -7$.

717. Use the greatest common factor method: $3(c^2 - 11c - 26) = 0$. Factor the trinomial expression using the trinomial factor form: $3(c - 13)(c + 2) = 0$. Ignore the factor 3 in the expression. Use the zero product property: $(c - 13) = 0$. Add 13 to both sides: $c - 13 + 13 = 0 + 13$. Combine like terms on each side: $c = 13$. Let the second factor equal zero: $(c + 2) = 0$. Subtract 2 from both sides: $c + 2 - 2 = 0 - 2$. Combine like terms on each side: $c = -2$. The solutions for the quadratic equation $3c^2 - 33c - 78 = 0$ are $c = 13$ and $c = -2$.

718. To transform the equation so that all terms are on one side and are equal to zero, first subtract 144 from both sides: $100r^2 - 144 = 144 - 144$. Combine like terms on each side: $100r^2 - 144 = 0$. The expression is the difference between two perfect squares. Factor the equation: $(10r + 12)(10r - 12) = 0$. Applying the zero product property [if $(a)(b) = 0$, then $a = 0$ or $b = 0$ or both $= 0$], the first factor or the second factor or both must equal zero: $(10r + 12) = 0$. Subtract 12 from both sides of the equation: $10r + 12 - 12 = 0 - 12$. Combine like terms on each side: $10r = -12$. Divide both sides by 10: $\frac{10r}{10} = \frac{-12}{10} = \frac{-6}{5}$. Simplify terms: $r = -1\frac{1}{5}$. Let the second factor equal zero: $(10r - 12) = 0$. Add 12 to both sides of the equation: $10r - 12 + 12 = 0 + 12$. Combine like terms on each side: $10r = 12$. Divide both sides by 10: $\frac{10r}{10} = \frac{12}{10} = \frac{6}{5}$. Simplify terms: $r = 1\frac{1}{5}$. The solution for the quadratic equation $100r^2 = 144$ is $r = \pm 1\frac{1}{5}$.

719. Use the greatest common factor method: $3(x^2 - 12x + 36) = 0$. Factor the trinomial expression using the trinomial factor form: $3(x - 6)(x - 6) = 0$. Ignore the factor 3 in the expression. Use the zero product property: $(x - 6) = 0$. Add 6 to both sides: $x - 6 + 6 = 0 + 6$. Combine like terms on each side: $x = 6$. Because both factors of the trinomial expression are the same, the solution for the quadratic equation $3x^2 - 36x + 108 = 0$ is $x = 6$.

720. Use the greatest common factor method: $7(a^2 - 3a - 4) = 0$. Factor the trinomial expression using the trinomial factor form: $7(a - 4)(a + 1) = 0$. Ignore the factor 7 in the expression. Use the zero product property: $(a - 4) = 0$. Add 4 to both sides: $a - 4 + 4 = 0 + 4$. Combine like terms on each side: $a = 4$. Let the second factor equal zero: $(a + 1) = 0$. Subtract 1 from both sides: $a + 1 - 1 = 0 - 1$. Combine like terms on each side: $a = -1$. The solutions for the quadratic equation $7a^2 - 21a - 28 = 0$ are $a = 4$ and $a = -1$.

721. Use the greatest common factor method: $8(y^2 + 7y + 12) = 0$. Factor the trinomial expression using the trinomial factor form: $8(y + 4)(y + 3) = 0$. Ignore the factor 8 in the expression. Use the zero product property: $(y + 4) = 0$. Subtract 4 from both sides: $y + 4 - 4 = 0 - 4$. Combine like terms on each side: $y = -4$. Let the second factor equal zero: $(y + 3) = 0$. Subtract 3 from both sides: $y + 3 - 3 = 0 - 3$. Simplify: $y = -3$. The solutions for the quadratic equation $8y^2 + 56y + 96 = 0$ are $y = -4$ and $y = -3$.

722. To transform the equation into the familiar trinomial equation form, first add 10 to both sides of the equation: $2x^2 + 9x + 10 = -10 + 10$. Combine like terms on each side: $2x^2 + 9x + 10 = 0$. Factor the trinomial expression using the trinomial factor form: $(2x + 5)(x + 2) = 0$. Use the zero product property: $(2x + 5) = 0$. Subtract 5 from both sides: $2x + 5 - 5 = 0 - 5$. Combine like terms on each side: $2x = -5$. Divide both sides by 2: $\frac{2x}{2} = \frac{-5}{2}$. Simplify terms: $x = -2\frac{1}{2}$. Let the second term equal zero: $x + 2 = 0$. Subtract 2 from both sides: $x + 2 - 2 = 0 - 2$. Simplify: $x = -2$. The solutions for the quadratic equation $2x^2 + 9x = -10$ are $x = -2\frac{1}{2}$ and $x = -2$.

723. To transform the equation into the familiar trinomial equation form, first subtract 15 from both sides of the equation: $4x^2 + 4x - 15 = 15 - 15$. Combine like terms on each side: $4x^2 + 4x - 15 = 0$. Factor the trinomial expression using the trinomial factor form: $(2x - 3)(2x + 5) = 0$. Use the zero product property: $(2x + 5) = 0$. Subtract 5 from both sides: $2x + 5 - 5 = 0 - 5$. Simplify: $2x = -5$. Divide both sides by 2: $\frac{2x}{2} = \frac{-5}{2}$. Simplify terms: $x = -2\frac{1}{2}$. Let the second factor equal zero: $(2x - 3) = 0$. Add 3 to both sides: $2x - 3 + 3 = 0 + 3$. Simplify: $2x = 3$. Divide both sides by 2: $\frac{2x}{2} = \frac{3}{2}$. Simplify terms: $x = 1\frac{1}{2}$. The solutions for the quadratic equation $4x^2 + 4x = 15$ are $x = -2\frac{1}{2}$ and $x = 1\frac{1}{2}$.

724. To transform the equation into the familiar trinomial equation form, first add 4 to both sides of the equation: $9x^2 + 12x + 4 = -4 + 4$. Combine like terms on each side: $9x^2 + 12x + 4 = 0$. Factor the trinomial expression using the trinomial factor form: $(3x + 2)(3x + 2) = 0$. Use the zero product property: $(3x + 2) = 0$. Subtract 2 from both sides: $3x + 2 - 2 = 0 - 2$. Simplify: $3x = -2$. Divide both sides by 3: $\frac{3x}{3} = \frac{-2}{3}$. Simplify terms: $x = \frac{-2}{3}$. Because both factors of the trinomial are the same, the solution to the quadratic equation $9x^2 + 12x = -4$ is $x = \frac{-2}{3}$.

725. To transform the equation into the familiar trinomial equation form, first subtract $19x$ from both sides: $3x^2 - 19x = 19x - 19x - 20$. Combine like terms: $3x^2 - 19x = -20$. Add 20 to both sides: $3x^2 - 19x + 20 = -20 + 20$. Combine like terms on each side: $3x^2 - 19x + 20 = 0$. Factor the trinomial expression using the trinomial factor form: $(3x - 4)(x - 5) = 0$. Use the zero product property: $(3x - 4) = 0$. Add 4 to both sides: $3x - 4 + 4 = 0 + 4$. Simplify: $3x = 4$. Divide both sides by 3: $\frac{3x}{3} = \frac{4}{3}$. Simplify terms: $x = 1\frac{1}{3}$. Now, let the second term equal zero: $x - 5 = 0$. Add 5 to both sides: $x - 5 + 5 = 0 + 5$. Simplify: $x = 5$. The solutions for the quadratic equation $3x^2 = 19x - 20$ are $x = 1\frac{1}{3}$ and $x = 5$.

726. To transform the equation into the familiar trinomial equation form, first subtract 42 from both sides of the equation: $8b^2 + 10b - 42 = 42 - 42$. Simplify: $8b^2 + 10b - 42 = 0$. Use the greatest common factor method to factor out 2: $2(4b^2 + 5b - 21) = 0$. Factor the trinomial expression using the trinomial factor form: $2(4b - 7)(b + 3) = 0$. Ignore the factor 2 in the expression. Use the zero product property: $(4b - 7) = 0$. Add 7 to both sides: $4b - 7 + 7 = 0 + 7$. Simplify: $4b = 7$. Divide both sides by 4: $\frac{4b}{4} = \frac{7}{4}$. Simplify terms: $b = 1\frac{3}{4}$. Now, let the second term equal zero: $(b + 3) = 0$. Subtract 3 from both sides: $b + 3 - 3 = 0 - 3$. Simplify: $b = -3$. The solutions for the quadratic equation $8b^2 + 10b = 42$ are $b = 1\frac{3}{4}$ and $b = -3$.

727. To transform the equation into the familiar trinomial equation form, first subtract $7n$ from both sides of the equation: $14n^2 - 7n = 7n - 7n + 21$. Simplify and subtract 21 from both sides: $14n^2 - 7n - 21 = 21 - 21$. Simplify the equation: $14n^2 - 7n - 21 = 0$. Factor the greatest common factor from each term: $7(2n^2 - n - 3) = 0$. Factor the trinomial expression using the trinomial factor form: $7(2n - 3)(n + 1) = 0$. Ignore the factor 7 in the expression. Use the zero product property: $(2n - 3) = 0$. Add 3 to both sides: $2n - 3 + 3 = 0 + 3$. Simplify: $2n = 3$. Divide both sides by 2: $\frac{2n}{2} = \frac{3}{2}$. Simplify terms: $n = 1\frac{1}{2}$. Now, set the second factor equal to zero: $n + 1 = 0$. Subtract 1 from both sides: $n + 1 - 1 = 0 - 1$. Simplify: $n = -1$. The solutions for the quadratic equation $14n^2 = 7n + 21$ are $n = 1\frac{1}{2}$ and $n = -1$.

728. To transform the equation into the familiar trinomial equation form, first add $9b$ to both sides of the equation: $6b^2 + 20b + 9b = -9b + 9b - 20$. Simplify and add 20 to both sides of the equation: $6b^2 + 29b + 20 = 20 - 20$. Simplify: $6b^2 + 29b + 20 = 0$. Factor the trinomial expression using the trinomial factor form: $(6b + 5)(b + 4) = 0$. Use the zero product property: $(6b + 5) = 0$. Subtract 5 from both sides: $6b + 5 - 5 = 0 - 5$. Simplify: $6b = -5$. Divide both sides by 6: $\frac{6b}{6} = \frac{-5}{6}$. Simplify terms: $b = \frac{-5}{6}$. Now, set the second factor equal to zero: $b + 4 = 0$. Subtract 4 from both sides: $b + 4 - 4 = 0 - 4$. Simplify: $b = -4$. The solutions for the quadratic equation $6b^2 + 20b = -9b - 20$ are $b = \frac{-5}{6}$ and $b = -4$.

729. Factor the greatest common factor from each term: $5(3x^2 - 14x - 24) = 0$. Now factor the trinomial expression using the trinomial factor form: $5(3x + 4)(x - 6) = 0$. Use the zero product property: $(3x + 4) = 0$. Subtract 4 from both sides: $3x + 4 - 4 = 0 - 4$. Simplify: $3x = -4$. Divide both sides by 3: $\frac{3x}{3} = \frac{-4}{3}$. Simplify terms: $x = \frac{-4}{3}$. Now, set the second factor equal to zero: $x - 6 = 0$. Add 6 to both sides: $x - 6 + 6 = 0 + 6$. Simplify: $x = 6$. The solutions for the quadratic equation $15x^2 - 70x - 120 = 0$ are $x = \frac{-4}{3}$ and $x = 6$.

730. To transform the equation into the familiar trinomial equation form, first subtract $52x$ from both sides of the equation: $7x^2 - 52x = 52x - 52x - 21$. Simplify and add 21 to both sides of the equation: $7x^2 - 52x + 21 = 21 - 21$. Simplify: $7x^2 - 52x + 21 = 0$. Factor the trinomial expression using the trinomial factor form: $(7x - 3)(x - 7) = 0$. Use the zero product property: $(7x - 3) = 0$. Add 3 to both sides: $7x - 3 + 3 = 0 + 3$. Simplify: $7x = 3$. Divide both sides by 7: $\frac{7x}{7} = \frac{3}{7}$. Simplify terms: $x = \frac{3}{7}$. Now, set the second factor equal to zero: $x - 7 = 0$. Add 7 to both sides of the equation: $x - 7 + 7 = 0 + 7$. Simplify: $x = 7$. The solutions for the quadratic equation $7x^2 = 52x - 21$ are $x = \frac{3}{7}$ and $x = 7$.

731. To transform the equation into the familiar trinomial equation form, first add 36 to both sides of the equation: $36z^2 + 78z + 36 = -36 + 36$. Combine like terms: $36z^2 + 78z + 36 = 0$. Factor out the greatest common factor from each term: $6(6z^2 + 13z + 6) = 0$. Factor the trinomial expression into two factors: $6(2z + 3)(3z + 2) = 0$. Ignore the numerical factor and set the first factor equal to zero: $(2z + 3) = 0$. Subtract 3 from both sides: $2z + 3 - 3 = 0 - 3$. Simplify terms: $2z = -3$. Divide both sides by 2: $\frac{2z}{2} = \frac{-3}{2}$. Simplify terms: $z = -1\frac{1}{2}$. Now, let the second factor equal zero: $(3z + 2) = 0$. Subtract 2 from both sides: $3z + 2 - 2 = 0 - 2$. Simplify terms: $3z = -2$. Divide both sides by 3: $\frac{3z}{3} = \frac{-2}{3}$. Simplify terms: $z = \frac{-2}{3}$. The solutions for the quadratic equation $36z^2 + 78z = -36$ are $z = -1\frac{1}{2}$ and $z = \frac{-2}{3}$.

732. To transform the equation into the familiar trinomial equation form, first add $(40r - 192)$ to both sides of the equation: $12r^2 + 40r - 192 = 192 - 40r + 40r - 192$. Combine like terms: $12r^2 + 40r - 192 = 0$. Factor the greatest common factor, 4, out of each term: $4(3r^2 + 10r - 48) = 0$. Now, factor the trinomial expression: $4(r + 6)(3r - 8) = 0$. Ignoring the numerical factor, set one factor equal to zero: $(r + 6) = 0$. Subtract 6 from both sides: $r + 6 - 6 = 0 - 6$. Simplify: $r = -6$. Now, set the second factor equal to zero: $3r - 8 = 0$. Add 8 to both sides: $3r - 8 + 8 = 0 + 8$. Simplify: $3r = 8$. Divide both sides by 3: $\frac{3r}{3} = \frac{8}{3}$. Simplify terms: $r = 2\frac{2}{3}$. The solutions for the quadratic equation $12r^2 = 192 - 40r$ are $r = -6$ and $r = 2\frac{2}{3}$.

733. Divide both sides of the equation by 3: $\frac{24x^2}{3} = \frac{3(43x - 15)}{3}$. Simplify terms: $8x^2 = 43x - 15$. Add $(15 - 43x)$ to both sides of the equation: $8x^2 + 15 - 43x = 43x - 15 + 15 - 43x$. Combine like terms: $8x^2 + 15 - 43x = 0$. Use the commutative property to move terms: $8x^2 - 43x + 15 = 0$. Factor the trinomial expression: $(8x - 3)(x - 5) = 0$. Use the zero product property: $8x - 3 = 0$. Add 3 to both sides: $8x - 3 + 3 = 0 + 3$. Divide both sides by 8: $\frac{8x}{8} = \frac{3}{8}$. Simplify terms: $x = \frac{3}{8}$. Now, let the second factor equal zero: $x - 5 = 0$. Add 5 to both sides: $x = 5$. The solutions for the quadratic equation $24x^2 = 3(43x - 15)$ are $x = \frac{3}{8}$ and $x = 5$.

734. The equation is in the proper form. First, list the values for a, b, and c: $a = 1$, $b = 2$, $c = -8$. Substitute the values into the quadratic formula: $x = \frac{-2 \pm \sqrt{(2)^2 - 4(1)(-8)}}{2}$. Simplify the expression under the radical sign: $x = \frac{-2 \pm \sqrt{4 - -32}}{2} = \frac{-2 \pm \sqrt{36}}{2}$. Evaluate the square root of 36: $x = \frac{-2 \pm 6}{2}$. Find the two solutions for x by simplifying terms. First, add the terms in the numerator, and then subtract them: $x = \frac{-2 + 6}{2} = \frac{4}{2} = 2$ and $x = \frac{-2 - 6}{2} = \frac{-8}{2} = -4$. The two solutions for the variable x are $x = 2$ and $x = -4$.

735. The equation is in the proper form. First, list the values for a, b, and c: $a = 2$, $b = -7$, $c = -30$. Substitute the values into the quadratic formula: $x = \frac{-(-7) \pm \sqrt{(-7)^2 - 4(2)(-30)}}{2(2)}$. Simplify the expression: $x = \frac{7 \pm \sqrt{49 - (-240)}}{4} = \frac{7 \pm \sqrt{289}}{4} = \frac{7 \pm 17}{4}$. Find the two solutions for x by adding and then subtracting in the numerator: $x = \frac{7 + 17}{4} = \frac{24}{4} = 6$ and $x = \frac{7 - 17}{4} = \frac{-10}{4} = -2.5$. The two solutions for the variable x are $x = 6$ and $x = -2.5$.

736. The equation is in the proper form. First, list the values for a, b, and c: $a = 6$, $b = 13$, $c = -28$. Substitute the values into the quadratic formula: $x = \frac{-(13) \pm \sqrt{(13)^2 - 4(6)(-28)}}{2(6)}$. Simplify the expression: $x = \frac{-13 \pm \sqrt{169 + 672}}{12} = \frac{-13 \pm \sqrt{841}}{12} = \frac{-13 \pm 29}{12}$. Find the two solutions for x by adding and then subtracting in the numerator: $x = \frac{-13 + 29}{12} = \frac{16}{12} = 1\frac{1}{3}$ and $x = \frac{-13 - 29}{12} = \frac{-42}{12} = -3\frac{1}{2}$. The two solutions for the variable x are $x = 1\frac{1}{3}$ and $-3\frac{1}{2}$.

737. The equation is in the proper form. First, list the values for a, b, and c: $a = 18$, $b = 9$, $c = 1$. Substitute the values into the quadratic formula: $x = \frac{-(9) \pm \sqrt{(9)^2 - 4(18)(1)}}{2(18)}$. Simplify the expression: $x = \frac{-9 \pm \sqrt{81 - 72}}{36} = \frac{-9 \pm \sqrt{9}}{36} = \frac{-9 \pm 3}{36}$. Find the two solutions for x by adding and then subtracting in the numerator: $x = \frac{-9 + 3}{36} = \frac{-6}{36} = \frac{-1}{6}$ and $x = \frac{-9 - 3}{36} = \frac{-12}{36} = \frac{-1}{3}$. The two solutions for the variable x are $x = \frac{-1}{6}$ and $\frac{-1}{3}$.

738. First transform the equation into the proper form. Subtract 28 from both sides of the equation: $6x^2 + 17x - 28 = 28 - 28$. Combine like terms on both sides: $6x^2 + 17x - 28 = 0$. Now, list the values for a, b, and c: $a = 6$, $b = 17$, $c = -28$. Substitute the values into the quadratic formula: $x = \frac{-(17) \pm \sqrt{(17)^2 - 4(6)(-28)}}{2(6)}$. Simplify the expression: $x = \frac{-17 \pm \sqrt{289 + 672}}{12} = \frac{-17 \pm \sqrt{961}}{12} = \frac{-17 \pm 31}{12}$. Find the two solutions for x by adding and then subtracting in the numerator: $x = \frac{-17 + 31}{12} = \frac{14}{12} = 1\frac{1}{6}$ and $x = \frac{-17 - 31}{12} = \frac{-48}{12} = -4$. The two solutions for the variable x are $x = 1\frac{1}{6}$ and -4.

739. First transform the equation into the proper form. Add $(-12x - 32)$ to both sides of the equation: $14x^2 - 12x - 32 = 12x + 32 - 12x - 32$. Combine like terms: $14x^2 - 12x - 32 = 0$. Now, list the values for a, b, and c: $a = 14$, $b = -12$, $c = -32$. Substitute the values into the quadratic formula: $x = \frac{-(-12) \pm \sqrt{(-12)^2 - 4(14)(-32)}}{2(14)}$. Simplify the expression: $x = \frac{12 \pm \sqrt{144 + 1,792}}{28} = \frac{12 \pm \sqrt{1,936}}{28} = \frac{12 \pm 44}{28}$. Find the two solutions for x by adding and then subtracting in the numerator: $x = \frac{12 + 44}{28} = \frac{56}{28} = 2$ and $x = \frac{12 - 44}{28} = \frac{-32}{28} = -1\frac{1}{7}$. The two solutions for the variable x are $x = 2$ and $x = -1\frac{1}{7}$. Yes, you could have divided both sides of the equation by 2 before listing your a, b, and c values. However, the solution would have been the same. Try it yourself.

740. The equation may not appear to be in proper form because there is no value for c. But you could write it as $4x^2 + 5x + 0 = 0$, and then your values would be: $a = 4$, $b = 5$, $c = 0$. Substitute the values into the quadratic formula: $x = \frac{-(5) \pm \sqrt{(5)^2 - 4(4)(0)}}{2(4)}$. Simplify the expression: $x = \frac{-5 \pm \sqrt{25 - 0}}{8} = \frac{-5 \pm 5}{8}$. Find the two solutions for x by adding and then subtracting in the numerator: $x = \frac{-5 + 5}{8} = \frac{0}{8} = 0$ and $x = \frac{-5 - 5}{8} = \frac{-10}{8} = -1\frac{1}{4}$. The two solutions for the variable x are $x = 0$ and $x = -1\frac{1}{4}$.

741. Subtract 27 from both sides of the equation: $5x^2 - 27 = 27 - 27$ or $5x^2 - 27 = 0$. In this equation, there appears to be no x term unless you realize that $0x = 0$, so you could write the equation in the proper form like this: $5x^2 + 0x - 27 = 0$. Now, list the values for a, b, and c: $a = 5$, $b = 0$, $c = -27$. Substitute the values into the quadratic formula: $x = \frac{-(0) \pm \sqrt{(0)^2 - 4(5)(-27)}}{2(5)}$. Simplify the expression: $x = \frac{\pm\sqrt{4 \times 5 \times 27}}{10} = \frac{\pm\sqrt{4 \times 9 \times 5 \times 3}}{10} = \frac{\pm 2 \times 3\sqrt{5 \times 3}}{10}$ $= \frac{\pm 6\sqrt{15}}{10} = \frac{\pm 3\sqrt{15}}{5}$. Simplify the radical: $\frac{\pm 3\sqrt{6}}{2} = \pm 1\frac{1}{2}\sqrt{6}$. The two solutions for the variable x are $x = \pm 1\frac{1}{2}\sqrt{6}$.

742. To transform the equation into the desired form, first subtract $18x$ from and add 17 to both sides: $5x^2 - 18x + 17 = 18x - 18x - 17 + 17$. Combine like terms on both sides: $5x^2 - 18x + 17 = 0$. Now, list the values for a, b, and c: $a = 5$, $b = -18$, $c = 17$. Substitute the values into the quadratic formula: $x = \frac{-(-18) \pm \sqrt{(-18)^2 - 4(5)(17)}}{2(5)}$. Simplify the expression: $x = \frac{18 \pm \sqrt{324 - 340}}{10} = \frac{18 \pm \sqrt{-16}}{10}$. There is no rational number equal to the square root of a negative number, so there are no solutions for this equation.

743. List the values of a, b, and c: $a = 3$, $b = 11$, $c = -7$. Substitute the values into the quadratic formula: $x = \frac{-(11) \pm \sqrt{(11)^2 - 4(3)(-7)}}{2(3)}$. Simplify the expression: $x = \frac{-11 \pm \sqrt{121 + 84}}{6} = \frac{-11 \pm \sqrt{205}}{6} = -1\frac{5}{6} \pm \frac{1}{6}\sqrt{205}$. The solutions for the variable x are $x = -1\frac{5}{6} \pm \frac{1}{6}\sqrt{205}$.

744. List the values of a, b, and c: $a = 5$, $b = 52$, $c = 20$. Substitute the values into the quadratic formula: $x = \frac{-(52) \pm \sqrt{(52)^2 - 4(5)(20)}}{2(5)}$. Simplify the expression: $x = \frac{-52 \pm \sqrt{2,704 - 400}}{10} = \frac{-52 \pm \sqrt{2,304}}{10} = \frac{-52 \pm 48}{10}$. Find the two solutions for x by adding and then subtracting in the numerator. $x = \frac{-52 + 48}{10} = \frac{-4}{10} = -0.4$ and $x = \frac{-52 - 48}{10} = \frac{-100}{10} = -10$. The two solutions for the variable x are $x = -0.4$ and $x = -10$.

745. First, multiply both sides of the equation by 2: $2x^2 = -5x - 2$. Then, add $5x + 2$ to both sides of the equation: $2x^2 + 5x + 2 = 0$. List the values of a, b, and c: $a = 2$, $b = 5$, $c = 2$. Substitute the values into the quadratic formula: $x = \frac{-(5) \pm \sqrt{(5)^2 - 4(2)(2)}}{2(2)} = \frac{-5 \pm \sqrt{25 - 16}}{2(2)} = \frac{-5 \pm 3}{4}$. Find the two solutions for x by adding and then subtracting in the numerator: $x = \frac{-5 + 3}{4} = \frac{-2}{4} = \frac{-1}{2}$ and $x = \frac{-5 - 3}{4} = \frac{-8}{4} = -2$. The two solutions for the variable x are $x = \frac{-1}{2}$ and $x = -2$.

746. Transform the equation by subtracting 5 from both sides: $x^2 + 8x - 5 = 0$. List the values of a, b, and c: $a = 1$, $b = 8$, $c = -5$. Substitute the values into the quadratic formula: $x = \frac{-8 \pm \sqrt{8^2 - 4(1)(-5)}}{2(1)}$. Simplify the expression: $x = \frac{-8 \pm \sqrt{64 + 20}}{2} = \frac{-8 \pm \sqrt{84}}{2} = \frac{-8 \pm \sqrt{4 \times 21}}{2} = \frac{-8 \pm 2\sqrt{21}}{2} = -4 \pm \sqrt{21}$. The solution for the variable x is $x = -4 \pm \sqrt{21}$.

747. Transform the equation by subtracting $20x$ from and adding 19 to both sides of the equation: $x^2 - 20x + 19 = 0$. List the values of a, b, and c: $a = 1$, $b = -20$, $c = 19$. Substitute the values into the quadratic formula: $x = \frac{-(-20) \pm \sqrt{(-20)^2 - 4(1)(19)}}{2(1)}$. Simplify the expression: $x = \frac{20 \pm \sqrt{400 - 76}}{2} = \frac{20 \pm \sqrt{324}}{2} = \frac{20 \pm 18}{2}$. Find the two solutions for x by adding and then subtracting in the numerator: $x = \frac{20 + 18}{2} = \frac{38}{2} = 19$ and $x = \frac{20 - 18}{2} = \frac{2}{2} = 1$. The two solutions for the variable x are $x = 19$ and $x = 1$.

748. Transform the equation into proper form. Use the distributive property of multiplication on the right side of the equation: $23x^2 = 16x - 2$. Subtract $16x$ and add 2 to both sides: $23x^2 - 16x + 2 = 0$. List the values of a, b, and c: $a = 23$, $b = -16$, $c = 2$. Substitute the values into the quadratic formula: $x = \frac{-(-16) \pm \sqrt{(-16)^2 - 4(23)(2)}}{2(23)}$. Simplify the expression: $x = \frac{16 \pm \sqrt{256 - 184}}{2(23)} = \frac{16 \pm \sqrt{72}}{2(23)} = \frac{16 \pm \sqrt{4 \times 9 \times 2}}{2(23)} = \frac{2 \times 8 \pm 2 \times 3\sqrt{2}}{2(23)} = \frac{8 \pm 3\sqrt{2}}{23} = \frac{8}{23} \pm \frac{3}{23}\sqrt{2}$. The solution for the variable x is $x = \frac{8}{23} \pm \frac{3}{23}\sqrt{2}$.

749. List the values of a, b, and c: $a = 1$, $b = 10$, $c = 11$. Substitute the values into the quadratic formula: $x = \frac{-(10) \pm \sqrt{(10)^2 - 4(1)(11)}}{2(1)}$. Simplify the expression: $x = \frac{-10 \pm \sqrt{100 - 44}}{2} = \frac{-10 \pm \sqrt{56}}{2} = \frac{-10 \pm \sqrt{4 \times 14}}{2} = \frac{-10 \pm 2\sqrt{14}}{2} = -5 + \sqrt{14}$. The two solutions for the variable x are $x = -5 + \sqrt{14}$ and $x = -5 - \sqrt{14}$.

750. List the values of a, b, and c: $a = 24$, $b = 18$, $c = -6$. Substitute the values into the quadratic formula: $x = \frac{-(18) \pm \sqrt{(18)^2 - 4(24)(-6)}}{2(24)}$. Simplify the expression: $x = \frac{-18 \pm \sqrt{324 + 576}}{48} = \frac{-18 \pm \sqrt{900}}{48} = \frac{-18 \pm 30}{48}$. Find the two solutions for x by adding and then subtracting in the numerator: $x = \frac{-18 + 30}{48} = \frac{12}{48} = \frac{1}{4}$ and $x = \frac{-18 - 30}{48} = \frac{-48}{48} = -1$. The two solutions for the variable x are $x = \frac{1}{4}$ and $x = -1$.

751. Transform the equation into the proper form. First use the distributive property of multiplication: $7x^2 = 4(3x) + 4(1) = 12x + 4$. Then subtract $(12x + 4)$ from both sides: $7x^2 - 12x - 4 = 0$. List the values of a, b, and c: $a = 7$, $b = -12$, $c = -4$. Substitute the values into the quadratic formula: $x = \frac{-(-12) \pm \sqrt{(-12)^2 - 4(7)(-4)}}{2(7)}$. Simplify the expression: $x = \frac{12 \pm \sqrt{144 + 112}}{14} = \frac{12 \pm \sqrt{256}}{14} = \frac{12 \pm 16}{14}$. Find the two solutions for x by adding and then subtracting in the numerator: $x = \frac{12 + 16}{14} = \frac{28}{14} = 2$ and $x = \frac{12 - 16}{14} = \frac{-4}{14} = \frac{-2}{7}$. The two solutions for the variable x are $x = 2$ and $x = \frac{-2}{7}$.

752. You could use the fractions as values for a and b, but it might be easier to first transform the equation by multiplying it by 12: $12(\frac{1}{3}x^2 + \frac{3}{4}x - 3 = 0)$. Use the distributive property: $12(\frac{1}{3}x^2) + 12(\frac{3}{4}x) - 12(3) = 12(0)$. Simplify the terms: $4x^2 + 9x - 36 = 0$. List the values of a, b, and c: $a = 4$, $b = 9$, $c = -36$. Substitute the values into the quadratic formula: $x = \frac{-(9) \pm \sqrt{(9)^2 - 4(4)(-36)}}{2(4)}$. Simplify the expression: $x = \frac{-9 \pm \sqrt{81 + 576}}{8} = \frac{-9 \pm \sqrt{657}}{8} = \frac{-9 \pm \sqrt{9 \times 73}}{8} = \frac{-9 \pm 3\sqrt{73}}{8}$. The two solutions for the variable x are $x = \frac{-9 + 3\sqrt{73}}{8}$ and $x = \frac{-9 - 3\sqrt{73}}{8}$.

753. List the values of a, b, and c: $a = 5$, $b = -12$, $c = 1$. Substitute the values into the quadratic formula: $x = \frac{-(-12) \pm \sqrt{(-12)^2 - 4(5)(1)}}{2(5)}$. Simplify the expression: $x = \frac{12 \pm \sqrt{144 - 20}}{2(5)} = \frac{12 \pm \sqrt{124}}{2(5)} = \frac{12 \pm \sqrt{4 \times 31}}{2(5)} = \frac{12 \pm 2\sqrt{31}}{2(5)} = \frac{6 \pm \sqrt{31}}{5}$. The two solutions for the variable x are $x = \frac{6 + \sqrt{31}}{5}$ and $x = \frac{6 - \sqrt{31}}{5}$.

754. List the values of a, b, and c: $a = 11$, $b = -4$, $c = -7$. Substitute the values into the quadratic formula: $r = \frac{-(-4) \pm \sqrt{(-4)^2 - 4(11)(-7)}}{2(11)}$. Simplify the expression: $r = \frac{4 \pm \sqrt{16 + 308}}{22} = \frac{4 \pm \sqrt{324}}{22} = \frac{4 \pm 18}{22}$. Find the two solutions for r by adding and then subtracting in the numerator: $r = \frac{4 + 18}{22} = \frac{22}{22} = 1$ and $r = \frac{4 - 18}{22} = \frac{-14}{22} = -0.64$. The two solutions for the variable r are $r = 1$ and $r = -0.64$.

755. List the values of a, b, and c: $a = 3$, $b = 21$, $c = -8$. Substitute the values into the quadratic formula: $m = \frac{-(21) \pm \sqrt{(21)^2 - 4(3)(-8)}}{2(3)}$. Simplify the expression: $m = \frac{-21 \pm \sqrt{441 + 96}}{6} = \frac{-21 \pm \sqrt{537}}{6}$. The square root of 537 rounded to the nearest hundredth is 23.17. Substitute into the expression and simplify: $m = \frac{-21 + 23.17}{6} = \frac{2.17}{6} = 0.36$ and $m = \frac{-21 - 23.17}{6} = \frac{-44.17}{6} = -7.36$. The two solutions for the variable m are $m = 0.36$ and $m = -7.36$.

756. Transform the equation by subtracting $16y$ from and adding 5 to both sides: $4y^2 - 16y + 5 = 16y - 5 - 16y + 5$. Combine like terms: $4y^2 - 16y + 5 = 0$. List the values of a, b, and c: $a = 4$, $b = -16$, $c = 5$. Substitute the values into the quadratic formula: $y = \frac{-(-16) \pm \sqrt{(-16)^2 - 4(4)(5)}}{2(4)}$. Simplify the expression: $y = \frac{16 \pm \sqrt{256 - 80}}{8} = \frac{16 - \sqrt{176}}{8} = \frac{16 \pm \sqrt{16 \times 11}}{8} = \frac{16 \pm 4\sqrt{11}}{8}$. The square root of 11 rounded to the nearest hundredth is 3.32. Substitute into the expression and simplify: $y = \frac{16 + 4(3.32)}{8} = \frac{29.28}{8} = 3.66$ and $y = \frac{16 - 4(3.32)}{8} = \frac{2.72}{8} = 0.34$. The two solutions for the variable y are $y = 3.66$ and $y = 0.34$.

757. List the values of a, b, and c: $a = 5$, $b = 12$, $c = -1$. Substitute the values into the quadratic formula: $s = \frac{-(12) \pm \sqrt{(12)^2 - 4(5)(-1)}}{2(5)}$. Simplify the expression: $s = \frac{-12 \pm \sqrt{144 + 20}}{10} = \frac{-12 \pm \sqrt{164}}{10}$. The square root of 164 rounded to the nearest hundredth is 12.81. Substitute into the expression and simplify: $s = \frac{-12 + 12.81}{10} = \frac{0.81}{10} = 0.081$ and $s = \frac{-12 - 12.81}{10} = \frac{-24.81}{10} = -2.48$. The two solutions for the variable s to the nearest hundredth are $s = 0.08$ and $s = -2.48$.

758. List the values of a, b, and c: $a = 4$, $b = -11$, $c = 2$. Substitute the values into the quadratic formula: $c = \frac{-(-11) \pm \sqrt{(-11)^2 - 4(4)(2)}}{2(4)}$. Simplify the expression: $c = \frac{11 \pm \sqrt{121 - 32}}{8} = \frac{11 \pm \sqrt{89}}{8} = \frac{11 \pm 9.43}{8}$. $c = \frac{11 + 9.43}{8} = \frac{20.43}{8} = 2.55$ and $c = \frac{11 - 9.43}{8} = \frac{1.57}{8} = 0.20$. The two solutions for the variable c are $c = 2.55$ and $c = 0.20$.

759. List the values of a, b, and c: $a = 11$, $b = -32$, $c = 10$. Substitute the values into the quadratic formula: $k = \frac{-(-32) \pm \sqrt{(-32)^2 - 4(11)(10)}}{2(11)}$. Simplify the expression: $k = \frac{32 \pm \sqrt{1,024 - 440}}{22} = \frac{32 \pm \sqrt{584}}{22} = \frac{32 \pm 24.17}{22}$. $k = \frac{32 + 24.17}{22} = \frac{56.17}{22} = 2.55$ and $k = \frac{32 - 24.17}{22} = \frac{7.83}{22} = 0.36$. The two solutions for the variable k are $k = 2.55$ and $k = 0.36$.

CHAPTER

12▶ Triangles

ANY CLOSED FIGURE made up of line segments is called a **polygon**. A **triangle** is a three-sided polygon with three angles. Before we look at triangles, let's review the different types of angles.

► **Types of Angles**

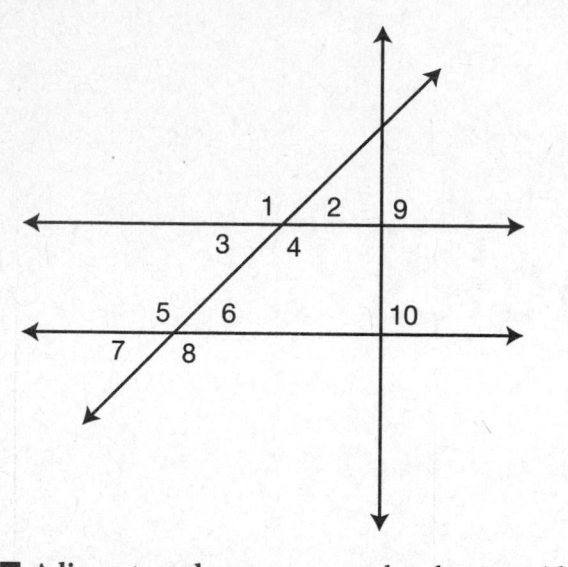

■ **Adjacent angles** are two angles that are side-by-side, sharing a vertex and a side. In the diagram, angles 1 and 2 are adjacent, as are angles 1 and 3.

■ **Corresponding angles** are pairs of congruent angles. (If two angles have the same measure, they are **congruent**. When parallel lines are cut by a transversal (a third line that intersects the parallel lines), alternating angles that are on the same side of the transversal are corresponding angles. Angles 1 and 5 are alternating (corresponding) angles. Angles 2 and 6, 3 and 7, and 4 and 8 are also corresponding angles.

■ **Vertical angles**, or opposite angles, are angles that are not adjacent, but made by the intersection of two lines. Vertical angles are congruent. Angles 1 and 4 are vertical angles, and they are equal in measure.

■ **Complementary angles** are two angles whose measures total 90 degrees. There are 90 degrees in a right angle, so any two angles that form a right angle are complementary angles.

■ **Supplementary angles** are two angles whose measures total 180 degrees. There are 180 degrees in a line, so any two angles that form a line

are supplementary angles. Angles 1 and 2 are supplementary angles, as are angles 6 and 8.

■ **Acute angles** are angles that measure less than 90 degrees. Angles 2, 3, 6, and 7 are acute angles.

■ **Right angles** are angles that measure exactly 90 degrees. Angles 9 and 10 are right angles.

■ **Obtuse angles** are angles that measure greater than 90 degrees. Angles 1, 4, 5, and 8 are obtuse angles.

■ **Straight angles** are angles that measure exactly 180 degrees.

■ **Reflex angles** are angles that measure more than 180 degrees.

As you can see, there are many ways to classify the same angle. In that diagram, angle 1 is an adjacent angle (to angles 2 and angle 3), a corresponding angle (to angle 5), a vertical angle (to angle 4), a supplementary angle (to angles 2 and 3), and an obtuse angle.

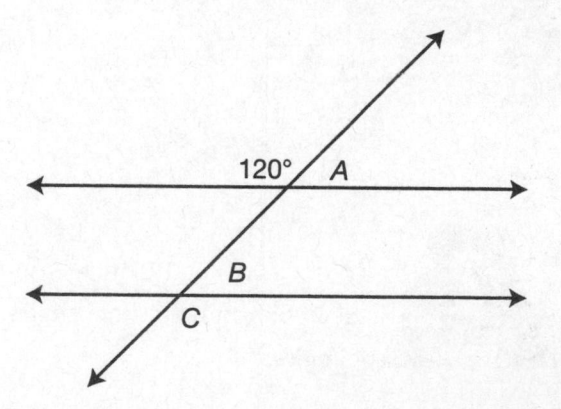

Example

What is the measure of angle *A*?

The angle labeled 120 degrees and angle *A* are supplementary angles because they form a line. Since there are 180 degrees in a line, the measure of angle *A* is 180 − 120 = 60 degrees.

Example

What is the measure of angle *B*?

Angles *A* and *B* are corresponding angles, so they are congruent. Angle *B* is also 60 degrees.

Example

What kind of angle is angle *C*?

Angle *C* is a corresponding angle, congruent to the angle labeled 120 degrees, which means that angle *C* is an obtuse angle.

▶ **Practice Questions**

Label each angle measurement as acute, right, obtuse, straight, or reflex.

760. 13.5°

761. 91°

762. 46°

763. 179.3°

764. 355°

765. 180.2°

766. 90°

Use the following diagram to answer questions 767 through 769.

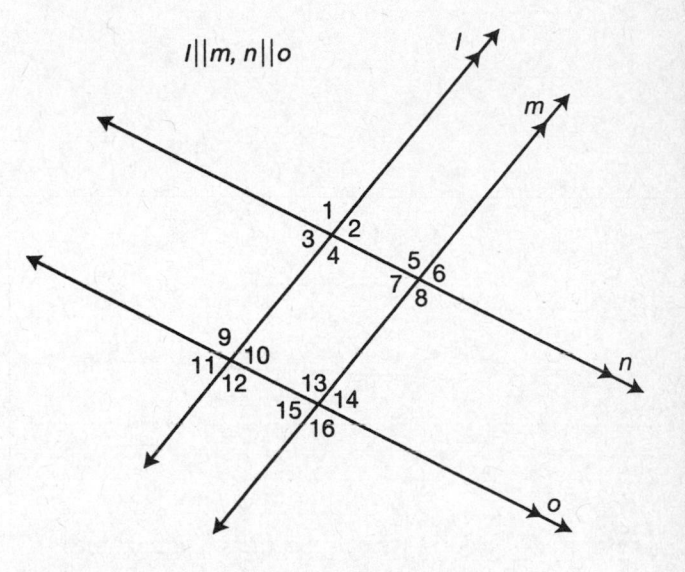

767. In sets, name all the congruent angles.

768. In pairs, name all the vertical angles.

769. In pairs, name all the corresponding angles.

Use the following diagram and information to determine whether lines *o* and *p* are proven to be parallel.

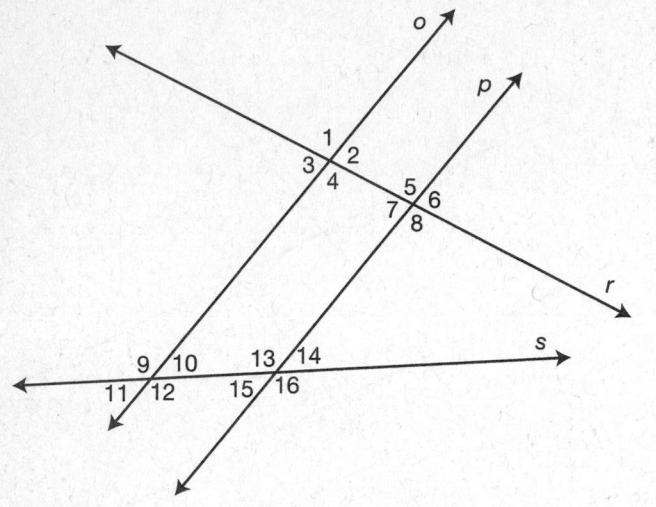

770. If ∠5 and ∠4 are congruent and equal, does this prove lines *o* and *p* are parallel?

771. If ∠1 and ∠2 are congruent and equal, does this prove lines *o* and *p* are parallel?

Circle the correct answer True or False.

772. Angles supplementary to the same angle or to angles with the same measure are also equal in measure.   **True or False**

773. Adjacent angles that are also congruent are always right angles.   **True or False**

774. Supplementary angles that are also congruent are right angles.   **True or False**

775. If vertical angles are acute, the angle adjacent to them must be obtuse.   **True or False**

776. When two lines intersect, all four angles formed are never congruent to each other. **True or False**

Choose the best answer for questions 777 through 781 based on the figure shown here.

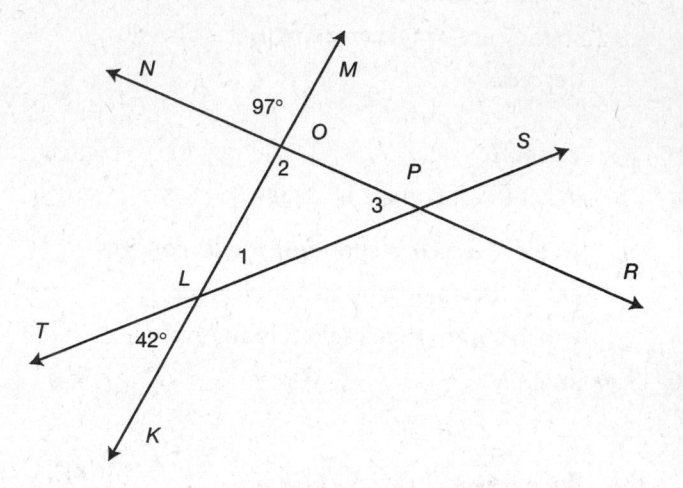

777. Name the angle vertical to ∠NOM.
  a. ∠NOL
  b. ∠KLP
  c. ∠LOP
  d. ∠MOP

778. Name the angle vertical to ∠TLK.
  a. ∠MOR
  b. ∠NOK
  c. ∠KLT
  d. ∠MLS

779. Name the pair of angles supplementary to ∠NOM.
  a. ∠MOR and ∠NOK
  b. ∠SPR and ∠TPR
  c. ∠NOL and ∠LOP
  d. ∠TLK and ∠KLS

780. ∠1, ∠2, and ∠3 respectively measure
  a. 90°, 40°, 140°.
  b. 139°, 41°, 97°.
  c. 42°, 97°, 41°.
  d. 41°, 42°, 83°.

781. The measure of exterior ∠OPS is
    a. 139°.
    b. 83°.
    c. 42°.
    d. 41°.

Choose the best answer.

782. If ∠LKN and ∠NOP are complementary angles, then
    a. they are both acute.
    b. they must both measure 45°.
    c. they are both obtuse.
    d. one is acute and the other is obtuse.
    e. No determination can be made.

783. If ∠KAT and ∠GIF are supplementary angles,
    a. they are both acute.
    b. they must both measure 90°.
    c. they are both obtuse.
    d. one is acute and the other is obtuse.
    e. No determination can be made.

784. If ∠DEF and ∠IPN are congruent, they are
    a. complementary angles.
    b. supplementary angles.
    c. right angles.
    d. adjacent angles.
    e. No determination can be made.

785. If ∠ABE and ∠GIJ are congruent supplementary angles, they are
    a. acute angles.
    b. obtuse angles.
    c. right angles.
    d. adjacent angles.
    e. No determination can be made.

786. If ∠EDF and ∠HIJ are supplementary angles, and ∠SUV and ∠EDF are also supplementary angles, then ∠HIJ and ∠SUV are
    a. acute angles.
    b. obtuse angles.
    c. right angles.
    d. congruent angles.
    e. No determination can be made.

Fill in the blanks based on your knowledge of angles and the figure shown here.

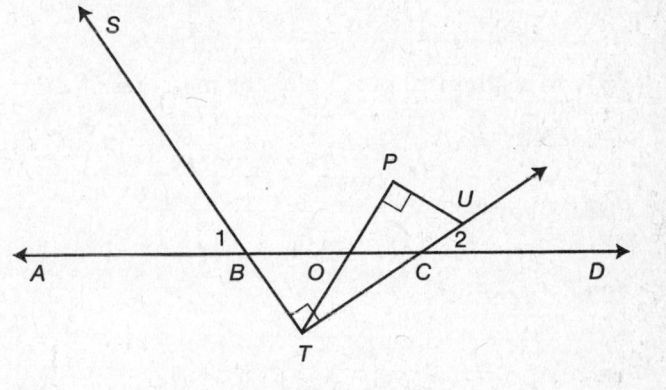

787. If ∠ABT is obtuse, ∠TBO is _____.

788. ∠BTO and ∠OTC are _____.

789. If ∠POC is acute, ∠POB is _____.

790. If ∠1 is congruent to ∠2, then _____.

State the relationship or sum of the angles given based on the figure shown here. If a relationship cannot be determined, then indicate this as your answer.

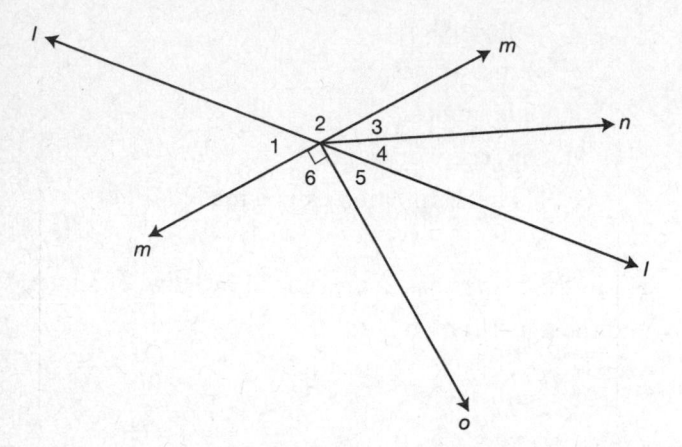

791. Measurement of ∠2 plus the measures of ∠6 and ∠5.

792. ∠1 and ∠3.

793. ∠1 and ∠2.

794. The sum of ∠5, ∠4, and ∠3.

795. ∠6 and ∠2.

796. The sum of ∠1, ∠6, and ∠5.

Now that we've reviewed the different types of angles, we're ready to look at triangles.

▶ Types of Triangles

A triangle that has three congruent sides and three congruent angles is called an **equilateral triangle**. Every angle in the triangle measures 60 degrees.

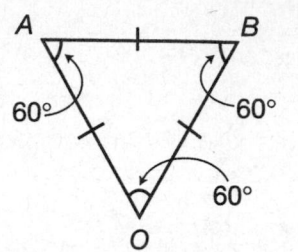

△ABO $AB \cong BO \cong OA$ ∠ABO ≅ ∠BOA ≅ ∠BAO

A triangle that has two congruent sides and two congruent angles is called an **isosceles triangle**. An isosceles triangle always has at least two acute angles, but its third angle can be acute, obtuse, or right.

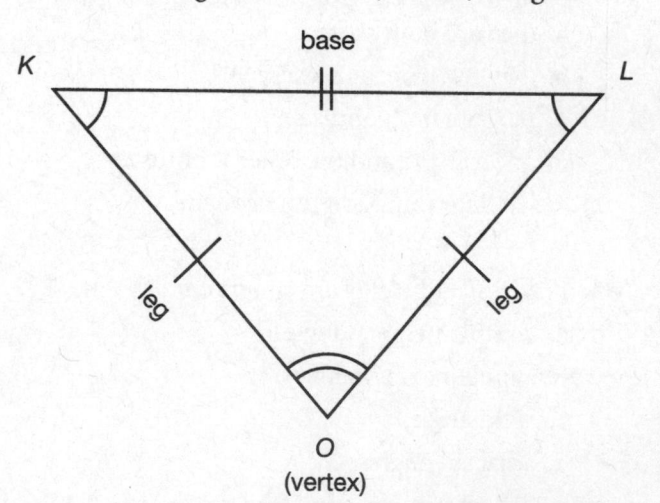

△KLO $KO \cong LO$ ∠LKO ≅ ∠KLO

A triangle that has no congruent sides or angles is called a **scalene triangle**. Scalene triangles can also be acute, obtuse, or right triangles.

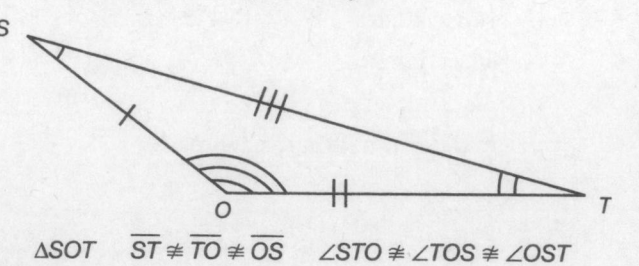

△SOT $\overline{ST} \not\cong \overline{TO} \not\cong \overline{OS}$ ∠STO ≠ ∠TOS ≠ ∠OST

A triangle with three angles that all measure less than 90 degrees is called an **acute triangle**. An isosceles triangle can be acute, and all equilateral triangles are acute.

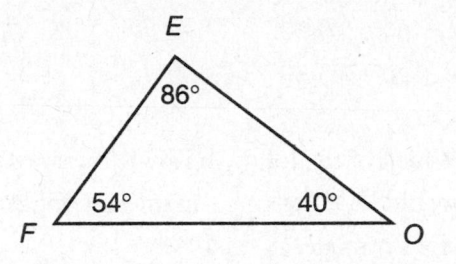

Scalene triangle *EOF*   ∠EOF, ∠OEF, and ∠FEO < 90

A triangle with one angle that measures exactly 90 degrees is called a **right triangle**. An isosceles triangle can be a right triangle if the two angles that are not the right angle both measure 45 degrees.

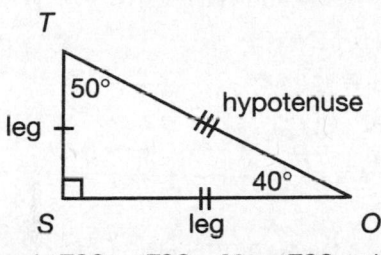

Scalene triangle *TOS*   ∠TSO = 90   ∠TOS and ∠STO < 90

A triangle with one angle that measures more than 90 degrees is called an **obtuse triangle**. An isosceles triangle can be obtuse.

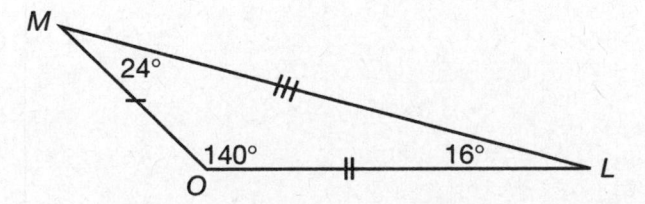

Scalene triangle *LMO*   ∠LOM > 90   ∠OLM and ∠LMO < 90

Example

What kind of triangle is the triangle shown here?

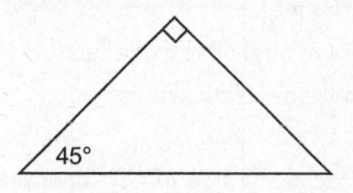

This triangle has a right angle, which means that it is a right triangle. It also has one angle that measures 45 degrees, which means that the third angle must also measure 45 degrees, because 180 − 90 − 45 = 45. There are two congruent angles in this triangle, so it must be an isosceles right triangle.

▶ Practice Questions

Fill in the blanks based on your knowledge of triangles and angles.

797. In right triangle *ABC*, if ∠*C* measures 31° and ∠*A* measures 90°, then ∠*B* measures

_____.

798. In scalene triangle *QRS*, if ∠*R* measures 134° and ∠*Q* measures 16°, then ∠*S* measures

_____.

799. In isosceles triangle *TUV*, if vertex ∠*T* is supplementary to an angle in an equilateral triangle, then base ∠*U* measures _____.

800. In obtuse isosceles triangle *EFG*, if the base ∠*F* measures 12°, then the vertex ∠*E* measures

_____.

801. In acute triangle *ABC*, if ∠*B* measures 45°, can ∠*C* measure 30°? _____.

Choose the best answer.

802. Which of the following sets of interior angle measures would describe an acute isosceles triangle?
a. 90°, 45°, 45°
b. 80°, 60°, 60°
c. 60°, 60°, 60°
d. 60°, 50°, 50°

803. Which of the following sets of interior angle measures would describe an obtuse isosceles triangle?
a. 90°, 45°, 45°
b. 90°, 90°, 90°
c. 100°, 50°, 50°
d. 120°, 30°, 30°

804. Which of the following angle measurements would NOT describe an interior angle of a right triangle?
a. 30°
b. 60°
c. 90°
d. 100°

805. If △*JNM* is equilateral and equiangular, which condition would NOT exist?
a. $\overline{JN} = \overline{MN}$
b. $\overline{JM} \cong \overline{JN}$
c. ∠*N* = ∠*J*
d. ∠*M* = $\overline{NM}$

806. In isosceles △*ABC*, if vertex ∠*A* is twice the measure of base ∠*B*, then ∠*C* measures
a. 30°.
b. 33°.
c. 45°.
d. 90°.

Using the obtuse triangle diagram shown here, determine which of the pair of angles given has a greater measure.

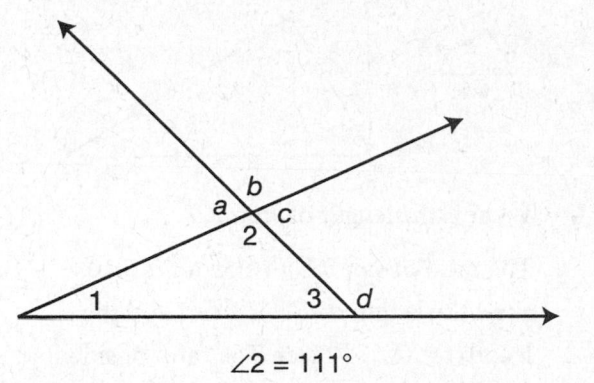

∠2 = 111°

807. ∠1 or ∠2

808. ∠3 or ∠d

809. ∠a or ∠b

810. ∠1 or ∠c

811. ∠a or ∠c

812. ∠3 or ∠b

813. ∠2 or ∠d

▶ Similar Triangles

Two triangles are **similar triangles** if they are both comprised of the same angles. For example, if one triangle has angles that measure 45 degrees, 65 degrees, and 70 degrees, and another triangle has angles that measure 45 degrees, 65 degrees, and 70 degrees, then these triangles are similar.

The two triangles shown here are similar. Although their angles are congruent, their sides are not. Each side of triangle *DEF* is three times the length of the corresponding side of triangle *ABC*.

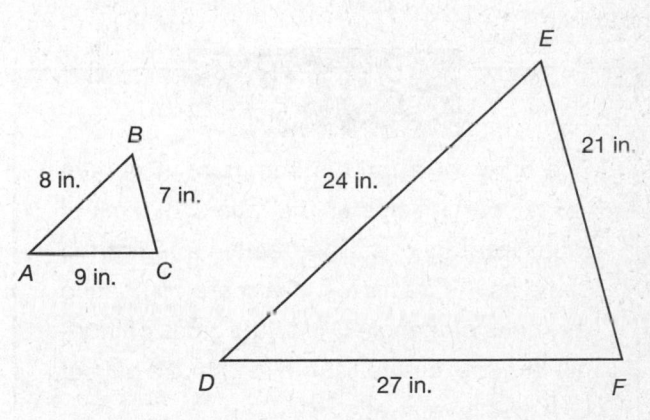

The ratio of the length of a side of triangle *ABC* to the length of a side of triangle *DEF* is 1:3. A **ratio** is a comparison between two quantities. We say that the sides of triangle *DEF* are in proportion to the sides of *ABC*. That is, each side of *DEF* is larger than the corresponding side of *ABC* according to a ratio.

We can use a proportion to compare the lengths of the sides of the two triangles. That's all a **proportion** is—a comparison between two ratios. The ratio of side *AB* to side *DE* is equal to the ratio of side *AC* to side *DF*: $\frac{8}{24} = \frac{9}{27}$. How can we check that these two ratios are equal? There are two ways: We can reduce them, or we can cross multiply and check that the products are equal. First, let's reduce the fractions.

The greatest common factor of 8 and 24 is 8, so we can divide the top and bottom of this fraction by 8: $\frac{8}{24} = \frac{1}{3}$. The greatest common factor of 9 and 27 is 9, so we can divide the top and bottom of this fraction by 9: $\frac{9}{27} = \frac{1}{3}$. Since both fractions reduce to the same fraction, these ratios are equal.

We also could cross multiply to check that the ratios are equal. Multiply the numerator of the first fraction, 8, by the denominator of the second fraction, 27, and see whether that product is equal to the product of the denominator of the first fraction, 24, and the numerator of the second fraction, 9: $(8)(27) = 216$, and $(9)(24) = 216$, which also confirms that the ratios are equal.

PITFALL

The order of a ratio is important. The ratio 1:3 is not the same as the ratio 3:1. Think of ratios as fractions. The fraction $\frac{1}{3}$ is not the same as the fraction $\frac{3}{1}$. If you are comparing the sides of triangle *GHI* to the sides of triangle *JKL*, be sure to list the length of a side of *GHI* first in your ratio.

How can ratios help us? If we know that two triangles are similar, and we know the lengths of some of the sides of the two triangles, we can use the ratio of their side measures to set up a proportion. That proportion can help us find the lengths of the other sides of the triangles.

Example

Triangles *TUV* and *XYZ*, shown here, are similar.

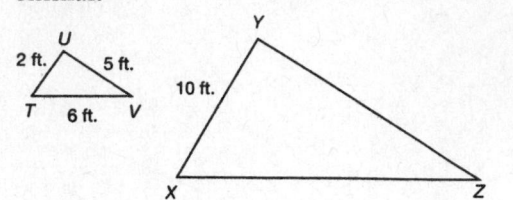

What is the length of side *YZ*?

The ratio of side *TU* to side *XY* is 2:10, since the length of *TU* is 2 feet and the length of *XY* is 10 feet. The ratio of side *UV* to *YZ* must also be 2:10. Since the length of *UV* is known but the length of *YZ* is not known, we can set up a proportion to find the length of *YZ*. Our proportion will compare the ratio of *TU* to *XY* to the ratio of *UV* to *YZ*: $\frac{2}{10} = \frac{5}{YZ}$.

To solve a proportion, cross multiply and see whether the products are equal to each other. Multiply the numerator of the first fraction, 2, by the denominator of the second fraction, *YZ*, and set that product equal to the product of the denominator of the first fraction, 10, and the numerator of the second fraction, 5: $(2)(YZ) = (10)(5)$, and $2(YZ) = 50$. Solve this equation by dividing both sides by 2: $YZ = 25$ feet.

TIP

It's often easier to reduce a ratio before working with it. The ratio 2:10 in the preceding example reduces to 1:5. Reducing a ratio is just like reducing a fraction (because a ratio is a fraction). Set up the proportion to find the length of *YZ* using the reduced ratio. $\frac{1}{5} = \frac{5}{YZ}$. Cross multiplying shows that $YZ = 25$— we have our answer in one step!

Example

What is the length of side *XZ*?

We can find the length of *XZ* in the same way. Set the ratio of *TU* to *XY* equal to the ratio of *TV* to *XZ*: $\frac{2}{10} = \frac{6}{XZ}$; $2(XZ) = (10)(6)$; $2(XZ) = 60$, $XZ = 30$ feet.

We know that two triangles are similar if they are comprised of the same three angles, but if we know that two triangles even have two identical angles, then the third angle of each triangle must be identical, too. This is called the Angle-Angle postulate. Knowing that two triangles share two congruent angles is enough to prove that the triangles are similar. We have two other postulates we can use to prove similarity.

$\overline{AB} : \overline{EF} = 3:12$

$\overline{BC} : \overline{FG} = 1:4$

$3:12 = 1:4$

Reduce each ratio:

$1:4 = 1:4$

Side-Angle-Side (SAS) Postulate: If the lengths of two pairs of corresponding sides of two triangles are proportional and the corresponding included angles are congruent, then the triangles are similar.

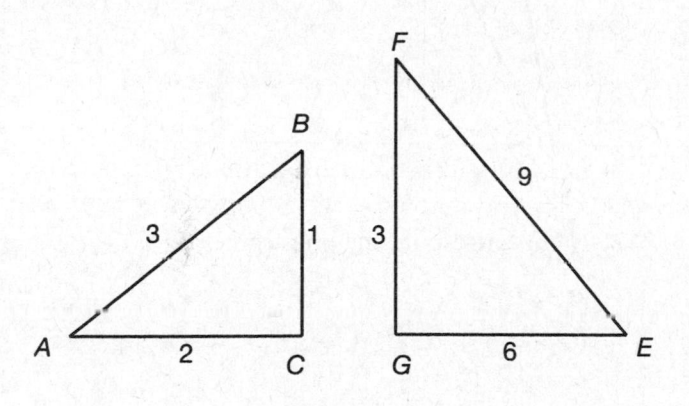

$\overline{AB} : \overline{EF} = 3:9$

$\overline{BC} : \overline{FG} = 1:3$

$\overline{CA} : \overline{GE} = 2:6$

$3:9 = 2:6 = 1:3$

Reduce each ratio:

$1:3 = 1:3 = 1:3$

Side-Side-Side (SSS) Postulate: If the lengths of the corresponding sides of two triangles are proportional, then the triangles are similar.

▶ **Practice Questions**

Choose the best answer.

814. In △*ABC*, side *AB* measures 16 inches. In similar △*EFG*, corresponding side *EF* measures 24 inches. State the ratio of side *AB* to side *EF*.

a. 2:4

b. 2:3

c. 2:1

d. 8:4

815. Use the following figure to find a proportion to solve for *x*.

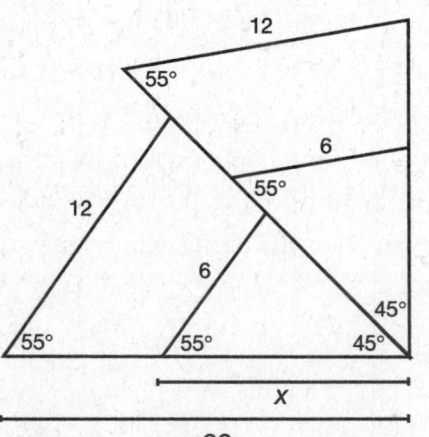

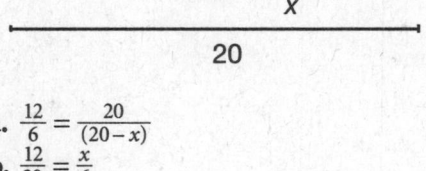

a. $\frac{12}{6} = \frac{20}{(20-x)}$

b. $\frac{12}{20} = \frac{x}{6}$

c. $\frac{20}{12} = \frac{6}{x}$

d. $\frac{12}{6} = \frac{20}{x}$

816. In similar triangles *UBE* and *ADF*, $\overline{UB}$ measures 10 inches while corresponding $\overline{AD}$ measures 2 inches. If $\overline{BE}$ measures 30 inches, then corresponding $\overline{DF}$ measures

a. 150 inches.

b. 60 inches.

c. 12 inches.

d. 6 inches.

Use the figure shown to answer question 817.

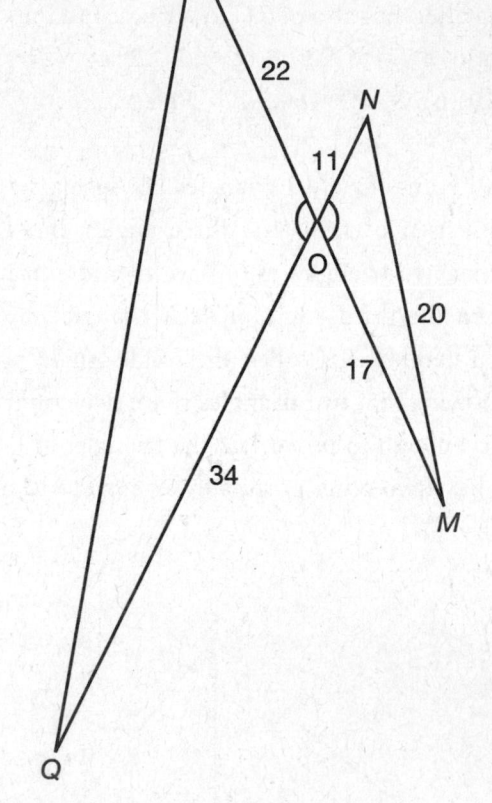

817. Name corresponding line segments.

Use the figure shown here to answer questions 818 and 819.

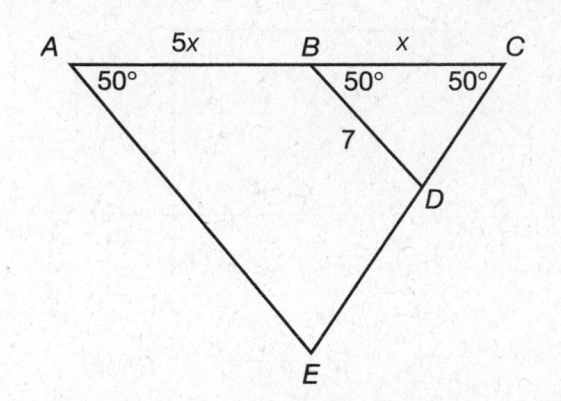

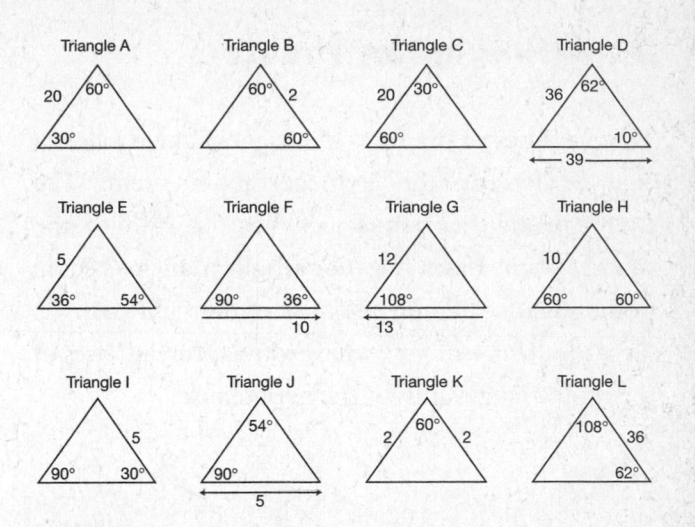

818. Name corresponding line segments.

819. Find $\overline{AE}$.

820. Name the triangle that is congruent to ΔA.

821. Which triangle is similar to ΔA?

822. Name the triangle that is congruent to ΔB.

823. Which triangle is similar to ΔB?

824. Name the triangle that is congruent to ΔE.

825. Which triangle is similar to ΔE?

826. Name the triangle that is congruent to ΔD.

827. Which triangle is similar to ΔD?

828. Name the triangles that are right triangles.

829. Which triangles are equilateral triangles?

▶ Pythagorean Theorem

The measures of the sides of a right triangle follow a rule known as the Pythagorean theorem. The **Pythagorean theorem** states that the sum of the squares of the bases (legs) of a right triangle is equal to the square of the hypotenuse of the right triangle. Simply put, $a^2 + b^2 = c^2$, where a and b are the bases of the right triangle and c is the hypotenuse.

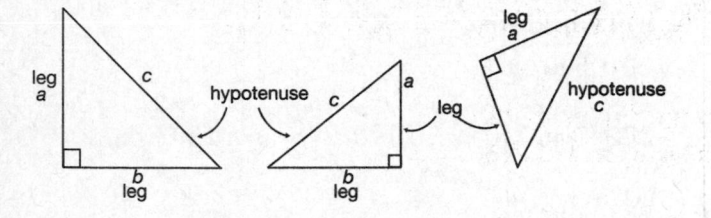

The longest side is always the hypotenuse; therefore the longest side is always c.

Find hypotenuse $\overline{QR}$.

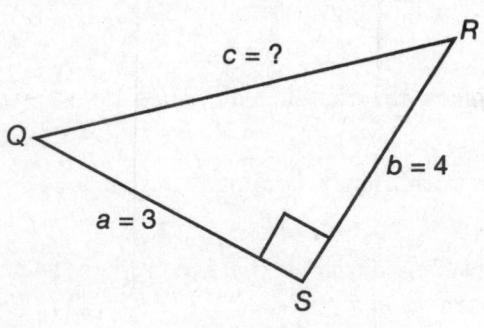

$$a^2 + b^2 = c^2$$
$$3^2 + 4^2 = c^2$$
$$9 + 16 = c^2$$
$$25 = c^2$$

Take the square root of each side:

$$\sqrt{25} = \sqrt{c^2}$$
$$5 = c$$

Find $\overline{KL}$.

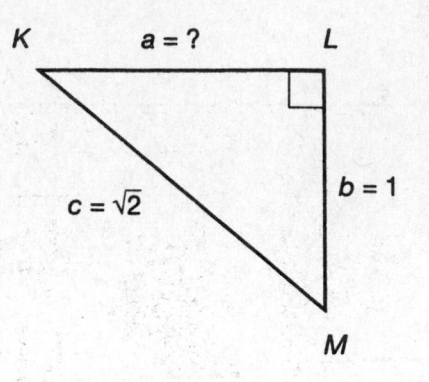

$$a^2 + b^2 = c^2$$
$$a^2 + 1^2 = (\sqrt{2})^2$$
$$a^2 + 1 = 2$$
$$a^2 = 1$$

Take the square root of each side:

$$\sqrt{a^2} = \sqrt{1}$$
$$a = 1$$

Find $\overline{CD}$.

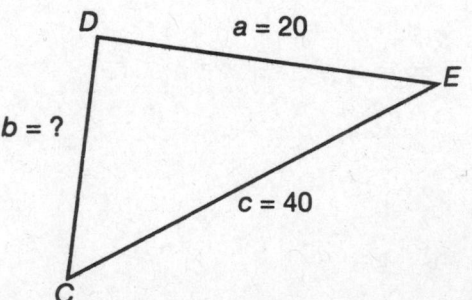

$$a^2 + b^2 = c^2$$
$$20^2 + b^2 = 40^2$$
$$400 + b^2 = 1{,}600$$
$$b^2 = 1{,}200$$

Take the square root of each side:

$$\sqrt{b^2} = \sqrt{1{,}200}$$
$$b = 20\sqrt{3}$$

The Pythagorean theorem can only find a side of a right triangle. However, if all the sides of any given triangle are known but none of the angles are known, the Pythagorean theorem can tell you whether that triangle is obtuse or acute based on how the longest side compares with the other two sides.

Is $\triangle GHI$ obtuse or acute?

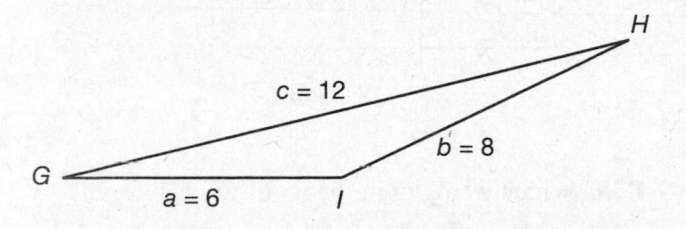

$a^2 + b^2$		c^2
$6^2 + 8^2$		12^2
$36 + 64$		144
100	$<$	144

The square of the longest side is greater than the sum of the squares of the other two sides. Therefore, $\triangle GHI$ is obtuse.

Is $\triangle JKL$ obtuse or acute?

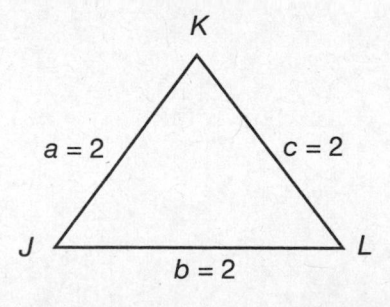

$a^2 + b^2$		c^2
$2^2 + 2^2$		2^2
$4 + 4$		4
8	$>$	4

The square of the longest side is less than the sum of the squares of the other two sides. Therefore, $\triangle JKL$ is obtuse.

TIP

Remember the order of operations. Exponents come before addition, so always find the squares before adding or subtracting when using the Pythagorean theorem.

▶ Practice Questions

Choose the best answer.

830. If the sides of a triangle measure 3, 4, and 5, then the triangle is
   a. acute.
   b. right.
   c. obtuse.
   d. It cannot be determined.

831. If the sides of a triangle measure 12, 16, and 20, then the triangle is
   a. acute.
   b. right.
   c. obtuse.
   d. It cannot be determined.

Choose the best answer.

832. Eva and Carr meet at a corner. Eva turns 90°
left and walks five paces; Carr continues
straight and walks six paces. If a line segment
connected them, it would measure
  a. $\sqrt{22}$ paces.
  b. $\sqrt{25}$ paces.
  c. $\sqrt{36}$ paces.
  d. $\sqrt{61}$ paces.

833. The legs of a table measure 3 feet long and the
top measures 4 feet long. If the legs are connected to the table at a right angle, then what is
the distance between the bottom of each leg
and the end of the tabletop?
  a. 5 feet
  b. 7 feet
  c. 14 feet
  d. 25 feet

834. Dorothy sees a plane flying 300 meters directly
above her, and another plane flying straight
behind the first that is 500 meters away from
her. How far apart are the planes from each
other?
  a. 40 meters
  b. 400 meters
  c. 4,000 meters
  d. 40,000 meters

835. Timmy arranges the walls of his shed on the
ground. The base of the first side measures 10
feet. The base of the second side measures 15
feet. If the walls are at a right angle to each
other, the measure from the end of one side to
the end of the second side equals
  a. 35 feet.
  b. 50 feet.
  c. $\sqrt{225}$ feet.
  d. $\sqrt{325}$ feet.

Use the figure shown here to answer questions 836
through 841.

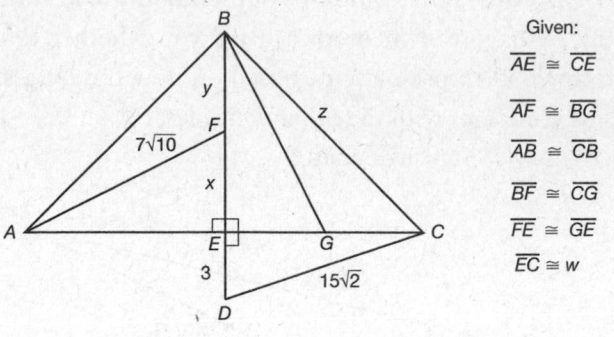

836. Which triangles in the figure are congruent
and/or similar?

837. Find the value of w.

838. Find the value of x.

839. Find the value of y.

840. Find the value of z.

841. Is $\triangle BGC$ acute or obtuse?

TERMS TO REVIEW

polygon

triangle

adjacent angles

corresponding angles

congruent

vertical angles

complementary angles

supplementary angles

acute angles

right angles

obtuse angles

straight angles

reflex angles

equilateral triangle

isosceles triangle

scalene triangle

acute triangle

right triangle

obtuse triangle

similar triangles

ratio

proportion

Pythagorean theorem

FORMULA TO REVIEW

Pythagorean theorem: $a^2 + b^2 = c^2$, where a and b are the lengths of the legs of a right triangle and c is the hypotenuse

▶ **Answers**

760. $0° < 13.5° < 90°$; acute

761. $90° < 91° < 180°$; obtuse

762. $0° < 46° < 90°$; acute

763. $90° < 179.3° < 180°$; obtuse

764. $180° < 355° < 360°$; reflex

765. $180° < 180.2° < 360°$; reflex

766. $90° = 90°$; right

767. $\angle 1 \cong \angle 4 \cong \angle 5 \cong \angle 8 \cong \angle 9 \cong \angle 12 \cong \angle 13 \cong \angle 16$; $\angle 2 \cong \angle 3 \cong \angle 6 \cong \angle 7 \cong \angle 10 \cong \angle 11 \cong \angle 14 \cong \angle 15$

768. $\angle 1, \angle 4$; $\angle 2, \angle 3$; $\angle 5, \angle 8$; $\angle 6, \angle 7$; $\angle 9, \angle 12$; $\angle 10, \angle 11$; $\angle 13, \angle 16$; $\angle 14, \angle 15$

769. $\angle 1, \angle 9$; $\angle 2, \angle 10$; $\angle 3, \angle 11$; $\angle 4, \angle 12$; $\angle 5, \angle 13$; $\angle 6, \angle 14$; $\angle 7, \angle 15$; $\angle 8, \angle 16$

770. Yes, because there are three congruent angle pairs that can prove a pair of lines cut by a transversal are parallel: alternate interior angles, alternate exterior angles, and corresponding angles. Angles 5 and 4 are alternate interior angles—notice the Z figure.

771. No, because $\angle 1$ and $\angle 2$ are adjacent angles. Their measurements combined must equal 180°, but they do not determine parallel lines.

772. True.

773. False. Adjacent congruent angles do not always form straight lines; to be adjacent, angles just need to share a vertex, a side, and no interior points. However, adjacent congruent angles that do form a straight line are always right angles.

774. **True.** A pair of supplementary angles must measure 180°. If the angles are also congruent, they must measure 90° each. An angle that measures 90° is a right angle.

775. **True.** When two lines intersect, they create four angles. The two angles opposite each other are congruent. Adjacent angles are supplementary. If vertical angles are acute, angles adjacent to them must be obtuse in order to measure 180°.

776. **False.** Perpendicular lines form all right angles.

777. **c.** ∠NOM and ∠LOP are opposite angles formed by intersecting lines NR and MK; thus, they are vertical angles.

778. **d.** ∠TLK and ∠MLS are opposite angles formed by intersecting lines TS and MK; thus, they are vertical angles.

779. **a.** ∠MOR and ∠NOK are both adjacent to ∠NOM along two different lines. The measure of each angle added to the measure of ∠NOM equals that of a straight line, or 180°. Each of the other answer choices is a pair of angles that are supplementary to each other, but not to ∠NOM.

780. **c.** ∠1 is the vertical angle to ∠TLK, which is given. ∠2 is the vertical pair to ∠NOM, which is also given. Since vertical angles are congruent, ∠1 and ∠2 measure 42° and 97°, respectively. To find the measure of ∠3, subtract the sum of ∠1 and ∠2 from 180° (the sum of the measure of a triangle's interior angles): $180 - (42 + 97) = ∠3$, and $41 = ∠3$.

781. **a.** There are two ways to find the measure of exterior angle OPS. The first method subtracts the measure of ∠3 from 180°. The second method adds the measures of ∠1 and ∠2 together because the measure of an exterior angle equals the sum of the two nonadjacent interior angles. ∠OPS measures 139°.

782. **a.** The sum of any two complementary angles must equal 90°. Any angle less than 90° is acute. Only the measures of two acute angles could add to 90°. Choice **b** assumes both angles are also congruent; however, that information is not given. If the measure of one obtuse angle equals more than 90°, then two obtuse angles could not possibly measure exactly 90° together. Choices **c** and **d** are incorrect.

783. **e.** Unlike the preceding question, where every complementary angle must also be acute, supplementary angles can be acute, right, or obtuse. If an angle is obtuse, its supplement is acute. If an angle is right, its supplement is also right. Two obtuse angles can never be a supplementary pair, and two acute angles can never be a supplementary pair. Without more information, the answer to this question cannot be determined.

784. **e.** Complementary angles that are also congruent measure 45° each. Supplementary angles that are also congruent measure 90° each. Without more information, the answer to this question cannot be determined.

785. **c.** Congruent supplementary angles always measure 90° each: $\angle ABE = x$; $\angle GIJ = x$; $\angle ABE + \angle GIJ = 180$; replace each angle with its measure: $x + x = 180$; $2x = 180$; divide each side by 2: $x = 90$. Any 90° angle is a right angle.

786. **d.** When two angles are supplementary to the same angle, they are congruent to each other: $\angle EDF + \angle HIJ = 180$; $\angle EDF + \angle SUV = 180$; $\angle EDF + \angle HIJ = \angle EDF + \angle SUV$; subtract $\angle EDF$ from each side: $\angle HIJ = \angle SUV$.

787. $\angle TBO$ is an acute angle. $\angle ABT$ and $\angle TBO$ are adjacent angles on the same line. As a supplementary pair, the sum of their measures must equal 180°. If one angle is more than 90°, the other angle must compensate by being less than 90°. Thus, if one angle is obtuse, the other angle is acute.

788. $\angle BTO$ and $\angle OTC$ are adjacent complementary angles. $\angle BTO$ and $\angle OTC$ share a side, a vertex, and no interior points; they are adjacent. The sum of their measures must equal 90° because they form a right angle; thus, they are complementary.

789. $\angle POB$ is an obtuse angle. $\angle POC$ and $\angle POB$ are adjacent angles on the same line. As a supplementary pair, the sum of their measures must equal 180°. If one angle is less than 90°, the other angle must compensate by being more than 90°. Thus, if one angle is acute, the other angle is obtuse.

790. $\angle SBO$ and $\angle OCU$ are congruent. When two angles are supplementary to the same angle or angles that measure the same, then they are congruent.

791. The measurements are equal. Together $\angle 5$ and $\angle 6$ form the vertical angle pair to $\angle 2$. Consequently, the angles are congruent and their measurements are equal.

792. A determination cannot be made. $\angle 1$ and $\angle 3$ may look like vertical angles, but do not be deceived. Vertical angle pairs are formed when lines intersect. The vertical angle to $\angle 1$ is the full angle that is opposite and between lines m and l.

793. $\angle 1$ and $\angle 2$ are adjacent supplementary angles. They share a side, a vertex, and no interior points; they are adjacent. The sum of their measures must equal 180° because they form a straight line; thus, they are supplementary.

794. $\angle 5$, $\angle 4$, and $\angle 3$ total 90°. $\angle 6$, $\angle 5$, $\angle 4$, and $\angle 3$ are on a straight line. All together, they measure 180°. If $\angle 6$ is a right angle, it equals 90°. The remaining three angles must equal 180° minus 90°, or 90°.

795. A determination cannot be made. $\angle 6$ and $\angle 2$ may look like vertical angles, but vertical pairs are formed when lines intersect. The vertical angle to $\angle 2$ is the full angle that is opposite and between lines m and l.

796. $\angle 1$, $\angle 6$, and $\angle 5$ total 180°.

797. $\angle B$ measures 59°. $180 - (\angle C + \angle A) = \angle B$; $180 - 121 = \angle B$; $59 = \angle B$.

798. $\angle S$ measures 30°. $180 - (\angle R + \angle Q) = \angle S$; $180 - 150 = \angle S$; $30 = \angle S$.

799. Base $\angle U$ measures 30°. Step 1: $180 - 60 = \angle T$; $120 = \angle T$. Step 2: $180 - \angle T = \angle U + \angle V$; $180 - 120 = \angle U + \angle V$; $60 = \angle U + \angle V$. Step 3: 60° shared by two congruent base angles equals two 30° angles.

800. Vertex $\angle E$ measures 156°. $180 - (\angle F + \angle G) = \angle E$; $180 - 24 = \angle E$; $156 = \angle E$.

801. No, because the sum of the measures of $\angle B$ and $\angle C$ equals only 75°. Subtract 75° from 180°, and $\angle A$ measures 105°. $\triangle ABC$ cannot be acute if one of its interior angles measures 90° or more.

802. c. Choice **a** is not an acute triangle because it has one right angle. In choice **b**, the sum of interior angle measures exceeds 180°. Choice **d** suffers the reverse problem; its sum does not make 180°. Though choice **c** describes an equilateral triangle, it also describes an acute isosceles triangle.

803. d. Choice **a** is not an obtuse triangle; it is a right triangle. In choice **b** and choice **c** the sum of the interior angle measures exceeds 180°.

804. d. A right triangle has a right angle and two acute angles; it does not have any obtuse angles.

805. d. Angles and sides are measured in different units; 60 inches (or other linear units) is not the same as 60°.

806. c. Let $\angle A = 2x$, $\angle B = x$, and $\angle C = x$. $2x + x + x = 180°$; $4x = 180°$; $x = 45°$.

807. $\angle 2$ is greater. If $\angle 2$ is the obtuse angle in an obtuse triangle, $\angle 1$ and $\angle 3$ must be acute.

808. $\angle d$ is greater. If $\angle 3$ is acute, its supplement is obtuse.

809. $\angle b$ is greater. $\angle b$ is vertical to obtuse angle 2, which means $\angle b$ is also obtuse. The supplement to an obtuse angle is always acute.

810. $\angle c$ is greater. The measure of an exterior angle equals the measure of the sum of nonadjacent interior angles, which means the measure of $\angle c$ equals the measure of $\angle 1$ plus the measure of $\angle 3$. Therefore, the measure of $\angle c$ is greater than the measure of $\angle 1$ all by itself.

811. Neither angle is greater; $\angle a$ equals $\angle c$. $\angle a$ and $\angle c$ are a vertical pair. They are congruent and equal.

812. $\angle b$ is greater. $\angle b$ is the vertical angle to obtuse $\angle 2$, which means $\angle b$ is also obtuse. Just as the measure of $\angle 2$ exceeds the measure of $\angle 3$, so too does the measure of $\angle b$.

813. $\angle d$ is greater. The measure of an exterior angle equals the measure of the sum of nonadjacent interior angles, which means the measure of $\angle d$ equals the measure of $\angle 1$ plus the measure of $\angle 2$. Therefore, the measure of $\angle d$ is greater than the measure of $\angle 2$ all by itself.

814. b. A ratio is a comparison. If one side of a triangle measures 16 inches, and a corresponding side in another triangle measures 24 inches, then the ratio is 16:24. This ratio can be simplified by dividing each side of the ratio by the common factor 8. The comparison now reads 2:3 or 2 to 3. Choice **a** simplifies into an incorrect ratio of 1:2, and choices **c** and **d** are or simplify into the incorrect ratio of 2:1.

815. **d.** When writing a proportion, corresponding parts must parallel each other. The proportions in choices **b** and **c** are misaligned. Choice **a** looks for the line segment $20 - x$, not x.

816. **d.** First, state the ratio between similar triangles; that ratio is 10:2 or 5:1. The ratio means that a line segment in the larger triangle is always 5 times more than the corresponding line segment in the similar smaller triangle. If the first line segment measures 30 inches, it is 5 times more than the corresponding line segment. Create the equation: $30 = 5x$, so $x = 6$.

817. The corresponding line segments are $\overline{OQ}$ and $\overline{OM}$; $\overline{QR}$ and $\overline{MN}$; $\overline{RO}$ and $\overline{NO}$. Always coordinate corresponding endpoints.

818. The corresponding line segments are $\overline{AE}$ and $\overline{BD}$; $\overline{EC}$ and $\overline{DC}$; $\overline{CA}$ and $\overline{CB}$. Always coordinate corresponding endpoints.

819. $\overline{AE} = 42$. This is a little tricky. When you state the ratio between triangles, remember that corresponding sides $\overline{AC}$ and $\overline{BC}$ share part of a line segment. $\overline{AC}$ actually measures $5x + x$, or $6x$. The ratio is $6x{:}1x$, or 6:1. If the side of the smaller triangle measures 7, then the corresponding side of the larger triangle will measure 6 times 7, or 42.

820. Triangle C is congruent to ΔA. Because the two angles given in ΔA are 30° and 60°, the third angle in ΔA is 90°. Like ΔA, ΔC and ΔI also have angles that measure 30°, 60°, and 90°. According to the Angle-Angle postulate, at least two congruent angles prove similarity. To be congruent, an included side must also be congruent. ΔA and ΔC have congruent hypotenuses, so they are congruent.

821. Triangle I is similar to ΔA. In the previous answer, ΔC was determined to be congruent to ΔA because of congruent angles and a congruent side. ΔI's hypotenuse measures 5 units; it has the same shape as ΔA but is smaller. Consequently, they are not congruent triangles; they are only similar triangles.

822. Triangle K is congruent to ΔB. ΔB is an equilateral triangle. ΔH and ΔK are also equilateral triangles (an isosceles triangle whose vertex measures 60° must also have base angles that measure 60°). However, only ΔK and ΔB are congruent because of congruent sides.

823. Triangle H has the same equilateral shape as ΔB, but they are different sizes. They are not congruent; they are only similar.

824. Triangle J is congruent to ΔE. The three angles in ΔE measure 36°, 54°, and 90°. ΔF and ΔJ also have angles that measure 36°, 54°, and 90°. According to the Angle-Angle postulate, at least two congruent angles prove similarity. To be congruent, an included side must also be congruent. The line segments between the 36° and 90° angles in ΔJ and ΔE are congruent.

825. Triangle F has the same right scalene shape as ΔE, but they are not congruent; they are only similar.

826. Triangle L is congruent to ΔD. The three angles in ΔD respectively measure 62°, 10°, and 108°. ΔL has a set of corresponding and congruent angles, which proves similarity; but it also has an included congruent side, which proves congruency.

827. Triangle G has only one given angle; the Side-Angle-Side postulate proves it is similar to ΔD. The sides on either side of the 108° angle are proportional, and the included angle is obviously congruent.

828. Triangles A, C, E, F, I, and J are right triangles. Any triangle with a 90° interior angle is a right triangle.

829. Triangles B, H, and K are equilateral triangles. Any triangle with congruent sides and congruent angles is an equilateral, equiangular triangle.

830. **b.** This is a popular triangle, so know it well. A 3-4-5 triangle is a right triangle. Apply the Pythagorean theorem: $a^2 + b^2 = c^2$: $3^2 + 4^2 = 5^2$; $9 + 16 = 25$.

831. **b.** This is also a 3-4-5 triangle. Simplify the measurement of each side by dividing 12, 16, and 20 by 4: $\frac{12}{4} = 3$; $\frac{16}{4} = 4$; $\frac{20}{4} = 5$.

832. **d.** The corner forms the right angle of this triangle; Eva and Carr walk the distance of each leg, and we want to know the hypotenuse. Plug the known measurements into the Pythagorean theorem: $5^2 + 6^2 = c^2$; $25 + 36 = c^2$; $61 = c^2$; $\sqrt{61} = c$.

833. **a.** The connection between the table leg and the tabletop forms the right angle of this triangle. The length of the leg and the length of the top are the legs of the triangle, and we want to know the distance of the hypotenuse. Plug the known measurements into the Pythagorean theorem: $3^2 + 4^2 = c^2$; $9 + 16 = c^2$; $25 = c^2$; $5 = c$. If you chose answer **d**, you forgot to take the square root of the 25. If you chose answer **b**, you added the table leg and tabletop together without squaring them first.

834. **b.** The first plane is actually this triangle's right vertex. The distance between Dorothy and the second plane is the hypotenuse. Plug the known measurements into the Pythagorean theorem: $300^2 + b^2 = 500^2$; $90,000 + b^2 = 250,000$; $b^2 = 160,000$; $b = 400$. Notice that if you divide each side by 100, this is another 3-4-5 triangle.

835. **d.** The bases of Timmy's walls form the legs of this right triangle. The hypotenuse is unknown. Plug the known measurements into the Pythagorean theorem: $10^2 + 15^2 = c^2$; $100 + 225 = c^2$; $325 = c^2$. $\sqrt{325} = c$.

836. $\triangle AFE$ and $\triangle BGE$ are congruent, and $\triangle ABF$ and $\triangle BCG$ are congruent (Side-Side-Side postulate).

837. Plug the measurements of $\triangle ECD$ into the Pythagorean theorem: $3^2 + w^2 = (15\sqrt{2})^2$; $9 + w^2 = 450$; $w^2 = 441$; $w = 21$.

838. Corresponding parts of congruent triangles are congruent (CPCTC). If $\overline{EC}$ is 21, then $\overline{EA}$ is also 21. Plug the measurements of $\triangle AFE$ into the Pythagorean theorem: $21^2 + x^2 = (7\sqrt{10})^2$; $441 + x^2 = 490$; $x^2 = 49$; $x = 7$.

839. Because of CPCTC, $\overline{AE}$ is also congruent to $\overline{BE}$. If $\overline{BE}$ is 21 and $\overline{FE}$ is 7, subtract 7 from 21 to find $\overline{BF}$: $21 - 7 = 14$; $y = 14$.

840. $z = 21\sqrt{2}$. Plug the measurements of $\triangle BEC$ into the Pythagorean theorem: $21^2 + 21^2 = z^2$; $441 + 441 = z^2$; $882 = z^2$; $21\sqrt{2} = z$.

841. $\triangle BGC$ is obtuse. You could just guess that $\angle BGC > 90°$. However, you can also use the Pythagorean theorem to show $(7\sqrt{10})^2 + 14^2 < (21\sqrt{2})^2$, so $\triangle BGC$ is obtuse.

Polygons and Quadrilaterals

N OW THAT WE'VE LEARNED about different triangles and their properties, we'll look at some other polygons, including quadrilaterals. Quadrilaterals are four-sided figures, and just as there are certain types of triangles with specific properties, there are certain types of quadrilaterals, each with its own properties.

Remember, a polygon is any closed figure made up of line segments. The following chart shows the names of some common polygons:

NAME OF POLYGON	NUMBER OF SIDES
triangle	3
quadrilateral	4
pentagon	5
hexagon	6
heptagon	7
octagon	8
nonagon	9
decagon	10
dodecagon	12

A **regular polygon** is a polygon whose sides are all congruent and whose angles are all congruent. This hexagon is a regular hexagon:

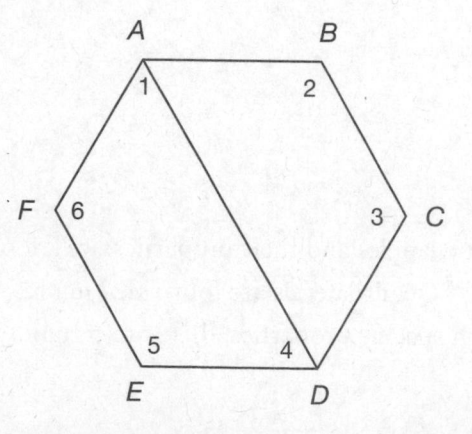

A polygon is named by its angles, in order, without skipping over a vertex. You can begin naming a polygon at any angle, but you must continue to name the figure in order of its vertices. A **vertex** is an angle of the polygon. This polygon could be named *ABCDEF* or *FEDCBA*. It could also be named *CDEFAB* or *EFABCD*. All of those names start at one vertex and continue around the figure in order. Sides *AB* and *BC* are consecutive sides, because they are con-

nected, and angles *A* and *B* are consecutive vertices, because one follows the other. Line *AD* is a **diagonal**, which is a line drawn within a polygon connecting two nonconsecutive vertices.

This hexagon, like all regular polygons, is a **convex polygon**, because all of its interior angles are no greater than 180 degrees. It is a **strictly convex polygon** if every interior angle is *less than* 180 degrees. What's the difference? A convex polygon can have an interior 180-degree angle, but a strictly convex polygon cannot. If any interior angle is greater than 180 degrees, the polygon is a **concave polygon**.

> ### TIP
>
> Note that *all* interior angles must be 180 degrees or less for a polygon to be a convex polygon, but only *one* interior angle has to be greater than 180 degrees for the polygon to be a concave polygon.

Angles 1 through 6 are on the inside of the hexagon, so they are called **interior angles**.

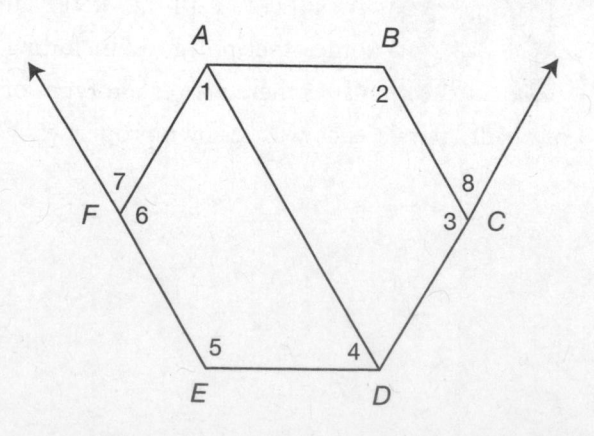

If we were to extend two sides of the hexagon, we would form angles 7 and 8. These angles are on the outside of the hexagon, so they are called **exterior**

angles. The sum of the interior angles of a polygon is equal to $180(n-2)$, where n is the number of sides of the polygon. A hexagon has six sides, so there are $180(6-2) = 180(4) = 720$ degrees inside a hexagon.

Example

A polygon is named *ABCDE*. What type of polygon is this?

A five-sided polygon is called a pentagon.

Example

What is the sum of the interior angles of *ABCDE*?

Use the formula: sum of interior angles = $180(n-2) = 180(5-2) = 180(3) = 540$ degrees.

Example

How many diagonals can be drawn in regular polygon *ABCDE*?

This polygon has five vertices, and each vertex has two vertices opposite it to which diagonals can be drawn. However, because every diagonal starts at one vertex and ends at another, we must be careful not to count diagonal *AC* and diagonal *CA* (for example) as two different diagonals. Therefore, we count only one diagonal for each vertex. Five diagonals can be drawn in polygon *ABCDE*.

▶ **Practice Questions**

State whether the object is or is not a polygon and why. (Envision each of these objects as simply as possible; otherwise, there will always be exceptions.)

842. a rectangular city block

843. Manhattan's grid of city blocks

844. branches of a tree

845. the block letter *M* carved into the tree

846. outline of a television screen

847. a human face on the TV screen

848. an ergonomic chair

849. lace

Use the diagram shown here to answer questions 850 through 852.

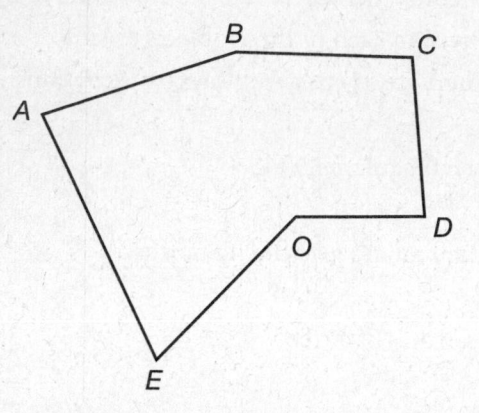

850. Name the polygon. Is it convex or concave?

851. How many diagonals can be drawn from vertex *O*?

852. How many sides does the polygon have? Based on its number of sides, what type of polygon is it?

Use the diagram shown here to answer questions 853 through 855.

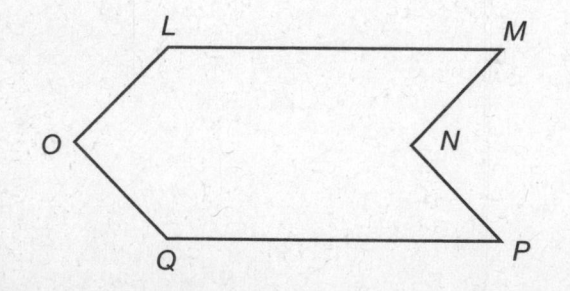

853. Name the polygon. Is it convex or concave?

854. How many diagonals can be drawn from vertex *O*?

855. How many sides does the polygon have? Based on its number of sides, what type of polygon is it?

Use the following diagram to answer questions 856 through 858.

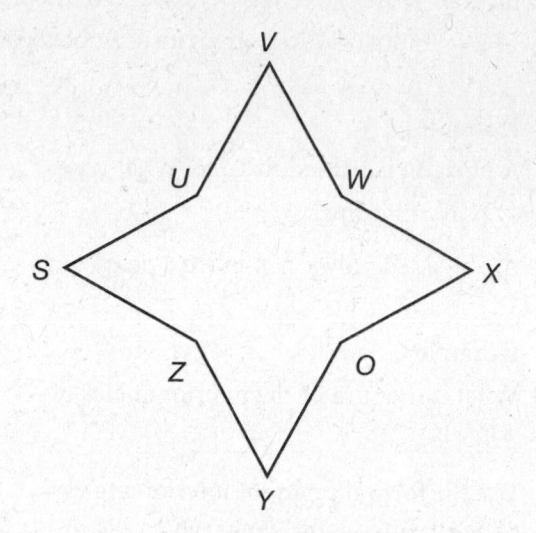

856. Name the polygon. Is it convex or concave?

857. How many diagonals can be drawn from vertex *O*?

858. How many sides does the polygon have? Based on its number of sides, what type of polygon is it?

Use the diagram shown here to answer questions 859 through 861.

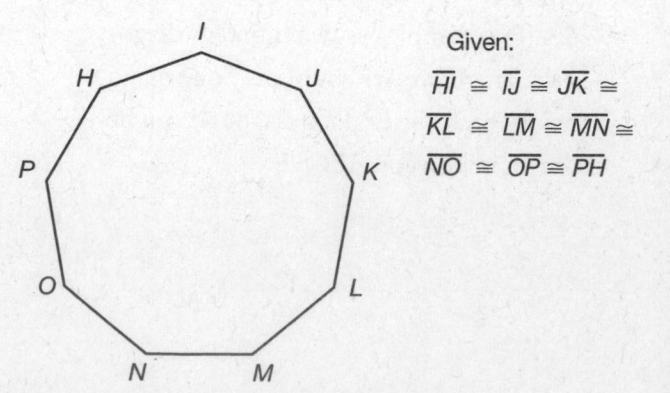

Given:

$\overline{HI} \cong \overline{IJ} \cong \overline{JK} \cong$
$\overline{KL} \cong \overline{LM} \cong \overline{MN} \cong$
$\overline{NO} \cong \overline{OP} \cong \overline{PH}$

859. Name the polygon. Is it convex or concave?

860. How many diagonals can be drawn from vertex *O*?

861. How many sides does the polygon have? Based on its number of sides, what type of polygon is it?

Draw a diagram of polygon *CDEFG*. Use your knowledge of polygons to fill in the blanks.

862. In polygon *CDEFG*, $\overline{CD}$ and $\overline{DE}$ are _____.

863. In polygon *CDEFG*, $\overline{CE}$, $\overline{DF}$, and $\overline{EG}$ are _____.

864. In polygon *CDEFG*, $\angle EFG$ is also _____.

865. In polygon *CDEFG*, $\angle DEF$ and $\angle EFG$ are _____.

Use diagonals to draw triangles within the polygons in questions 866 through 868.

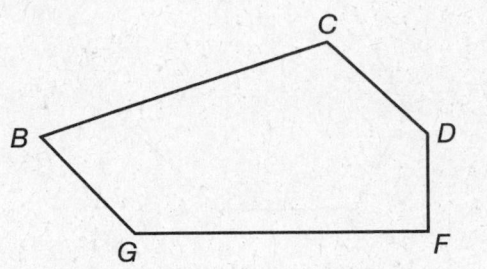

866. Determine the sum of the polygon's interior angles using the number of triangles; verify your answer by using the formula $s = 180(n - 2)$, where *s* is the sum of the interior angles and *n* is the number of sides the polygon has.

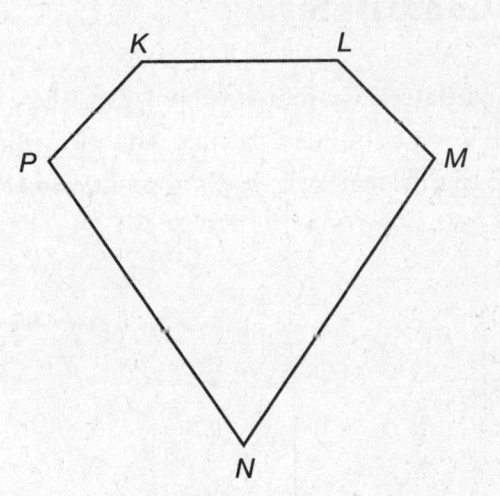

867. Determine the sum of the polygon's interior angles using the number of triangles; then apply the formula $s = 180(n - 2)$ to verify your answer.

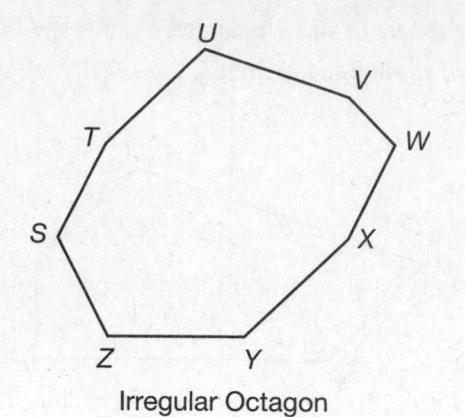

Irregular Octagon

868. Determine the sum of the polygon's interior angles using the number of triangles; then apply the formula $s = 180(n - 2)$ to verify your answer.

▶ Quadrilaterals

A **quadrilateral** is a four-sided polygon. There are six different types of quadrilaterals with properties that are so specific that each type is given its own name.

▶ Parallelogram

A **parallelogram** is a quadrilateral whose opposite sides are parallel and congruent. Opposite angles are also congruent, consecutive angles are supplementary, and the diagonals bisect each other.

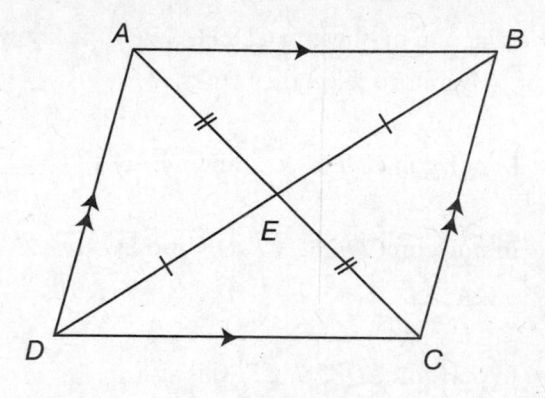

In the parallelogram shown here, sides AB and CD are congruent and parallel. Sides AD and BC are also congruent and parallel. Angles A and C are congruent, as are angles B and D. Angles A and B are supplementary, as are angles C and D. Diagonals AC and BD bisect each other, which means that they cut each other into two equal pieces. AE and EC are congruent, as are BE and ED.

▶ Rectangle

A **rectangle** is a type of parallelogram whose angles are *all* congruent and whose diagonals are congruent. Wait a minute—how could a rectangle be a type of parallelogram? The reason is simple: A rectangle has all of the properties of a parallelogram, plus two additional properties. It has opposite sides that are parallel and congruent, opposite angles that are congruent, consecutive angles that are supplementary, and diagonals that bisect each other. However, all of its angles are 90-degree angles. That's why not only are opposite angles congruent, but *all* angles are congruent, and *any* two angles are supplementary. We can say that all rectangles are types of parallelograms, because all rectangles have the properties of parallelograms.

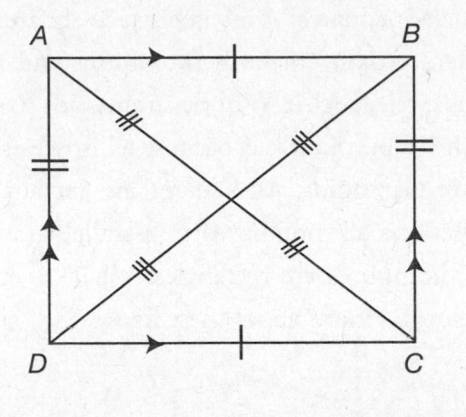

Angles *A*, *B*, *C*, and *D* are all 90 degrees. Sides *AB* and *CD* are parallel and congruent, as are sides *AD* and *BC*. Diagonals *AC* and *BD* are congruent, as are their respective halves.

Rhombus

A **rhombus** is a type of parallelogram with *four* congruent sides. Rhombuses have perpendicular diagonals that bisect their angles. Perpendicular diagonals are diagonals that form right angles at their intersection point. If a rhombus is a type of parallelogram, is a rhombus a rectangle? Not necessarily. A rectangle must have four 90-degree angles; a rhombus, like the one shown here, does not. Is a rectangle a rhombus? Not necessarily. A rhombus must have four congruent sides; a rectangle, like the one shown earlier, does not.

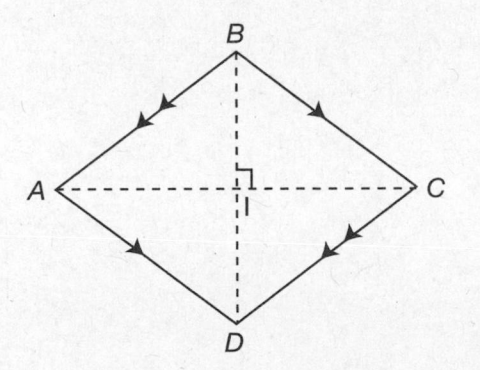

Sides *AB*, *BC*, *CD*, and *DA* are all congruent, and diagonals *AC* and *BD* form right angles. Those diagonals bisect angles *A*, *B*, *C*, and *D*.

Could a rhombus be a rectangle? Could a rectangle be a rhombus? Yes—the key word is *could*. If a quadrilateral has all the properties of a rhombus *and* a rectangle, that quadrilateral could be referred to as a rhombus or a rectangle. But we have a special name for that kind of quadrilateral—a square.

Square

A **square** has all of the properties of a rhombus and a rectangle: four congruent sides; four congruent, 90-degree angles; two pairs of parallel sides; consecutive angles that are supplementary; and diagonals that are congruent, are perpendicular, bisect each other, and bisect their angles. All squares are parallelograms, all squares are rectangles, and all squares are rhombuses.

Trapezoid

A **trapezoid** is a four-sided figure, but it is *not* a parallelogram. A trapezoid has only one pair of parallel sides, whereas a parallelogram has two pairs. Sides *AB* and *CD* of the trapezoids shown here are parallel. If a trapezoid also has congruent legs, base angles, and diagonals, it is called an **isosceles trapezoid**, shown on the right. Legs *DA* and *BC* of the isosceles trapezoid are congruent, as are diagonals *AC* and *BD*, angles *A* and *B*, and angles *C* and *D*.

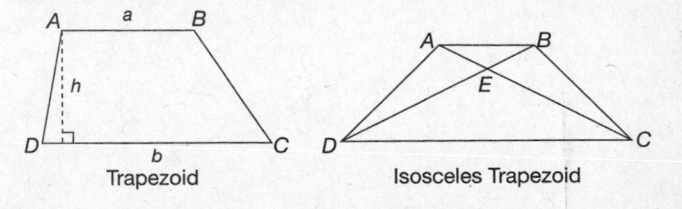

Trapezoid   Isosceles Trapezoid

Example

Tommy's quadrilateral has no congruent angles. What kind of quadrilateral could it be?

Parallelograms, rectangles, rhombuses, squares, and isosceles trapezoids all have two pairs of congruent angles, so we know those are not the answer. A trapezoid could have four different angles, so Tommy's quadrilateral must be a trapezoid.

Example

Becky's quadrilateral has an angle that is 120 degrees. Of the six types of quadrilaterals we've learned about, which could be Becky's quadrilateral? Which could not be?

If Becky's quadrilateral has an angle that is 120 degrees, then it cannot have four 90-degree angles, so it cannot be a square or a rectangle. It could be a parallelogram, rhombus, trapezoid, or isosceles trapezoid, since all of these quadrilaterals could contain an angle of 120 degrees.

The Venn diagram can be used to describe quadrilaterals. Notice that the oval labeled "rhombuses" and the oval labeled "rectangles" overlap in the area labeled "squares." This is because squares combine characteristics of both rhombuses and rectangles. The circle labeled "isosceles trapezoids" is wholly inside the trapezoid circle, because all isosceles trapezoids are trapezoids. All squares are rectangles, all rectangles and all rhombuses are parallelograms, but not all rhombuses are rectangles. What other statements can you make about these figures?

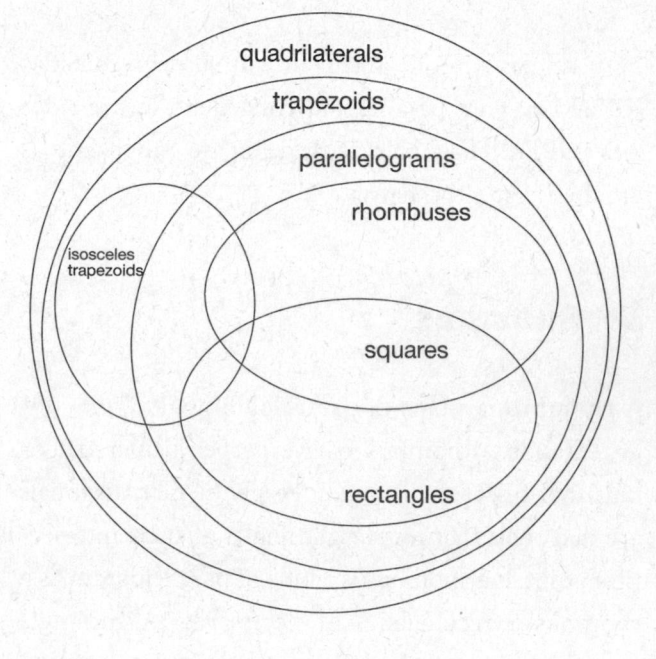

▶ **Practice Questions**

Choose the best answer.

869. The sides of Mary's chalkboard consecutively measure 9 feet, 5 feet, 9 feet, and 5 feet. Without any other information, you can determine that Mary's chalkboard is a
   a. rectangle.
   b. rhombus.
   c. parallelogram.
   d. square.

870. Four line segments connected end to end will always form
   a. an open figure.
   b. four interior angles that measure 360°.
   c. a square.
   d. It cannot be determined.

871. A square whose vertices are the midpoints of another square is
   a. congruent to the other square.
   b. half the size of the other square.
   c. twice the size of the other square.
   d. It cannot be determined.

872. The sides of a square measure 2.5 feet each. If three squares fit perfectly side by side in one rectangle, what are the minimum dimensions of the rectangle?
   a. 5 feet, 2.5 feet
   b. 7.5 feet, 7.5 feet
   c. 7.5 feet, 3 feet
   d. 7.5 feet, 2.5 feet

873. A rhombus, a rectangle, and an isosceles trapezoid all have
   a. congruent diagonals.
   b. opposite congruent sides.
   c. interior angles that measure 360°.
   d. opposite congruent angles.

874. A figure with four sides and four congruent angles could be a
   a. rhombus or square.
   b. rectangle or square.
   c. trapezoid or rhombus.
   d. rectangle or trapezoid.

875. A figure with four sides and perpendicular diagonals could be a
   a. rhombus or square.
   b. rectangle or square.
   c. trapezoid or rhombus.
   d. rectangle or trapezoid.

876. A figure with four sides and diagonals that bisect each angle could be a
   a. rectangle.
   b. rhombus.
   c. parallelogram.
   d. trapezoid.

877. A figure with four sides and diagonals that bisect each other could NOT be a
   a. rectangle.
   b. rhombus.
   c. parallelogram.
   d. trapezoid.

Fill in the blanks based on your knowledge of quadrilaterals. More than one answer may be correct.

878. If quadrilateral *ABCD* has two sets of parallel lines, it could be _____.

879. If quadrilateral *ABCD* has four congruent sides, it could be _____.

880. If quadrilateral *ABCD* has exactly one set of opposite congruent sides, it could be _____.

881. If quadrilateral *ABCD* has opposite congruent angles, it could be _____.

882. If quadrilateral *ABCD* has consecutive angles that are supplementary, it could be _____.

883. If quadrilateral *ABCD* has congruent diagonals, it could be _____.

884. If quadrilateral *ABCD* can be divided into two congruent triangles, it could be _____.

885. If quadrilateral *ABCD* has diagonals that bisect each vertex angle in two congruent angles, it is _____.

Choose the best answer.

886. If an angle in a rhombus measures 21°, then the other three angles consecutively measure
   a. 159°, 21°, 159°
   b. 21°, 159°, 159°
   c. 69°, 21°, 69°
   d. 21°, 69°, 69°
   e. It cannot be determined.

887. In an isosceles trapezoid, the angle opposite an angle that measures 62° measures
   a. 62°.
   b. 28°.
   c. 118°.
   d. 180°.
   e. It cannot be determined.

888. In rectangle *WXYZ*, ∠*WXZ* and ∠*XZY*
   a. are congruent.
   b. are alternate interior angles.
   c. form complementary angles with ∠*WZX* and ∠*YXZ*.
   d. all of the above
   e. It cannot be determined.

889. In square *ABCD*, ∠*ABD*
   a. measures 45°.
   b. is congruent with ∠*ADC*.
   c. forms a supplementary pair with ∠*ADB*.
   d. all of the above
   e. It cannot be determined.

890. In parallelogram *KLMN*, if diagonal *KM* measures 30 inches, then
   a. $\overline{KL}$ measures 18 inches.
   b. $\overline{LM}$ measures 24 inches.
   c. diagonal *LN* is perpendicular to diagonal *KM*.
   d. all of the above
   e. It cannot be determined.

Use the following figure to answer questions 891 through 893.

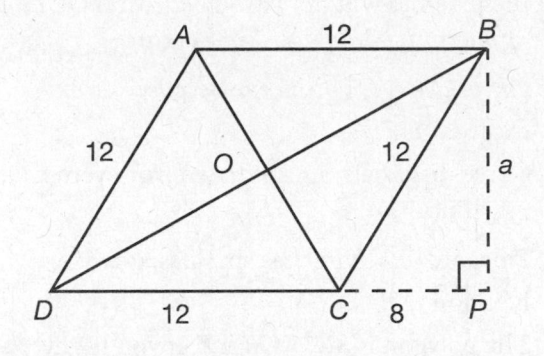

$\angle BCA = 72$
$\angle BDA = 18$

891. Using your knowledge of triangles and quadrilaterals, show that diagonals *AC* and *BD* intersect perpendicularly.

892. Using your knowledge of triangles and quadrilaterals, what is the length of imaginary side *BP*?

893. Using your knowledge of triangles and quadrilaterals, what is the length of diagonal *DB*?

TERMS TO REVIEW

regular polygon

vertex

diagonal

convex polygon

strictly convex polygon

concave polygon

interior angles

exterior angles

quadrilateral

parallelogram

rectangle

rhombus

square

trapezoid

isosceles trapezoid

FORMULA TO REVIEW

sum of interior angles: $s = 180(n - 2)$ (n = number of sides of polygon)

▶ **Answers**

842. A single city block is a polygon, because it is a closed four-sided figure; each of its corners is a vertex.

843. A grid is not a polygon, because its lines intersect at points that are not endpoints.

844. Tree branches do not form a polygon. Branches are open, and they fork at points that are not endpoints.

845. The block letter *M* is a polygon. Block letters are closed multisided figures in which the edges of the line segments begin and end at endpoints.

846. A television screen is a rectangular polygon; it has four sides and four vertices.

847. A face is not a polygon. The human face is very complex, but primarily it has few if any straight line segments.

848. An ergonomic chair is a chair designed to contour to your body. It is not a polygon. It is usually curved to support the natural curves of the hip and spine.

849. Lace is not a polygon. Like the human face, lace is very intricate. Unlike the human face, lace has lots of line segments that meet at lots of different points.

850. The polygon is *ABCDOE*. As long as you list the vertices in consecutive order, any one of these names will do: *BCDOEA*, *CDOEAB*, *DOEABC*, *OEABCD*, *EABCDO*. Also, polygon *ABCDOE* is concave because the measure of vertex *O* exceeds 180°.

851. Three diagonals can be drawn from vertex *O*: $\overline{OA}$, $\overline{OB}$, $\overline{OC}$. $\overline{OD}$ and $\overline{OE}$ are not diagonals; they are sides.

852. Polygon *ABCDOE* has six sides; it is a hexagon.

853. The polygon is *OLMNPQ*. As long as you list the vertices in consecutive order, any one of these names will do: *LMNPQO*, *MNPQOL*, *NPQOLM*, *PQOLMN*, *QOLMNP*. Also, polygon *OLMNPQ* is concave because vertex *N* exceeds 180°.

854. Three diagonals can be drawn from vertex *O*: $\overline{OM}$, $\overline{ON}$, $\overline{OP}$.

855. Polygon *OLMNPQ* has six sides; it is a hexagon.

856. The polygon is *SUVWXOYZ*. If you list every vertex in consecutive order, then your name for the polygon given is correct. Also, polygon *SUVWXOYZ* is concave. The measures of vertices *U*, *W*, *O*, and *Z* each exceed 180°.

857. Five diagonals can be drawn from vertex *O*: *OZ*, *OS*, *OU*, *OV*, *OW*.

858. Polygon *SUVWXOYZ* has eight sides; it is an octagon.

859. The polygon is *HIJKLMNOP*. List every vertex in consecutive order and your answer is correct. Also, polygon *HIJKLMNOP* is regular and convex.

860. Six diagonals can be drawn from vertex *O*: $\overline{OH}$, $\overline{OI}$, $\overline{OJ}$, $\overline{OK}$, $\overline{OL}$, and $\overline{OM}$.

861. Polygon *HIJKLMNOP* has nine sides; it is a nonagon.

862. $\overline{CD}$ and $\overline{DE}$ are consecutive sides. Draw polygon CDEFG to see that yes, $\overline{CD}$ and $\overline{DE}$ are consecutive sides.

863. $\overline{CE}$, $\overline{DF}$, and $\overline{EG}$ are diagonals. When a line segment connects nonconsecutive endpoints in a polygon, it is a diagonal.

864. ∠*EFG* is also ∠*GFE* or ∠*F*.

865. ∠*DEF* and ∠*EFG* are consecutive vertices. Look back at the drawing you made of polygon *CDEFG*. You can see that ∠*E* and ∠*F* are consecutive vertices.

For the solution to 866, refer to the image shown.

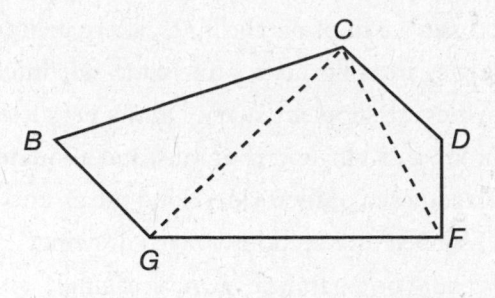

866. At any one time, three triangles can be drawn in polygon *BCDFG*. Remember when drawing your triangles that a diagonal must go from endpoint to endpoint. If the interior angles of a triangle measure 180° together, then three sets of interior angles measure 180 × 3, or 540. Apply the formula $s = 180(n - 2)$: $s = 180(5 - 2)$; $s = 180(3)$; $s = 540$. The interior angles of a convex pentagon will always measure 540° together.

For the solution to 867, refer to the image shown.

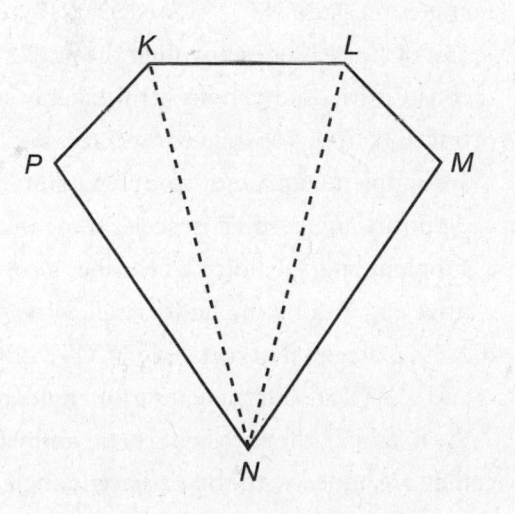

867. At any one time, three triangles can be drawn in polygon *KLMNP*. If the interior angles of a triangle measure 180° together, then three sets of interior angles measure 180 × 3 = 540. Apply the formula $s = 180(n - 2)$. Again, $s = 540$. You have again confirmed that the interior angles of a convex pentagon will always measure 540° together.

For the solution to 868, refer to the image shown.

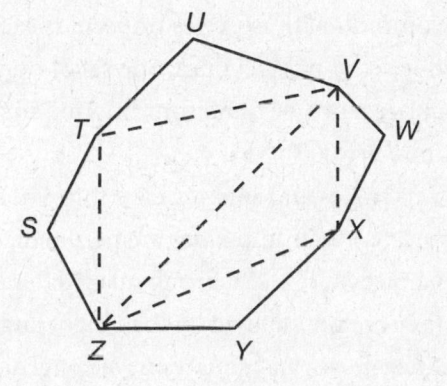

868. At any one time, six triangles can be drawn in polygon *STUVWXYZ*. If the interior angles of a triangle measure 180° together, then six sets of interior angles measure 180 × 6 = 1,080. Apply the formula $s = 180(n - 2)$: $s = 180(8 - 2)$; $s = 180(6)$; $s = 1,080$.

869. c. All parallelograms have opposite congruent sides, including rectangles, rhombuses, and squares. However, without more information, you cannot be any more specific than a parallelogram.

870. b. The interior angles of a quadrilateral total 360°. Choices **a** and **c** are incorrect because the question states that the line segments connect end to end; this is a closed figure, but it is not necessarily a square.

871. b. Find the point along each of the larger square's line segments that would divide that line segment into two equal pieces. That is the line segment's midpoint. Connect the midpoints together and you have another square that is half the existing square.

872. d. Three squares in a row will have three times the length of one square, or 2.5 inches × 3 = 7.5 inches. However, the width will remain the length of just one square, or 2.5 inches.

873. c. Rectangles and rhombuses have very little in common with isosceles trapezoids except one set of parallel lines, one set of opposite congruent sides, and four interior angles that measure 360°.

874. b. Rectangles and squares have four 90° angles because their four sides are perpendicular. Choices **a**, **c**, and **d** are all quadrilaterals, but they are not defined by their right angles.

875. a. Rhombuses and squares have congruent sides and diagonals that are perpendicular. Rectangles and trapezoids do not have diagonals that cross perpendicularly, because their sides are not congruent.

876. b. A rhombus's diagonals bisect its vertices.

877. d. Diagonals of a trapezoid are not congruent unless the trapezoid is an isosceles trapezoid. Diagonals of any trapezoid do not bisect each other.

878. It could be a parallelogram, a rectangle, a rhombus, or a square. Two pairs of parallel lines define each of these four-sided figures.

879. It could be a rhombus or a square.

880. It could be an isosceles trapezoid.

881. It could be a parallelogram, a rectangle, a rhombus, or a square. When a transversal crosses a pair of parallel lines, alternate interior angles are congruent, while same-side interior angles are supplementary. Draw a parallelogram, a rectangle, a rhombus, and a square; extend each of their sides. Find the *Z*- and *C*-shaped intersections in each drawing.

882. It could be a parallelogram, a rectangle, a rhombus, or a square. Again, look at the drawings you made in the preceding answer to see why consecutive angles are supplementary.

883. It could be a rectangle, a square, or an isosceles trapezoid.

884. It could be a parallelogram, a rectangle, a rhombus, or a square.

885. It is a rhombus or a square.

886. a. The first consecutive angle must be supplementary to the given angle. The angle opposite the given angle must be congruent. Consequently, in consecutive order, the angles measure 180 − 21, or 159°, 21°, and 159°. Choice **b** does not align the angles in consecutive order; choice **c** mistakenly subtracts 21 from 90 when consecutive angles are supplementary, not complementary.

887. c. Opposite angles in an isosceles trapezoid are supplementary. Choice **a** describes a consecutive angle along the same parallel line.

888. d. $\overline{XZ}$ is a diagonal in rectangle *WXYZ*. ∠*WXZ* and ∠*XZY* are alternate interior angles along the diagonal; they are congruent; and when they are added with their adjacent angle, the two angles form a 90° angle.

889. **a.** $\overline{BD}$ is a diagonal in square $ABCD$. It bisects vertices B and D, creating four congruent 45° angles. Choice **b** is incorrect because $\angle ABD$ is half of $\angle ADC$; they are not congruent. Also, choice **c** is incorrect because when two 45° angles are added together they measure 90°, not 180°.

890. **e.** It cannot be determined.

891. First, opposite sides of a rhombus are parallel, which means alternate interior angles are congruent. If $\angle BCA$ measures 72°, then $\angle CAD$ also measures 72°. The sum of the measures of all three interior angles of a triangle must equal 180°: $72 + 18 + \angle AOB = 180$, and $\angle AOD = 90$. Second, because AC and BD are intersecting straight lines, if one angle of the intersection measures 90°, all four angles of the intersection measure 90°, which means the lines perpendicularly meet.

892. Imaginary side BP or $a = 4\sqrt{5}$. $\overline{BP}$ is the height of rhombus $ABCD$ and the leg of $\triangle BPC$. Use the Pythagorean theorem: $a^2 + 8^2 = 12^2$; $a^2 + 64 = 144$; $a^2 = 80$; $a^2 = 16 \times 5$; therefore, $a = 4\sqrt{5}$.

893. Diagonal DB or $c = 4\sqrt{30}$. Use the Pythagorean theorem to find the hypotenuse (side c) of $\triangle BPD$, which is diagonal BD: $(4\sqrt{5})^2 + (12 + 8)^2 = c^2$; $80 + 400 = c^2$; $480 = c^2$; $16 \times 30 = c^2$; therefore, $= 4\sqrt{30} = c$.

14 ▶ Area and Perimeter

WHEN WORKING WITH TWO-DIMENSIONAL SHAPES, the most common quantities you'll be asked to find are area and perimeter. **Area** is the amount of surface, in square units, that an object covers. A rectangular carpet covers an area of a floor; a painting might cover an area of a wall. **Perimeter** is the distance around an object. If you were to place a fence around a yard, you would need to find the perimeter of the yard in order to know how much fencing to buy.

▶ Area of Parallelograms, Rectangles, Rhombuses, and Squares

Area is always measured in square units. Why? Because area is the product of two measurements, as area represents a two-dimensional quantity. This rectangle has a length of 8 units and a height of 4 units. It is cut into 32 unit squares, each of which measures 1 square unit. The area of the rectangle is 32 square units.

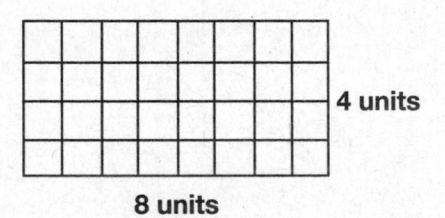

4 units

8 units

We don't need to count unit squares to find the areas of most shapes, though. The areas of parallelograms, rectangles, rhombuses, and squares can be found using either of two formulas: $A = bh$, where b is the base of the shape and h is its height, or $A = lw$, where l is length and w is width.

Area of a Parallelogram

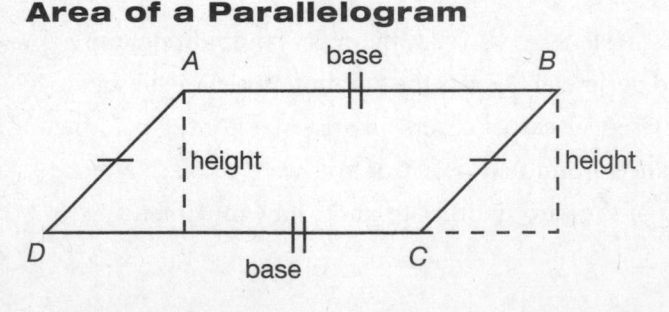

Area of parallelogram *ABCD* in square increments = base × height

Area of a Rectangle

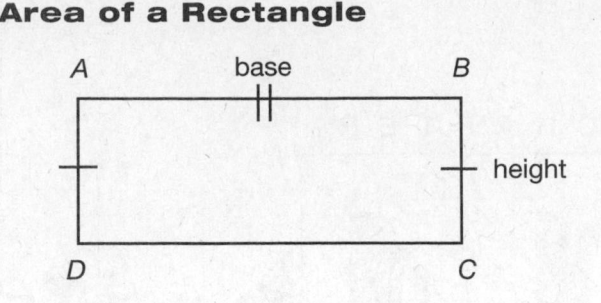

Area of rectangle *ABCD* in square increments = base × height

Area of a Rhombus

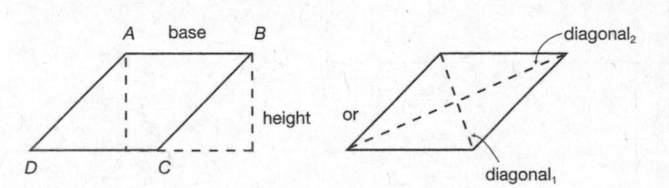

Note: A rhombus has an area like a rectangle, not a square.

Area of rhombus *ABCD* in square increments = base × height

Area = $\frac{1}{2}$(diagonal$_1$ × diagonal$_2$)

Area of a Square

Area of square *ABCD* in square increments = base × height

Area = side^2

Why are the main area formulas for these four shapes the same? Remember, a rectangle is a type of parallelogram, and a square is a type of rhombus, which, itself, is a type of parallelogram. All of these shapes are four-sided figures, with a base and a height.

The area of a square has a second formula. Because the base and height of a square are the same, the area can also be described as the length of a side squared (s^2).

PITFALL

Although a rhombus also has four congruent sides, the height of rhombus is usually *not* equal to the base (side) of the rhombus, so we cannot find the area of a rhombus by squaring the length of one side of the figure. Look at the rhombus shown here. The dashed line represents the height of the figure, and it forms a right triangle with the right side of the rhombus and part of the base. The longest side of a right triangle is the hypotenuse, which means that the dashed line, or the height, must be less than 10 units.

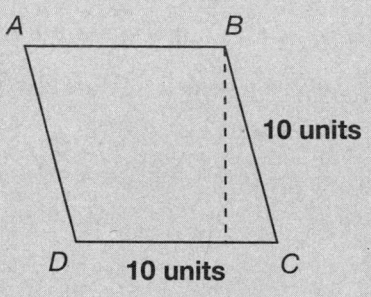

A square with a side of 10 units will have an area of $(10)^2 = 100$ square units. Unless a rhombus *is* a square, a rhombus with a side of 10 units will have an area that is less than 100 square units. In fact, if the rhombus looks like the one shown here, as angle C decreases in size, the area of the rhombus decreases in size, because although its base doesn't change, its height is decreasing. The greater the difference in size between the obtuse angles and the acute angles of a rhombus, the smaller the area of the rhombus.

Example

What is the area of a parallelogram with a base of 12 units and a height of 7 units?

Use the formula $A = bh$: $A = (12)(7) = 84$ square units.

Example

If the area of a square is 81 square inches, what is the length of one side of the square?

Use the formula for the area of a square to find the length of one side of the square: $A = s^2$; $s^2 = 81$; $s = 9$ inches.

TIP

Geometrical measures (length, area, volume, etc.) can never have negative values. If you solve a geometry problem by taking the square root of a number, the answer is always the positive square root of that number—never the negative square root.

Example

A square has a base of 3 feet and a rhombus has a base of 3 feet. If the measure of one angle of the rhombus is 95 degrees, which figure has the larger area?

The square has the larger area. If one angle of the rhombus measures 95 degrees, then the rhombus is not a square, and its height is less than its base (and therefore, its height is less than the height of the square). Therefore, the product of its base and height will be less than the product of the base and height of the square, and the area of the rhombus will be smaller.

▶ Area of Triangles

A rectangle or square can be cut into two triangles by drawing a diagonal through either shape. Therefore, it's no surprise that the area of a triangle is equal to half the area of a rectangle that has the same base and height. The area of a triangle is equal to $\frac{1}{2}bh$.

Area of a Triangle

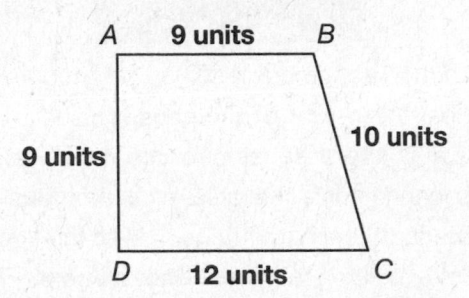

Triangle$_1$ $\cong$ triangle$_2$; therefore area $\Delta_1 \cong$ area Δ_2

Area $\Delta_1 = \frac{1}{2}$ area of polygon $ABCD$

Area $\Delta_1 = \frac{1}{2} b \times h$

Area of triangle ABC in square increments $= \frac{1}{2}$ base $\times$ height

Example

A triangle has a base of 16 centimeters and a height of 5 centimeters. What is the area of the triangle?

Use the formula $A = \frac{1}{2}bh$: $A = \frac{1}{2}(16)(5) =$ 40 square centimeters.

Example

If a triangle has an area of 32 square inches and a base of 8 inches, what is the height of the triangle?

Use the formula $A = \frac{1}{2}bh$ and solve for h: $32 = \frac{1}{2}(8)(h)$; $32 = 4h$; $h = 8$ inches.

▶ Area of a Trapezoid

Remember, a trapezoid is a four-sided figure with one set of parallel sides. We can find the area of a trapezoid by multiplying the average of its bases by its height. The parallel sides of a trapezoid are the bases of the trapezoid. Area $= \frac{1}{2}(b_1 + b_2)h$.

The bases of this trapezoid are 9 units and 12 units, respectively. The height of the figure is also 9 units. The length of BC, 10 units, is not used in finding the area of the shape.

Example

What is the area of the trapezoid shown?

Plug the given measurements into the formula for area of a trapezoid:
$A = \frac{1}{2}(b_1 + b_2)h$; $A = \frac{1}{2}(9 + 12)9 =$
$\frac{1}{2}(21)(9) = (10.5)(9) = 94.5$ square units.

> **TIP**
>
> If you are asked to find the area of an irregular shape, try to divide the shape into common figures: squares, rectangles, and triangles. You know the formulas for the areas of these shapes, and you can add the areas of each piece of the irregular shape together to find the total area of the shape.

Example

What is the area of the following figure?

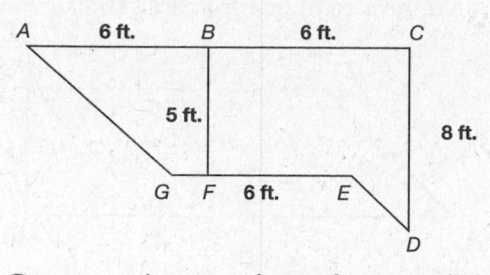

Connect points *B* and *D* to form parallelogram *ABEG* and right triangle *BCD*:

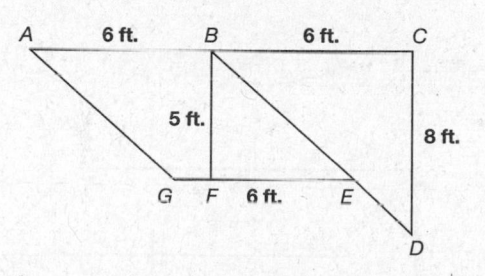

The area of *ABEG* is equal to $(6)(5) = 30$ square feet, and the area of $BCD = \frac{1}{2}(6)(8) = 24$ square feet, for a total area of $30 + 24 = 54$ square feet.

▶ **Practice Questions**

Choose the best answer.

894. Area is
   a. the negative space inside a polygon.
   b. a positive number representing the interior space of a polygon.
   c. all the space on a plane.
   d. no space at all.

895. Two congruent figures have
   a. equal areas.
   b. disproportional perimeters.
   c. no congruent parts.
   d. dissimilar shapes.

Circle whether the following statements are true or false.

896. A rhombus with opposite sides that measure 5 feet has the same area as a square with opposite sides that measure 5 feet.
   True or False

897. A rectangle with opposite sides that measure 5 feet and 10 feet has the same area as a parallelogram with opposite sides that measure 5 feet and 10 feet.   **True or False**

898. A rectangle with opposite sides that measure 5 feet and 10 feet has twice the area of a square with opposite sides that measure 5 feet.
   True or False

899. A parallelogram with opposite sides that measure 5 feet and 10 feet has twice the area of a rhombus whose height is equal to the height of the parallelogram and whose opposite sides measure 5 feet.   **True or False**

900. A triangle with a base of 10 and a height of 5 has a third the area of a trapezoid with base lengths of 10 and 20 and a height of 5.
   True or False

901. Find the area of △*DEF*.

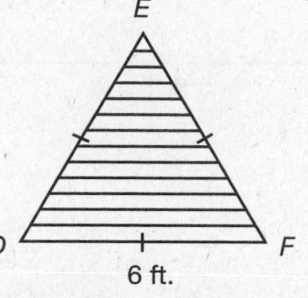

902. Find the area of quadrilateral *ABCD*.

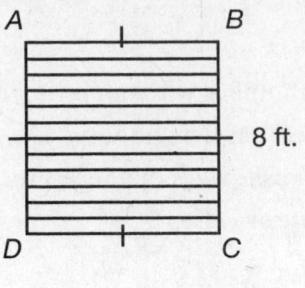

8 ft.

903. Find the shaded area of the following figure.

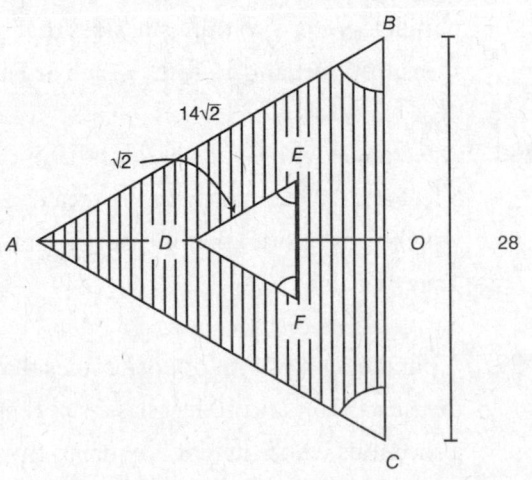

$14\sqrt{2}$

$\sqrt{2}$

28

904. Find the shaded area of the following figure.

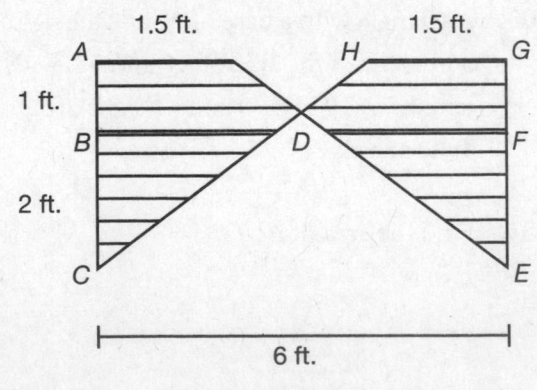

1.5 ft.   1.5 ft.

1 ft.

2 ft.

6 ft.

Find the area of each of the following figures.

905. Find the area of quadrilateral *ABCD*.

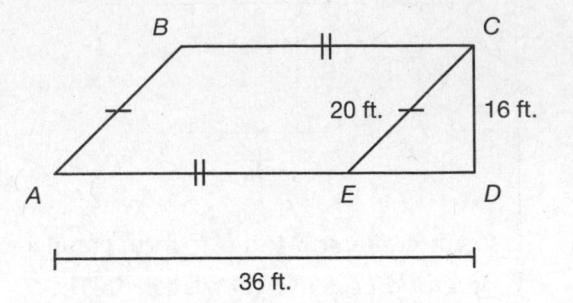

20 ft.   16 ft.

36 ft.

906. Find the area of polygon *RSTUV*.

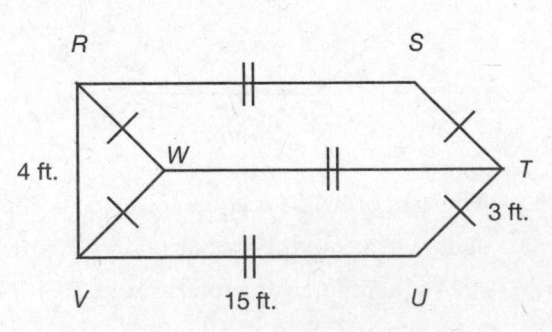

4 ft.

3 ft.

15 ft.

907. Find the area of concave polygon *KLMNOPQR*.

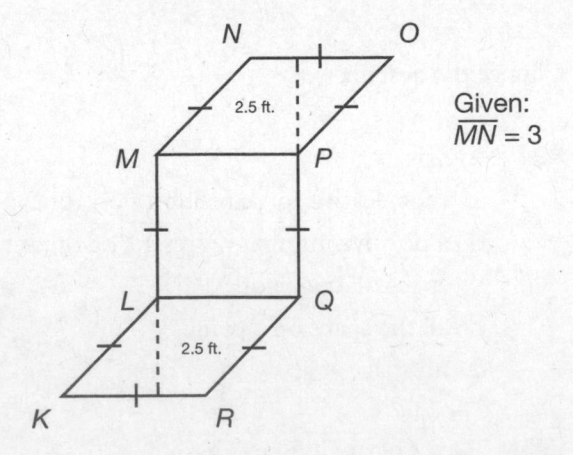

2.5 ft.

Given:
$\overline{MN} = 3$

2.5 ft.

908. Find the area of polygon *BCDEFGHI*.

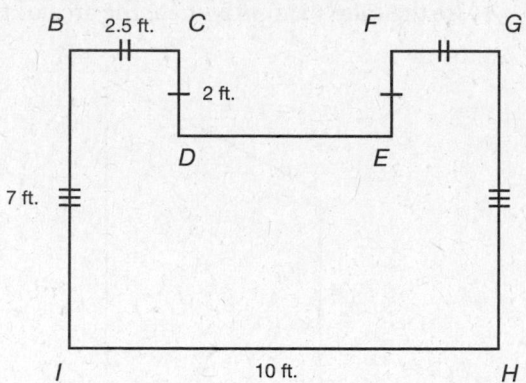

909. Find the area of concave polygon *MNOPQR*.

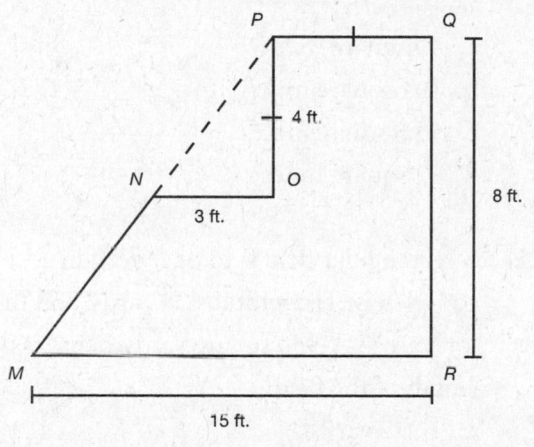

Use the following figure and information to answer questions 910 through 912.

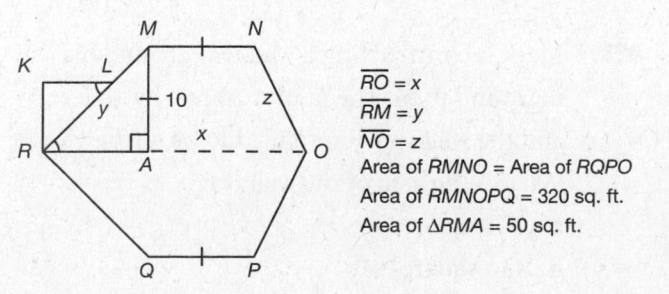

$\overline{RO} = x$
$\overline{RM} = y$
$\overline{NO} = z$
Area of *RMNO* = Area of *RQPO*
Area of *RMNOPQ* = 320 sq. ft.
Area of $\triangle$*RMA* = 50 sq. ft.

910. Find the measure of side *x*.

911. Find the measure of side *y*.

912. Find the measure of side *z*.

913. Charlie wants to know the area of his property, which measures 120 feet by 150 feet. Which formula will he use?
   a. $A = s^2$
   b. $A = \pi r^2$
   c. $A = \frac{1}{2}bh$
   d. $A = lw$

914. Rick is ordering a new triangular sail for his boat. He needs to know the area of the sail. Which formula will he use?
   a. $A = lw$
   b. $A = \frac{1}{2}bh$
   c. $A = bh$
   d. $A = \frac{1}{2}h(b_1 + b_2)$

915. Cathy wants to know the total area of her square-shaped quilt. Which formula will she use?
   a. $A = s^2$
   b. $A = \frac{1}{2}bh$
   c. $A = \neq r^2$
   d. $A = \frac{1}{2}h(b_1 + b_2)$

916. A racquetball court is 40 feet by 20 feet. What is the area of the court?
   a. 60 square feet
   b. 80 square feet
   c. 800 square feet
   d. 120 square feet

917. Allan has been hired to mow the school soccer field, which is 180 feet wide by 330 feet long. If his mower mows strips that are 2 feet wide, how many times must he mow across the width of the field?
   a. 90
   b. 165
   c. 255
   d. 60

918. Erin is painting a bathroom with four walls each measuring 9 feet by 5.5 feet. Ignoring the doors or windows, what is the area to be painted?
a. 198 square feet
b. 66 square feet
c. 49.5 square feet
d. 160 square feet

919. Brittney would like to carpet her bedroom. If her room is 11 feet by 14 feet, what is the area to be carpeted in square feet?
a. 121 square feet
b. 25 square feet
c. 169 square feet
d. 154 square feet

920. If a triangular sail has a vertical height of 83 feet and horizontal length of 40 feet, what is the area of the sail?
a. 1,660 square feet
b. 1,155 square feet
c. 201 square feet
d. 3,320 square feet

921. Stuckeyburg is a small town in rural America. Use the map to approximate the area of the town.

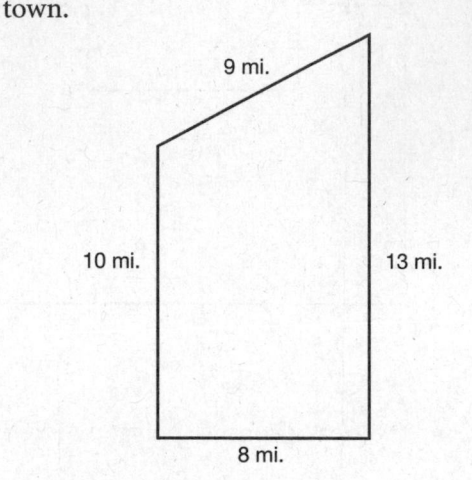

a. 40 square miles
b. 104 square miles
c. 93.5 square miles
d. 92 square miles

922. A rectangular field is to be fenced in completely. The width is 28 yards, and the total area is 1,960 square yards. What is the length of the field?
a. 1,932 yards
b. 70 yards
c. 31 yards
d. 473 yards

923. Mark is constructing a walkway around his inground pool. The pool is 20 feet by 40 feet, and the walkway is intended to be 4 feet wide. What is the area of the walkway?
a. 224 square feet
b. 416 square feet
c. 256 square feet
d. 544 square feet

924. The picture frame shown here has outer dimensions of 8 inches by 10 inches and inner dimensions of 6 inches by 8 inches. Find the area of section *A* of the frame.

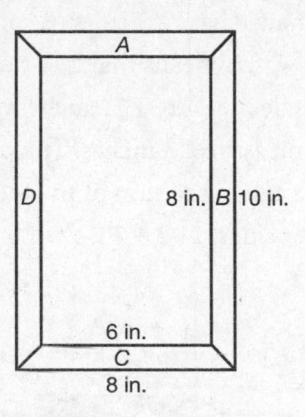

a. 18 square inches
b. 14 square inches
c. 7 square inches
d. 9 square inches

For question 925, use the following illustration.

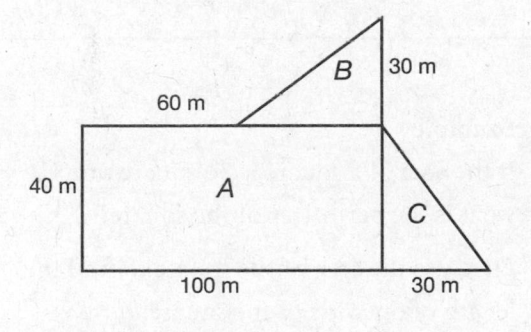

925. John is planning to purchase an irregularly shaped plot of land. Referring to the diagram, find the total area of the land.
a. 6,400 square meters
b. 5,200 square meters
c. 4,500 square meters
d. 4,600 square meters

926. The base of a triangle is four times as long as its height. If together they measure 95 centimeters, what is the area of the triangle?
a. 1,444 square centimeters
b. 100 square centimeters
c. 722 square centimeters
d. 95 square centimeters

927. The length and width of a rectangle together measure 130 yards. Their difference is 8 yards. What is the area of the rectangle?
a. 4,209 squarc yards
b. 130 square yards
c. 3,233 square yards
d. 4,270 square yards

928. Find the area of the region.

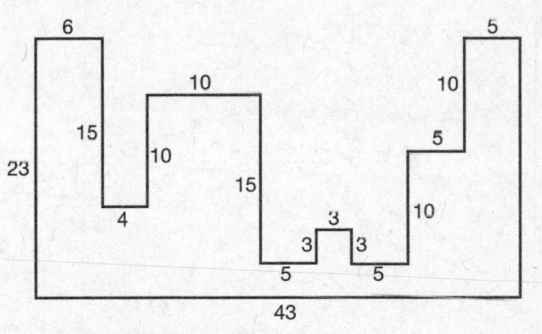

a. 478 square units
b. 578 square units
c. 528 square units
d. 428 square units

► Perimeter

Perimeter, the distance around a two-dimensional shape, is a bit easier to find because there are fewer formulas to remember. To find the perimeter of a polygon, just add the lengths of every side of the polygon. Because a square has four equal sides, the perimeter of a square is equal to $4s$ (where s is the length of one side of the square). In fact, a similar formula can be used to find the area of any regular (equal-sided) polygon. Perimeter = ns, where n is the number of sides of the regular polygon. Be careful, though, because this formula can be used only when every side of the polygon is equal in length.

Example

What is the perimeter of the figure shown here?

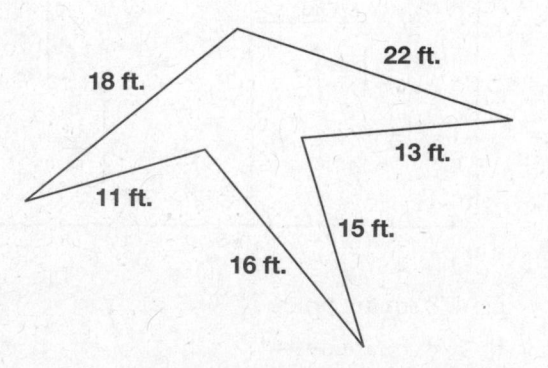

18 ft.   22 ft.   13 ft.   11 ft.   15 ft.   16 ft.

Add the lengths of each side of the figure:
$18 + 22 + 13 + 15 + 16 + 11 = 95$ feet.

TIP

Unlike area, perimeter is measured in linear units, because it is the sum of one measurement (length).

Example

What is the perimeter of a rectangle that has a width of 11.2 inches and length of 4.3 inches?

A rectangle has two pairs of congruent sides, which means that two sides of this rectangle measure 11.2 inches and two sides measure 4.3 inches. The perimeter of a rectangle is the sum of the measures of its four sides: $11.2 + 11.2 + 4.3 + 4.3 = 31$ inches.

TIP

The perimeter of a rectangle can also be found by doubling both the width and the length, and then adding those two totals. In other words, the perimeter of a rectangle is equal to $2l + 2w$.

Example

If the area of a square is 36 square units, what is the perimeter of the square?

First, use the area of the square to find the length of one side of the square: $A = s^2$; $s^2 = 36$; $s = 6$. If the length of one side of the square is 6 units, then the perimeter of the square must be $(4)(6) = 24$ units.

▶ **Practice Questions**

Choose the best answer.

929. A regular octagonal gazebo is added to a
Victorian lawn garden. Each side of the
octagon measures 5 feet. The formula for the
gazebo's perimeter is
a. $P = 8 \times 5$.
b. $8 = n \times 5$.
c. $5 = n \times 8$.
d. $s = n \times P$.

930. Timmy randomly walks ten steps to the left.
He does this nine more times. His path never
crosses itself, and he returns to his starting
point. The perimeter of the figure Timmy
walked equals
a. 90 steps.
b. 90 feet.
c. 100 steps.
d. 100 feet.

931. The perimeter of the Periwinkle High School
building is 1,600 feet. It has four sides of equal
length. Each side measures
a. 4 feet.
b. 40 feet.
c. 400 feet.
d. 4,000 feet.

932. Roberta draws two similar pentagons. The
perimeter of the larger pentagon is 93 feet;
one of its sides measures 24 feet. If the
perimeter of the smaller pentagon equals 31
feet, then the corresponding side of the
smaller pentagon can be found using which of
these calculations?
a. $5s = 31$
b. $93s = 24 \times 31$
c. $93 \times 24 = 31s$
d. $5 \times 31 = s$

933. Isadora wants to know the perimeter of the
face of a building; however, she does not have
a ladder. She knows that the building's
rectangular facade casts a 36-foot shadow at 5
o'clock, while a nearby mailbox casts a 12-foot
shadow. The mailbox is 4.5 feet tall. If the
façade is 54 feet long, the facade's perimeter
measures
a. 13.5×4.
b. 54×4.
c. $4.5(2) + 12(2)$.
d. $13.5(2) + 54(2)$.

Choose the best answer.

934. Which perimeter is not the same?

a.

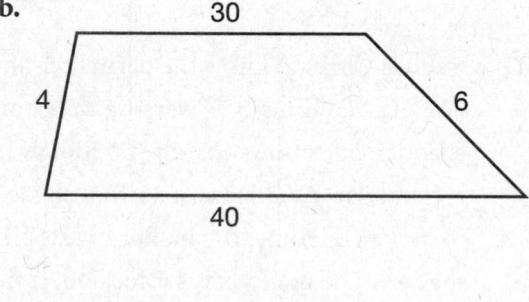

35

7 ⊨   ⊨ 7

35

b.

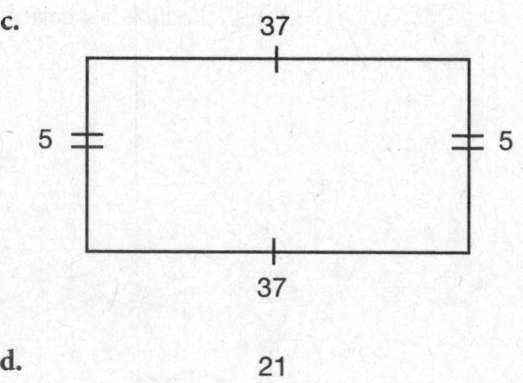

30

4   6

40

c.

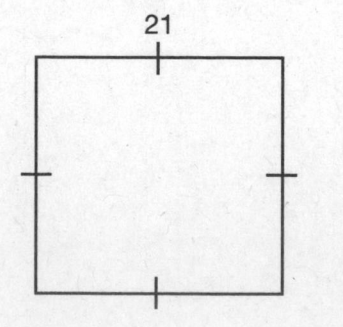

37

5 ⊨   ⊨ 5

37

d.

21

935. Which perimeter is not the same?

a. a 12-foot regular square backyard

b. an 8-foot regular hexagonal pool

c. a 6-foot regular octagonal patio

d. a 4-foot regular decagon Jacuzzi

e. It cannot be determined.

936. Which of the following choices has a different perimeter than the others?

a.

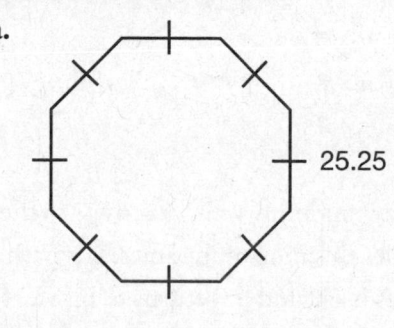

25.25

b.

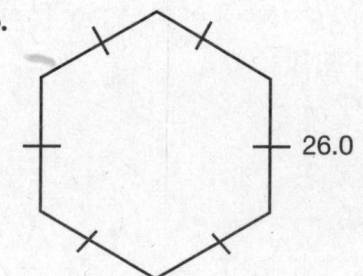

26.0

c.

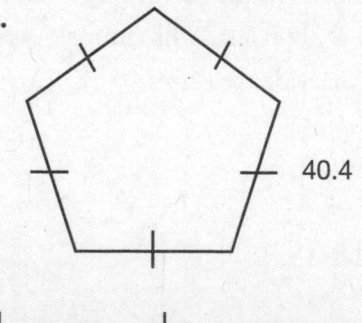

40.4

d.

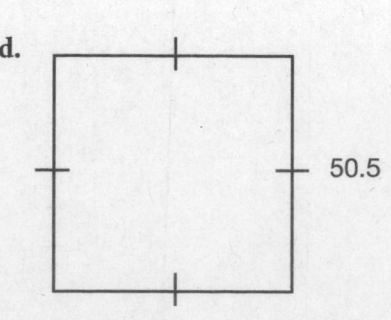

50.5

937. The measure of which figure's side is different from the other four figures?

    a. a regular nonagon whose perimeter measures 90 feet

    b. an equilateral triangle whose perimeter measures 27 feet

    c. a regular heptagon whose perimeter measures 63 feet

    d. a regular octagon whose perimeter measures 72 feet

    e. It cannot be determined.

938. Which figure does not have 12 sides?

    a. Regular figure A with sides that measure 4.2 in. and a perimeter of 50.4 in.

    b. Regular figure B with sides that measure 1.1 in. and a perimeter of 13.2 in.

    c. Regular figure C with sides that measure 5.1 in. and a perimeter of 66.3 in.

    d. Regular figure D with sides that measure 6.0 in. and a perimeter of 72.0 in.

    e. It cannot be determined.

Find the perimeters of the following figures.

939.

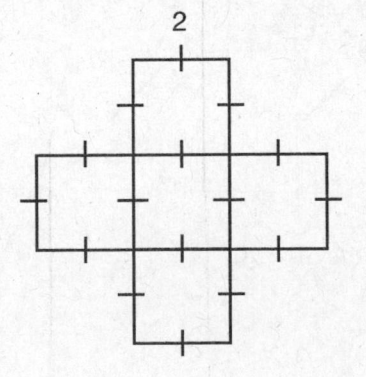

940.

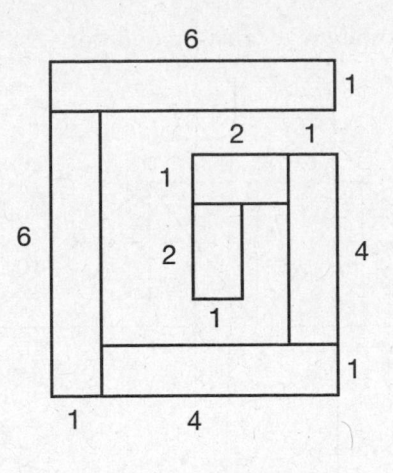

941.

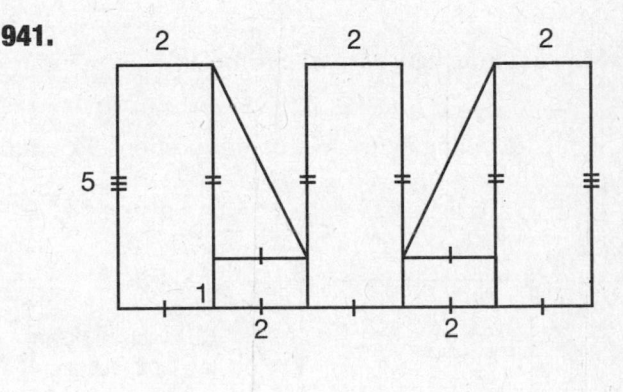

942.

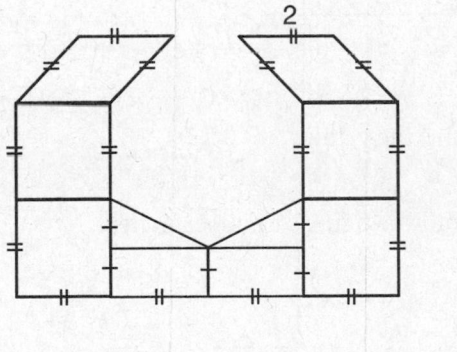

Use the figure shown to answer questions 943 and 944.

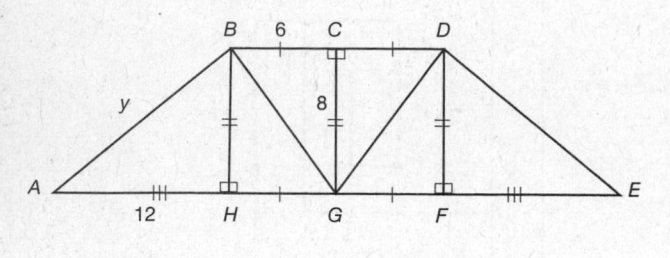

943. Find the value of y.

944. Find the figure's total perimeter.

Use the following figure to answer questions 945 and 946.

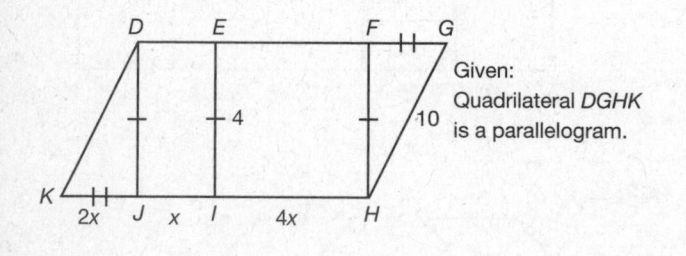

Given:
Quadrilateral *DGHK*
is a parallelogram.

945. Find the value of x.

946. Find the figure's total perimeter.

Use the following figure to answer questions 947 through 949.

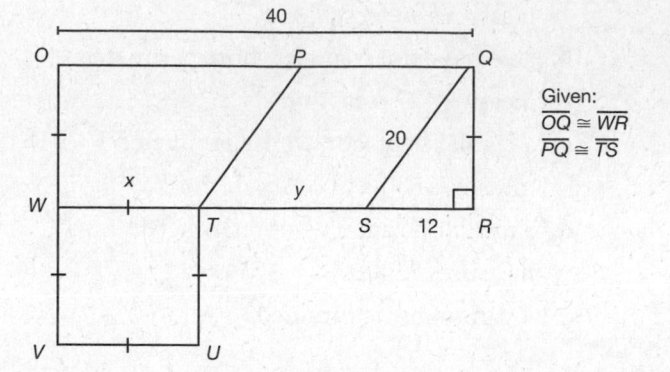

Given:
$\overline{OQ} \cong \overline{WR}$
$\overline{PQ} \cong \overline{TS}$

947. Find the value of x.

948. Find the value of y.

949. Find the figure's total perimeter.

Use the figure shown to answer questions 950 through 952.

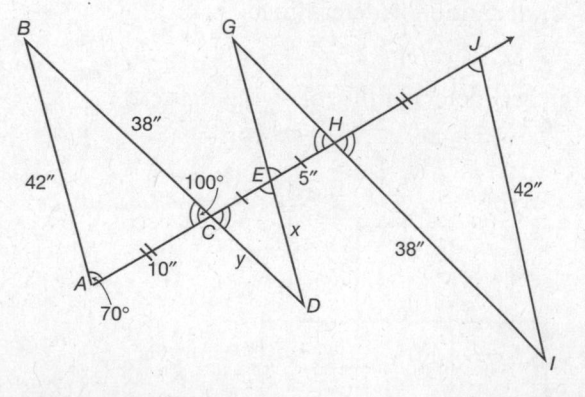

950. Find the value of x.

951. Find the value of y.

952. Find the figure's total perimeter.

953. Audrey is creating a raised flowerbed that is 4.5 feet by 4.5 feet. She needs to calculate how much lumber to buy. Which formula is easiest to use to find the distance around the flowerbed?
   a. $P = a + b + c$
   b. $A = lw$
   c. $P = 4s$
   d. $C = 2\pi r$

954. Danielle needs to know the distance around a basketball court. Which geometry formula will she use?
   a. $P = 2l + 2w$
   b. $P = 4s$
   c. $P = a + b + c$
   d. $P = b_1 + b_2 + h$

955. To find the perimeter of a triangular region, which formula would you use?
   a. $P = a + b + c$
   b. $P = 4s$
   c. $P = 2l + 2w$
   d. $C = 2\pi r$

956. Using the following illustration, determine the perimeter of the plot of land.
   a. 260 meters
   b. 340 meters
   c. 360 meters
   d. 320 meters

957. The longer base of a trapezoid is three times the shorter base. The nonparallel sides are congruent, and are 5 centimeters longer than the shorter base. The perimeter of the trapezoid is 40 centimeters. What is the length of the longer base?
   a. 15 centimeters
   b. 5 centimeters
   c. 10 centimeters
   d. 21 centimeters

958. The perimeter of the parallelogram is 32 centimeters. What is the length of the longer side?

   a. 9 centimeters
   b. 10 centimeters
   c. 6 centimeters
   d. 12 centimeters

TERMS TO REVIEW

area

perimeter

FORMULAS TO REVIEW

(A is area, b is base, h is height, s is side, P is perimeter, l is length, and w is width.)

area of parallelogram, rectangle, and rhombus:
$A = bh$ and $A = lw$

area of square: $A = s^2$ and $A = bh$

area of triangle: $A = \frac{1}{2}bh$

area of trapezoid: $A = \frac{1}{2}(b_1 + b_2)h$

perimeter of rectangle: $P = 2l + 2w$

perimeter of square: $P = 4s$

perimeter of regular polygon: $P = ns$

 Answers

894. **b.** All areas are positive numbers. Choice **a** is incorrect because if an area represented negative space, then it would be a negative number, which area cannot be. Choice **c** is incorrect because the area of a plane is infinite; when you measure area, you are measuring only a part of that plane inside a polygon. Points, lines, and planes do not occupy space, but figures do. The area of a figure is how much space that figure occupies.

895. **a.** Congruent figures have congruent parts, perimeters, and areas.

896. **False.** If the rhombus is not a square, it is a tilted square, which makes its height less than 5 feet. Consequently, the area of the square is 25 square feet, but the area of the rhombus is less than 25 square feet.

897. **False.** If the parallelogram is not a rectangle, it is a tilted rectangle, which makes its height less than 5 feet. Conseqently, the area of the rectangle is 50 square feet, but the area of the parallelogram is less than 50 square feet.

898. **True.** If two squares can fit into one rectangle, then the rectangle has twice the area of one square.

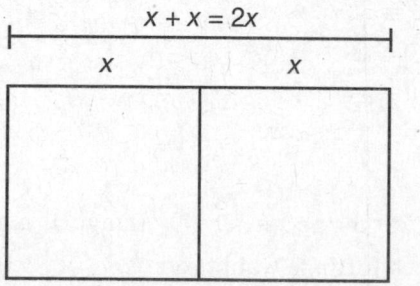

899. **True.** Like the squares and rectangle in the preceding answer, if two rhombuses can fit into one parallelogram, then the parallelogram has twice the area of one rhombus.

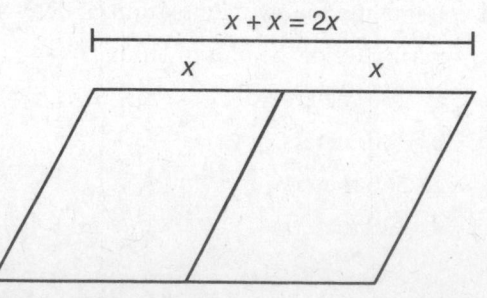

900. **True.** The triangle has an area of 25 square feet. The trapezoid has an area that measures 75 square feet. Three triangles of this size fit into one trapezoid, or the area of one triangle is a third of the area of the trapezoid.

901. The area of $\triangle DEF$ is $9\sqrt{3}$ square feet. To find the height of equilateral $\triangle DEF$, draw a perpendicular line segment from vertex E to the midpoint of $\overline{DF}$. This line segment divides $\triangle DEF$ into two congruent right triangles. Plug the given measurement into the Pythagorean theorem: $(\frac{1}{2} \times 6)^2 + b^2 = 6^2$; $9 + b^2 = 36$; $b = \sqrt{27}$; $b = 3\sqrt{3}$. To find the area, multiply the height by the base: $3\sqrt{3}$ feet $\times$ 6 feet $= 18\sqrt{3}$ square feet. Then, take half of $18\sqrt{3}$ to get $9\sqrt{3}$ square feet.

902. The area of $ABCD$ is 64 square feet. If one side of the square measures 8 feet, the other three sides of the square each measure 8 feet. Multiply two sides of the square to find the area: 8 feet $\times$ 8 feet $= 64$ square feet.

903. The area is 195 square feet. The area of this shaded figure requires the dual use of the Pythagorean theorem and the ratio of areas between similar triangles. First, find half the area of $\triangle ABC$. Perpendicularly extend a line segment from vertex A to the midpoint of $\overline{CB}$. The height of right triangle ABO is 14^2 ft. $+ b^2 = (14\sqrt{2})^2$ ft.; 196 sq. ft. $+ b^2 = 392$ sq. ft.; $b^2 = 196$ sq. ft.; $b = 14$ ft. Using the height, find the area of $\triangle ABC$: $\frac{1}{2}(14$ ft. $\times 28$ ft.$) = 196$ sq. ft. Within $\triangle ABC$ is a void, $\triangle DEF$. The area of the void must be subtracted from 196 square feet. Because $\triangle ABC$ is similar to $\triangle DEF$ (by the Angle-Angle postulate), $(\frac{14\sqrt{2}}{\sqrt{2}})^2 = \frac{196}{x}$. Therefore, $x = 1$ square foot; 196 square feet $-$ 1 square foot $= 195$ square feet.

904. The area is 10.5 square feet. Find the area of a rectangle with sides 6 feet and 3 feet: $A = 6$ ft. $\times 3$ ft. $= 18$ sq. ft. Find the area of both triangular voids: Area of the smaller triangular void $= \frac{1}{2}(3$ ft. $\times 1$ ft.$) = 1.5$ sq. ft. Area of the larger triangular void $= \frac{1}{2}(6$ ft. $\times 2$ ft.$) = 6$ sq. ft. Subtract 7.5 sq. ft. from 18 sq. ft., and 10.5 square feet remain.

905. The area of $ABCD$ is 480 square feet. You can treat figure $ABCD$ either like a trapezoid or like a parallelogram and a triangle. However you choose to work with the figure, you must begin by finding the measure of $\overline{ED}$ using the Pythagorean theorem: $16^2 + a^2 = 20^2$; $256 + a^2 = 400$; $a^2 = 144$; $a = 12$. Subtract 12 feet from 36 feet to find the measure of $\overline{BC}$: $36 - 12 = 24$. Should you choose to treat the figure as the sum of two polygons, to find the area of the entire figure, you find the area of each polygon separately and add them together. Parallelogram $ABCE$: 16 ft. $\times 24$ ft. $= 384$ sq. ft.; $\triangle ECD$: $\frac{1}{2} \times 16$ ft. $\times 12$ ft. $= 96$ sq. ft.; 384 sq. ft. $+ 96$ sq. ft. $= 480$ sq. ft. Should you choose to treat the figure like a trapezoid and need to find the area, simply plug in the appropriate measurements: $\frac{1}{2} \times 16$ ft.$(24$ ft. $+ 36$ ft.$) = 480$ square feet.

906. The area of *RSTUV* is $60 + 2\sqrt{5}$ square feet. Extend $\overline{TW}$ to $\overline{RV}$. Let's call this $\overline{XW}$. $\overline{XW}$ perpendicularly bisects $\overline{RV}$; as a perpendicular bisector, it divides isosceles triangle *RWV* into two congruent right triangles and establishes the height for parallelograms *RSTW* and *VUTW*. Solve the area of parallelogram *VUTW*: 2 ft. × 15 ft. = 30 sq. ft. Find the height of $\triangle RWV$ using the Pythagorean theorem: $a^2 + 2^2 = 3^2$; $a^2 + 4 = 9$; $a^2 = 5$; $a = \sqrt{5}$. Solve the area of $\triangle RWV$: $\frac{1}{2} \times \sqrt{5}$ ft. × 4 ft. = $2\sqrt{5}$ sq. ft. Add all the areas together: $2\sqrt{5}$ sq. ft. + 30 sq. ft. + 30 sq. ft. = $60 + 2\sqrt{5}$ square feet.

907. The area is 24.0 square feet. Rhombuses *KLQR* and *MNOP* are congruent. Their areas each equal 2.5 ft. × 3 ft. = 7.5 sq. ft. The area of square *LMPQ* equals the product of two sides: 3 ft. × 3 ft. = 9 sq. ft. The sum of all the areas is 9 sq. ft. + 7.5 sq. ft. + 7.5 sq. ft. = 24 square feet.

908. The area is 60.0 square feet. The simplest way to find the area of polygon *BCDEFGHI* is first to find the area of rectangle *BGHI*: 10 ft. × 7 ft. = 70 sq. ft. Subtract the area of rectangle *CFED*: 5 ft. × 2 ft. = 10 sq. ft.; 70 sq. ft. − 10 sq. ft. = 60 square feet.

909. The area is 70 square feet. Again, the simplest way to the find the area of polygon *MNOPQR* is first to find the area of trapezoid *MPQR*: $\frac{1}{2} \times$ 8 ft.(4 ft. + 15 ft.) = $\frac{1}{2} \times 8(19)$ = 76 sq. ft. Subtract the area of $\triangle NPO$: $\frac{1}{2} \times$ 3 ft. × 4 ft. = 6 sq. ft.; 76 sq. ft. − 6 sq ft. = 70 square feet.

910. Side $x = 22$ feet. The area of trapezoid *RMNO* plus the area of trapezoid *RQPO* equals the area of figure *RMNOPQ*. Because trapezoid *RMNO* and trapezoid *RQPO* are congruent, their areas are equal: $\frac{1}{2}$(320 sq. ft.) = 160 sq. ft. The congruent height of each trapezoid is known, and one congruent base length is known. Using the equation to find the area of a trapezoid, create the equation: 160 sq. ft. = $\frac{1}{2}$(10 ft.)(10 ft. + x); 160 sq. ft. = 50 sq. ft. + 5x ft.; 110 sq. ft. = 5x ft.; 22 feet = x.

911. Side $y = 10\sqrt{2}$ feet. Work backwards using the given area of $\triangle RMA$: 50 sq. ft. = $\frac{1}{2}b$(10 ft.); 50 sq. ft. = 5 ft. × b; 10 ft. = b. Once the base and height of $\triangle RMA$ are established, use the Pythagorean theorem to find $\overline{RM}$ (side y) $10^2 + 10^2 = c^2$; $100 + 100 = c^2$; $200 = c^2$; $10\sqrt{2} = c$; $\overline{RM} = 10\sqrt{2}$ feet.

912. Side $z = 2\sqrt{26}$ feet. Imagine a perpendicular line from vertex *N* to the base of trapezoid *RMNO*. This imaginary line divides $\overline{RO}$ into another 10-foot segment. The remaining portion of line $\overline{RO}$ is 2 feet long. Use the Pythagorean theorem to find the length of $\overline{NO}$ (side z): (10 ft.)^2 + (2 ft.)^2 = z^2; 100 sq. ft. + 4 sq. ft. = z^2; 104 sq. ft. = z^2; $2\sqrt{26}$ feet = z.

913. **d.** The area of a rectangle is length × width.

914. **b.** The area of a triangle is $\frac{1}{2}$ times the length of the base times the length of the height.

915. **a.** The area of a square is side squared or side times side.

916. **c.** The area of a rectangle is length times width. Therefore, the area of the racquetball court is equal to 40 feet times 20 feet or 800 square feet. If you chose answer **d**, you found the perimeter or distance around the court.

917. **a.** The width of the field, 180 feet, must be divided by the width of the mower, 2 feet. The result is that he must mow across the field 90 times. If you chose **b**, you calculated as if he were mowing the length of the field. If you chose **c**, you combined length and width, which would result in mowing the field twice.

918. **a.** The area of the room is the sum of the areas of four rectangular walls. Each wall has an area of length times width, or (9)(5.5), which equals 49.5 square feet. Multiply this by 4, which equals 198 square feet. If you chose **b**, you added 9 feet and 5.5 feet instead of multiplying.

919. **d.** The area of a rectangle is length times width. Using the dimensions described, area = (11)(14) or 154 square feet.

920. **a.** The area of a triangle is $\frac{1}{2}$(base)(height). Using the dimensions given, area = $\frac{1}{2}$(40)(83) or 1,660 square feet. If you chose **d**, you omitted $\frac{1}{2}$ from the formula.

921. **d.** The area of Stuckeyburg can be found by dividing the region into a rectangle and a triangle. Find the area of the rectangle ($A = lw$) and add the area of the triangle ($\frac{1}{2}bh$) for the total area of the region. Referring to the diagram, the area of the rectangle is (10)(8) = 80 square miles. The area of the triangle is $\frac{1}{2}$(8)(3) = 12 square miles. The sum of the two regions is 80 + 12 square miles = 92 square miles. If you chose **a**, you found the perimeter. If you chose **b**, you found the area of the rectangular region but did not include the triangular region.

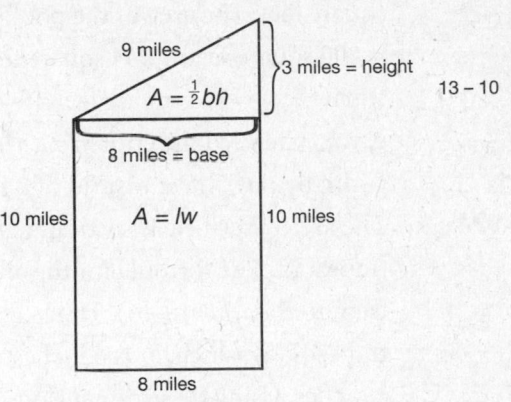

922. **b.** The area of a rectangle is length times width. Using the formula 1,960 square yards = (l)(28), solve for l by dividing both sides by 28; l = 70 yards.

923. d. The area of the walkway is equal to the entire area (area of the walkway and pool) minus the area of the pool. The area of the entire region is length times width. The pool is 20 feet wide and the walkway adds 4 feet onto each side, so the width of the rectangle formed by the walkway and pool is $20 + 4 + 4 = 28$ feet. Because the pool is 40 feet long and the walkway adds 4 feet onto each side, the length of the rectangle formed by the walkway and pool is $40 + 4 + 4 = 48$ feet. Therefore, the area of the walkway and pool is $(28)(48) = 1,344$ square feet. The area of the pool is $(20)(40) = 800$ square feet; 1,344 square feet $- 800$ square feet $= 544$ square feet. If you chose **c**, you extended the entire area's length and width by only 4 feet instead of 8 feet.

924. c. The area described as section A is a trapezoid. The formula for the area of a trapezoid is $\frac{1}{2}h(b_1 + b_2)$. The height of the trapezoid is 1 inch, b_1 is 6 inches, and b_2 is 8 inches. Using these dimensions, area $= \frac{1}{2}(1)(6 + 8)$ or 7 square inches. If you chose **b**, you used a height of 2 inches rather than 1 inch. If you chose **d**, you found the area of section B or D.

925. b. To find the total area, add the area of region A plus the area of region B plus the area of region C. The area of region A is length times width or $(100)(40) = 4,000$ m². Area of region B is $\frac{1}{2}bh$ or $\frac{1}{2}(40)(30) = 600$ m². The area of region C is $\frac{1}{2}bh$ or $\frac{1}{2}(30)(40) = 600$ square meters. The combined area is the sum of the previous areas or $4,000 + 600 + 600 = 5,200$ square meters. If you chose **a**, you miscalculated the area of a triangle as bh instead of $\frac{1}{2}bh$. If you chose **c**, you found only the area of the rectangle. If you chose **d**, you found the area of the rectangle and only one of the triangles.

926. c. This problem requires two steps. First, determine the base and height of the triangle. Second, determine the area of the triangle. To determine the base and height we will use the equation $x + 4x = 95$. Simplifying, $5x = 95$. Divide both sides by 5: $x = 19$. By substitution, the height is 19 and the base is $4(19)$ or 76. The area of the triangle is found by using the formula area $= \frac{1}{2}$ base $\times$ height. Therefore, the area $= \frac{1}{2}(76)(19)$ or 722 square centimeters. If you chose **a**, the area formula was incorrect. Area $= \frac{1}{2}$ base $\times$ height, not base $\times$ height. If you chose **b**, the original equation $x + 4x = 95$ was simplified incorrectly as $4x^2 = 95$.

927. a. There are two ways to solve this problem. The first method requires a linear equation with one variable. The second method requires a system of equations with two variables. Let the length of the rectangle equal x. Let the width of the rectangle equal $x + 8$. Together they measure 130 yards. Therefore, $x + x + 8 = 130$. Simplify: $2x + 8 = 130$. Subtract 8 from both sides: $2x = 122$. Divide both sides by 2: $x = 61$. The length of the rectangle is 61, and the width of the rectangle is $61 + 8$ or 69; $61 \times 69 = 4{,}209$. The second method is to develop a system of equations using x and y. Let x = the length of the rectangle, and let y = the width of the rectangle. Since the sum of the length and width of the rectangle is 130, we have the equation $x + y = 130$. The difference is 8, so we have the equation $x - y = 8$. If we add the two equations vertically, we get $2x = 122$. Divide both sides by 2: $x = 61$. The length of the rectangle is 61. Substitute 61 into either equation: $61 + y = 130$. Subtract 61 from both sides, giving you $y = 130 - 61 = 69$. To find the area of the rectangle, we use the formula length $\times$ width or $(61)(69) = 4{,}209$ square yards. If you chose **b**, you added 61 to 69 rather than multiplied. If you chose **c**, the length is 61 but the width was decreased by 8 to 53.

928. b. Refer to the diagram to find the area of the shaded region. One method is to enclose the figure into a rectangle, and subtract the area of the unwanted regions from the area of the rectangle. The unwanted regions have been labeled A through F. The area of region A is $(15)(4) = 60$. The area of region B is $(5)(10) = 50$. The area of region C is $(20)(5) = 100$. The area of region D is $(17)(3) = 51$. The area of region E is $(20)(5) = 100$. The area of region F is $(10)(5) = 50$. The area of the rectangle is $(23)(43) = 989$. The area of the shaded region is $989 - 60 - 50 - 100 - 51 - 100 - 50 = 578$ square units. If you chose **a**, **c**, or **d**, you omitted one or more of the regions A through F.

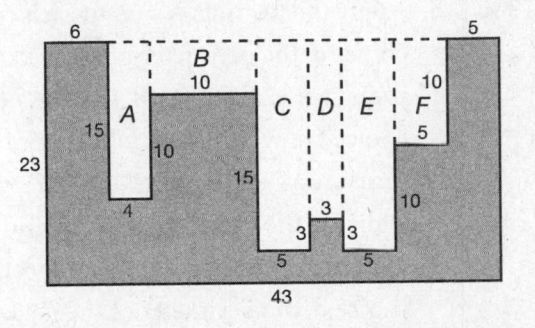

929. a. To find the perimeter, multiply the number of sides by the measure of one side. The perimeter of this Victorian gazebo is $P = 8 \times 5$.

930. c. Timmy walked ten ten-step sets. To find the perimeter of the figure Timmy walked, multiply 10 by 10 and remember that each side of that figure was measured in steps, not feet. Choice **a** forgot to count the first ten steps and turn that Timmy made. Choices **b** and **d** use the wrong increment, feet.

931. **c.** Plug the numbers into the formula: $P = ns$; $1,600 = 4s$; $400 = s$.

932. **b.** A proportion can find an unknown side of a figure using known sides of a similar figure; a proportion can also find an unknown side using known perimeters: $\frac{93}{24} = \frac{31}{s}$. Cross multiply: $93s = 24 \times 31$; $s = 8$.

933. **d.** Using a proportion, find x: $\frac{12}{36} = \frac{4.5}{x}$. Cross multiply: $12x = 36(4.5)$, and $x = 13.5$. The face of the building is a rectangle whose sides measure 13.5, 54, 13.5, and 54. To find its perimeter, add the measures of its sides together.

934. **b.** Each figure except the trapezoid has a perimeter of 84 feet; its perimeter measures only 80 feet.

935. **d.** Apply the formula $P = ns$ to each choice. In choice **a**, the perimeter of the backyard measures 12 feet $\times$ 4 sides, or 48 feet. In choice **b**, the perimeter of the pool measures 8 feet $\times$ 6 sides, or 48 feet. In choice **c**, the perimeter of the patio measures 6 feet $\times$ 8 sides, or 48 feet. In choice **d**, the perimeter of the Jacuzzi measures only 4 feet by 10 sides, or 40. Consequently, the Jacuzzi has a different perimeter.

936. **b.** Each figure has a perimeter of 202 feet except the hexagon; its perimeter measures 156 feet.

937. **a.** To find the measure of each side, change the formula $P = ns$ to $\frac{P}{n} = s$. Plug each choice into this formula. In choice **a**, the sides of the nonagon measure $\frac{90 \text{ feet}}{9 \text{ sides}}$, or 10 feet per side. In choice **b**, the sides of the triangle measure $\frac{27 \text{ feet}}{3 \text{ sides}}$, or 9 feet per side. In choice **c**, the sides of the heptagon measure $\frac{63 \text{ feet}}{7 \text{ sides}}$, or 9 feet per side. In choice d, the sides of the octagon measure $\frac{72 \text{ feet}}{8 \text{ sides}}$, or 9 feet per side.

938. **c.** To find the number of sides a figure has, change the formula $P = ns$ to $\frac{P}{s} = n$. Plug each choice into this formula. In choice **a**, figure A has 12 sides. In choice **b**, figure B has 12 sides. In choice **c**, figure C has 13 sides.

939. The perimeter equals 24 units. You can find this perimeter by either adding the measure of each side, or by using the formula $P = ns$. If you choose to add each side, your solution looks like this: $4(2 + 2 + 2) = 24$. If you choose to use the formula, $P = 12s$; $P = 12(2)$; $P = 24$.

940. The perimeter equals 50 units. Using your knowledge of rectangles and their congruent sides, you find the measure of each exterior side not given. To find the perimeter, you add the measures of the exterior sides together: $1 + 6 + 1 + 6 + 1 + 4 + 1 + 4 + 1 + 2 + 1 + 2 + 1 + 2 + 1 + 3 + 3 + 5 + 5 = 50$.

941. The perimeter equals $34 + 4\sqrt{5}$ units. First, find the hypotenuse of at least one of the two congruent triangles using the Pythagorean theorem: $2^2 + 4^2 = c^2$; $4^2 + 16^2 = c^2$; $20 = c^2$; $2\sqrt{5} = c$. Add the measures of the exterior sides together: $2 + 5 + 2 + 2 + 2 + 2 + 2 + 5 + 2 + 2\sqrt{5} + 4 + 2 + 4 + 2\sqrt{5} = 34 + 4\sqrt{5}$.

942. The perimeter equals $32 + 2\sqrt{5}$ units. First find the hypotenuse of at least one of the two congruent triangles using the Pythagorean theorem: $1^2 + 2^2 = c^2$; $1 + 4 = c^2$; $\sqrt{5} = c$. Add the measures of the exterior sides together: $2(2 + 2 + 2) + 4(2 + 2) + 2(2) + 2\sqrt{5} = 32 + 2\sqrt{5}$.

943. $y = 4\sqrt{13}$. $\overline{CG}$ and $\overline{BH}$ are congruent because the opposite sides of a rectangle are congruent. Plug the measurements of $\triangle ABH$ into the Pythagorean theorem: $12^2 + 8^2 = y^2$; $144 + 64 = y^2$; $208 = y^2$; $4\sqrt{13} = y$.

944. The perimeter is $48 + 8\sqrt{13}$ units. Figure $ABDE$ is an isosceles trapezoid; $\overline{AB}$ is congruent to $\overline{ED}$. Add the measures of the exterior line segments together: $6 + 6 + 4\sqrt{13} + 12 + 6 + 6 + 12 + 4\sqrt{13} = 48 + 8\sqrt{13}$.

945. $x = \sqrt{21}$. In parallelogram $DGHK$, opposite sides are congruent, so $\triangle KDJ$ and $\triangle GFH$ are also congruent (Side-Side-Side postulate or Side-Angle-Side postulate). Plug the measurements of $\triangle KDJ$ and $\triangle GFH$ into the Pythagorean theorem: $(2x)^2 + 4^2 = 10^2$; $4x^2 + 16 = 100$; $4x^2 = 84$; $x^2 = 21$; $x = \sqrt{21}$.

946. The perimeter is $14\sqrt{21} + 20$ units. Replace each x with $\sqrt{21}$ and add the exterior line segments together: $2\sqrt{21} + \sqrt{21} + 4\sqrt{21} + 10 + 2\sqrt{21} + 4\sqrt{21} + \sqrt{21} + 10 = 14\sqrt{21} + 20$.

947. $x = 16$. The hatch marks indicate that $\overline{WT}$ and $\overline{QR}$ are congruent. Plug the measurements of $\triangle SQR$ into the Pythagorean theorem: $12^2 + x^2 = 20^2$; $144 + x^2 = 400$; $x^2 = 256$; $x = 16$.

948. $y = 12$. Opposite sides of a rectangle are congruent. $\overline{OQ}$ equals the sum of $\overline{WT}, \overline{TS}$, and $\overline{SR}$. Create the equation: $40 = 16 + y + 12$; $40 = 28 + y$; $12 = y$.

949. The perimeter is 144 units. Add the measures of the exterior line segments together: $40 + 16 + 12 + 12 + 16 + 16 + 16 + 16 = 144$.

950. $x = 21$ inches. $\triangle ABC$ and $\triangle JIH$ are congruent (Side-Side-Side postulate). $\triangle EDC$ and $\triangle EGH$ are also congruent because three angles and a side are congruent. However, $\triangle ABC$ and $\triangle JIH$ are only similar to $\triangle EDC$ and $\triangle EGH$ (Angle-Angle postulate). A comparison of side $\overline{AC}$ to side $\overline{EC}$ reveals a 10:5 or 2:1 ratio between similar triangles. If $\overline{AB}$ measures 42 inches, then corresponding line segment $\overline{ED}$ measures half as much, or 21 inches.

951. $y = 19$ inches. Using the same ratio determined for x, if $\overline{BC}$ measures 38 inches, then corresponding line segment $\overline{DC}$ measures half as much, or 19 inches.

952. The perimeter is 270 inches. Add the measures of the exterior line segments together: $2(42 + 38 + 10) + 2(21 + 19 + 5) = 180 + 90 = 270$ inches.

953. c. The perimeter of a square is four times the length of one side.

954. a. The perimeter of a rectangle is two times the length plus two times the width.

955. **a.** The perimeter of a triangle is length of side a plus length of side b plus length of side c.

956. **c.** To find the perimeter, we must know the lengths of all sides. According to the diagram, we must find the length of the hypotenuse for the triangular regions B and C. Use the Pythagorean theorem for triangular region B: $30^2 + 40^2 = c^2$; $900 + 1,600 = c^2$; $2,500 = c^2$; 50 m $= c$. The hypotenuse for triangular region C is also 50 m, because the legs are 30 m and 40 m as well. Now adding the lengths of all sides, 40 m $+ 100$ m $+ 30$ m $+ 50$ m $+ 30$ m $+ 50$ m $+ 60$ m $= 360$ meters, the perimeter of the plot of land. If you chose **a**, you did not calculate in the hypotenuse on either triangle. If you chose **b**, you miscalculated the hypotenuse as having a length of 40 m. If you chose **d**, you miscalculated the hypotenuse as having a length of 30 m.

957. **a.** The two bases of the trapezoid are represented by x and $3x$. The nonparallel sides are each $x + 5$. Setting up the equation for the perimeter will allow us to solve for x; $x + 3x + x + 5 + x + 5 = 40$. Simplify to $6x + 10 = 40$. Subtract 10 from both sides; $6x = 30$. Divide both sides by 6; $x = 5$. The longer base is represented by $3x$. Using substitution, $3x$ or $(3)(5)$ equals 15 centimeters, the longer base. If you chose **b**, you solved for the shorter base. If you chose **c**, you solved for the nonparallel side. If you chose **d**, the original equation was incorrect, $x + x + 5 + 3x = 40$.

958. **b.** The perimeter of a parallelogram is the sum of the lengths of all four sides. Using this information and the fact that opposite sides of a parallelogram are equal, we can write the following equation: $x + x + \frac{3x+2}{2} + \frac{3x+2}{2} = 32$. Simplify to $2x + 3x + 2 = 32$. Simplify again: $5x + 2 = 32$. Subtract 2 from both sides: $5x = 30$. Divide both sides by 5: $x = 6$. The longer base is represented by $\frac{3x+2}{2}$. Using substitution, $\frac{3(6)+2}{2}$ equals 10 centimeters. If you chose **c**, you solved for the shorter side.

15 ▶ Circles

A CIRCLE IS A CLOSED FIGURE with no sides; it is a set of points that are equidistant from a single point, the center of the circle.

▶ Center Point, Radius, Central Angle

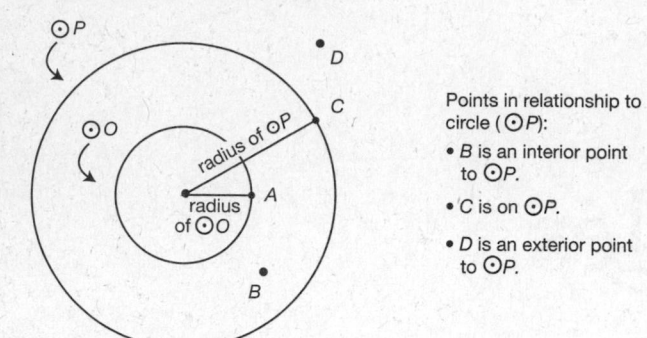

Points in relationship to circle (⊙*P*):

- *B* is an interior point to ⊙*P*.
- *C* is on ⊙*P*.
- *D* is an exterior point to ⊙*P*.

A **center point** is a stationary point at the center of a circle. All the points that lie on the curve of the circle are equidistant from the center point.

A **radius** is a line segment that extends from the center of the circle and meets exactly one point on the circle.

Circles with the same center point but different radii are **concentric circles**.

A **central angle** is an angle formed by two radii.

▶ Chords and Diameters

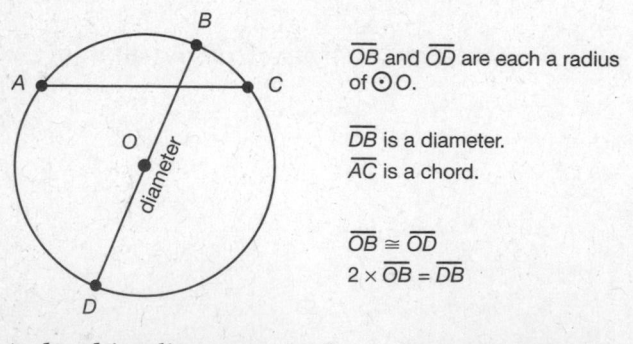

$\overline{OB}$ and $\overline{OD}$ are each a radius of ⊙*O*.

$\overline{DB}$ is a diameter.
$\overline{AC}$ is a chord.

$\overline{OB} \cong \overline{OD}$
$2 \times \overline{OB} = \overline{DB}$

A **chord** is a line segment that joins two points on a circle.

A **diameter** is a chord that joins two points on a circle and passes through the center point.

Note: A diameter is twice the length of a radius, and a radius is half the length of a diameter.

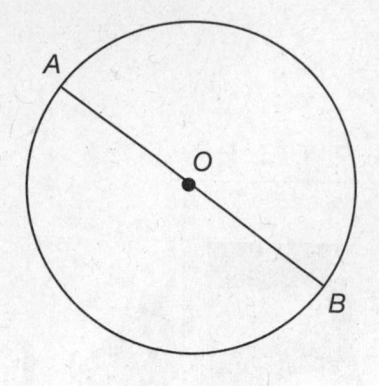

Example

If *AO* measures 5 units, what is the measure of *OB*?

Since point *A* is on the curve of the circle and point *O* is the center of the circle, *AO* is a radius of the circle. Point *B* is also on the circle and connects to the center of the circle, so *OB* is also a radius of the circle. If *AO* is 5 units, *OB* must also be 5 units.

Example

If *AB* is 14 units, what is the measure of *AO*?

AB is a diameter of the circle and *AO* is a radius. The radius of a circle is half the length of the diameter, so the length of *AO* is $\frac{14}{2} = 7$ units.

▶ Arcs

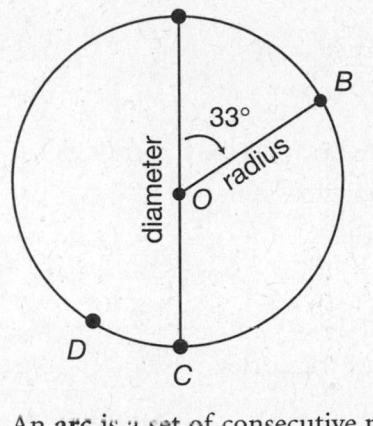

$\overarc{AB} = 33°$
$\overarc{AB} = 10.1$ inches

$\overarc{AB}$ is a minor arc.
$\overarc{ABC}$ is a semicircle.
$\overarc{ABD}$ is a major arc.

An **arc** is a set of consecutive points on a circle. Arcs can be measured by their rotation and by their length.

A **minor arc** is an arc that measures less than 180°.

A **semicircle** is an arc that measures exactly 180°. The endpoints of a semicircle are the endpoints of a diameter.

A **major arc** is an arc that measures greater than 180°.

Note: An arc formed by a central angle has the same rotation as that angle.

▶ Other Lines and Circles

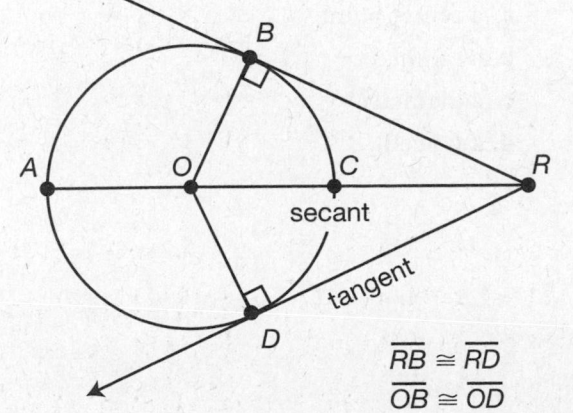

$\overline{RB} \cong \overline{RD}$
$\overline{OB} \cong \overline{OD}$

A **tangent** is a ray or line segment that intercepts a circle at exactly one point. The angle formed by a radius and a tangent at the point where it meets a circle is a right angle.

Note: Two tangents from the same exterior point are congruent.

A **secant** is a ray or line segment that intercepts a circle at two points.

▶ Congruent Arcs and Circles

Congruent circles have congruent radii and diameters. Congruent central angles form congruent arcs in congruent circles.

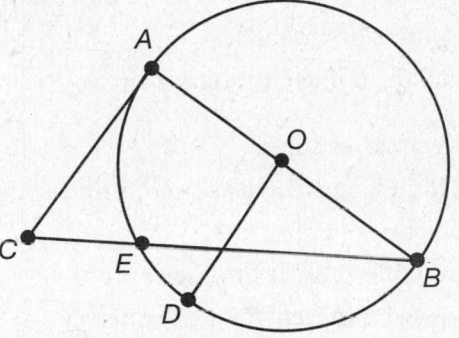

Example

If angle *DOB* measures 90 degrees, what is the length of arc *DB*?

Since angle *DOB* is a central angle, its measure and the measure of its intercepted arc are equal. The length of arc *DB* is also 90 degrees.

Example

AC is a tangent and *CB* is a secant. If the length of *OB* is 8 units and the length of *AC* is 12 units, what is the length of *CB*?

A tangent and a radius form a right angle, which means that triangle *ABC* is a right triangle. Since *OB* is a radius, *AB* is a diameter, which is equal to twice the length of *OB*: $8 \times 2 = 16$ units. We use the Pythagorean theorem to find the length of the hypotenuse of the triangle, *CB*: $12^2 + 16^2 = CB^2$; $144 + 256 = CB^2$; $400 = CB^2$; *CB* = 20 units.

▶ Practice Questions

Choose the best answer.

959. Which points of a circle are on the same plane?
   a. only the center point and points on the curve of the circle
   b. points on the circle but no interior points
   c. the center point, interior points, but no points on the circle
   d. all the points in and on a circle

960. In a circle, a radius
   a. is the same length as a radius in a congruent circle.
   b. extends outside the circle.
   c. is twice the length of a diameter.
   d. determines an arc.

961. Congruent circles
   a. have the same center point.
   b. have diameters of the same length.
   c. have radii of the same length.
   d. **b** and **c**

Use the following figure to answer question 962.

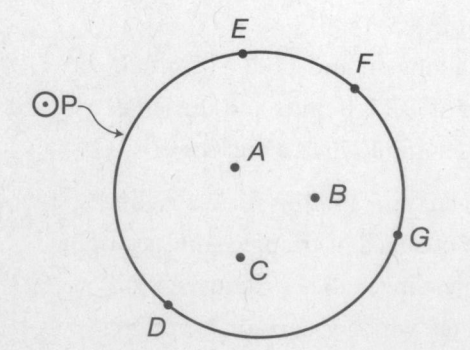

962. Name the exterior point(s).
   a. *A*, *B*, *C*
   b. *D*, *E*, *F*, *G*
   c. *H*
   d. *A*, *E*, *G*, *H*

963. Point *A* lies 12 inches from the center of ⊙*O*. If ⊙*O* has a 1-foot radius, •*A* lies
   a. inside the circle.
   b. on the circle.
   c. outside the circle.
   d. between concentric circles.

964. A diameter is also
   a. a radius.
   b. an arc.
   c. a chord.
   d. a line.

965. Both tangents and radii
   a. extend from the center of a circle.
   b. are half a circle's length.
   c. meet a circle at exactly one point.
   d. are straight angles.

966. From a stationary point, Billy throws four balls in four directions. Where each ball lands determines the radius of a different circle. What do the four circles have in common?
   a. a center point
   b. a radius
   c. a diameter
   d. a tangent

967. From a stationary point, Kim aims two arrows at a bull's-eye. The first arrow nicks one point on the edge of the bull's-eye; the other strikes the center of the bull's-eye. Kim knows the first arrow traveled 100 miles. If the bull's-eye is 200 miles wide, how far is Kim from the center of the bull's-eye?

   a. 100 miles

   b. $2\sqrt{100}$ miles

   c. 1,000 miles

   d. $100\sqrt{2}$ miles

Use the following figure to answer question 968.

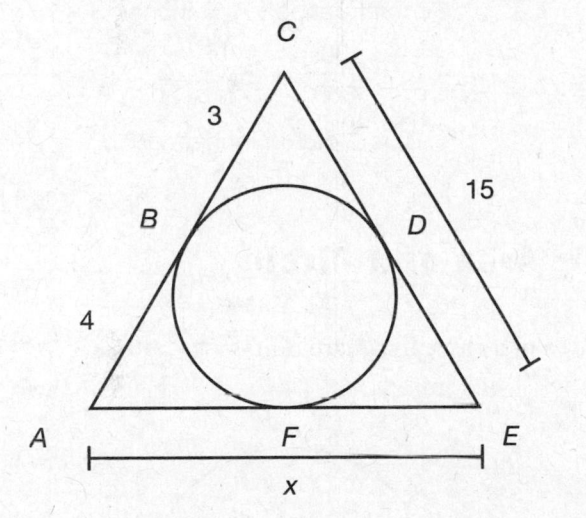

968. What is the value of x?

Use the figure shown here to answer question 969.

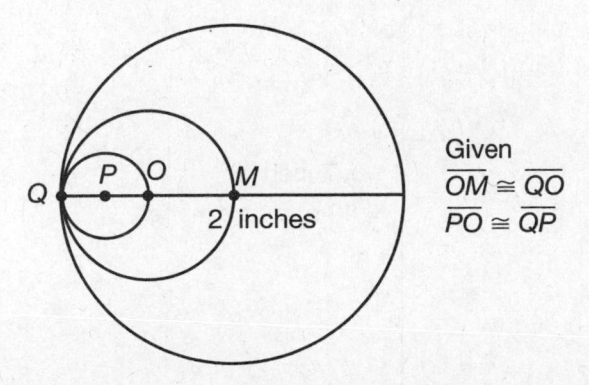

Given
$\overline{OM} \cong \overline{QO}$
$\overline{PO} \cong \overline{QP}$

2 inches

969. If the diameter of $\odot M$ is 2 inches, then what is the diameter of $\odot P$?

Use the figures shown here to answer question 970.

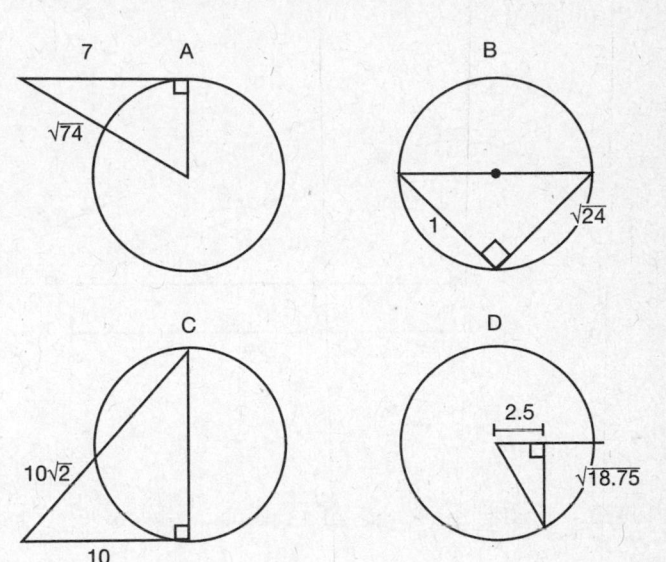

970. Which circle is NOT congruent?

Use the figures shown here to answer question 971.

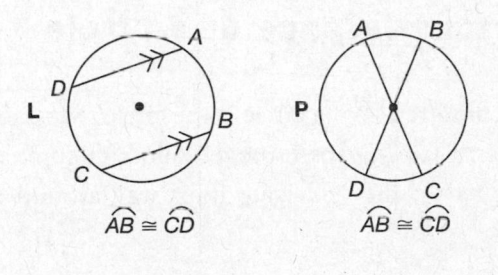

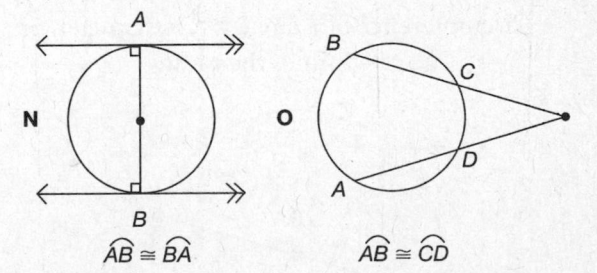

971. In which figure (L, N, P, O) is the set of arcs not congruent to each other even though so labeled?

Use the following figure to answer questions 972 through 974.

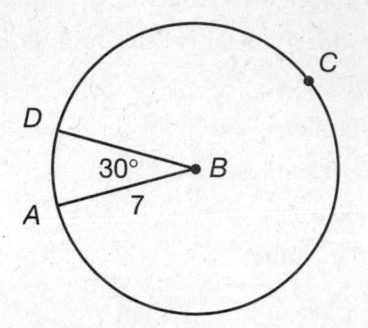

972. What is the length of a radius in the circle?

973. What is the area of $\triangle DEF$?

974. Is $\overset{\frown}{DHG}$ a major or minor arc?

▶ Circumference of a Circle

The **circumference** of a circle is the circle's version of perimeter. *Circa* means *around*. Sailors circumnavigate the earth; they navigate their way around the earth.

Circumference of a circle = π × diameter, or
π × 2 times the radius

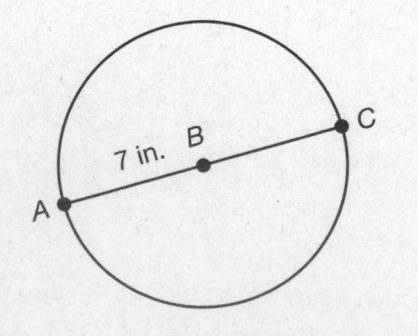

$2 \times \overline{AB} = \overline{AC}$
Circumference = π2r
= π2 × 7 inches
= 14π inches

▶ Measure of an Arc

Using the circumference of a circle, you can find the measure of an arc.

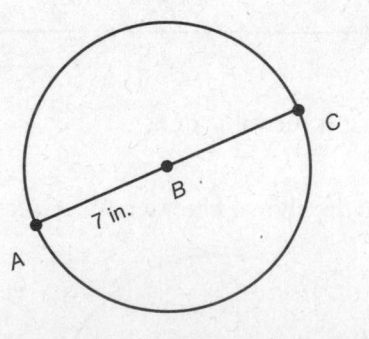

Circumference = 14π inches

$$\frac{30°}{360°} = \frac{1}{12}$$

$\overset{\frown}{AD}$ is $\frac{1}{12}$ of 14π inches, or $\frac{7}{6}\pi$ inches

▶ Area of a Circle

Area of a circle in square units = π × radius²

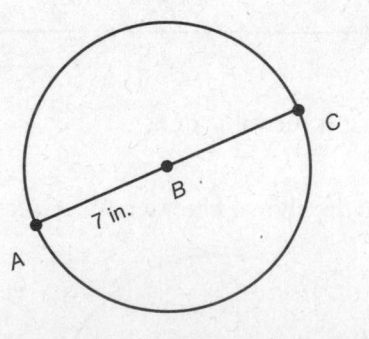

Area = πr^2
= π(7 inches)²
= 49π square inches

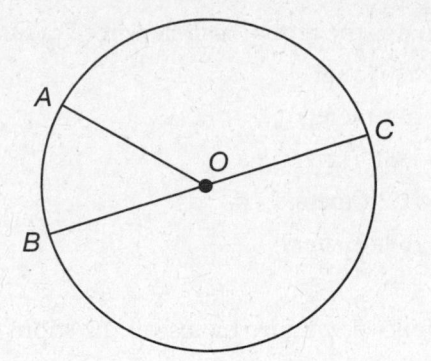

Example

If the circumference of the circle is 20π centimeters, what is the diameter of the circle? What is the radius? What is the area?

The circumference of a circle is equal to π multiplied by the diameter, which means that the diameter of the circle is 20 centimeters. Since the diameter of the circle is 20 cm, the radius of the circle is $\frac{20}{2} = 10$ centimeters. The area of a circle is equal to πr^2, which means that the area of the circle is $\pi(10)^2$, or 100π square centimeters.

Example

If the radius of the circle is 2 inches, what is the circumference of the circle? What is the area?

The diameter of a circle is twice the radius, so the diameter is 4 inches and the circumference is 4π inches. The area of a circle is equal to πr^2: $\pi(2)^2 = 4\pi$ square inches.

Example

If angle *AOC* is 120 degrees and diameter *BC* is 9 feet, what is the length of arc *AC* in feet?

If diameter *BC* is 9 feet, then the circumference of the circle is 9π feet; 120 degrees is $\frac{120}{360}$ or $\frac{1}{3}$ of the circle, which means that arc *AC* is equal to one-third of the circumference of the circle, or $\frac{9\pi}{3} = 3\pi$ feet.

Choose the best answer.

975. What is the circumference of the following figure?

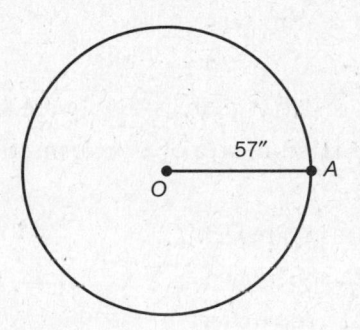

a. 57π inches
b. 114π inches
c. 26.5π inches
d. $\sqrt{57}\pi$ inches

976. What is the area of the following figure?

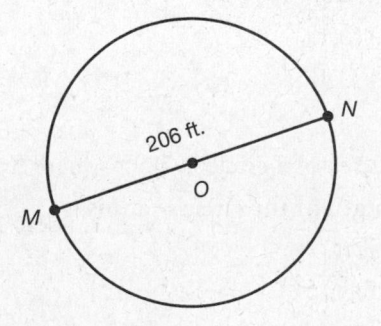

a. 51.5π square feet
b. 103π square feet
c. 206π square feet
d. 10,609π square feet

977. What is the radius of the following figure?

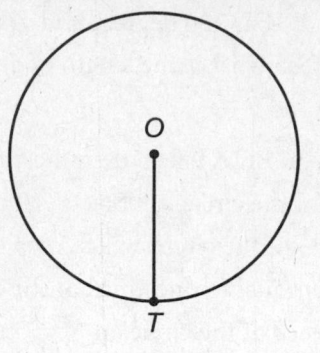

Circumference of ⊙O = 64π centimeters

a. 8 centimeters
b. 16 centimeters
c. 32 centimeters
d. 64 centimeters

978. The area of a square is 484 square feet. What is the maximum area of a circle inscribed in the square?

a. 11π square feet
b. 22π square feet
c. 121π square feet
d. 122π square feet

979. If the circumference of a circle is 192π feet, then the length of the circle's radius is

a. $16\sqrt{6}$ feet.
b. 96 feet.
c. 192 feet.
d. 384 feet.

980. If the area of a circle is 289π square feet, then the length of the circle's radius is

a. 17 feet.
b. 34 feet.
c. 144.5 feet.
d. 289 feet.

981. What is the area of a circle whose radius is 13 meters long?

a. 26π meters
b. 156π meters
c. 42.2π meters
d. 169π meters

Use the following figure to answer questions 982 and 983.

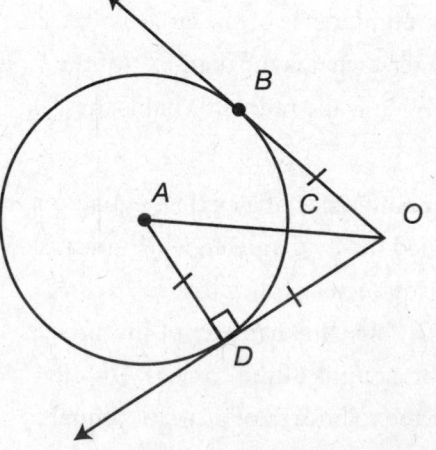

Circumference = 64π feet

982. $\overset{\frown}{BD}$ is a quarter of the circumference of the circle. If the total circumference is 64π feet, then what is the length of $\overset{\frown}{BD}$?

a. 16π feet
b. 32π feet
c. 48π feet
d. 90π feet

983. What type of angle is the central angle that intercepts $\overset{\frown}{BD}$?

a. an acute angle
b. a right angle
c. an obtuse angle
d. a straight angle

Use the following figure to answer question 984.

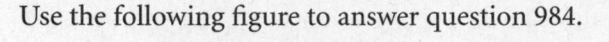

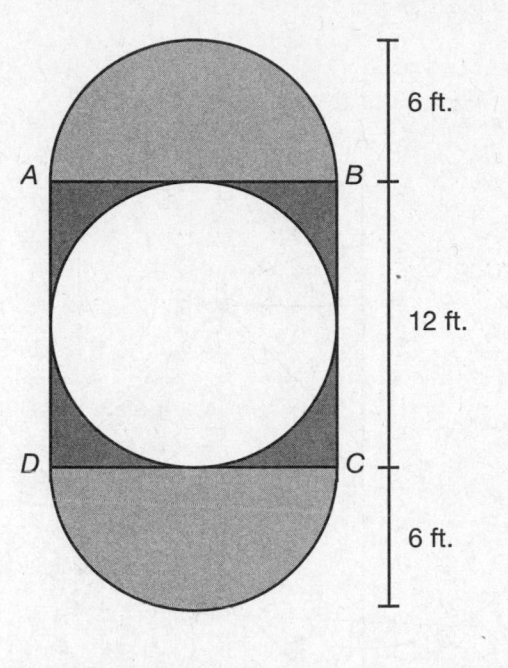

6 ft.

12 ft.

6 ft.

984. What is the area of the shaded figure (light and dark combined)?

   a. 144 square feet – 12π square feet
   b. 12 square feet – 144π square feet
   c. 144 square feet
   d. 144 square feet – 24π square feet + 12π square feet

Use the following figure to answer questions 985 and 986.

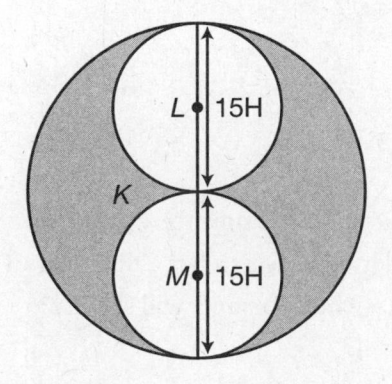

985. What is the area of the shaded figure?

   a. 56.25π square feet
   b. 112.5π square feet
   c. 225π square feet
   d. 337.4π square feet

986. What is the ratio of the area of ⊙M and the area of ⊙K?

   a. 1:8
   b. 1:4
   c. 1:2
   d. 1:1

Use the following figure to answer questions 987 and 988.

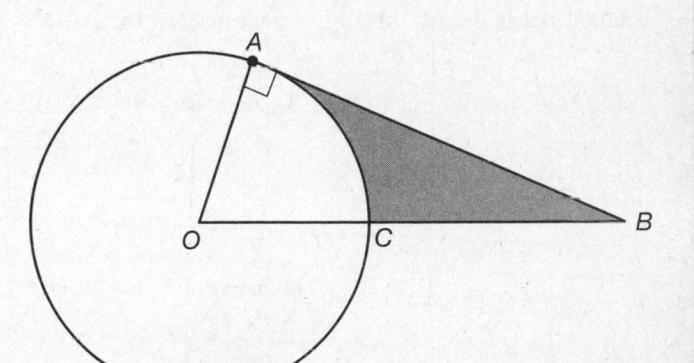

987. If $\overline{AB}$ = 60 and $\overline{OB}$ = 75, what is the measure of $\overline{OA}$?

988. If central angle AOC measures 60°, what is the area of the shaded figure?

Use the following figure to answer questions 989 and 990.

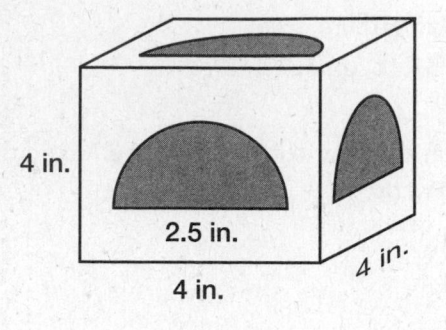

4 in.

2.5 in.

4 in.

4 in.

989. If each side of a cube has an identical semicircle carved into it, what is the total carved area of the cube?

990. What is the remaining surface area of the cube?

Using the following figure, answer questions 991 through 994.

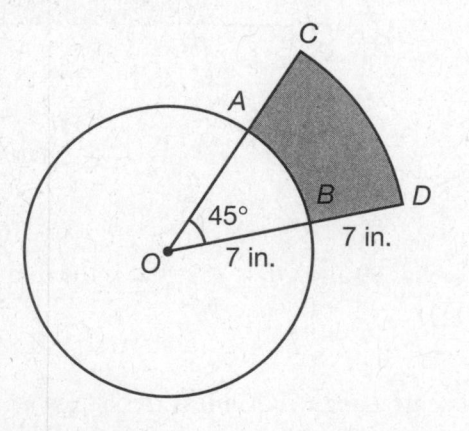

991. Find the shaded area of the figure.

992. Find the length of $\overarc{AB}$.

993. Find the length of $\overarc{CD}$.

994. Are $\overarc{AB}$ and $\overarc{CD}$ the same length?

Use the following figure to answer questions 995 and 996.

995. What is the area of trapezoid *ABDE*?

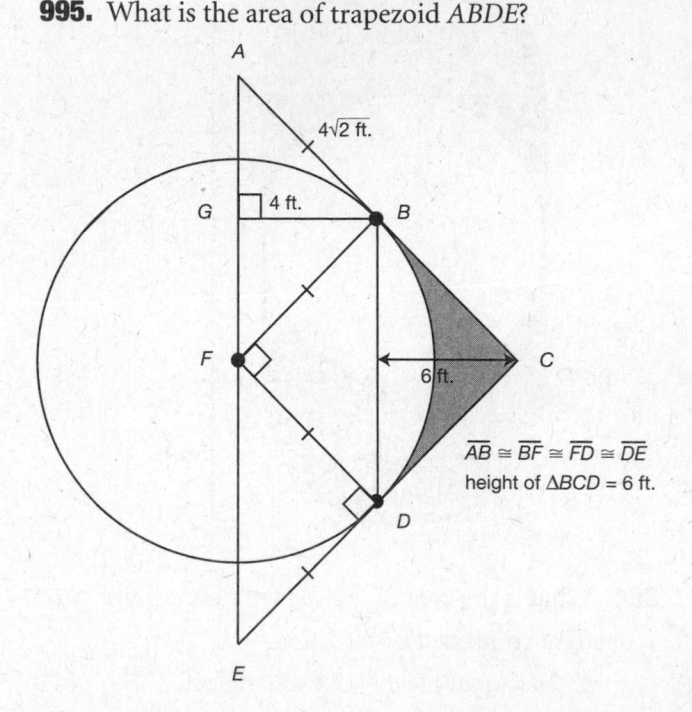

$\overline{AB} \cong \overline{BF} \cong \overline{FD} \cong \overline{DE}$
height of $\triangle BCD = 6$ ft.

996. What is the shaded area?

997. Aaron is installing a ceiling fan in his bedroom. He needs to know the area the fan will cover once it is in motion. Which formula will he use?
 a. $A = bh$
 b. $A = s^2$
 c. $A = \frac{1}{2}bh$
 d. $A = \pi r^2$

998. If Lisa wants to know the distance around her circular table, which has a diameter of 42 inches, which formula will she use?
 a. $P = 4s$
 b. $P = 2l + 2w$
 c. $C = \pi d$
 d. $P = a + b + c$

999. The arm of a ceiling fan measures a length of 25 inches. What is the area covered by the motion of the fan blades when turned on? (π = 3.14)

   a. 246.49 square inches

   b. 78.5 square inches

   c. 1,962.5 square inches

   d. 157 square inches

1000. How far will a bowling ball roll in one rotation if the ball has a diameter of 10 inches? (π = 3.14)

   a. 31.4 inches

   b. 78.5 inches

   c. 15.7 inches

   d. 62.8 inches

1001. A water sprinkler sprays in a circular pattern a distance of 10 feet. What is the circumference of the spray? (π = 3.14)

   a. 31.4 feet

   b. 314 feet

   c. 62.8 feet

   d. 628 feet

1002. How many degrees does a minute hand move in 25 minutes?

   a. 25°

   b. 150°

   c. 60°

   d. 175°

1003. A circular print is being matted in a square frame. If the frame is 18 inches by 18 inches and the radius of the print is 7 inches, what is the area of the matting? (π = 3.14)

   a. 477.86 square inches

   b. 170.14 square inches

   c. 280.04 square inches

   d. 288 square inches

1004. Pat is making a Christmas tree skirt. She needs to know how much fabric to buy. Using the illustration provided, determine the area of the skirt to the nearest square foot.

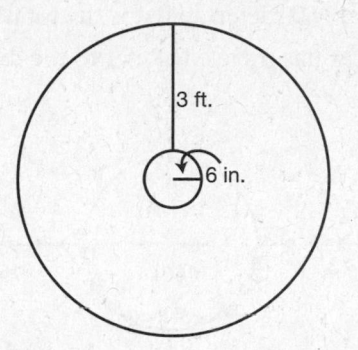

   a. 37.7 square feet

   b. 27 square feet

   c. 75 square feet

   d. 38 square feet

1005. A Norman window is to be installed in a new home. Using the dimensions marked on the illustration, find the area of the window to the nearest tenth of an inch. (π = 3.14)

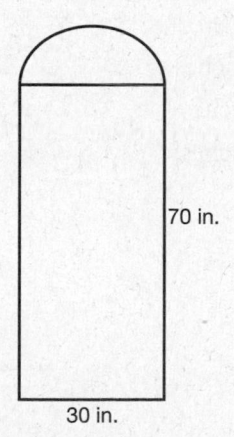

   a. 2,453.3 square inches

   b. 2,806.5 square inches

   c. 147.1 square inches

   d. 2,123.6 square inches

1006. A car is initially 200 meters due west of a roundabout (traffic circle). If the car travels to the roundabout, continues halfway around the circle, exits due east, then travels an additional 160 meters, what is the total distance the car has traveled? Refer to the diagram.

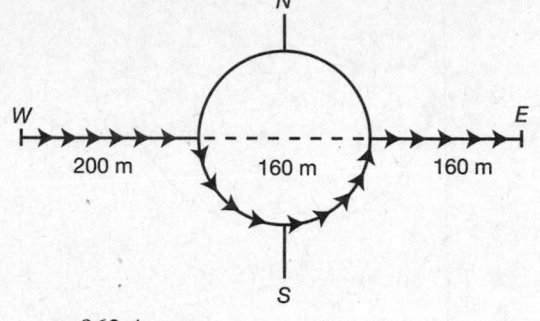

a. 862.4 meters
b. 611.2 meters
c. 502.4 meters
d. 451.2 meters

1007. If Gretta's bicycle has a 25-inch-diameter wheel, how far will she travel in two turns of the wheel? ($\pi = 3.14$)
a. 491 inches
b. 78.5 inches
c. 100 inches
d. 157 inches

1008. The following figure represents the cross section of a pipe $\frac{1}{2}$ inch thick that has an inside diameter of 3 inches. Find the area of the shaded region in terms of π.

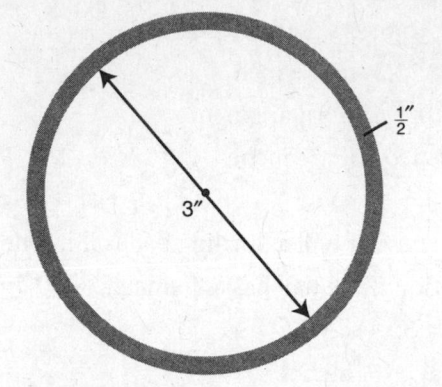

a. 8.75 π square inches
b. 3.25 π square inches
c. 7 π square inches
d. 1.75 π square inches

1009. Using the illustration provided, find the area of the shaded region in terms of π.

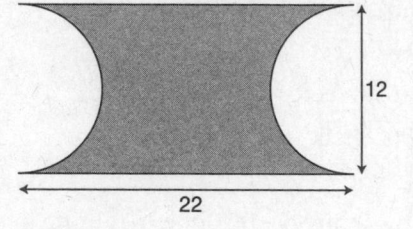

a. $264 - 18\pi$
b. $264 - 36\pi$
c. $264 - 12\pi$
d. $18\pi - 264$

1010. Find the area of the shaded portions, where $\overline{AB} = 6$ and $\overline{BC} = 10$. Give the answer in terms of π.

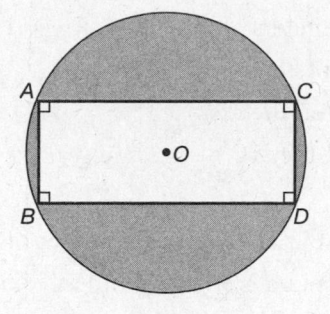

- **a.** $25\pi - 72$
- **b.** $25\pi - 48$
- **c.** $25\pi - 8$
- **d.** $100\pi - 48$

1011. On a piece of machinery, the centers of two pulleys are 3 feet apart, and the radius of each pulley is 6 inches. How long a belt (in feet) is needed to wrap around both pulleys?

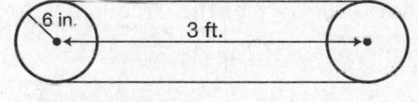

- **a.** $(6 + 0.5\pi)$ feet
- **b.** $(6 + 0.25\pi)$ feet
- **c.** $(6 + 12\pi)$ feet
- **d.** $(6 + \pi)$ feet

1012. Find the total area of the shaded regions, if the radius of each circle is 5 centimeters. Give the answer in terms of π.

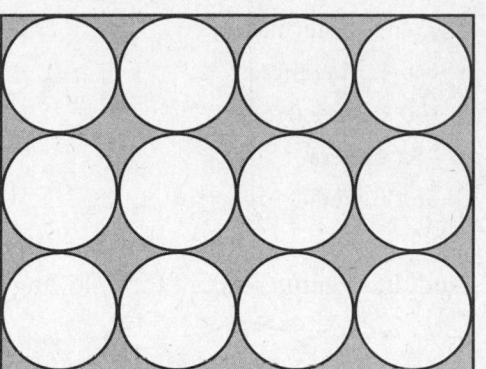

- **a.** $1,200 - 300\pi$ cm^2
- **b.** $300 - 300\pi$ cm^2
- **c.** $300\pi - 1,200$ cm^2
- **d.** $300\pi - 300$ cm^2

1013. If the radius of a circle is tripled, the circumference is
- **a.** multiplied by 3.
- **b.** multiplied by 6.
- **c.** multiplied by 9.
- **d.** multiplied by 12.

1014. Find the area of the shaded region. Give the answer in terms of π.

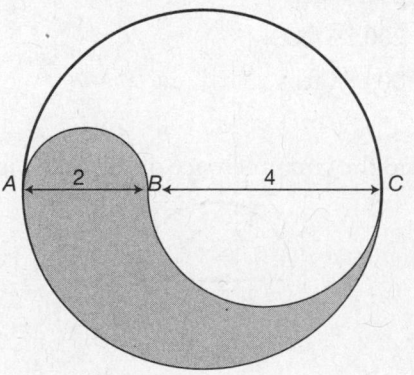

- **a.** 16.5π
- **b.** 30π
- **c.** 3π
- **d.** 7.5π

1015. A round tower with a 40-meter circumference is surrounded by a security fence that is 8 meters from the tower. How long is the security fence in terms of π?
   a. $(40 + 16\pi)$ meters
   b. $(40 + 8\pi)$ meters
   c. 48π meters
   d. 56π meters

1016. Find the circumference of the following circle.

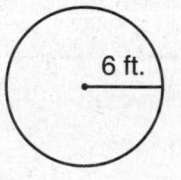

6 ft.

   a. 12π feet
   b. 6π feet
   c. 3π feet
   d. 36π feet
   e. 48π feet

1017. The diameter of a circular swimming pool is 16 feet. What is the circumference of the pool?
   a. 50.24 feet
   b. 100.48 feet
   c. 25.12 feet
   d. 200.96 feet
   e. 803.84 feet

1018. Find the circumference of the following circle.

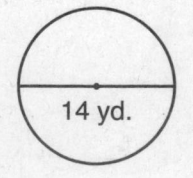

14 yd.

   a. 14π yards
   b. 7π yards
   c. 49π yards
   d. 196π yards
   e. 28π yards

1019. If the area of a circle is 42.25π square meters, what is the length of the radius of the circle?
   a. 10.5 meters
   b. 6.5 meters
   c. 7.5 meters
   d. 21.125 meters
   e. 2.54 meters

1020. At the Pizza Place, the diameter of a small round pizza is 9 inches and the diameter of a large round pizza is 15 inches. Approximately how much more pizza do you get in a large pizza than in a small pizza? Use $\pi = 3.14$.
   a. 113 square inches
   b. 144 square inches
   c. 6 square inches
   d. 19 square inches
   e. 452 square inches

1021. What is the area of the irregular figure shown here? Use $\pi = 3.14$.

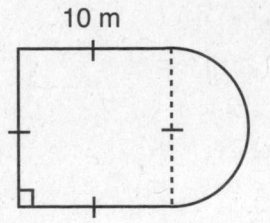

10 m

   a. 139.25 square meters
   b. 178.5 square meters
   c. 78.5 square meters
   d. 414 square meters
   e. 109.5 square meters

1022. What is the area of the shaded region?

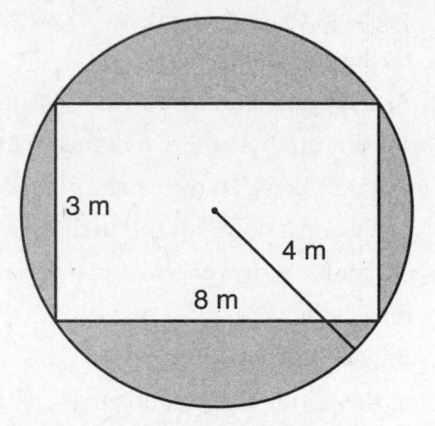

a. $(64\pi - 12)$ square meters

b. $(16\pi - 24)$ square meters

c. $(16\pi - 12)$ square meters

d. $(24 - 16\pi)$ square meters

e. $(24 - 64\pi)$ square meters

1023. $\angle ACB$ is a central angle. What is the measure of $\widehat{AB}$?

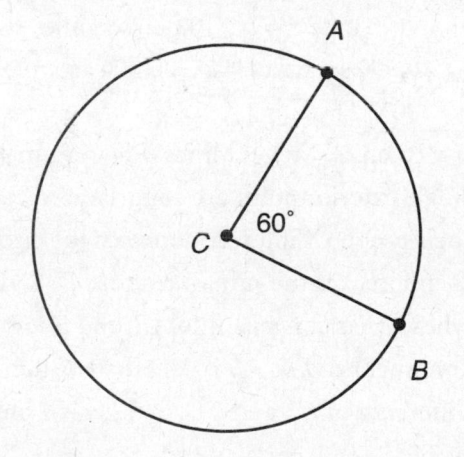

a. 15°

b. 30°

c. 60°

d. 90°

e. 120°

1024. In the following figure, $\overline{CD}$ is a diameter of the circle and $\overline{BA}$ is a radius. If the measure of $\widehat{AC}$ is 125°, what is the measure of $\widehat{AD}$?

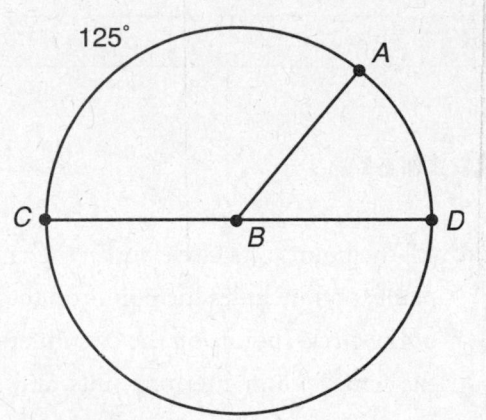

a. 125°

b. 62.5°

c. 27.5°

d. 55°

e. 180°

TERMS TO REVIEW

circle

center point

radius

concentric circles

central angle

chord

diameter

arc

minor arc

semicircle

major arc

tangent

secant

circumference

FORMULAS TO REVIEW

area of a circle = πr^2

circumference = $2\pi r$ or πd

diameter = $2r$

▶ Answers

959. **d.** All the points of a circle are on the same plane; that includes the points on the curve of the circle (points on the circumference), the center point, interior points, and exterior points (unless otherwise stated).

960. **a.** A circle is a set of points equidistant from a center point. Congruent circles have points that lie the same distance from two different center points. Consequently, the radii (the line segments that connect the center point to the points on a circle) of congruent circles are congruent. Choice **d** describes a chord.

961. **d.** Congruent circles have congruent radii; if their radii are congruent, then their diameters are also congruent. Choice **a** describes concentric circles, not congruent circles.

962. **c.** An exterior point is a point that lies outside a circle. Choice **a** represents a set of interior points. Choice **b** represents a set of points on ⊙P; and choice **d** is a mix of points in, on, and outside of ⊙P.

963. **b.** 12 inches is a foot, so •A lies on ⊙O. If the distance from •A to the center point measured less than the radius, then •A would rest inside ⊙O. If the distance from •A to the center point were greater than the radius, then •A would rest outside ⊙O.

964. **c.** A diameter is a special chord; it is a line segment that bridges a circle *and passes through the center point.*

965. **c.** As a tangent skims by a circle, it intercepts a point on that circle. A radius spans the distance between the center point of a circle and a point on the circle; like a tangent, a radius meets exactly one point on a circle.

966. **a.** Billy acts as the central fixed point of each of these four circles, and circles with a common center point are concentric.

967. **d.** A bull's-eye is a circle; the flight path of each arrow is a line. The first arrow is a tangent that also forms the leg of a right triangle. The path of the second arrow forms the hypotenuse. Use the Pythagorean theorem to find the distance between Kim and the center of the bull's-eye: 100 miles^2 + 100 miles^2 = c^2. 10,000 sq. miles + 10,000 sq. miles = c^2. 20,000 sq. miles = c^2. $100\sqrt{2} = c$.

968. $x = 16$ units. Tangent lines drawn from a single exterior point are congruent to each of their points of interception with the circle; therefore, x is the sum of lengths $\overline{AF}$ and $\overline{EF}$ where $\overline{AF}$ is congruent to $\overline{AB}$, and $\overline{EF}$ is congruent to $\overline{ED}$. $\overline{AB}$ is 4, and $\overline{DE}$ is the difference of $\overline{CE}$ and $\overline{CD}$, or 12; x is 4 plus 12, or 16.

969. The diameter of ⊙P = 0.5 inch. The diameter of ⊙O is half the diameter of ⊙M. The diameter of ⊙O is 1 inch. The diameter of ⊙P is half the diameter of ⊙O. The diameter of ⊙P is 0.5 inch.

970. Use the Pythagorean theorem to find the length of each circle's radius: $\odot A$: $7^2 + b^2 = \sqrt{74}^2$; $49 + b^2 = 74$; $b^2 = 25$; $b = 5$. Radius = 5. $\odot B$: $1^2 + \sqrt{24}^2 = c^2$; $1 + 24 = c^2$; $25 = c^2$; $5 = c$. Radius = $\frac{1}{2}(5) = 2.5$. $\odot C$: $10^2 + b^2 = 10\sqrt{2}^2$; $100 + b^2 = 200$; $b^2 = 100$; $b = 10$. Radius = $\frac{1}{2}(10) = 5$. $\odot D$: $2.5^2 + \sqrt{18.75}^2 = c^2$; $6.25 + 18.75 = c^2$; $25 = c^2$; $5 = c$. Radius = 5. Only $\odot B$ is not congruent to $\odot A$, $\odot C$, and $\odot D$.

971. The arcs in $\odot O$ are not congruent to each other. Parallel lines form congruent arcs. Two diameters form congruent arcs. Parallel tangent lines form congruent semicircles. Secants extending from a fixed exterior point form noncongruent arcs.

972. The radius = 15 units. Use the Pythagorean theorem to find the length of $\overline{DF}$ or a: $a^2 + 20^2 = 25^2$; $a^2 + 400 = 625$; $a^2 = 225$; $a = 15$.

973. The area of $\triangle DEF = 150$ square inches. The length of $\overline{ED}$ is the height of $\triangle DEF$. To find the area of $\triangle DEF$, plug the measures of the radius and the height into $\frac{1}{2}bh$: $\frac{1}{2}(15 \text{ in.} \times 20 \text{ in.}) = 150$ square inches.

974. $\overset{\frown}{DHG}$ is a major arc.

975. **b.** The perimeter of a circle is twice the radius times π: $(2 \times 57 \text{ inches})\pi$.

976. **d.** The area of a circle is the radius squared times π: $\pi(103 \text{ feet})^2$.

977. **c.** If the circumference of a circle is 64π centimeters, then the radius of that circle is half of 64, or 32 centimeters.

978. **c.** If the area of a square is 484 square feet, then the sides of the square must measure 22 feet each. The diameter of an inscribed circle has the same length as one side of the square. The maximum area of an inscribed circle is $\pi(11 \text{ feet})^2$, or 121π square feet.

979. **b.** The circumference of a circle is *pi* times twice the radius. 192 feet is twice the length of the radius; therefore half of 192 feet, or 96 feet, is the actual length of the radius.

980. **a.** The area of a circle is *pi* times the square of its radius. If 289 feet is the square of the circle's radius, then 17 feet is the length of its radius. Choice **c** is not the answer because 144.5 is half of 289, not the square root of 289.

981. **d.** The area of a circle is *pi* times the square of its radius. The area of the circle is $\pi(13 \text{ meters})^2$, or 169π square meters.

982. **a.** The length of arc BD is a quarter of the circumference of the circle, or 16π feet.

983. **b.** A quarter of 360° is 90°; it is a right angle.

984. **c.** This question is much simpler than it seems. The half circles that cap square $ABCD$ form the same area as the circular void in the center. Find the area of square $ABCD$, and that is your answer. 12 feet × 12 feet = 144 feet. Choices **a** and **d** are the same answer. Choice **b** is a negative area and is incorrect.

985. **b.** The radii of $\odot L$ and $\odot M$ are half the radius of $\odot K$. Their areas equal $\pi(7.5 \text{ feet})^2$, or 56.25π square feet each. The area of $\odot K$ is $\pi(15^2)$, or 225π square feet. Subtract the areas of circles L and M from the area of $\odot K$: 225π sq. ft. $- 112.5\pi$ sq. ft. $= 112.5\pi$ square feet.

986. **b.** Though $\odot M$ has half the radius of $\odot K$, it has a fourth of the area of $\odot K$: 56.25π square feet: 225π square feet or 1:4.

987. The radius $OA = 45$ feet. Use the Pythagorean theorem: $a^2 + 60^2 = 75^2$; $a^2 + 3,600 = 5,625$; $a^2 = 2,025$; $a = 45$ feet.

988. The area of $\odot O$ is $\pi(45 \text{ feet})^2$, or $2,025\pi$ square feet. If central angle AOC measures $60°$, then the area inside the central angle is $\frac{1}{6}$ the total area of $\odot O$, or 337.5π square feet. The area of $\triangle ABO$ is $\frac{1}{2}(45 \text{ feet} \times 60 \text{ feet})$, or $1,350$ square feet. Subtract the area inside the central angle from the area of the triangle: shaded area = $1,350$ square feet – 337.5π square feet.

989. The area of one semicircle is $A = \frac{1}{2}\pi(1.25 \text{ in}^2)$. $A \approx 0.78125\pi$ square inches. Multiply the area of one semicircle by 6: $6 \times 0.78125\pi$ square inches $\approx 4.6875\pi$ square inches.

990. The surface area of a cube is $6(4 \text{ inches}^2)$, or 96 square inches. Subtract the area of six semicircles from the surface area of the cube: remaining surface area = 96 square inches – 18.75π square inches.

991. The shaded area = 18.4π square inches. $\overset{\frown}{CD}$ is part of a concentric circle outside $\odot O$. Its area is $\pi(14 \text{ inches})^2$, or 196π square inches. A $45°$ slice of that area is one-eighth the total area, or 24.5π square inches. This is still not the answer. The area of $\odot O$ is $\pi(7 \text{ inches})^2$, or 49π square inches. Again, a $45°$ slice of that area is one-eighth the total area, or 6.1π square inches. Subtract the smaller wedge from the larger wedge, and the shaded area is 18.4π square inches.

992. $\overset{\frown}{AB}$ is 1.8π inches. The circumference of $\odot O$ is 14π inches. A $45°$ slice of that circumference is one-eighth the circumference, or 1.8π inches.

993. $\overset{\frown}{CD}$ is 3.5π inches. The circumference of concentric $\odot O$ is 28π inches. An eighth of that circumference is 3.5π inches.

994. $\overset{\frown}{AB}$ and $\overset{\frown}{CD}$ may have the same rotation ($45°$), but they do not have the same length.

995. The area of trapezoid $ABDE = 48$ square feet. Use the Pythagorean theorem to find $\overline{AG}$: $(4\sqrt{2} \text{ ft.})^2 = (4 \text{ ft.})^2 + b^2$; 32 sq. ft. = 16 sq. ft. + b^2; $b = 4$ ft. If $\overline{AG}$ equals 4 feet, then $\overline{AF}$ and $\overline{EF}$ equal 8 feet, and $\overline{AE}$ equals 16 feet. The area of a trapezoid is its height times half the sum of its bases: $\frac{1}{2}(8 \text{ ft.} + 16 \text{ ft.})(4 \text{ ft.}) = 12(4) = 48$ square feet.

996. The shaded area ≈ 14.88 square feet. The shaded area is the difference of $\triangle BCD$'s area and the area between chord BD and arc BD. The height of $\triangle BCD$ is 6 feet. Its area is $\frac{1}{2}(6 \text{ ft.} \times 8 \text{ ft.}) = 24$ sq. ft. The area of $\overset{\frown}{BD}$ is tricky. It is the portion of circle F contained within $\angle BFD$ minus the area of inscribed $\triangle BFD$. Central angle BFD is a right angle; it is a quarter of a circle's rotation and a quarter of its area. The circle's radius is $4\sqrt{2}$ feet. The area of circle F is $\pi(4\sqrt{2} \text{ ft.})^2$, or 32π square feet. A quarter of that area is 8π square feet. The area of $\triangle BFD$ is $\frac{1}{2}(4\sqrt{2} \text{ ft.} \times 4\sqrt{2} \text{ ft.}) = 16$ sq. ft. Subtract 16 square feet from 8π square feet; then subtract that answer from 24 square feet, and your answer is approximately 14.88 square feet.

997. **d.** The area of a circle is π times the radius squared.

998. **c.** The circumference or distance around a circle is π times the diameter.

999. **c.** The ceiling fan follows a circular pattern; therefore, area = πr^2. Area = $(3.14)(25)^2 = 1,962.5$ square inches. If you chose **a**, the incorrect formula you used was $\pi^2 r$. If you chose **d**, the incorrect formula you used was πd.

1000. a. The circumference of a circle is πd. Using the diameter of 10 inches, the circumference is equal to $(3.14)(10)$ or 31.4 inches. If you chose **b**, you found the area of a circle. If you chose **c**, you mistakenly used πr for circumference rather than $2\pi r$ or πd. If you chose **d**, you used the formula $2\pi d$.

1001. c. The circumference of a circle is πd. Since 10 feet represents the radius, the diameter is 20 feet. The diameter of a circle is twice the radius. Therefore, the circumference is $(3.14)(20)$ or 62.8 feet. If you chose **a**, you used πr rather than $2\pi r$ or πd. If you chose **b**, you found the area rather than circumference.

1002. b. A minute hand moves 180 degrees in 30 minutes. Use the following proportion: $\frac{30 \text{ minutes}}{180 \text{ degrees}} = \frac{25 \text{ minutes}}{x \text{ degrees}}$. Cross multiply: $30x = 4,500$. Solve for x; $x = 150$ degrees.

1003. b. To find the area of the matting, subtract the area of the print from the area of the frame. The area of the print is found using πr^2 or $(3.14)(7)^2$, which equals 153.86 sq. in. The area of the frame is length of side times length of side or $(18)(18)$, which equals 324 square inches. The difference, 324 square inches − 153.86 square inches, or 170.14 square inches, is the area of the matting. If you chose **c**, you mistakenly used the formula for the circumference of a circle, $2\pi r$, instead of the area of a circle, πr^2.

1004. d. To find the area of the skirt, find the area of the outer circle minus the area of the inner circle. The area of the outer circle is $\pi(3.5)^2$ or 38.465 sq. ft. The area of the inner circle is $\pi(0.5)^2$ or 0.785 sq. ft. The difference is $38.465 − 0.785$ square feet or 37.68 square feet. The answer, rounded to the nearest square foot, is 38 square feet. If you chose **a**, you rounded to the nearest tenth of a foot. If you chose **b**, you miscalculated the radius of the outer circle as being 3 feet instead of 3.5 feet.

1005. a. To find the area of the rectangular region, multiply length times width, or $(30)(70)$, which equals 2,100 sq. in. To find the area of the semicircle, multiply $\frac{1}{2}$ times πr^2 or $\frac{1}{2}\pi(15)^2$, which equals 353.25 sq. in. Add the two areas together: 2,100 plus 353.25 equals 2,453.3 square inches, rounded to the nearest tenth, for the area of the entire window. If you chose **b**, you included the area of a circle, not a semicircle.

1006. b. The question requires us to find the distance around the semicircle. This distance will then be added to the distance traveled before entering the roundabout, 200 meters, and the distance traveled after exiting the roundabout, 160 meters. According to the diagram, the diameter of the roundabout is 160 meters. The distance or circumference of half a circle is $\frac{1}{2}\pi d$, $\frac{1}{2}(3.14)(160)$ or 251.2 m. The total distance or sum is 200 m + 160 m + 251.2 m = 611.2 meters. If you chose **a**, you included the distance around the entire circle. If you chose **c**, you found the distance around the circle. If you chose **d**, you did not include the distance after exiting the circle, 160 m.

1007. d. To find how far Gretta will travel, find the circumference of the wheel multiplied by 2. The formula for the circumference of the wheel is πd. Since the diameter of the wheel is 25 inches, the circumference of the wheel is 25π. Multiply this by 2: $(2)(25\pi)$ or 50π. Finally, substitute 3.14 for π; $50(3.14) = 157$ inches, the distance the wheel covers in two turns. If you chose **a**, you used the formula for area of a circle rather than circumference. If you chose **b**, the distance traveled was one rotation, not two.

1008. d. To find the area of the shaded region, we must find the area of the outer circle minus the area of the inner circle. To find the area of the outer circle, we will use the formula area $= \pi r^2$. The outer circle has a diameter of 4 $(3 + \frac{1}{2} + \frac{1}{2})$ and a radius of 2; therefore, its area $= \pi 2^2$ or 4π. The inner circle has a radius of 1.5; therefore, its area $= \pi(1.5)^2$ or 2.25π. The difference, $4\pi - 2.25\pi$, or 1.75π square inches, is the area of the shaded region. If you chose **a**, you gave the outer circle a radius of 3 and the inner circle a radius of $\frac{1}{2}$. If you chose **b**, you gave the outer circle a radius of $\frac{7}{2}$ and the inner circle a radius of 3. If you chose **c**, you gave the outer circle a radius of 4 and the inner circle a radius of 3.

1009. b. The area of the shaded region is the area of a rectangle, 22 by 12, minus the area of a circle with a diameter of 12. The area of the rectangle is $(22)(12) = 264$. The area of a circle with a diameter of 12 and a radius of 6 is $\pi(6)^2 = 36\pi$. Therefore, the area of the shaded region is $264 - 36\pi$. If you chose **a**, the formula for area of a circle was incorrect, $\frac{1}{2}\pi r^2$. If you chose **c**, the formula for area of a circle was incorrect, πd. If you chose **d**, this was the reverse of choice **a**— area of the circle minus area of the rectangle.

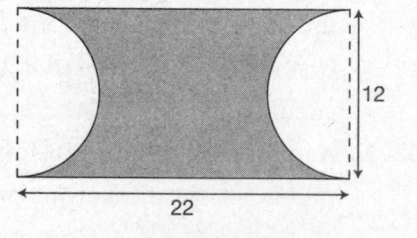

1010. b. To find the area of the shaded region, we must find the area of the circle minus the area of the rectangle. The formula for the area of a circle is πr^2. The radius is $\frac{1}{2}\overline{BC}$ or $\frac{1}{2}(10)$, which is 5. The area of the circle is $\pi(5^2)$ or 25π. The formula for the area of a rectangle is length × width. Using the fact that the rectangle is divided into two triangles with width of 6 and hypotenuse of 10, and using the Pythagorean theorem, we find the length: $a^2 + b^2 = c^2$; $a^2 + 6^2 = 10^2$; $a^2 + 36 = 100$; $a^2 = 64$; $a = 8$. The area of the rectangle is length × width or $6 \times 8 = 48$. Finally, to answer the question, the area of the shaded region is the area of the circle minus the area of the rectangle, or $25\pi - 48$. If you chose **a**, the error was in the use of the Pythagorean theorem, $6^2 + 10^2 = c^2$. If you chose **c**, the error was in finding the area of the rectangle. If you chose **d**, you used the wrong formula for area of a circle, πd^2.

1011. d. To solve for the length of the belt, begin with the distance between the centers of the pulleys, 3 feet, and multiply by 2: $(3)(2)$ or 6 feet. Second, you need to know that the circumference of two semicircles with the same radius is equivalent to the circumference of one circle. Therefore $C = \pi d$ or 12π inches. Since the units are in feet, and not inches, convert 12π inches to feet or 1π feet. Now, add these two values together, $(6 + \pi)$ feet, to determine the length of the belt around the pulleys. If you chose **a** or **b**, you used an incorrect formula for circumference of a circle. Recall that circumference $= \pi d$. If you chose **c**, you forgot to convert the unit from inches to feet.

1012. a. To find the total area of the shaded region, we must find the area of the rectangle minus the sum of the areas of all the circles. The area of the rectangle is base × height. Since the rectangle is four circles wide and three circles high, and each circle has a diameter of 10 centimeters (radius of 5 cm × 2), the rectangle is 40 cm wide and 30 cm high; $(40)(30) = 1{,}200$ cm^2. The area of one circle is πr^2 or $\pi(5)^2 = 25\pi$. Multiply this value times 12, since we are finding the area of 12 circles: $(12)(25)\pi = 300\pi$. The difference is $1{,}200 - 300\pi$ cm^2, the area of the shaded region. If you chose **b**, the area of the rectangle was incorrectly calculated as $(20)(15)$. If you chose **c**, you reversed the calculation to be the area of the circles minus the area of the rectangle. If you chose **d**, you reversed choice **b** as the area of the circles minus the area of the rectangle.

1013. a. The formula for finding the circumference of a circle is πd or $2\pi r$. If the radius is tripled, the diameter is also tripled. The new circumference is $\pi 3d$ or $\pi 6r$. Compare this expression to the original formula; with a factor of 3, the circumference is multiplied by 3.

1014. **c.** To find the area of the shaded region, we must find $\frac{1}{2}$ the area of the circle with diameter AC, minus $\frac{1}{2}$ the area of the circle with diameter BC, plus $\frac{1}{2}$ the area of the circle with diameter AB. To find $\frac{1}{2}$ the area of the circle with diameter AC, we use the formula area $= \frac{1}{2}\pi r^2$. Since the diameter is 6, the radius is 3; therefore, the area is $\frac{1}{2}\pi 3^2$ or 4.5π. To find $\frac{1}{2}$ the area of the circle with diameter BC, we again use the formula area $= \frac{1}{2}\pi r^2$. Since the diameter is 4, the radius is 2; therefore the area is $\frac{1}{2}\pi 2^2$ or 2π. To find $\frac{1}{2}$ the area of the circle with diameter AB, we again use the formula area $= \frac{1}{2}\pi r^2$. Since the diameter is 2, the radius is 1; therefore the area is $\frac{1}{2}\pi$. Finally, $4.5\pi - 2\pi + 0.5\pi = 3\pi$, the area of the shaded region. If you chose **a** or **b**, in the calculations you mistakenly used πd^2 as the area formula rather than πr^2.

1015. **a.** This problem has three parts. First, we must find the diameter of the existing tower. Second, we will increase the diameter by 16 meters for the purpose of the fence. Finally, we will find the circumference using this new diameter; this will be the length of the fence. The formula for circumference of a circle is πd. This formula, along with the fact that the tower has a circumference of 40 meters, gives us the following formula: $40 = \pi d$. To solve for d, the diameter, divide both sides by π. $D = \frac{40}{\pi}$, the diameter of the existing tower. Now increase the diameter by 16 meters, $\frac{40}{\pi} + 16$, to determine the diameter of the fenced-in section. Finally, use this value for d in the equation πd or $\pi(\frac{40}{\pi} + 16)$ meters. Simplify by distributing π through the expression: $(40 + 16\pi)$ meters. This is the length of the security fence. If you chose **b**, you added 8 to the circumference of the tower rather than 16. If you chose **c**, you merely added 8 to the circumference of the tower.

1016. **a.** The formula for the circumference of a circle is $C = 2\pi r$ or $C = \pi d$. The radius (halfway across the circle) is given. Double the radius to 12 feet: $C = \pi(12)$, which is equivalent to 12π feet.

1017. **a.** The formula for circumference is $C = \pi d$ or $C = 2\pi r$. Use 3.14 for π; $C = (3.14)(16) = 50.24$ feet.

1018. **a.** The formula for circumference is $C = \pi d$ or $C = 2\pi r$. The diameter of the circle is 14 yards: $C = \pi(14)$, which is equivalent to 14π yards.

1019. b. Use the formula $A = \pi r^2$ where r is the radius of the circle. Since the area is given, substitute into the formula to find the radius: $42.25\pi = \pi r^2$. Divide each side of the equation by π to get $42.25 = r^2$, so $r = \sqrt{42.25} = 6.5$ meters.

1020. a. To find the difference, find the area of the small pizza and subtract it from the area of the large pizza. Be sure to use the radius of each pizza in the formula, which is half of the diameter: $A_{large} - A_{small} = \pi r^2 - \pi r^2 = (3.14)(7.5)^2 - (3.14)(4.5)^2$. Evaluate exponents: $(3.14)(56.25) - (3.14)(20.25)$. Multiply within each term: $176.625 - 63.585$. Subtract to get 113.04 sq. in. There are approximately 113 more square inches in the large pizza.

1021. a. The figure is made up of a square and a half-circle. The square has sides that are 10 meters long and the diameter of the semicircle is a side of the square. Therefore, the radius of the circle is 5 meters. To find the area of the figure, find the area of each part and add the areas together: $A_{square} + A_{\frac{1}{2} circle} = bh + \frac{1}{2}\pi r^2 = (10)(10) + \frac{1}{2}(3.14)(5)^2$. Simplify the expression: $100 + \frac{1}{2}(3.14)(25) = 100 + \frac{1}{2}(78.5) = 100 + 39.25$. Therefore the total area is 139.25 square meters.

1022. b. The area of the shaded region is the area of the inner figure subtracted from the area of the outer figure; $A_{shaded} = A_{outer} - A_{inner}$. In this figure, the outer region is a circle and the inner region is a rectangle: $A_{shaded} = A_{circle} - A_{rectangle} = \pi r^2 - bh$. Substitute into the formula; $\pi(4)^2 - (3)(8) = (16\pi - 24)$ square meters.

1023. c. A central angle of a circle is an angle whose vertex is the center of the circle and whose sides are radii of the circle. Since $\angle ACB$ is a central angle, the measure of the intercepted $\overset{\frown}{AB}$ is equal to the measure of the angle. Therefore, the measure of $\overset{\frown}{AB}$ is 60°.

1024. d. Since $\overline{CD}$ is a diameter of the circle, it divides the circle into two congruent parts. Thus, the measure of $\overset{\frown}{CAD}$ is one-half of 360°, or 180°. Since the measure of $\overset{\frown}{AC}$ is 125°, subtract 125 from 180 to find the measure of $\overset{\frown}{AD}$: $180 - 125 = 55°$.

CHAPTER

16▶ Volume and Surface Area

WHEN WORKING WITH THREE-DIMENSIONAL SOLIDS, we are often asked to find volume and surface area. **Volume** is the amount of space taken up by an object. Volume is measured in cubic units, such as cubic centimeters or cubic inches. **Surface area** is the total area of every face of a solid. Surface area is measured in square units.

▶ Volume of a Right Prism

The sides of a right prism perpendicularly meet the base. The base is the polygon that defines the shape of the solid. The volume of a right prism is equal to the area of its base multiplied by its height.

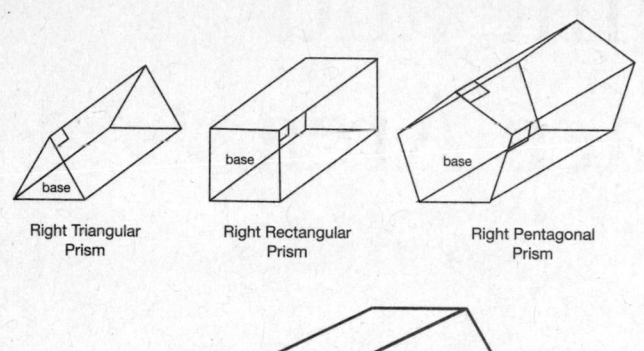

Right Triangular Prism   Right Rectangular Prism   Right Pentagonal Prism

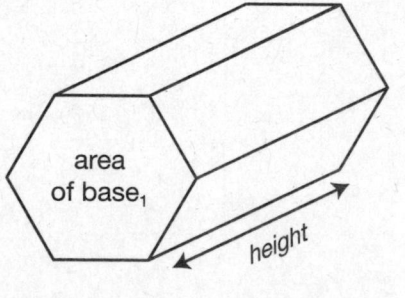

area of base₁

height

Example

The length of the base of a right triangular prism is 4 inches, and the height of the base is 3 inches. If the height of the prism is 12 inches, what is the volume of the triangular prism?

First, we find the area of the base of the prism. The area of a triangle is equal to half the product of its base and its height: $\frac{4\times3}{2} = \frac{12}{2} = 6$ sq. in. Now, multiply the area of the base by the height of the prism: $6 \times 12 = 72$ cu. in. The volume of the triangular prism is 72 cubic inches.

▶ Volume of a Right Rectangular Prism

A rectangular prism is a right prism that has a rectangle as its base. The volume of a rectangular prism is equal to the area of its base (length times width) times its height. A shoebox might be a right rectangular prism.

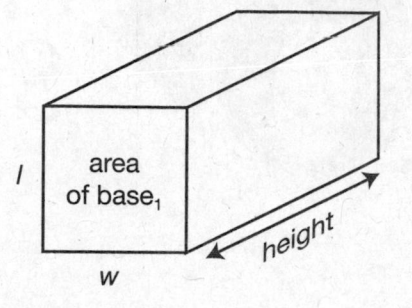

l   area of base₁   *w*   height

Example

What is the volume of a rectangular prism that has a length of 8 meters, a width of 6 meters, and a height of 7 meters?

Multiply the length by the width by the height: $8 \times 6 \times 7 = 336$ square meters.

► Volume of a Right Cube

A right cube is a type of rectangular prism—it's a rectangular prism whose base is a square and whose height has the same measure, which means that the measures of the length, width, and height of the prism are identical. The volume of a cube is equal to the length multiplied by the width multiplied by the height, or the cube of any one of those measures: $V = lwh$, or $V = s^3$. Numbered dice are cubes.

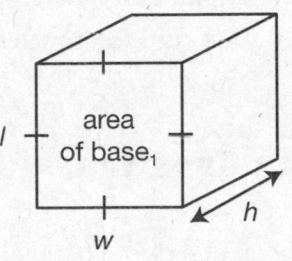

Example

If the volume of a cube is 343 square feet, what is the length of the cube?

Since the volume of a cube is equal to the length of any one of its sides cubed, we can take the cube root of the volume to find the length of one side of the cube. The cube root of 343, $\sqrt[3]{343}$, is 7. The length of the cube is 7 feet.

► Volume of a Pyramid

The volume of a pyramid is equal to one-third the area of its base multiplied by its height. The following diagram shows how three rectangular pyramids form a rectangular prism, which is why the volume of a pyramid is equal to one-third the volume of a rectangular prism that has the same base and height measurements.

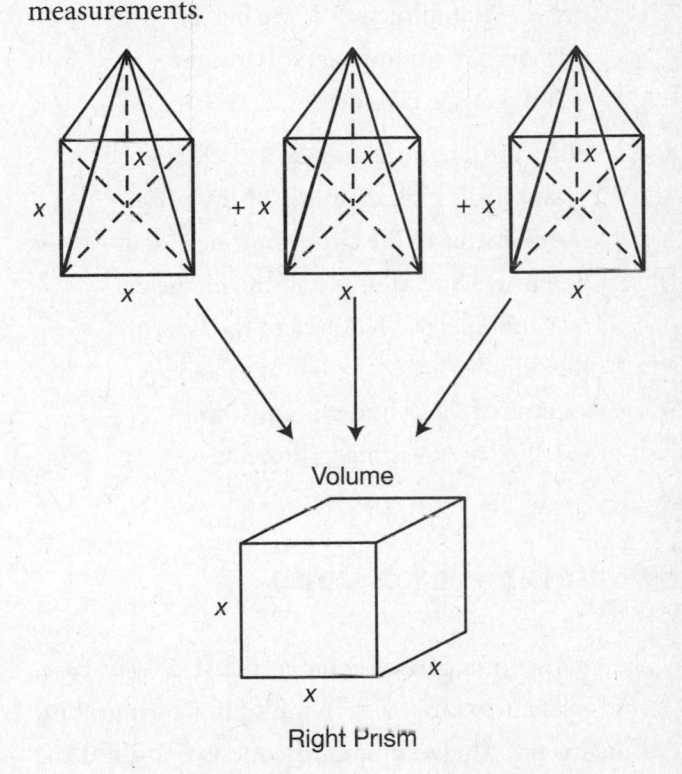

Example

A pyramid has a base area of 24 square feet and a height of 10 feet. What is the volume of the pyramid?

Multiply the area of the base by the height of the pyramid and divide by 3: $\frac{24 \times 10}{3} = \frac{240}{3} = 80$ cubic feet.

▶ Volume of a Cylinder

The volume of a cylinder is equal to the area of its base, which is a circle (area = πr^2), multiplied by its height. A roll of paper towels might be a cylinder.

Example

The circumference of the base of a cylinder is 16π millimeters. If the height of the cylinder is 6 millimeters, what is the volume of the cylinder?

Since the circumference of a circle is π multiplied by the diameter of the circle, the diameter of the circle must be 16 millimeters. The radius is half the diameter, or 8 millimeters. The area of the base of the cylinder is $\pi(8)^2 = 64\pi$ mm^2, so the volume of the cylinder is equal to $(64\pi)(6) = 384$ cubic millimeters.

▶ Volume of a Cone

The volume of a cone is equal to the area of its base, which is a circle (area = πr^2), multiplied by one-third of its height. The volume of a cone is one-third the volume of a cylinder that has the same base and height. A party hat might be a cone.

Example

The radius of the base of a cone is 14 inches and the height of the cone is 9 inches. What is the volume of the cone?

The base of a cone is a circle, which means that the area of the base is $\pi(14)^2 = 196\pi$ sq. in. One-third of the height of the cone is $\frac{9}{3} = 3$, so the volume of the cone is $(196\pi)(3) = 588\pi$ cubic inches.

▶ Volume of a Sphere

The volume of a sphere is equal to $\frac{4}{3}(\pi r^3)$. A basketball is a sphere.

Example

What is the volume of a ball that has a radius of 9 centimeters?

Plug the radius of the ball into the formula for the volume of a sphere: $\frac{4}{3}[\pi (9)^3]$ $= \frac{4}{3}(729\pi) = 972\pi$ cubic centimeters.

▶ Practice Questions

1025. Which choice describes a figure that has a third of the volume of the figure shown?

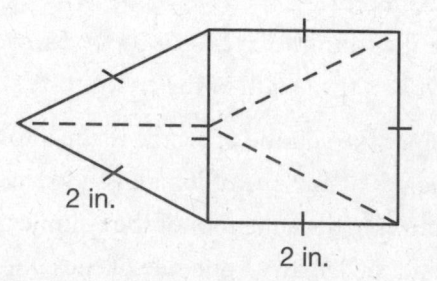

2 in.

2 in.

a. a right triangular prism with base sides that measure 2 inches and a height that measures 2 inches.
b. a cube with base sides that measure 2 inches and a height that measures 2 inches.
c. a triangular pyramid with base sides that measure 2 inches and a height that measures 2 inches.
d. a square pyramid with base sides that measure 2 inches and a height that measures 2 inches.

1026. Which figure has a third of the volume of a 3-inch cube?

a.
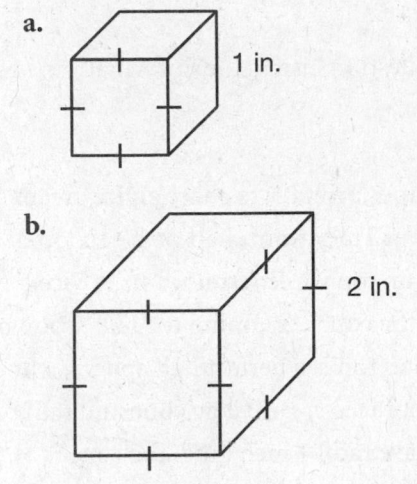
1 in.

b.

2 in.

c.
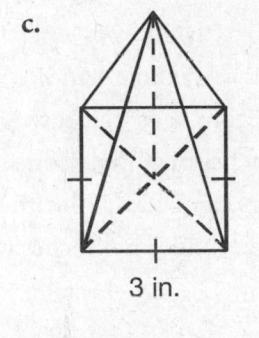
3 in.

d.
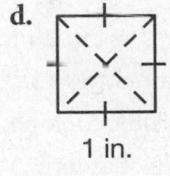
1 in.

Find the volume of each solid.

1027. Find the volume of a pyramid with four congruent base sides. The length of each base side and the prism's height each measure 2.4 feet.

1028. Find the volume of a pyramid with an eight-sided base that measures 330 square inches and a height that measures 10 inches.

Find each unknown element using the information provided.

1029. Find the height of a right rectangular prism with a volume of 295.2 cubic inches and a base area that measures 72 square inches.

1030. Find the measure of a triangular pyramid's base side if its volume measures $72\sqrt{3}$ cubic meters and its height measures 6 meters. The base of the pyramid forms an equilateral triangle.

Use the solid figure to answer question 1031.

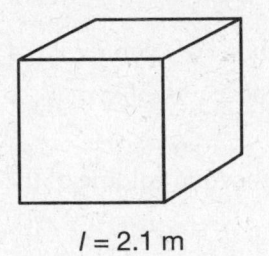

$l = 2.1$ m

1031. What is the volume?

Use the figure to answer questions 1032 and 1033.

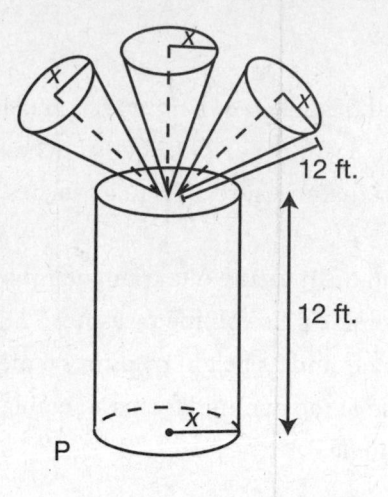

P

Volume of cylinder P = 432π cu. ft.

1032. If the volume of cylinder P is 432π cubic feet, what is the length of x?

1033. What is the total volume of the solid?

Use the following figure to answer questions 1034 and 1035.

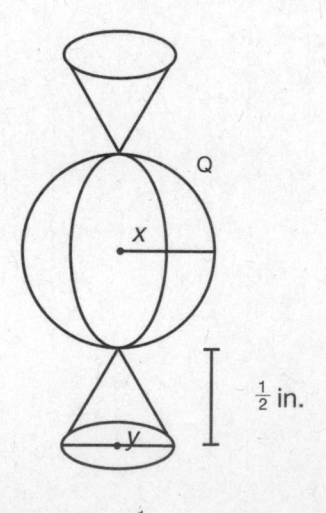

Volume Q = $\frac{1}{6}$ π cu. in.

1034. If the volume of a candy wrapper Q is $\frac{1}{6}$π cubic inches, what is the length of x?

1035. If the conical ends of candy wrapper Q have $\frac{1}{96}$π cubic inch volume each, what is the length of y?

Solve each question using the information in each word problem.

1036. Tracy and Jarret try to share an ice cream cone, but Tracy wants half of the scoop of ice cream on top while Jarret wants the ice cream inside the cone. Assuming the half scoop of ice cream on top is a perfect half sphere, who will have more ice cream? The cone and scoop both have radii 1 inch long; the cone is 3 inches high.

1037. Dillon fills the cylindrical coffee grind containers each day from bags that each contain 32π cubic inches of coffee. How many cylindrical containers can Dillon fill with two bags of coffee if each cylinder is 4 inches wide and 4 inches high?

1038. Before dinner, Jen measures the circumference and length of her roast. It measures 12π inches around and 4 inches long. After cooking, the roast is half its volume but just as long. What is the new circumference of the roast?

1039. Mike owns many compact discs (CDs), which he has to organize. If his CD holder is 5 inches wide by 4.5 inches high by 10 inches long and his CDs measure 4 inches wide by an eighth of an inch thick, how many CDs fit back-to-back in Mike's CD case?

1040. Munine is trying to carry her new 24-inch-tall cylindrical speakers through her front door. Unfortunately, they do not fit, either upright or sideways, through the width of the doorway. If each speaker is 2,400π cubic inches, how narrow is her doorway?

1041. Tory knows the space within a local cathedral dome that is a half sphere is 13,122π cubic feet. Using her knowledge of geometry, what does Tory calculate the height of the dome to be?

1042. Joe carves a perfect 3-meter-wide sphere inside a right prism. How much material did he remove? If the volume of the prism was 250 cubic meters, how much material remains?

1043. Theoretically, how many spherical candies should fit into a cylindrical jar if the diameter of each candy is 0.5 inch, and the jar is 4.5 inches wide and 6 inches high?

1044. A sphere with a 2-foot radius rests inside a cube with 4.5-foot edges. What is the volume of the space between the sphere and the cube, assuming π ≈ 3.14?

Use Puppet Dan to answer questions 1045 through 1053.

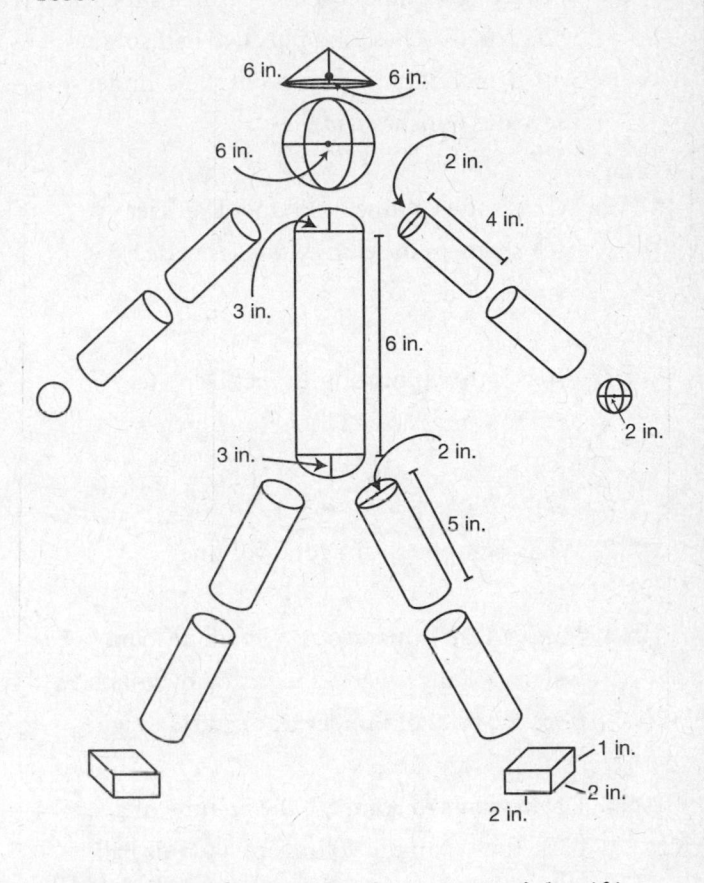

1045. What is the volume of Puppet Dan's hat if it measures 6 inches wide by 6 inches high?

1046. What is the volume of Puppet Dan's head if it measures 6 inches wide?

1047. What is the volume of Puppet Dan's arms if one segment measures 2 inches wide by 4 inches long?

1048. What is the volume of Puppet Dan's hands if each one measures 2 inches wide?

1049. What is the volume of Puppet Dan's body if it consists of a cylinder that measures 6 inches wide and 6 inches long plus two half spheres with 3-inch radii? Each end of the cylinder measures 6 inches wide.

1050. What is the volume of Puppet Dan's legs if each segment measures 2 inches wide by 5 inches long?

1051. What is the volume of Puppet Dan's feet if each foot measures 2 inches × 2 inches × 1 inch?

1052. What is puppet Dan's total volume?

1053. Puppet Dan is made out of foam. If foam weighs 3 ounces per cubic inch, how much does the total of Puppet Dan's parts weigh?

1054. Dawn wants to compare the volume of a basketball with the volume of a tennis ball. Which formula will she use?
- a. $V = \pi r^2 h$
- b. $V = \frac{4}{3}\pi r^3$
- c. $V = \frac{1}{3}\pi r^2 h$
- d. $V = s^3$

1055. Mimi is filling a tennis ball can with water. She wants to know the volume of the cylinder-shaped can. What formula will she use?
- a. $V = \pi r^2 h$
- b. $V = \frac{4}{3}\pi r^3$
- c. $V = \frac{1}{3}\pi r^2 h$
- d. $V = s^3$

1056. To find the volume of a cube that measures 3 centimeters by 3 centimeters by 3 centimeters, which formula would you use?
- a. $V = \pi r^2 h$
- b. $V = \frac{4}{3}\pi r^3$
- c. $V = \frac{1}{3}\pi r^2 h$
- d. $V = s^3$

1057. A circular pool is filling with water. Assuming the water will be 4 feet deep and the diameter is 20 feet, what is the volume of the water needed to fill the pool? ($\pi = 3.14$)
- a. 251.2 cubic feet
- b. 1,256 cubic feet
- c. 5,024 cubic feet
- d. 3,140 cubic feet

1058. What is the volume of a ball whose radius is 4 inches? Round to the nearest inch. ($\pi = 3.14$)
- a. 201 cubic inches
- b. 268 cubic inches
- c. 804 cubic inches
- d. 33 cubic inches

1059. A plum has a radius of 1.5 inches. Find the volume of the plum. ($\pi = 3.14$)
- a. 9.42 cubic inches
- b. 113.04 cubic inches
- c. 28.26 cubic inches
- d. 14.13 cubic inches

1060. Safe-deposit boxes are rented at the bank. The dimensions of a box are 22 inches by 5 inches by 5 inches. What is the volume of the box?
- a. 220 cubic inches
- b. 550 cubic inches
- c. 490 cubic inches
- d. 360 cubic inches

1061. An inground pool is filling with water. The shallow end is 3 feet deep and gradually slopes to the deepest end, which is 10 feet deep. The width of the pool is 15 feet and the length is 30 feet. What is the volume of the pool?

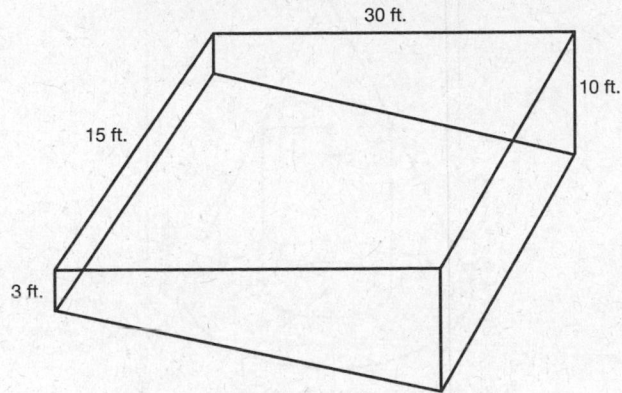

30 ft.

10 ft.

15 ft.

3 ft.

a. 1,575 cubic feet
b. 4,500 cubic feet
c. 2,925 cubic feet
d. 1,350 cubic feet

1062. A sphere has a volume of 288π cubic centimeters. Find its radius.
a. 9.5 centimeters
b. 7 centimeters
c. 14 centimeters
d. 6 centimeters

1063. Using the same cross section of pipe as in question 1008, answer the following question: If the pipe is 18 inches long, what is the volume of the half-inch-thick shaded region in terms of π?
a. 31.5π cubic inches
b. 126π cubic inches
c. 157.5 cubic inches
d. 58.5 cubic inches

1064. A sand pile is shaped like a cone as illustrated here. How many cubic yards of sand are in the pile? Round to the nearest tenth. ($\pi = 3.14$)

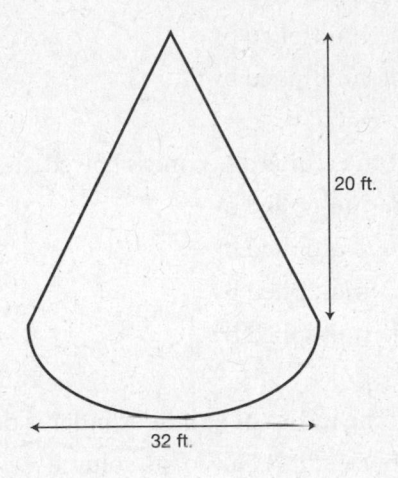

20 ft.

32 ft.

a. 5,358.9 cubic yards
b. 595.4 cubic yards
c. 198.5 cubic yards
d. 793.9 cubic yards

1065. If the lengths of all sides of a box are doubled, how much is the volume increased?
a. 2 times
b. 4 times
c. 6 times
d. 8 times

1066. If the diameter of a sphere is tripled, the volume is
a. multiplied by 3.
b. multiplied by 27.
c. multiplied by 9.
d. multiplied by 6.

1067. If the radius of a cone is doubled, the volume is
  a. multiplied by 2.
  b. multiplied by 4.
  c. multiplied by 6.
  d. multiplied by 8.

1068. If the radius of a cone is halved, the volume is
  a. multiplied by $\frac{1}{4}$.
  b. multiplied by $\frac{1}{2}$.
  c. multiplied by $\frac{1}{8}$.
  d. multiplied by $\frac{1}{16}$.

1069. If the radius of a right cylinder is doubled and the height is halved, its volume
  a. remains the same.
  b. is multiplied by 2.
  c. is multiplied by 4.
  d. is multiplied by $\frac{1}{2}$.

1070. If the radius of a right cylinder is doubled and the height is tripled, its volume is
  a. multiplied by 12.
  b. multiplied by 2.
  c. multiplied by 6.
  d. multiplied by 3.

1071. A solid is formed by cutting the top off of a cone with a slice parallel to the base, and then cutting a cylindrical hole into the resulting solid. Find the volume of the remaining hollowed solid in terms of π.

  a. 834π cubic centimeters
  b. $2,880\pi$ cubic centimeters
  c. 891π cubic centimeters
  d. $1,326\pi$ cubic centimeters

1072. A rectangular container is 5 centimeters wide and 15 centimeters long, and contains water to a depth of 8 centimeters. An object is placed in the water and the water rises 2.3 centimeters. What is the volume of the object?
  a. 92 cubic centimeters
  b. 276 cubic centimeters
  c. 172.5 cubic centimeters
  d. 312.5 cubic centimeters

1073. A spherical holding tank whose diameter to the outer surface is 20 feet is constructed of 1-inch-thick steel. How many cubic feet of steel is needed to construct the holding tank? Round to the nearest integer value. ($\pi = 3.14$)

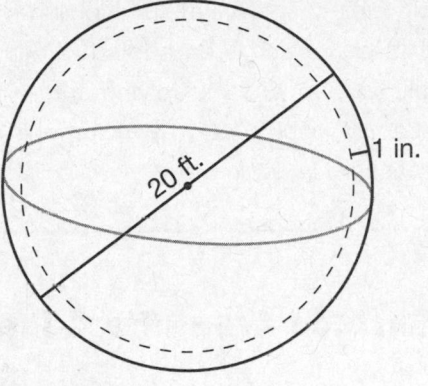

a. 78 cubic feet
b. 104 cubic feet
c. 26 cubic feet
d. 125 cubic feet

1074. How many cubic inches of lead are there in the pencil? Round to the nearest thousandth. ($\pi = 3.14$)

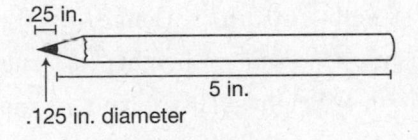

a. 0.061 cubic inch
b. 0.060 cubic inch
c. 0.062 cubic inch
d. 0.063 cubic inch

1075. A cylindrical hole with a diameter of 4 inches is cut through a cube. The edge of the cube is 5 inches. Find the volume of the hollowed solid in terms of π.

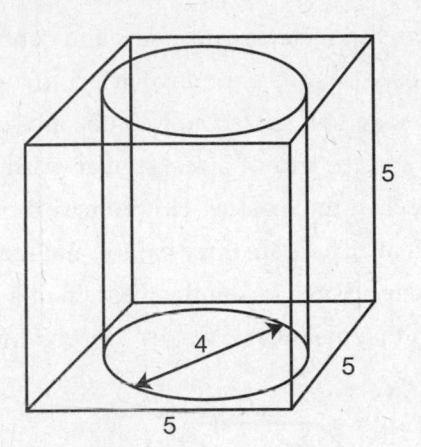

a. $125 - 80\pi$
b. $125 - 20\pi$
c. $80\pi - 125$
d. $20\pi - 125$

▶ Surface Area

We've reviewed how to find the areas of two-dimensional shapes such as triangles, rectangles, squares, and circles. Three-dimensional figures can have one, two, three, four, five, six, or more faces. To find the total surface area of a solid, we must find the area of each face, and then add the areas of those faces together.

► Surface Area of a Rectangular Prism

Not all prisms have six faces (a triangular prism has five faces), but rectangular prisms and cubes have six faces. All six faces of a rectangular prism are types of rectangles, and all six faces of a cube are squares. To find the surface area of a rectangular prism, find the area of each of the six faces. However, we can take a bit of a shortcut. There are three pairs of unique faces to a rectangular prism. Look at the diagram of a rectangular prism that measures 6 feet by 5 feet by 1 foot:

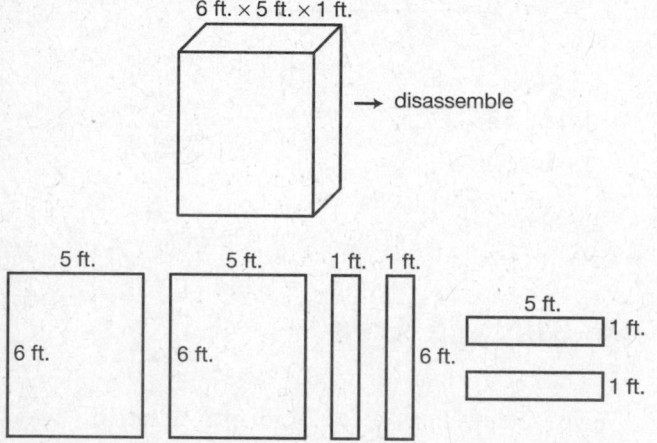

The surface area of a prism is the sum of the areas of its face areas, or $SA = $ (length $\times$ width) $+$ (length $\times$ height) $+$ (width $\times$ height) $+$ (width $\times$ height) $+$ (length $\times$ height) $+$ (length $\times$ width). This formula simplifies into $SA = 2(lw + wh + lh)$.

Example

If a rectangular prism has a length of 7 inches, a width of 3 inches, and a height of 2 inches, what is the surface area of the prism?

Plug the measurements into the formula for the surface area of a rectangular prism: $SA = 2(lw + wh + lh) =$ $2[(7)(3) + (3)(2) + (7)(2)] = 2(21 + 6 + 14)$ $= 2(41) = 82$ square inches.

► Surface Area of a Cube

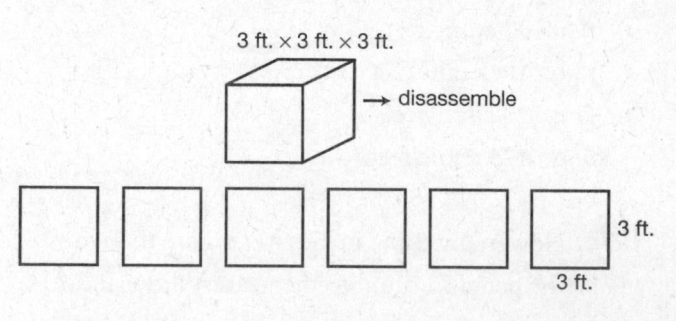

The surface area of a cube is the sum of its face areas, or $SA = $ (length $\times$ width) $+$ (length $\times$ width) $+$ (width $\times$ height) $+$ (width $\times$ height) $+$ (length $\times$ height) $+$ (length $\times$ height). This formula simplifies into: $SA = 6s^2$, where s is the length of one side.

Example

The width of one side of a cube is 1.3 centimeters. What is the surface area of the cube?

Use the formula for surface area of a cube: $SA = 6s^2$; $6(1.3)^2 = 6(1.69) = 10.14$ cm^2.

Example

A cube has a surface area of 150 square feet. What is the volume of the cube?

Work backward from the formula for surface area of a cube: $SA = 6s^2$; $150 = 6s^2$; $25 = s^2$; $s = 5$ feet. The formula for volume of a cube is s^3, so the volume of the cube is $(5)^3 = 125$ cubic feet.

▶ Surface Area of a Cylinder

A cylinder is made up of three faces: two identical circles and one curved surface that can be flattened into a rectangle. The area of one circle is πr^2, so the total area of the two circles is $2\pi r^2$. The length of the rectangle is equal to the circumference of the circle, $2\pi r$, and the width of the rectangle is the height of the cylinder, h. Therefore, the area of the rectangle is $2\pi rh$. The total surface area of a cylinder is equal to the sum of the areas of its three faces: $2\pi r^2 + 2\pi rh$.

Example

The area of one circular face of a cylinder is 49π square millimeters. If the height of the cylinder is 5 millimeters, what is the surface area of the cylinder?

If the area of one circular face is 49π square millimeters, then the area of both circular faces is $(2)(49) = 98\pi$ mm^2. Since the area of a circle is πr^2, and the area of one circular face is 49π square millimeters, the radius of that circle (and the radius of the cylinder) must be equal to the square root of 49, which is 7. The area of the rectangular face of the cylinder is $2\pi rh$: $2\pi(7)(5) = 70\pi$ mm^2. The total surface area of the cylinder is 70π mm^2 + 98π mm^2 = 168π square millimeters.

▶ Surface Area of a Sphere

A sphere has one single, round face. The formula for the surface area of a sphere is $4\pi r^2$, which is four times the area of one circle that could be found by slicing the sphere through its center.

Example

If the radius of a sphere is 14 inches, what is the surface area of the sphere?

The surface area of a sphere is $4\pi r^2$, so $4\pi(14)^2 = 4\pi(196) = 784\pi$ square inches.

▶ Practice Questions

Choose the best answer.

1076. A rectangular prism has
   a. one set of congruent faces.
   b. two pairs of congruent faces.
   c. three pairs of congruent faces.
   d. four pairs of congruent faces.

1077. How many faces of a cube have equal areas?
   a. two
   b. three
   c. four
   d. six

Find the surface area.

1078. Mark plays a joke on Tom. He removes the bottom from a box of bookmarks. When Tom lifts the box, all the bookmarks fall out. What is the surface area of the empty box Tom is holding if the box measures 5.2 inches wide by 17.6 inches long and 3.7 inches deep?

1079. Crafty Tara decides to make each of her friends a light box. To let the light out, she removes a right triangle from each side of the box such that the area of each face of the box is the same. What is the remaining surface area of the box if each edge of the box measures 3.3 feet and the area of each triangle measures 6.2 square feet?

1080. Jimmy gives his father the measurements of a table he wants built. If this drawing represents that table, how much veneer does Jimmy's father need to buy in order to cover all the exterior surfaces of his son's table?

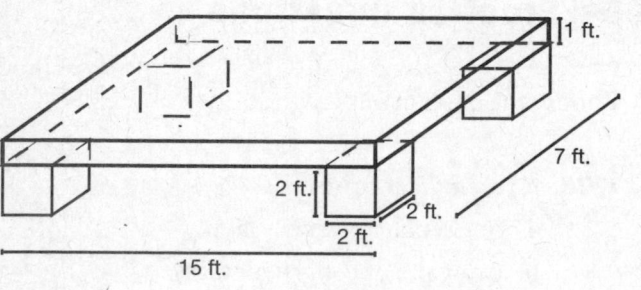

1081. The 25th Annual Go-Kart Race is just around the corner, and Dave still needs to build a platform for the winner. In honor of the tradition's longevity, Dave wants the platform to be special; therefore, he will cover all the exposed surfaces of his platform in red velvet. If the base step measures 15 feet by 7 feet by 1 foot, and each consecutive step is uniformly 1 foot from the edge of the last step, how much exposed surface area must Dave cover?

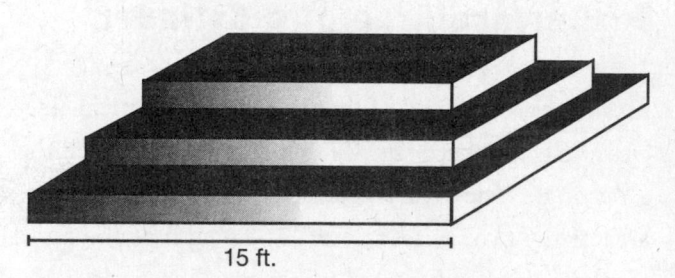

1082. Sarah cuts three identical blocks of wood and joins them end to end. How much exposed surface area remains?

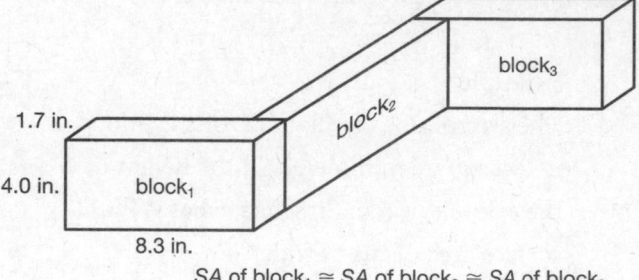

SA of block$_1$ ≅ SA of block$_2$ ≅ SA of block$_3$

Find each value of *x* using the figures and information provided.

1083. *Surface area* = 304 square feet

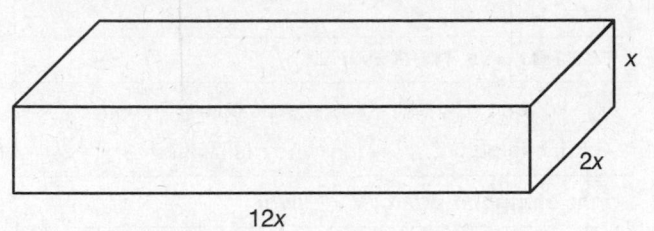

1084. *Surface area* = 936 square meters

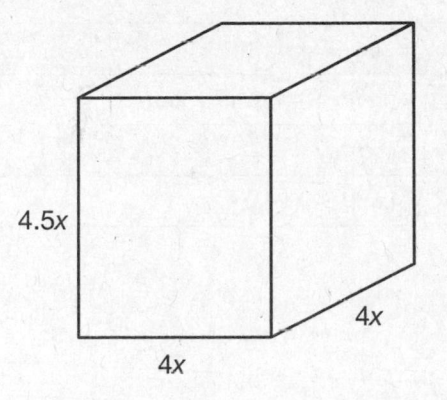

1085. *Surface area* = 864 square yards

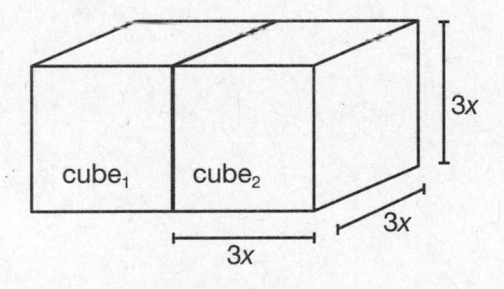

cube₁ ≅ cube₂

$\text{cube}_1 \cong \text{cube}_2$

1086. Using the same cylinder diagram as in questions 1032 and 1033 (page 332), what is the surface area of cylinder P?

1087. Using the same candy wrapper diagram as in questions 1034 and 1035 (page 332), what is the surface area of the candy inside the wrapper?

1088. In art class, Billy adheres 32 identical half spheres to canvas. What is their total surface area, not including the flat side adhered to the canvas, if the radius of one sphere is 8 centimeters?

1089. Keith wants to know the surface area of a basketball. Which formula will he use?
   a. $SA = 6s^2$
   b. $SA = 4\pi r^2$
   c. $SA = 2\pi r^2 + 2\pi rh$
   d. $SA = \pi r^2 + 2\pi rh$

1090. Al is painting a right cylinder storage tank. In order to purchase the correct amount of paint, he needs to know the total surface area to be painted. Which formula will he use if he does not paint the bottom of the tank?
   a. $SA = 2\pi r^2 + 2\pi rh$
   b. $SA = 4\pi r^2$
   c. $SA = \pi r^2 + 2\pi rh$
   d. $SA = 6s^2$

1091. The formula for the surface area of a sphere is $4\pi r^2$. What is the surface area of a ball with a diameter of 6 inches? Round to the nearest square inch. ($\pi = 3.14$)
   a. 452 square inches
   b. 113 square inches
   c. 38 square inches
   d. 28 square inches

1092. Barbara is wrapping a wedding gift that is contained within a rectangular box 20 inches by 18 inches by 4 inches. How much wrapping paper will she need?
   a. 512 square inches
   b. 1,440 square inches
   c. 1,024 square inches
   d. 92 square inches

1093. A publishing company is designing a book jacket for a soon-to-be-published textbook. Find the area of the book jacket, given that the front cover is 8 inches wide by 11 inches high, the spine is 1.5 inches by 11 inches, and the jacket will extend 2 inches inside the front and back covers.
   a. 236.5 square inches
   b. 192.5 square inches
   c. 188 square inches
   d. 232 square inches

1094. If the diameter of a sphere is doubled, the surface area is
   a. multiplied by 4.
   b. multiplied by 2.
   c. multiplied by 3.
   d. multiplied by 8.

1095. A surveyor is hired to measure the width of a river. Using the illustration provided, determine the width of the river.

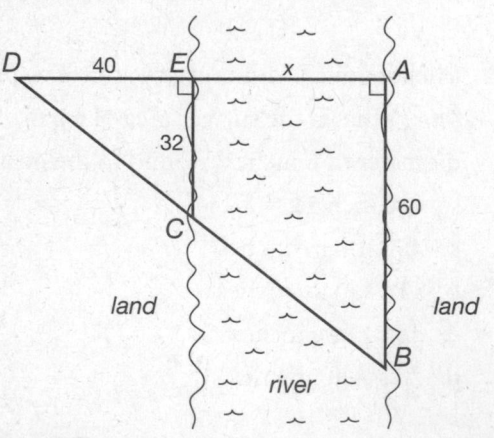

   a. 48 ft.
   b. 8 ft.
   c. 35 ft.
   d. 75 ft.

TERMS TO REVIEW

volume

surface area

FORMULAS TO REVIEW

(*V* is volume, *SA* is surface area, *l* is length, *w* is width, *h* is height, *s* is side, and *r* is radius)

right triangular prism: $V = \frac{1}{2}(lwh)$

rectangular prism: $V = lwh$, $SA = 2(lw + wh + lh)$

cube: $V = s^3$, $SA = 6s^2$

pyramid: $V = \frac{1}{3}(lwh)$

cylinder: $V = (\pi r^2)h$, $SA = 2\pi r^2 + 2\pi rh$

cone: $V = \frac{1}{3}(\pi r^2)h$

sphere: $V = \frac{4}{3}(\pi r^3)$, $SA = 4\pi r^2$

▶ Answers

1025. c. If their base measurements are congruent, a pyramid's volume is a third of a prism's volume. Choices **a** and **b** are eliminated because they are not pyramids. Choice **d** is also eliminated because its base polygon is not equivalent to the given base polygon, an equilateral triangle.

1026. c. Again, you are looking for a pyramid with the same base measurements as the given cube. Twenty-seven choice **a**'s can fit into the given cube; meanwhile, 81 choice **d**'s fit into that same cube. Only three choice **c**'s fit into the given cube; it has one-third the volume.

1027. The volume of the pyramid = 4.6 cubic feet. This is a square-based pyramid; its volume is a third of a cube's volume with the same base measurements, or $\frac{1}{3}$(area of its base × height). Plug its measurements into the formula: $\frac{1}{3}$(2.4 ft.)^2 × 2.4 ft. Volume of square pyramid = $\frac{1}{3}$(5.76 sq. ft.) × 2.4 ft. = $\frac{1}{3}$(13.824 cu. ft.) = 4.608 cubic feet.

1028. The volume of the pyramid = 1,100 cubic inches. Unlike the preceding example, this pyramid has an octagonal base. However, it is still a third of a right octagonal prism with the same base measurements, or $\frac{1}{3}$(area of its base × height). Conveniently, the area of the base has been given to you: area of octagonal base = 330 square inches. Volume of octagonal pyramid = $\frac{1}{3}$(330 sq. in.) × 10 in. = $\frac{1}{3}$(3,300 cu. in.) = 1,100 cubic inches.

1029. The prism's height = 4.1 inches. If the volume of a right rectangular prism measures 295.2 cubic inches, and the area of one of its two congruent bases measures 72 square inches, then its height measures 4.1 inches: 295.2 cu. in. = 72 sq. in. × h; h = 4.1 inches.

1030. The pyramid's base side = 12 meters. If the volume of a triangular pyramid is $72\sqrt{3}$ cubic meters, work backwards to find the area of its triangular base and then the length of a side of that base (remember, you are working with regular polygons, so the base will be an equilateral triangle). $72\sqrt{3}$ cubic meters = $\frac{1}{3}$ area of base × 6 meters; $72\sqrt{3}$ cubic meters = a × 2 meters; $36\sqrt{3}$ square meters = a. Divide both sides by $6\sqrt{3}$ meters: $36\sqrt{3}$ square meters = $\frac{1}{2}$ side of base × $6\sqrt{3}$ meters; 6 meters = $\frac{1}{2}b$; b = 12 meters.

1031. The cube's volume = 9.3 cubic meters. The volume of a cube is its length multiplied by its width multiplied by its height; $V = lwh$, or $V = s^3$: V = 2.1 meters × 2.1 meters × 2.1 meters; V = 9.261 cubic meters.

1032. x = 6 feet. The radius of cylinder P is represented by x; it is the only missing variable in the volume formula. Plug in and solve: 432π cu. ft. = (πx^2)12 ft.; 36 sq. ft. = x^2; x = 6 feet.

1033. The total volume of the solid = 864π cubic feet. This problem is easier than you might think. Each cone has exactly the same volume. The three cones together equal the volume of the cylinder. Multiply the volume of the cylinder by 2, and you have the combined volume of all three cones and the cylinder.

1034. $x = \frac{1}{2}$ inch. The volume of a sphere is $\frac{4}{3}\pi r^3$, where x is the value of r. Plug the variables in and solve: $\frac{1}{6}\pi$ cu. in. $= \frac{4}{3}\pi x^3$; $\frac{1}{8}$ cu. in. $= x^3$; $\frac{1}{2}$ inch $= x$.

1035. $y = \frac{1}{4}$ inch. The volume of a cone is $\frac{1}{3}\pi r^2 h$, where y is the value of r. Plug in the variables and solve: $\frac{1}{96}\pi$ cu. in. $= \frac{1}{3}\pi y^2 \frac{1}{2}$ in.; $\frac{1}{96}\pi$ cu. in. $= \frac{1}{6}\pi y^2$; $\frac{1}{16}\pi$ sq. in. $= y^2$; $\frac{1}{4}$ inch $= y$.

1036. The volume of a half sphere is $\frac{1}{2}(\frac{4}{3}\pi r^3)$. Tracy's half scoop is then $\frac{1}{2}(\frac{4}{3}\pi \times 1 \text{ inch}^3)$, or $\frac{2}{3}\pi$ cubic inches. The volume of a cone is $\frac{1}{3}\pi r^2 h$. The ice cream in the cone is $\frac{1}{3}\pi(1 \text{ inch}^2 \times 3 \text{ inches})$, or π cubic inches. Jarret has $\frac{1}{3}\pi$ cubic inches more ice cream than Tracy.

1037. The volume of each container is $\pi(2 \text{ in.})^2$ (4 in.), or 16π cubic inches. One bag fills two containers, and two bags fill four containers.

1038. The roast's new circumference $= 6\sqrt{2}\,\pi$ inches. This is a multistep problem. Find the radius of the roast: $2\pi r = 12\pi$ inches; $r = 6$ inches. The volume of the roast is $\pi(6 \text{ in.})^2$ (4 in.), or 144π cubic inches. After cooking, the roast is half is original volume, or 72π cubic inches. Its new radius is 72π cubic inches $= \pi r^2 \times 4$ inches; $r = 3\sqrt{2}$ inches. The new circumference of the roast is $2\pi r$, or $6\sqrt{2}\,\pi$ inches.

1039. Mike's CD case will hold 80 discs. This problem is not as hard as it might seem. A 4-inch-wide CD's diameter is 4 inches. It will fit snugly in a box with a 5-by-4.5-inch face. To find how many CDs will sit back-to-back in this container, divide the length of the container by the thickness of each CD: $\frac{10 \text{ in.}}{0.125 \text{ in. per CD}} = 80$ CDs:

1040. The radius of a single speaker is $\pi(r^2 \times 24$ inches$) = 2,400\pi$ cubic inches; $r^2 = 100$ square inches; $r = 10$ inches. The width of each speaker is twice the radius, or 20 inches. Munine's door is less than 20 inches wide!

1041. The dome is 27 feet high. Half the volume of a sphere is $\frac{1}{2}(\frac{4}{3}\pi r^3)$, or $\frac{2}{3}\pi r^3$. If the volume is $13,122\pi$ cubic feet, then the radius is 27 feet. The height of the dome is equal to the radius of the dome; therefore the height is also 27 feet.

1042. Joe removed the same amount of material as volume in the sphere, or $\frac{4}{3}\pi(1.5 \text{ meters})^3$, which simplifies to 4.5π cubic meters. The remaining volume is 250 cubic meters minus 4.5π cubic meters, or approximately 235.9 cubic meters.

1043. The jar holds 1,518 candies. The volume of each candy is $\frac{4}{3}\pi(0.25 \text{ inches})^3$, or 0.02π cubic inches. The volume of the jar is $\pi(2.25 \text{ inches}^2 \times 6)$ inches, or 30.375π cubic inches. Divide the volume of the jar by the volume of a candy $(\frac{30.375\pi \text{ cu. in.}}{0.02\pi \text{ cu. in.}})$, and 1,518 candies can theoretically fit into the given jar (not including the space between candies).

1044. First, find the volume of the cube, which is $(4.5 \text{ ft.})^3$, or approximately 91.1 cubic feet. The volume of the sphere within is only $\frac{4}{3}\pi(2 \text{ ft.})^3$, or approximately 33.5 cubic feet. Subtract the volume of the sphere from the volume of the cube. The remaining volume is approximately 57.6 cubic feet.

1045. Volume of a cone $= \frac{1}{3}\pi r^2 h$; $V = \frac{1}{3}\pi (3 \text{ in.})^2 (6$ in.$)$; $V = 18\pi$ cubic inches.

1046. Volume of a sphere $= \frac{4}{3}\pi r^3$; $V = \frac{4}{3}\pi (3 \text{ in.})^3$; $V = 36\pi$ cubic inches.

1047. Volume of a cylinder $= \pi r^2 h$; $V = \pi (1 \text{ in.}^2 \times 4 \text{ in.})$; $V = 4\pi$ cubic inches. There are four arm segments, so four times the volume $= 16\pi$ cubic inches.

1048. Volume of a sphere $= \frac{4}{3}\pi r^3$; $V = \frac{4}{3}\pi (1 \text{ in.}^3)$; $V = \frac{4}{3}\pi$ cubic inches. There are two hands, so two times the volume $= \frac{8}{3}\pi$ cubic inches.

1049. The body is the sum of two congruent half spheres, which total one sphere, and a cylinder. Volume of sphere $= \frac{4}{3}\pi r^3$; $V = \frac{4}{3}\pi (3 \text{ in.})^3$; $V = 36\pi$ cubic inches. Volume of cylinder $= \pi r^2 h$; $V = \pi (3 \text{ in.})^2 (6 \text{ in.})$; $V = 54\pi$ cubic inches. Total volume $= 90\pi$ cubic inches.

1050. Volume of a cylinder $= \pi r^2 h$; $V = \pi (1 \text{ in.}^2 \times 5 \text{ in.})$; $V = 5\pi$ cubic inches. There are four leg segments, so four times the volume $= 20\pi$ cubic inches.

1051. Each foot is a rectangular prism. Volume of a prism $=$ length $\times$ width $\times$ height; $V = 2 \text{ in.} \times 2 \text{ in.} \times 1 \text{ in}$; $V = 4$ cubic inches. There are two feet, so two times the volume $= 8$ cubic inches.

1052. The sum of the volumes of Puppet Dan's parts equals the total volume: 18π cubic inches $+ 36\pi$ cubic inches $+ 16\pi$ cubic inches $+ \frac{8}{3}\pi$ cubic inches $+ 90\pi$ cubic inches $+ 20\pi$ cubic inches $\approx 182.6\pi$ cubic inches $+ 8$ cubic inches. If $\pi \approx 3.14$, then $V \approx 581.36$ cubic inches.

1053. Multiply: $\frac{3 \text{ oz.}}{1 \text{ cu. in.}} \times 581.36$ cubic inches $= 1,744.08$ ounces or 109 pounds. Puppet Dan is surprisingly light for all his volume!

1054. b. The volume of a sphere is $\frac{4}{3}$ times π times the radius cubed.

1055. a. The volume of a cylinder is π times the radius squared, times the height of the cylinder.

1056. d. The volume of a cube is the length of the side cubed, or the length of the side times the length of the side times the length of the side.

1057. b. The volume of a cylinder is $\pi r^2 h$. Using a depth of 4 feet and radius of 10 feet, the volume of the pool is $(3.14)(10)^2(4)$ or 1,256 cubic feet. If you chose **a**, you used πdh instead of $\pi r^2 h$. If you chose **c**, you used the diameter squared instead of the radius squared.

1058. b. The volume of a sphere is $\frac{4}{3}\pi r^3$. Using the dimensions given, volume $= \frac{4}{3}(3.14)(4)^3$ or 267.9. Rounding this answer to the nearest inch is 268 cubic inches. If you chose **a**, you found the surface area rather than volume. If you chose **c**, you miscalculated surface area by using the diameter.

1059. d. To find the volume of a sphere, use the formula volume $= \frac{4}{3}\pi r^3$. Volume $= \frac{4}{3}(3.14)(1.5)^3 = 14.13$ cubic inches. If you chose **a**, you squared the radius instead of cubing the radius. If you chose **b**, you cubed the diameter instead of the radius. If you chose **c**, you found the surface area of the sphere, not the volume.

1060. b. The volume of a rectangular solid is length times width times depth. Using the dimensions in the question, volume $= (22)(5)(5)$ or 550 cubic inches. If you chose **c**, you found the surface area of the box.

1061. c. The volume of a rectangular solid is length times width times height. First, calculate what the volume would be if the entire pool had a depth of 10 feet. The volume would be (10)(30)(15) or 4,500 cu. ft. Now subtract the area under the sloped plane, a triangular solid. The volume of the region is $\frac{1}{2}$(base)(height)(depth) or $\frac{1}{2}$(7)(30)(15) or 1,575 cu. ft. Subtract: 4,500 cu. ft. minus 1,575 cu. ft. results in 2,925 cubic feet as the volume of the pool. If you chose **a**, this is the volume of the triangular solid under the sloped plane in the pool. If you chose **b**, you did not calculate the slope of the pool, but rather a pool that is consistently 10 feet deep.

1062. d. The volume of a sphere is found by using the formula $\frac{4}{3}\pi r^3$. Since the volume is 288π cm^3 and we are asked to find the radius, we will set up the following equation: $\frac{4}{3}\pi r^3 = 288\pi$. To solve for r, multiply both sides by 3: $4\pi r^3 = 864\pi$. Divide both sides by π: $4r^3 = 864$. Divide both sides by 4: $r^3 = 216$. Take the cube root of both sides: $r = 6$ centimeters. If you chose **a**, the formula for volume of a sphere was incorrect; $\frac{1}{3}\pi r^3$ was used instead of $\frac{4}{3}\pi r^3$. If you chose **c**, near the end of calculations you mistakenly took the square root of 216 rather than the cube root.

1063. a. To find the volume of the pipe with a known cross section and length of 18 inches, simply multiply the area of the cross section times the length of the pipe. The area of the cross section obtained from question 1008 was 1.75π square inches. The length is 18 inches. Therefore, the volume is 1.75 square inches times 18 inches, or 31.5π cubic inches. If you chose **b**, you multiplied choice **c** from question 1008 by 18. If you chose **c**, you multiplied choice **a** from question 1008 by 18. If you chose **d**, you multiplied choice **b** from question 1008 by 18.

1064. c. To find how many cubic yards of sand are in the pile, we must find the volume of the pile in cubic feet and convert the answer to cubic yards. The formula for volume of a cone is $V = \frac{1}{3}$(height)(area of the base). The area of the base is found by using the formula area $= \pi r^2$. The area of the base of the sand pile is $\pi(16)^2$ or 803.84 sq. ft. The height of the pile is 20 feet. The volume of the pile in cubic feet is (803.84)(20) or 5,358.93 cu. ft. To convert to cubic yards, divide 5,358.93 by 27, because 1 yard = 3 feet and 1 cu. yd. means 1 yd. $\times$ 1 yd. $\times$ 1 yd., which equals 3 ft. $\times$ 3 ft. $\times$ 3 ft. or 27 cu. ft. The answer is 198.5 cubic yards. If you chose **a**, you did not convert to cubic yards. If you chose **b**, you converted incorrectly by dividing 5,358.93 by 9 rather than 27. If you chose **d**, the area of the base formula was incorrect. The area of a circle does not equal πd^2.

1065. d. The volume of a box is found by multiplying length × length × length or $l \times l \times l = l^3$. If the length is doubled, the new volume is $(2l) \times (2l) \times (2l)$ or $8(l^3)$. When we compare the two expressions, we can see that the difference is a factor of 8. Therefore, the volume has been increased by a factor of 8.

1066. b. If the diameter of a sphere is tripled, the radius is also tripled. The formula for the volume of a sphere is $\frac{4}{3}\pi r^3$. If the radius is tripled, volume $= \frac{4}{3}\pi(3r)^3$, which equals $\frac{4}{3}\pi(27r^3)$ or $\frac{4}{3}(27)\pi r^3$. Compare this equation for volume with the original formula; with a factor of 27, the volume is now 27 times as great.

1067. b. The formula for the volume of a cone is $\frac{1}{3}\pi r^2 h$. If the radius is doubled, then volume $= \frac{1}{3}\pi(2r)^2 h$ or $\frac{1}{3}\pi 4r^2 h$. Compare this expression to the original formula; with a factor of 4, the volume is multiplied by 4.

1068. a. The formula for the volume of a cone is $\frac{1}{3}\pi r^2 h$. If the radius is halved, the new formula is $\frac{1}{3}\pi(\frac{1}{2}r)^2 h$ or $\frac{1}{3}\pi(\frac{1}{4})r^2 h$. Compare this expression to the original formula; with a factor of $\frac{1}{4}$, the volume is multiplied by $\frac{1}{4}$.

1069. b. The volume of a right cylinder is $\pi r^2 h$. If the radius is doubled and the height halved, the new volume is $\pi(2r)^2(\frac{1}{2}h)$ or $\pi 4r^2(\frac{1}{2}h)$ or $2\pi r^2 h$. Compare this expression to the original formula; with a factor of 2, the volume is multiplied by 2.

1070. a. The formula for finding the volume of a right cylinder is volume $= \pi r^2 h$. If the radius is doubled and the height is tripled, the formula has changed to $\pi(2r)^2(3h)$. Simplified, this is $\pi 4r^2 3h$ or $\pi 12r^2 h$. Compare this expression to the original formula; with a factor or 12, the volume is now multiplied by 12.

1071. a. To find the volume of the hollowed solid, we must find the volume of the original cone minus the volume of the smaller cone sliced from the original cone minus the volume of the cylindrical hole. The volume of the original cone is found by using the formula $V = \frac{1}{3}\pi r^2 h$. Using the values $r = 9$ and $h = 40$, substitute and simplify to find the volume $= \frac{1}{3}\pi(9)^2(40)$ or $1{,}080\pi$ cm^3. The volume of the smaller cone is found by using the formula $V = \frac{1}{3}\pi r^2 h$. Using the values $r = 3$ and $h = 19$, substitute and simplify to find the volume $= \frac{1}{3}\pi(3)^2(19)$ or 57π cm^3. The volume of the cylinder is found by using the formula $V = \pi r^2 h$. Using the values $r = 3$ and $h = 21$, substitute and simplify to find the volume $= \pi(3)^2(21)$ or 189π cm^3. Finally, calculate the volume of the hollowed solid: $1{,}080\pi - 57\pi - 189\pi$ or 834π cm^3. If you chose **b**, you used an incorrect formula for the volume of a cone, $V = \pi r^2 h$. If you chose **c**, you subtracted the volume of the cylinder from the volume of the large cone. If you chose **d**, you added the volumes of all three sections.

1072. c. To find the volume of the object, we must find the volume of the water that is displaced after the object is inserted. Since the container is 5 cm wide and 15 cm long, and the water rises 2.3 cm after the object is inserted, the volume of the displaced water can be found by multiplying length by width by depth: $(5)(15)(2.3) = 172.5$ cubic centimeters.

1073. b. To find the volume of steel needed, we must find the volume of the outer sphere minus the volume of the inner sphere. The formula for the volume of a sphere is $\frac{4}{3}\pi r^3$. The volume of the outer sphere is $\frac{4}{3}\pi(120)^3$. Here, the radius is 10 feet (half the diameter) multiplied by 12 (converted to inches), or 120 inches. The volume equals 7,234,560 cu. in. The volume of the inner sphere is $\frac{4}{3}\pi(119)^3$ or 7,055,199 cu. in. (This is rounded to the nearest integer value.) The difference between the volumes is 7,234,560 minus 7,055,199 or 179,361 cu. in. This answer is in cubic inches, and the question is asking for cubic feet. Since one cubic foot equals 1,728 cubic inches, we simply divide 179,361 by 1,728, which equals 104, rounded to the nearest integer value. As an alternative to changing units to inches, only to have to change them back into feet again, keep units in feet. The radius of the outer sphere is 10 feet, and the radius of the inner sphere is one inch less than 10 feet, which is $9\frac{11}{12}$ feet, or 9.917 feet. Use the formula for the volume of a sphere: $\frac{4}{3}\pi r^3$, and find the difference in the volumes. If you chose **a**, you used an incorrect formula for the volume of a sphere, $V = \pi r^3$. If you chose **c**, you also used an incorrect formula for the volume of a sphere, $V = \frac{1}{3}\pi r^3$. If you chose **d**, you found the correct answer in cubic inches; however, your conversion to cubic feet was incorrect.

1074. c. To solve this problem, we must find the volume of the sharpened tip and add this to the volume of the remaining lead that has a cylindrical shape. To find the volume of the sharpened point, we will use the formula for finding the volume of a cone, $\frac{1}{3}\pi r^2 h$. Using the values $r = 0.0625$ (half the diameter) and $h = 0.25$, the volume = $\frac{1}{3}\pi(0.0625)^2(0.25)$ or 0.002 cu. in. To find the volume of the remaining lead, we will use the formula for finding the volume of a cylinder, $\pi r^2 h$. Using the values $r = 0.0625$ and $h = 5$, the volume = $\pi(.0625)^2(5)$ or 0.0613. The sum is $0.001 + 0.0613$ or 0.0623 cubic inches, the volume of the lead. If you chose **a**, this is the volume of the lead without the sharpened tip. If you chose **b**, you subtracted the volumes calculated.

1075. b. To find the volume of the hollowed solid, we must find the volume of the cube minus the volume of the cylinder. The volume of the cube, found by multiplying length × width × height or (5)(5)(5), equals 125. The value of the cylinder is found by using the formula $\pi r^2 h$. In this question, the radius of the cylinder is 2 and the height is 5. Therefore, the volume is $\pi(2)^2(5)$ or 20π. The volume of the hollowed solid is $125 - 20\pi$. If you chose **a**, you made an error in the formula of a cylinder, using $\pi d^2 h$ rather than $\pi r^2 h$. If you chose **c**, this was choice **a** reversed: the volume of the cylinder minus the volume of the cube. If you chose **d**, you found the reverse of choice **b**.

1076. c. When the faces of a rectangular prism are laid side by side, you always have three pairs of congruent faces. That means every face of the prism (and there are six faces) has one other face that shares its shape, size, and area.

1077. d. A cube, like a rectangular prism, has six faces. If you have a small box nearby, pick it up and count its faces. It has six. In fact, if it is a cube, it has six congruent faces.

1078. The box's surface area = 260.24 square inches. Begin by finding the whole surface area: surface area = $2(lw + wh + lh)$. SA = $2[17.6 \text{ in.}(5.2 \text{ in.}) + 5.2 \text{ in.}(3.7 \text{ in.}) + 17.6 \text{ in.}(3.7 \text{ in.})]$; SA = 2(91.52 sq. in. + 19.24 sq. in. + 65.12 sq. in.); SA = 2(175.88 sq. in.); SA = 351.76 square inches. From the total surface area, subtract the area of the missing face: Remaining SA = 351.76 sq. in. − 91.52 sq. in. Remaining SA = 260.24 square inches.

1079. The surface area of the box = 28.14 square feet. You could use the formula to determine the surface area of a rectangular prism to also determine the surface area of a cube, or you could simplify the equation to 6 times the square of the length of one side: SA = $6(3.3 \text{ ft.})^2$; SA = 6(10.89 sq. ft.); SA = 65.34 square feet. Tara removes six triangular pieces, one from each face of the cube. It is given that each triangular cutout removes 6.2 square feet from the total surface area. 6 × 6.2 sq. ft. = 37.2 square feet. To find the remaining surface area, subtract the area removed from the surface area: 65.34 sq. ft. − 37.2 sq. ft. = 28.14 square feet.

1080. The table's surface area = 318 square feet. These next few problems are tricky: Look at the diagram carefully. Notice that the top of each cubed leg is not an exposed surface area, nor is the space where it touches the large rectangular prism. Let's find these surface areas first. The top of each cubed leg equals the square of the length of the cube, 2 feet, = 4 square feet. There are four congruent cubes, thus four congruent faces: 4×4 sq. ft. = 16 sq. ft. It is reasonable to assume that where the cubes meet the rectangular prism, an equal amount of area from the prism is also not exposed. Total area concealed = 16 sq. ft + 16 sq. ft. = 32 sq. ft. Now, find the total surface area of the table's individual parts.

SA of one cube = $6(2 \text{ ft.})^2 = 6(4 \text{ sq. ft.}) =$ 24 sq. ft.

SA of four congruent cubes = 4×24 sq. ft. = 96 sq. ft.

SA of one rectangular prism = $2[15 \text{ ft.}(7 \text{ ft.}) + 7 \text{ ft.}(1 \text{ ft.}) + 15 \text{ ft.}(1 \text{ ft.})] =$ $2(105 \text{ sq. ft.} + 7 \text{ sq. ft.} + 15 \text{ sq. ft.}) =$ $2(127 \text{ sq. ft.}) = 254$ sq. ft.

Total *SA* = 96 sq. ft. + 254 sq. ft. = 350 sq. ft.

Finally, subtract the concealed surface area from the total surface area = 350 sq. ft. – 32 sq. ft. = 318 square feet.

1081. The platform's surface area = 318 square feet. Like the preceding question, there are concealed surface areas in this question. However, let's only solve exposed areas this time around. Find the surface area for the base rectangular prism. Do not worry about any concealed parts; imagine the top plane rising with each step. *SA* of base rectangular prism = $2[15 \text{ ft.}(7 \text{ ft.}) + 7 \text{ ft.}(1 \text{ ft.}) + 15 \text{ ft.}(1 \text{ ft.})] =$ $2(105 \text{ sq. ft.} + 7 \text{ sq. ft.} + 15 \text{ sq. ft.}) = 2(127 \text{ sq. ft.}) = 254$ sq. ft. Of the next two prisms, only their sides are considered exposed surfaces (the lips of their top surfaces have already been accounted for). The new formula removes the top and bottom planes: *SA* of sides only = $2(lh + wh)$. Subtracting a foot from each side of the base prism, the second prism measures 13 feet by 5 feet by 1 foot. The last prism measures 11 feet by 3 feet by 1 foot. Plug the remaining two prisms into the formula:

SA of sides only = $2[13 \text{ ft.}(1 \text{ ft.})] +$ $[5 \text{ ft.}(1 \text{ ft.})] = 2(13 \text{ sq. ft.} + 5 \text{ sq. ft.}) =$ $2(18 \text{ sq. ft.}) = 36$ sq. ft.

SA of sides only = $2[11 \text{ ft.}(1 \text{ ft.}) +$ $3 \text{ ft.}(1 \text{ ft.})] = 2(11 \text{ sq. ft.} + 3 \text{ sq. ft.}) =$ $2(14 \text{ sq. ft.}) = 28$ sq. ft.

Add all the exposed surface areas together: 254 sq. ft. + 36 sq. ft. + 28 sq. ft = 318 square feet.

1082. The exposed surface area = 297.5 square inches. The three blocks are congruent; find the surface area of one block and multiply it by three: $SA = 2(8.3 \text{ in.}(4.0 \text{ in.}) + 4.0 \text{ in.}(1.7 \text{ in.}) + 8.3 \text{ in.}(1.7 \text{ in.}) = 2(33.2 \text{ sq. in.} + 6.8 \text{ sq. in.} + 14.11 \text{ sq. in.}) = 2(54.11 \text{ sq. in.}) = 108.22$ sq. in.; 108.22 sq. in. $\times 3 = 324.66$ sq. in. Look at the diagram: The ends of one block are concealed, and they conceal an equal amount of space on the other two blocks: $2 \times 2[4.0 \text{ in.}(1.7 \text{ in.})] = 27.2$ sq. in. Subtract the concealed surface area from the total surface area: 324.66 sq. in. − 27.2 sq. in. = 297.46 square inches.

1083. $x = 2$ feet. Plug the variables into the formula for the SA of a prism: 304 sq. ft. = $2[12x(2x) + 2x(x) + 12x(x)]$; 304 sq. ft. = $2(24 x^2 + 2x^2 + 12x^2)$; 304 sq. ft. = $2(38x^2)$; 304 sq. ft. = $76x^2$; 4 sq. ft. = x^2; 2 feet = x.

1084. $x = 3$ meters. Plug the variables into the formula for the SA of a prism: 936 square meters = $2[4.5x(4x) + 4x(4x) + 4.5x(4x)]$; 936 m^2 = $2(18x^2 + 16x^2 + 18x^2)$; 936 m^2 = $2(52x^2)$; 936 m^2 = $104x^2$; 9 m^2 = x^2; 3 meters = x.

1085. $x = 2\sqrt{2}$ yards. To find the area of one of the two congruent cubes, divide 864 square yards by 2: $\frac{864 \text{ sq. yd.}}{2} = 432$ sq. yd. Plug the measure of each edge into the formula $SA = 6s^2$: 432 sq. yd. = $6(3x^2)$; 432 sq. yd. = $6(9x^2)$; 432 sq. yd. = $54x^2$; 8 sq. yd. = x^2; $2\sqrt{2}$ yards = x.

1086. The cylinder's surface area = 216π square feet. The surface area of a cylinder is $2\pi r^2 + 2\pi rh$: Plug the variables in and solve: $SA = 2\pi(6 \text{ ft.})^2 + 2\pi(6 \text{ ft.} \times 12 \text{ ft.})$; 72π sq. ft. + 144π sq. ft. = 216π square feet.

1087. The surface area of the candy = 1.0π square inch. The candy inside the wrapper is a perfect sphere. Its surface area is $4\pi r^2$. Plug the variables in and solve: $SA = 4\pi(0.5 \text{ in.})^2$; $SA = 1.0\pi$ square inch.

1088. The exposed surface area is 4,096π square centimeters. Surface area of a whole sphere is $4\pi r^2$. The surface area of half a sphere is $2\pi r^2$. Each sphere's surface area is $2\pi(8 \text{ cm}^2)$, or 128π square centimeters. Now, multiply the surface area of one half sphere by 32, because there are 32 halves: $32 \times 128\pi \text{ cm}^2 = 4{,}096\pi$ square centimeters.

1089. b. The surface area of a sphere is four times π times the radius squared.

1090. c. The area of the circular region is π times radius squared. The area of the curved region is two times π times radius times height. Notice there is only one circular region, because the storage tank would be on the ground. This area would not be painted.

1091. b. If the diameter of a sphere is 6 inches, the radius is 3 inches; the radius of a circle is half the diameter. Using the radius of 3 inches, surface area equals $(4)(3.14)(3)^2$ or 113.04 sq. in. Rounded to the nearest inch, this is 113 square inches. If you chose **a**, you used the diameter rather than the radius. If you chose **c**, you did not square the radius. If you chose **d**, you omitted the value 4 from the formula for the surface area of a sphere.

1092. c. The surface area of the box is the sum of the areas of all six sides. Two sides are 20 inches by 18 inches, or $(20)(18) = 360$ sq. in. Two sides are 18 inches by 4 inches, or $(18)(4) = 72$ sq. in. The last two sides are 20 inches by 4 inches, or $(20)(4) = 80$ sq. in. Adding up all six sides: 360 sq. in. + 360 sq. in. + 72 sq. in. + 72 sq. in. + 80 sq. in. + 80 sq. in. = 1,024 square inches, which is the total area. If you chose **a**, you did not double all sides. If you chose **b**, you calculated the volume of the box.

1093. a. The area of the front cover is length times width or $(8)(11) = 88$ sq. in. The back cover has the same area as the front, 88 sq. in. The area of the spine is length times width or $(1.5)(11) = 16.5$ sq. in. The extension inside the front cover is length times width or $(2)(11) = 22$ sq. in. The extension inside the back cover is also 22 sq. in. The total area is the sum of all previous areas or 88 sq. in. + 88 sq. in. + 16.5 sq. in. + 22 sq. in. + 22 sq. in. or 236.5 sq. in. If you chose **b**, you did not calculate the extensions inside the front and back covers. If you chose **c**, you miscalculated the area of the spine as $(1.5)(8)$ and omitted the extensions inside the front and back covers. If you chose **d**, you miscalculated the area of the spine as $(1.5)(8)$ only.

1094. a. The formula for the surface area of a sphere is $4\pi r^2$. If the diameter is doubled, this implies that the radius is also doubled. The formula then becomes $4\pi(2r)^2$. Simplifying this expression, $4\pi(4r^2)$ equals $16\pi r^2$. Compare $4\pi r^2$ to $16\pi r^2$; $16\pi r^2$ is 4 times greater than $4\pi r^2$. Therefore, the surface area is four times as great.

1095. c. $\triangle ABD$ is similar to $\triangle ECD$. Using this fact, the following proportion is true: $\frac{DE}{EC} = \frac{DA}{AB}$ or $\frac{40}{32} = \frac{(40 + x)}{60}$. Cross multiply: $2,400 = 32(40 + x)$; $2,400 = 1,280 + 32x$. Subtract 1,280; $1,120 = 32x$; divide by 32; $x = 35$ feet.

CHAPTER

17 ▶ Conversions, Part II

I N CHAPTER 2, WE LEARNED HOW TO CONVERT between different units of time and money. Now that we're familiar with length, area, and volume, we'll convert between those units, as well as units of weight. But first, we'll learn how to express very large and very small decimals using scientific notation.

▶ Scientific Notation

Scientific notation is the representation of a number as the product of a decimal and the number 10 raised to an exponent. Any number can be written in scientific notation, but it is typically used for writing very large or very small numbers.

To write a number in scientific notation, move the decimal point to the left or to the right until there is only one digit to the left of the decimal point. Count the number of places that you have moved the decimal point. This number will be the exponent of 10, the number by which you will multiply your new decimal. If you moved the decimal point to the left, then the exponent will be positive. If you moved the decimal point to the right, then the exponent will be negative.

Example

Write 134 in scientific notation.

The number 134 is really 134.0. In order for there to be only one digit to the left of the decimal point, we must move the decimal point two places to the left. Because we moved the decimal point to the left, the exponent of 10 will be positive: $134 = 1.34 \times 10^2$.

Example

Write 0.0000056 in scientific notation.

In order for there to be only one digit to the left of the decimal point, we must move the decimal point six places to the right. Because we moved the decimal point to the right, the exponent of 10 will be negative: $0.0000056 = 5.6 \times 10^{-6}$.

To convert from scientific notation back to a decimal, move the decimal point in the decimal part of the number to the left or right according to the exponent of 10. If the exponent is positive, move the decimal point to the right; if the exponent is negative, move the decimal point to the left.

Example

Write 2.1×10^5 as a decimal.

Move the decimal point in 2.1 five places to the right, because the exponent of 10 is positive: $2.1 \times 10^5 = 210,000$.

▶ Practice Questions

1096. 541,000 in scientific notation is
- **a.** 5.41×10^{-5}.
- **b.** 541×10^4.
- **c.** 5.41×10^6.
- **d.** 5.41×10^5.
- **e.** 54.1×10^5.

1097. 580 converted to scientific notation form is
- **a.** 580×10^3.
- **b.** 5.80×10^{-2}.
- **c.** 5.8×10^2.
- **d.** 58.0×10^2.
- **e.** 58.0×10^1.

1098. $0.000045 =$
- **a.** 4.5×10^{-5}
- **b.** 4.5×10^5
- **c.** 45×10^{-6}
- **d.** 4.5×10^{-4}
- **e.** 0.45×10^5

1099. Five million =
  a. 5.0×10^5
  b. 0.5×10^4
  c. 5.0×10^6
  d. 5.0×10^{-6}
  e. 5.0×10^7

1100. 0.0908 in scientific notation form is
  a. 9.08×10^2.
  b. 9.08×10^{-3}.
  c. 90.8×10^{-2}.
  d. 90.8×10^3.
  e. 9.08×10^{-2}.

1101. Juan checked his odometer and found that his car had 46,230 miles on it. What is this number in scientific notation?
  a. 462.3×10^3 miles
  b. 4.623×10^5 miles
  c. 4.6×10^{-5} miles
  d. 4.623×10^4 miles
  e. 4.623×10^{-5} miles

1102. Four thousand plus nine thousand =
  a. 1.3×10^4
  b. 1.3×10^{-4}
  c. 4.9×10^4
  d. 13×10^4
  e. 1.3×10^3

1103. 0.00000002 =
  a. 2.0×10^9
  b. 20×10^8
  c. 2.0×10^{-8}
  d. 2.0×10^{-7}
  e. 2.0×10^8

Refer to the following chart for questions 1104 and 1105.

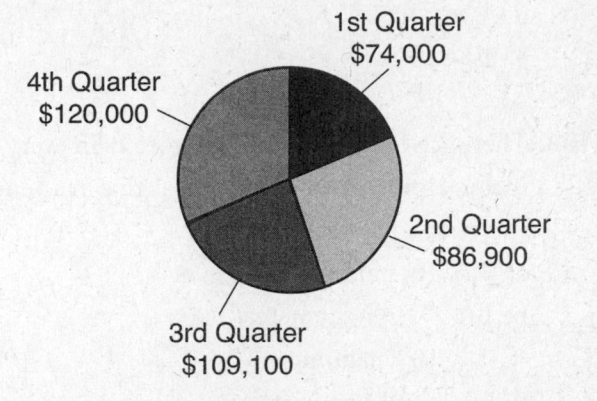

Company Earnings per Quarter

1st Quarter $74,000
4th Quarter $120,000
2nd Quarter $86,900
3rd Quarter $109,100

1104. What is the total amount of earnings in the second quarter in scientific notation?
  a. $\$8.69 \times 10^{-5}$
  b. $\$8.69 \times 10^3$
  c. $\$86.9 \times 10^4$
  d. $\$8.69 \times 10^4$
  e. $\$8.69 \times 10^5$

1105. What is the total amount of earnings in the first and fourth quarters together, expressed in scientific notation?
  a. $\$1.94 \times 10^5$
  b. $\$19.4 \times 10^4$
  c. $\$1.2 \times 10^6$
  d. $\$8.6 \times 10^5$
  e. $\$8.6 \times 10^{-5}$

1106. What is 0.000009001 in scientific notation?
  a. 9.0×10^{-6}
  b. 9.0×10^6
  c. 9.001×10^{-6}
  d. 9.001×10^6
  e. $9,001 \times 10^{-9}$

1107. In scientific notation, 50 =
  a. 5×10^{-1}
  b. 50×10^1
  c. 0.5×10^3
  d. 5×10^1
  e. 5×10^2

1108. There are 1,000,000,000 nanometers in every meter. How many nanometers is this in scientific notation?
  a. 1×10^9 nanometers
  b. 0.1×10^9 nanometers
  c. 1×10^{10} nanometers
  d. 10×10^{10} nanometers
  e. 1×10^8 nanometers

1109. Written in scientific notation, 2,340.25 is
  a. 2.34025×10^5.
  b. $2,340.25 \times 10^2$.
  c. 2.34025×10^3.
  d. 2.34025×10^2.
  e. 2.34025×10^4.

1110. What is −0.0063 written in scientific notation?
  a. 6.3×10^{-3}
  b. 6.3×10^4
  c. 63×10^4
  d. -6.3×10^{-3}
  e. -6.3×10^3

1111. $8.43 \times 10^4 =$
  a. 0.000843
  b. 0.00843
  c. 84,300
  d. 8,430,000
  e. 8,430

1112. $2 \times 10^{-3} =$
  a. 0.002
  b. 0.0002
  c. 2,000
  d. 200
  e. 0.020

1113. The moon is approximately 2.52×10^5 miles from the earth. What is the equivalent number of miles?
  a. 2,520 miles
  b. 25,200,000 miles
  c. 2,520,000 miles
  d. 25,200 miles
  e. 252,000 miles

1114. $1 \times 10^0 =$
  a. 10
  b. 0
  c. 1
  d. 0.1
  e. It cannot be determined.

1115. $9.0025 \times 10^{-6} =$
  a. 9,002,500
  b. 0.000090025
  c. 0.00090025
  d. 0.0000090025
  e. 0.00000090025

1116. The speed of sound is approximately 3.4×10^2 miles per second at sea level. This is equal to what number?
  a. 3,400 miles per second
  b. 340 miles per second
  c. 34 miles per second
  d. 0.0034 miles per second
  e. 0.034 miles per second

1117. $8 \times 10^1 =$

a. 8

b. 0.8

c. 80

d. 800

e. It cannot be determined.

1118. $2.11 \times 10^7 =$

a. 2,110,000

b. 21,100,000

c. 211,000,000

d. 211,000

e. 2,110,000,000

1119. $5.3 \times 10^{-1} =$

a. 530

b. 53

c. 5.3

d. 0.53

e. 0.053

1120. One μL is equal to 1×10^{-6} L. This number is equal to

a. 0.00001 L.

b. 0.0000001 L.

c. 1,000,000 L.

d. 1.000000 L.

e. 0.000001 L.

1121. A distribution manager ordered 9.35×10^4 items for the stores in the Pacific Northwest. This number is equal to

a. 93,500.

b. 9,350.

c. 935,000.

d. 935.

e. 9,350,000.

1122. $1.04825 \times 10^3 =$

a. 1,048.25

b. 10,482.5

c. 104.825

d. 104,825

e. 104,825,000

1123. $7.955 \times 10^{-5} =$

a. 0.000007955

b. 795,500

c. 79,550

d. 0.00007955

e. 0.0007955

▶ Metric and Standard Units

The following charts show the conversion between standard units and metric units for area, length, volume, and weight. To convert from a larger unit of measure, such as feet, to a smaller unit of measure, such as inches, you must multiply. To convert from a smaller unit of measure to a larger unit of measure, you must divide. Review the section on rounding decimals in Chapter 3, as some conversions, especially those with standard units, will require rounding.

Conversions between Standard Units

1 foot (ft.) = 12 inches (in.)

1 yard (yd.) = 3 feet = 36 inches

1 mile (mi.) = 1,760 yards = 5,280 feet

1 square foot (sq. ft., ft.^{2}) = 144 square inches (sq. in., in.^{2})

1 square yard (sq. yd., yd.^{2}) = 9 square feet

1 square mile (sq. mi., mi.^{2}) = 640 acres = 3,097,600 square yards

1 acre (A) = 4,840 square yards = 43,560 square feet

1 cubic foot (cu. ft., ft.^{3}) = 1,728 cubic inches (cu. in., in.^{3}) = 7.48 gallons

1 cubic yard (cu. yd., yd.^{3}) = 27 cubic feet

1 tablespoon (tbs.) = 0.5 fluid ounce (fl. oz.)

1 tablespoon = 3 teaspoons (tsp.)

1 cup (c.) = 8 fluid ounces

1 pint (pt.) = 2 cups = 16 fluid ounces

1 quart (qt.) = 2 pints = 32 fluid ounces

1 gallon (gal.) = 4 quarts = 128 fluid ounces

1 gallon $\approx$ 231 cubic inches $\approx$ 0.1337 cubic foot

Conversions between Metric Units

1 centimeter (cm) = 10 millimeters (mm)

1 meter (m) = 100 centimeters = 1,000 millimeters

1 kilometer (km) = 1,000 meters

1 square centimeter (cm^2) = 100 square millimeters (mm^2)

1 square meter (m^2) = 10,000 square centimeters = 1,000,000 square millimeters

1 square kilometer (km^2) = 1,000,000 square meters

1 cubic centimeter (cm^3, cc) = 1,000 cubic millimeters (mm^3)

1 cubic meter (m^3) = 1,000,000 cubic centimeters

1 liter (L) = 1,000 milliliters (mL) = 100 centiliters (cL)

1 kiloliter (kL) = 1,000 liters = 1,000,000 milliliters

Conversions between Standard and Metric Units

1 inch $\approx$ 25.4 millimeters $\approx$ 2.54 centimeters

1 foot $\approx$ 0.3048 meter $\approx$ 30.480 centimeters

1 yard $\approx$ 0.9144 meter

1 mile $\approx$ 1,609.34 meters $\approx$ 1.6093 kilometers

1 kilometer $\approx$ 0.6214 mile

1 meter $\approx$ 3.281 feet $\approx$ 39.37 inches

1 centimeter $\approx$ 0.3937 inch

1 square inch $\approx$ 645.16 square millimeters $\approx$ 6.4516 square centimeters

1 square foot $\approx$ 0.0929 square meter

1 square yard $\approx$ 0.8361 square meter

1 square mile $\approx$ 2,590,000 square meters $\approx$ 2.59 square kilometers

1 acre $\approx$ 4,046.8564 square meters $\approx$ 0.004047 square kilometers

1 cubic inch $\approx$ 16,387.064 cubic millimeters $\approx$ 16.3871 cubic centimeters

1 cubic foot $\approx$ 0.0283 cubic meter

1 cubic yard $\approx$ 0.7646 cubic meter

1 teaspoon $\approx$ 5 milliliters

1 tablespoon $\approx$ 15 milliliters

1 fluid ounce $\approx$ 29.57 milliliters $\approx$ 2.957 centiliters

1 fluid ounce $\approx$ 0.00002957 cubic meters

1 gallon $\approx$ 3.785 liters

1 liter $\approx$ 1.057 quarts $\approx$ 0.264 gallon

1 quart $\approx$ 0.946 liter

Weight Conversions between Standard Units

1 pound (lb.) = 16 ounces (oz.)

1 ton (t.) = 2,000 pounds

Weight Conversions between Metric Units

1 centigram (cg) = 10 milligrams (mg)

1 gram (g) = 100 centigrams = 1,000 milligrams

1 kilogram (kg) = 1,000 grams

Weight Conversions between Standard and Metric Units

1 gram ≈ 0.035 ounce

1 pound ≈ 0.454 kilogram

1 pound ≈ 454 grams

1 kilogram ≈ 2.205 pounds

1 ton ≈ 908 kilograms

Example

How many yards are 6 feet?

Because we are going from a smaller unit of measure, feet, to a larger unit of measure, yards, we must divide. Divide the number of feet by the number of feet in one yard. The preceding chart shows that there are three feet in one yard: $\frac{6}{3} = 2$. Six feet equal two yards.

Example

Convert 54.7 liters to milliliters.

Because we are going from a larger unit of measure, liters, to a smaller unit of measure, milliliters, we must multiply. Multiply the number of liters by the number of milliliters in one liter. The chart shows that there are 1,000 milliliters in 1 liter; 54.7 × 1,000 = 54,700 milliliters.

Example

Two and a half pounds are equal to approximately how many grams?

The chart shows that one pound is equal to approximately 454 grams. Because we are going from a larger unit to a smaller unit, we must multiply: 2.5 × 454 = 1,135 grams.

Example

Approximately how many cubic feet are in one cubic meter?

The chart shows that one cubic foot is approximately 0.0283 cubic meter. We can use a proportion to solve this problem. If one cubic foot is approximately 0.0283 meter, then x cubic feet are in one cubic meter: $\frac{1}{0.0283} = \frac{x}{1}$, 0.0283$x$ = 1, and $x = \frac{1}{0.0283}$, which is approximately 35.336 cubic feet.

▶ Practice Questions

1124. A recipe calls for three ounces of olive oil. Convert this measurement into cups.

   a. 0.3 cup

   b. 0.5 cup

   c. 0.375 cup

   d. 24 cups

   e. $2\frac{2}{3}$ cups

1125. 48 inches = _____ yard(s).

   a. $0.\overline{33}$

   b. 1

   c. $1.\overline{33}$

   d. 2

   e. 84

1126. A two-liter bottle of soda contains approximately how many fluid ounces?
a. 0.06 fluid ounce
b. 968.96 fluid ounces
c. 128.53 fluid ounces
d. 256 fluid ounces
e. 67.6 fluid ounces

1127. The perimeter of a room is measured and found to be 652 inches. Trim for the room is sold by the foot. How many feet of trim must be purchased so that the room can be trimmed?
a. 50 feet
b. 54 feet
c. 55 feet
d. 60 feet
e. 18 feet

1128. Thomas is 6 feet 1 inch in height. His son is 3 feet 3 inches tall. What is the difference in their heights?
a. 30 inches
b. 32 inches
c. 34 inches
d. 36 inches
e. 38 inches

1129. Martha walks to school, a distance of 0.85 mile. What is the distance she walks to school in feet?
a. 4,488 feet
b. 6,212 feet
c. 1,496 feet
d. 5,280 feet
e. 1,760 feet

1130. A road race is 33,000 feet long. How many miles long is the race?
a. 18.75 miles
b. 6.25 miles
c. 11,000 miles
d. 38,280 miles
e. 5 miles

1131. A child's sandbox is being constructed in Tony's backyard. The sandbox is 6 feet wide and 5 feet long. Tony wants the sand to be at least 1.5 feet deep. The volume of sand in the box is 6 feet $\times$ 5 feet $\times$ 1.5 feet = 45 cubic feet. Convert the volume into cubic yards.
a. 72 cubic yards
b. 15 cubic yards
c. 0.6 cubic yard
d. $1.\overline{66}$ cubic yards
e. 5 cubic yards

1132. Which of the following represents a method by which one could convert inches into miles?
a. Multiply by 12, then multiply by 5,280.
b. Divide by 12, then divide by 5,280.
c. Add 12, then multiply by 5,280.
d. Multiply by 12, then divide by 5,280.
e. Divide by 12, then multiply by 5,280.

1133. 4.5 miles = _____ feet.
a. 13.5
b. 5,275.5
c. 5,284.5
d. 7,920
e. 23,760

1134. How much is 2 pints 6 ounces + 1 cup 7 ounces?

a. 1 quart

b. 3 pints 1 cup 13 ounces

c. 2 pints 5 ounces

d. 3 pints 5 ounces

e. 3 pints 1 cup

1135. 35 mm = _____ cm.

a. 0.35

b. 3.5

c. 35

d. 350

e. 3,500

1136. Susan wishes to create bows from 12 yards of ribbon. Each bow requires 6 inches of ribbon to make. How many inches of ribbon does Susan have?

a. 18 inches

b. 432 inches

c. 48 inches

d. 144 inches

e. 24 inches

1137. The living room in Donna's home is 182 square feet. How many square yards of carpet should she purchase to carpet the room?

a. 9 square yards

b. 1,638 square yards

c. 61 square yards

d. 21 square yards

e. 546 square yards

1138. Sofie needed to take $\frac{3}{4}$ teaspoon of cough syrup three times a day. Convert $\frac{3}{4}$ teaspoon into milliliters.

a. 0.375 milliliter

b. 3.75 milliliters

c. 2.25 milliliters

d. 22.5 milliliters

e. 0.15 milliliter

1139. 3.9 kiloliters = _____ milliliter(s).

a. 0.00000039

b. 0.0000039

c. 0.0039

d. 3,900,000

e. 39,000,000

1140. The 1,500-meter race is a popular distance running event at track meets. How far is this distance in miles?

a. 0.6214 mile

b. 0.5 mile

c. 0.9321 mile

d. 2.4139 miles

e. 0.4143 mile

1141. 58.24 mm^3 = _____ cm^3.

a. 0.05824

b. 5.824

c. 58,240

d. 582.4

e. 5,824

1142. Abigail is 5.5 feet tall. Which of the following is closest to this height?

a. 17 meters

b. 170 centimeters

c. 1.7 kilometers

d. 170 millimeters

e. 0.17 kilometer

1143. 62.4 meters ≈ _____ feet.
- **a.** 1.58
- **b.** 2,456.7
- **c.** 65.7
- **d.** 19.01
- **e.** 205

1144. 0.16 acre = _____ square yards.
- **a.** 774.4
- **b.** 6,969.6
- **c.** 30,250
- **d.** 4,840.16
- **e.** 4,839.84

1145. 4.236 km + 23 m + 654 cm = _____ m.
- **a.** 681.236
- **b.** 33.776
- **c.** 4,913
- **d.** 453.14
- **e.** 4,265.54

1146. Murray owns five square kilometers of farmland in Pennsylvania. How many acres does Murray own?
- **a.** 0.020235 acre
- **b.** 1,235.48 acres
- **c.** 20.234 acres
- **d.** 1.235 acres
- **e.** 2.0234 acres

1147. A stretch limousine is 4.5 meters long. How many yards long is the limousine?
- **a.** 7.5 yards
- **b.** 44.307 yards
- **c.** 13.5 yards
- **d.** 14.769 yards
- **e.** 4.92 yards

1148. 12.59 m^2 ≈ _____ square yards.
- **a.** 15.06
- **b.** 10.53
- **c.** 13.43
- **d.** 11.75
- **e.** 1.17

1149. Sandi purchased three cubic yards of mulch for her garden. About how many cubic meters of mulch did she buy?
- **a.** 2.2304 cubic meters
- **b.** 3.7696 cubic meters
- **c.** 3.9236 cubic meters
- **d.** 2.2938 cubic meters
- **e.** 3.5621 cubic meters

1150. 15 cL ≈ _____ fl. oz.
- **a.** 44.36
- **b.** 17.96
- **c.** 5.1
- **d.** 4.3
- **e.** 12.04

1151. The average adult should drink eight 8-oz. glasses of water per day. Convert the daily water requirement to gallons.
- **a.** 0.64 gallon
- **b.** 6.4 gallons
- **c.** 0.5 gallon
- **d.** 5 gallons
- **e.** 3.2 gallons

1152. 19 quarts = _____ gallons.
- **a.** 76
- **b.** 4.75
- **c.** 9.5
- **d.** 2.375
- **e.** 1.1875

1153. A tree is measured and found to be 16.9 meters tall. How many centimeters tall is the tree?

a. 0.169 centimeters

b. 1.69 centimeters

c. 169 centimeters

d. 1,690 centimeters

e. 16,900 centimeters

1154. 1 quart + 1 pint =

a. 1.419 liters.

b. 1.586 liters.

c. 2.446 liters.

d. 1.5 liters.

e. 0.946 liter.

1155. A science experiment calls for the use of a porous sponge approximately 35 cubic centimeters in volume. About how big is the sponge in cubic inches?

a. 4.52 cubic inches

b. 5,735.49 cubic inches

c. 4,682.03 cubic inches

d. 573.55 cubic inches

e. 2.14 cubic inches

1156. The Johnsons' kitchen is 221 square feet in size. How many square meters of tile must they purchase to tile the kitchen floor?

a. 22.84 square meters

b. 20.53 square meters

c. 2,378.90 square meters

d. 221.09 square meters

e. 23.91 square meters

1157. 345.972 mL = _____ L.

a. 345,972

b. 3,459.72

c. 34.5972

d. 3.45972

e. 0.345972

1158. How many cubic inches are in a cubic yard?

a. 64 cubic inches

b. 46,656 cubic inches

c. 1,755 cubic inches

d. 108 cubic inches

e. 62,208 cubic inches

1159. _____ fl. oz. = $2\frac{3}{4}$ c.

a. 24

b. 16

c. $10\frac{3}{4}$

d. 18.72

e. 22

1160. Brenda measures her kitchen window so she can purchase blinds. She finds that the window is 40 inches long. Convert this dimension to centimeters.

a. 101.6 centimeters

b. 15.75 centimeters

c. 157.5 centimeters

d. 1,016 centimeters

e. 42.54 centimeters

1161. A photo has an area of 300 square centimeters. Convert this area into square millimeters.

a. 3,000 square millimeters

b. 30,000 square millimeters

c. 300,000 square millimeters

d. 3 square millimeters

e. 0.3 square millimeter

1162. A signpost is 4.3 meters tall. How tall is the signpost in centimeters?

a. 0.043 centimeter

b. 0.43 centimeter

c. 43 centimeters

d. 430 centimeters

e. 4,300 centimeters

1163. Use the diagram to answer the question:

◆————— 12.4 yds. —————◆

◆— ? —◆————— 8.34 m —————◆

Find the missing length in meters.

a. 4.06 meters

b. 1.356 meters

c. 1.134 meters

d. 9.12 meters

e. 3 meters

1164. A book cover has an area of 72 square inches. What is its area in square centimeters?

a. 46.45 square centimeters

b. 464.5 square centimeters

c. 11.16 square centimeters

d. 111.6 square centimeters

e. 4,645 square centimeters

1165. 11.8 in.^{3} = _____ mm^3.

a. 19,336.7

b. 16,398.86

c. 1,388.73

d. 193,367.36

e. 163,988.6

1166. Fred wants to replace the felt on his pool table. The table has an area of 29,728 square centimeters. About how many square meters of felt does he need?

a. 30 square meters

b. 20 square meters

c. 10 square meters

d. 3 square meters

e. 2 square meters

1167. A storage bin can hold up to 3.25 cubic meters of material. How many cubic centimeters of material can the bin accommodate?

a. 325,000 cubic centimeters

b. 3,250,000 cubic centimeters

c. 32,500,000 cubic centimeters

d. 32,500 cubic centimeters

e. 0.00000325 cubic centimeter

1168. William drank nine cups of juice on Saturday. How many fluid ounces of juice did he drink?

a. 17 fluid ounces

b. 11.25 fluid ounces

c. 7.2 fluid ounces

d. 1.125 fluid ounces

e. 72 fluid ounces

1169. A birdbath can hold about 226 cubic inches of water. Convert this amount into cubic feet.

a. 0.13 cubic foot

b. 1.3 cubic feet

c. 13 cubic feet

d. 7.65 cubic feet

e. 0.765 cubic foot

1170. A pitcher of punch contains 1.5 liters. Convert this amount to kiloliters.
  a. 15 kiloliters
  b. 1,500 kiloliters
  c. 0.0015 kiloliter
  d. 0.15 kiloliter
  e. 1.5 kiloliters

1171. A test tube is marked in cubic centimeters. If the test tube is filled to the 24 cubic centimeter mark with water, how many cubic millimeters of water are in the tube?
  a. 2,400 cubic millimeters
  b. 240 cubic millimeters
  c. 24,000 cubic millimeters
  d. 0.024 cubic millimeter
  e. 2.4 cubic millimeters

1172. 7.98 L = _____ mL.
  a. 0.798
  b. 0.00798
  c. 79,800
  d. 7,980
  e. 798,000

1173. A local charity has asked a water distributor to donate water for participants to drink during a fund-raising walk. The company donated 231 gallons of water. How many quarts is this?
  a. 57.75 quarts
  b. 462 quarts
  c. 115.5 quarts
  d. 1,848 quarts
  e. 924 quarts

1174. 6 lb. = _____ oz.
  a. 96
  b. 69
  c. 2.667
  d. 0.375
  e. 22

1175. Tommy drives a half-ton pickup truck. If *half-ton* refers to the amount of weight the truck can carry in its bed, how many pounds can the truck carry?
  a. 8 pounds
  b. 100 pounds
  c. 2,000 pounds
  d. 1,000 pounds
  e. 4,000 pounds

1176. 48 kg = _____ gram(s).
  a. 0.0048
  b. 0.48
  c. 4.8
  d. 4,800
  e. 48,000

1177. 3 kg ≈ _____ lb.
  a. 1.4
  b. 6.6
  c. 5.2
  d. 3.5
  e. 2.5

1178. 845.9 cg = _____ kg.
  a. 0.008459
  b. 0.8450
  c. 8.459
  d. 84.59
  e. 84,590

1179. 2.5 lb. ≈ _____ grams.
  a. 1.816
  b. 1.135
  c. 1,135
  d. 1,816
  e. 5.51

1180. A recipe calls for three ounces of cheese per serving. If Rebecca is making 12 servings, how many pounds of cheese will she need?
  a. 3.6 pounds
  b. 36 pounds
  c. 2.25 pounds
  d. 22.5 pounds
  e. 4.5 pounds

1181. Graham crackers come in a 13-ounce package. What is the approximate weight of the package in kilograms?
  a. 0.371 kilogram
  b. 371.4 kilograms
  c. 0.455 kilogram
  d. 455 kilograms
  e. 286.3 kilograms

1182. 432 ounces = _____ ton(s).
  a. 27
  b. 2.7
  c. 0.135
  d. 0.0135
  e. 0.216

1183. "Take 500 milligrams of ibuprofen if your headache acts up," the doctor ordered. How many grams of ibuprofen is this?
  a. 0.05 gram
  b. 5,000 grams
  c. 50 grams
  d. 5 grams
  e. 0.5 gram

1184. 4.2 tons ≈ _____ kilograms.
  a. 3,813.6
  b. 38.136
  c. 216.2
  d. 21.62
  e. 2,162

1185. A Galapagos tortoise can weigh up to 300 pounds. Convert this weight into tons.
  a. 60 tons
  b. 6 tons
  c. 6.67 tons
  d. 1.5 tons
  e. 0.15 ton

1186. 3.15 tons = _____ ounces.
  a. 100,800
  b. 10,080
  c. 50
  d. 6,300
  e. 393.75

1187. A box of Jell-O weighs about 85 grams. Convert this weight to kilograms.
  a. 85,000 kilograms
  b. 850 kilograms
  c. 0.85 kilogram
  d. 0.085 kilogram
  e. 0.0085 kilogram

1188. 1.4 kg ≈ _____ oz.
  a. 0.049
  b. 0.1929
  c. 22
  d. 11
  e. 49

1189. Susan bought a 25-gram container of ground cloves to use in cider. How many milligrams of cloves did she buy?

a. 250 milligrams

b. 2,500 milligrams

c. 25,000 milligrams

d. 2.5 milligrams

e. 0.025 milligram

1190. 58 g ≈ _____ oz.

a. 2.03

b. 1,657

c. 26.332

d. 127.75

e. 48.13

1191. Josephine has 70 ounces of meat to use in her annual batch of spaghetti sauce. How many pounds of meat does Josephine have?

a. 4.375 pounds

b. 1,120 pounds

c. 8.75 pounds

d. 3.5 pounds

e. 5.13 pounds

1192. Fill in the missing value in the following table so that the values are equivalent:

Kilograms	Milligrams
	1,254,600

a. 1,254,600 kilograms

b. 1.2546 kilograms

c. 1,254,600,000 kilograms

d. 1,254,600,000,000 kilograms

e. 12.546 kilograms

1193. Kelly's dog weighs 52 pounds. Convert the dog's weight to kilograms.

a. 23,608

b. 8.73

c. 236.1

d. 114.5

e. 23.608

1194. 12.71 kilograms = _____ milligram(s).

a. 0.01271

b. 0.00001271

c. 12,710

d. 1,271,000

e. 12,710,000

1195. A doctor prescribed 2 centigrams of medicine to his patient. How many milligrams of medicine did the doctor prescribe?

a. 2,000 milligrams

b. 200 milligrams

c. 20 milligrams

d. 2 milligrams

e. 0.2 milligram

1196. 0.24 kg = _____ cg.

a. 0.00024

b. 240

c. 24

d. 24,000

e. 0.0000024

1197. Michael weighed 12 pounds 9 ounces at his last checkup. Convert Michael's weight into ounces.

a. 148 ounces

b. 21 ounces

c. 9.75 ounces

d. 156 ounces

e. 201 ounces

1198. _____ cg = 564.9 g.

    a. 5.649

    b. 56,490

    c. 5,649

    d. 56.49

    e. 0.5649

1199. The empty container weighs 512 centigrams; 42,370 milligrams of sand are added to the cup. What is the total weight of the sand and the container in grams?

    a. 554.37 grams

    b. 42,882 grams

    c. 47.49 grams

    d. 93.57 grams

    e. 9.357 grams

1200. An African elephant can weigh as much as 5,448 kilograms. Convert the elephant's weight into tons.

    a. 2.7 tons

    b. 6.0 tons

    c. 3.5 tons

    d. 2.2 tons

    e. 1.8 tons

TERM TO REVIEW

scientific notation

▶ Answers

1096. d. In this problem, the value is greater than 10, so the exponent on the base of 10 will be positive. The original number, 541,000, has the decimal point to the right of the last zero. Move the decimal point five places to the left in order to achieve the correct form for scientific notation; $541,000 = 5.41 \times 10^5$.

1097. c. In order to get only one whole non-zero number (one number to the left of the decimal), move the decimal two places to the left; 580 is greater than 10; therefore, the exponent is positive 2.

1098. a. First, find the decimal place that will leave one digit to the left of the decimal (4.5). The decimal point needs to be moved five places to the right in order to get there. Because it is being moved to the right and the number is smaller than 1.0, the power will be negative: 4.5×10^{-5}.

1099. c. Five million is equal to 5,000,000. In order to convert this into scientific notation, the decimal needs to be moved six places to the left to get 5×10^6. The power is positive because the original number is greater than 10.

1100. e. Move the decimal two places to the right to get one whole number. Add a negative sign to the power of 10, because the decimal is moved to the right, and the number is smaller than 1.0: 9.08×10^{-2}.

1101. d. Move the decimal four places to the left in order to get only one digit to the left of the decimal: 4.623×10^4. The power is positive because the original number is greater than 10.

1102. a. The first step is to add up the two numbers: $4,000 + 9,000 = 13,000$. Now convert to scientific notation by moving the decimal four places to the left: 1.3×10^4; 13,000 is greater than 10, so the exponent is positive.

1103. c. Carefully count how many places the decimal must be moved to the right in order to end up with the number 2.0 (eight places to the right). The power sign is negative because the decimal is moved to the right: 2.0×10^{-8}. The original number is less than 1.0, so the exponent is negative.

1104. d. According to the chart, $86,900 was made in the second quarter. Move the decimal four places to the left: 8.69×10^4; 86,900 is greater than 10, so the exponent is positive.

1105. a. Add together the earnings from the first and fourth quarters; $74,000 + \$120,000 = \$194,000$. Then move the decimal point five places to the left: 1.94×10^5. The sum of the earnings is greater than 10, so the exponent is positive.

1106. c. Move the decimal point six places to the right in order to convert this number to scientific notation. Add the negative sign to the power, because the decimal is moved to the right, and the number is smaller than 1.0. So, the answer is 9.001×10^{-6}.

1107. d. Move the decimal one place to the left in order to have only one digit to the left of the decimal: 5×10^1. The exponent is positive because 50 is greater than 10.

1108. a. In order to get to the number 1.0, the decimal needs to be moved nine places to the left. Therefore, the number of nanometers in a meter is 1×10^9. The exponent is positive because the original number is greater than 10.

1109. c. In this problem, do not be confused by the fact that the decimal does not start off at the end of the number. The decimal needs to be moved three places to the left to get 2.34025×10^3. The original value is greater than 10, so the exponent is positive.

1110. d. Move the decimal three places to the right in order to get the correct scientific notation form. Add the negative sign to the power, because the decimal is moved to the right, and the number is smaller than 1.0. Finally, keep the negative sign in front of the original number because this has not changed; it is still a negative value. So, the answer is -6.3×10^{-3}.

1111. c. The power (exponent) is positive 4, so the number is a value greater than 10, and the decimal point must be moved four places to the right. Two trailing zeros must be added to get the number 84,300.

1112. a. The power sign is negative 3, so the number will be less than 1.0, and the decimal needs to be moved three places to the left. Do not forget that 2 is equal to 2.0 (although the decimal is not shown, it follows the last digit). The answer is 0.002.

1113. e. The power is positive 5, so the number will be greater than 10, and the decimal is moved five places to the right to become 252,000.

1114. c. Any number to the zero power is equal to 1. Therefore, $10^0 = 1$. Placing this back into the question gives 1×1, which equals 1.

1115. d. The power is negative 6, so the number will be less than 1.0, and the decimal is moved six places to the left to get 0.0000090025.

1116. b. The power is positive 2, so the number will be greater than 10, and the decimal is moved two places to the right to give the answer, 340 miles per second.

1117. c. The power is positive 1, so the number will be greater than 10, and the decimal is moved one place to the right to give the answer, 80.

1118. b. The power is positive 7, so the number will be greater than 10, and the decimal is moved seven places to the right in order to get 21,100,000.

1119. d. The power is negative 1, so the number will be less than 1.0, and the decimal is moved one place to the left in order to get 0.53.

1120. e. The power is negative 6, so the number will be less than 1.0, and the decimal is moved six places to the left. Therefore, one μL is equal to 0.000001 L.

1121. a. The power is positive 4, so the number will be greater than 10, and the decimal is moved four places to the right to give the answer 93,500.

1122. a. The power is positive 3, so the number will be greater than 10, and the decimal is moved three places to the right, even if this still leaves some digits to the right of the decimal point. The answer is 1,048.25.

1123. d. The power is negative 5, so the number will be less than 1.0, and the decimal is moved five places to the left in order to get the answer, 0.00007955.

1124. c. Every 8 ounces equals 1 cup, so divide: 3 oz. ÷ 8 oz./c. = 0.375 cup.

1125. c. Every 36 inches equals 1 yard, so divide: 48 in. ÷ 36 in./yd. = $1\frac{1}{3}$ yards, or $1.\overline{33}$ yards written as a decimal.

1126. e. Each liter equals 0.264 gallons, so for two liters: 2 L × 0.264 gal./L = 0.528 gal. A gallon equals 128 fluid ounces, so multiply: 0.528 gal. × 128 fl. oz./gal. = 67.584 fl. oz. ≈ 67.6 fluid ounces.

1127. c. Twelve inches equals 1 foot, so divide: 652 in. ÷ 12 in./ft. = 54.333 feet. Because trim is sold by the foot, round up; 55 feet must be purchased so that there is enough trim.

1128. c. One foot equals 12 inches, so Thomas's height is 6 ft. × 12 in./ft. = 72 in. + 1 in. = 73 inches. His son's height is 3 ft. × 12 in./ft. = 36 in. + 3 in. = 39 inches. The difference between their heights is 73 in. – 39 in. = 34 inches.

1129. a. A mile is 5,280 feet, so to find 0.85 mile, multiply: 0.85 mi. × 5,280 ft./mi. = 4,488 feet.

1130. b. It takes 5,280 feet to make a mile. To find how many miles are in 33,000 feet, divide: 33,000 ft. ÷ 5,280 ft./mi. = 6.25 miles.

1131. d. One cubic yard requires 27 cubic feet. To find the number of cubic yards in 45 cubic feet, divide: 45 ft.^{3} ÷ 27 ft.^{3}/yd.^{3} = $1.\overline{66}$ cubic yards.

1132. b. Because the conversion is from smaller units to larger units, division is required. Every 12 inches equals 1 foot, so to figure out the number of feet in the given number of inches, divide by 12. Then, to figure out how many miles are in the calculated number of feet, divide by the number of feet in a mile, 5,280.

1133. e. Each mile equals 5,280 feet. Because there are 4.5 miles, multiply: 4.5 mi. × 5,280 ft./mi. = 23,760 feet.

1134. d. The two amounts can be added as they are, but then the sum needs to be simplified; 2 pints 6 ounces + 1 cup 7 ounces = 2 pints 1 cup 13 ounces. Note that the sum is written with the largest units first and smallest units last. To simplify, start with the smallest unit, ounces, and work toward larger units. Every 8 ounces makes 1 cup, so 13 ounces = 1 cup 5 ounces. Replace the 13 ounces with the 1 cup 5 ounces, adding the cup portions together: 2 pints 2 cups 5 ounces. Now note that 2 cups = 1 pint, so the sum can be simplified again combining this pint with the 2 pints in the sum. The simplified total is 3 pints 5 ounces.

1135. b. There are 10 millimeters in every centimeter, so divide: 35 mm ÷ 10 mm/cm = 3.5 cm.

1136. b. There are 36 inches in a yard. To find the number of inches in 12 yards, multiply: 12 yd. × 36 in./yd. = 432 inches.

1137. d. It takes 9 square feet to make a square yard. To find out how many square yards are in 182 square feet, divide: 182 sq. ft. ÷ 9 sq. ft./sq. yd. = 20.22 sq. yd. Because Donna cannot purchase part of a square yard, she has to round up. She must purchase 21 square yards to have enough to carpet the room.

1138. b. One teaspoon equals 5 milliliters. Therefore, $\frac{3}{4}$ tsp. × 5 mL/tsp. = 3.75 mL.

1139. d. One kiloliter equals 1,000,000 milliliters: 3.9 kL × 1,000,000 mL/L = 3,900,000 milliliters.

1140. c. Each mile is about 1,609.34 meters. Divide to find out how many miles are equivalent to 1,500 meters: 1,500 m ÷ 1,609.34 m/mi. ≈ 0.9321 mile.

1141. a. There are 1,000 cubic millimeters in one cubic centimeter. Divide to determine how many cubic centimeters are in 58.24 mm³; 58.24 mm³ ÷ 1,000 mm³/cm³ = 0.05824 cubic centimeter.

1142. b. Because one foot equals about 0.3048 meter, 5.5 feet equals 5.5 ft. × 0.3048 m/ft. = 1.6764 meters. Answer **a** is much too large. Now convert 1.6764 m into the remaining units given in the answer choices; 1.6764 m × 100 cm/m = 167.64 cm, which is very close to answer **b**. 1.6764 m ÷ 1,000 m/km = .0016764 km, which eliminates answers **c** and **e**; 1.6764 m × 1,000 mm/m = 1,676.4 mm, which is much higher than answer **d**. The answer is **b**: 170 centimeters.

1143. e. One meter is approximately 3.281 feet. Therefore, 62.4 m × 3.281 ft./m ≈ 204.73 ft. ≈ 205 feet.

1144. a. One acre equals 4,840 square yards. Multiply: 0.16 acre × 4,840 sq. yd./acre = 774.4 square yards.

1145. e. Convert all units to meters first: 4.236 km × 1,000 m/km = 4,236 m; 654 cm ÷ 100 cm/m = 6.54 m. Now add: 4,236 m + 23 m + 6.54 m = 4,265.54 meters.

1146. b. One acre equals 0.004047 square kilometers, so divide: 5 km² ÷ 0.004047 km²/acre ≈ 1,235.48 acres.

1147. e. One yard equals about 0.9144 meter. Divide: 4.5 m ÷ 0.9144 m/yd. ≈ 4.92 yards.

1148. a. One square yard equals about 0.8361 square meter. Divide: 12.59 m² ÷ 0.8361 m²/yd.² ≈ 15.06 square yards.

1149. d. One cubic yard is equivalent to about 0.7646 cubic meters. Multiply: 3 yd.³ × 0.7646 m³/yd.³ ≈ 2.2938 cubic meters.

1150. c. One fluid ounce is equal to 2.957353 centiliters. Divide: 15 cL ÷ 2.957353 cL/fl. oz. ≈ 5.1 fluid ounces.

1151. c. Eight 8-oz. glasses account for 64 ounces of fluid: 8 × 8 = 64. There are 128 ounces in a gallon. Divide: 64 oz. ÷ 128 oz./gal. = 0.5 gallon.

1152. b. There are four quarts in a gallon. Divide: 19 qt. ÷ 4 qt./gal. = 4.75 gallons.

1153. d. One meter equals 100 centimeters. Multiply: 16.9 m × 100 cm/m = 1,690 centimeters.

1154. a. Two pints equal one quart, so one pint is equal to half a quart, or 0.5 quart. Therefore, 1 quart + 1 pint = 1 quart + 0.5 quart = 1.5 quarts. Each quart equals 0.946 liter, so multiply: 1.5 qt. × 0.946 L/qt. = 1.419 liters.

1155. e. One cubic inch is equivalent to 16.3871 cubic centimeters. Divide: 35 cm^3 ÷ 16.3871 cm^3/in.^{3} ≈ 2.14 cubic inches.

1156. b. One square foot equals 0.0929 square meter. Multiply: 221 ft.^{2} × 0.0929 m^2/ft.^{2} ≈ 20.53 square meters.

1157. e. There are 1,000 milliliters in one liter. Divide: 345.972 mL ÷ 1,000 mL/L = 0.345972 liter.

1158. b. One cubic yard is equal to 27 cubic feet. Each cubic foot is equal to 1,728 cubic inches. Multiply: 27 ft.^{3} × 1,728 in.^{3}/ft.^{3} = 46,656 cubic inches.

1159. e. One cup equals 8 fluid ounces; $2\frac{3}{4}$ cups = 2.75 cups. Multiply: 2.75 c. × 8 fl. oz./c. = 22 fluid ounces.

1160. a. One inch equals 2.54 centimeters. Multiply: 40 in. × 2.54 cm/in. = 101.6 centimeters.

1161. b. One square centimeter equals 100 square millimeters. Multiply: 300 cm^2 × 100 mm^2/cm^2 = 30,000 square millimeters.

1162. d. One meter is equal to 100 centimeters. Multiply: 4.3 m × 100 cm/m = 430 centimeters.

1163. e. To answer the question, subtract: 12.4 yds. − 8.34 m. Because the answer must be given in meters, and the subtraction cannot be done unless the units on the two quantities match, convert 12.4 yards into meters. One yard = 0.9144 meter; multiply: 12.4 yd. × 0.9144 m/yd. ≈ 11.34 m. Now subtract: 11.34 m − 8.34 m = 3 meters.

1164. b. One square inch equals 6.4516 square centimeters. Multiply: 72 in.^{2} × 6.4516 cm^2/in.^{2} ≈ 464.5 square centimeters.

1165. d. One cubic inch equals 16,387.064 cubic millimeters. Multiply: 11.8 in.^{3} × 16,387.064 mm^3/in.^{3} ≈ 193,367.36 cubic millimeters.

1166. d. One square meter is equal to 10,000 square centimeters. Divide: 29,728 cm^2 ÷ 10,000 cm^2/m^2 = 2.9728 m^2. Fred should purchase 3 square meters in order to have enough felt.

1167. b. One cubic meter equals 1,000,000 cubic centimeters. Multiply: 3.25 m^3 × 1,000,000 cm^3/m^3 = 3,250,000 cubic centimeters.

1168. e. One cup equals 8 fluid ounces. Multiply: 9 c. × 8 fl. oz./c. = 72 fluid ounces.

1169. a. One cubic foot equals 1,728 cubic inches. Divide: 226 in.^{3} ÷ 1,728 in.^{3}/ft.^{3} ≈ 0.13 cubic foot.

1170. c. There are 1,000 liters in one kiloliter. Divide: 1.5 L ÷ 1,000 L/kL = 0.0015 kiloliter.

1171. c. One cubic centimeter equals 1,000 cubic millimeters. Multiply: 24 cm^3 × 1,000 mm^3/cm^3 = 24,000 cubic millimeters.

1172. d. One liter equals 1,000 milliliters. Multiply: 7.98 L × 1,000 mL/L = 7,980 mL.

1173. e. One gallon equals 4 quarts. Multiply: 231 gal. × 4 qts./gal. = 924 quarts.

1174. a. Every pound equals 16 ounces, so multiply: 6 lb. × 16 oz./lb. = 96 ounces.

1175. d. One ton equals 2,000 pounds. Therefore, half of a ton is 2,000 ÷ 2 = 1,000 pounds.

1176. e. Every kilogram equals 1,000 grams, so multiply: 48 kg × 1,000 g/kg = 48,000 grams.

1177. b. Each kilogram equals about 2.205 pounds. Because there are 3 kilograms, multiply: 3 kg × 2.205 lb./kg = 6.615 lb. ≈ 6.6 pounds when rounded to the nearest tenth.

1178. a. Every 100,000 centigrams make a kilogram, so divide: 845.9 cg ÷ 100,000 cg/kg = 0.008459 kg.

1179. c. Every pound equals 454 grams. Therefore, to convert 2.5 pounds into grams, multiply: 2.5 lb. × 454 g/lb. = 1,135 grams.

1180. c. Because each serving calls for three ounces of cheese, and Rebecca is making 12 servings, she needs: 3 oz. × 12 = 36 oz. of cheese. Every 16 ounces is equal to one pound, so divide to convert ounces into pounds: 36 oz. ÷ 16 oz./lb. = 2.25 pounds.

1181. a. Every 0.035 ounce is equal to 1 gram, so first figure out how many grams are in 13 ounces by dividing: 13 oz. ÷ 0.035 oz./g ≈ 371.4 g. Because 1,000 g equals 1 kg, divide to determine the number of kilograms: 371.4 g ÷ 1,000g/kg ≈ 0.371 kilogram.

1182. d. There are 16 ounces in a pound, so divide: 432 oz. ÷ 16 oz./lb. = 27 lb. It takes 2,000 pounds to make 1 ton, so divide to discover what fraction of a ton 27 pounds is: 27 lb. ÷ 2,000 lb./t. = 0.0135 ton.

1183. e. 1,000 milligrams equals 1 gram, so divide: 500 mg ÷ 1,000 mg/g = 0.5 gram.

1184. a. Every ton equals 908 kilograms. Multiply: 4.2 t. × 908 kg/t. = 3,813.6 kilograms.

1185. e. There are 2,000 pounds in a ton, so divide: 300 lb. ÷ 2,000 lb./t. = 0.15 ton.

1186. a. Every ton equals 2,000 pounds, so multiply to find 3.15 tons: 3.15 t. × 2,000 lb./t. = 6,300 lb. Each pound equals 16 ounces, so multiply again: 6,300 lb. × 16 oz./lb. = 100,800 ounces.

1187. d. A thousand grams equals one kilogram, so divide to convert 85 grams: 85 g ÷ 1,000g/kg = 0.085 kilogram.

1188. e. Each kilogram is equivalent to 2.205 pounds, so multiply: 1.4 kg × 2.205 lb./kg = 3.087 lb. Each pound equals 16 ounces, so multiply again: 3.087 lb. × 16 oz./lb. = 49.392 ounces. Alternatively, you can first convert kilograms to grams, and then to pounds. Each kilogram is equal to 1,000 grams, so multiply: 1.4 × 1,000 = 1,400 grams. Each gram is 0.035 ounce, so multiply again: 1,400 × 0.035 = 49 ounces.

1189. c. Each gram equals 1,000 milligrams. Multiply: 25 g × 1,000 mg/g = 25,000 milligrams.

1190. a. Each gram equals about 0.035 ounce. Multiply: 58 g × 0.035 oz./g = 2.03 ounces.

1191. a. Every 16 ounces make one pound, so divide: 70 oz. ÷ 16 oz./lb. = 4.375 pounds.

1192. b. Every 1,000 milligrams equals 1 gram, so divide to determine the number of grams: 1,254,600 mg ÷ 1,000 mg/g = 1,254.6 g. Now, every 1,000 grams equals 1 kilogram, so divide again: 1,254.6 g ÷ 1,000 g/kg = 1.2546 kilograms.

1193. e. Each pound equals 0.454 kilogram, so multiply to find the number of kilograms in 52 pounds: 52 lb. × 0.454 kg/lb. = 23.608 kilograms.

1194. e. Each kilogram is 1,000 grams, so multiply to find the number of grams in 12.71 kilograms: 12.71 kg × 1,000 g/kg = 12,710 grams. Each gram equals 1,000 milligrams, so multiply again: 12,710 g × 1,000 mg/g = 12,710,000 milligrams.

1195. c. Each centigram equals 10 milligrams. Therefore, to find the number of milligrams in 2 centigrams, multiply: 2 cg × 10 mg/cg = 20 milligrams.

1196. d. Each kilogram is 1,000 grams, so multiply to find the number of grams in 0.24 kg: 0.24 kg × 1,000 g/kg = 240 g. Each gram equals 100 centigrams, so multiply again: 240 g × 100 cg/g = 24,000 centigrams.

1197. e. Every pound equals 16 ounces, so 12 pounds would equal 12 lb. × 16 oz./lb. = 192 oz. Since Michael weighed 12 pounds 9 ounces, add the 9 ounces to the number of ounces in 12 pounds: 192 oz. + 9 oz. = 201 ounces.

1198. b. Each gram is 100 centigrams, so multiply: 564.9 g $\times$ 100 cg/g = 56,490 centigrams.

1199. c. There are 100 centigrams in a gram, so 512 centigrams are equal to 5.12 grams: 512 cg $\div$ 100 cg/g = 5.12 g. Every 1,000 milligrams equals 1 gram, so 42,370 milligrams equals 42.370 grams: 42,370 mg $\div$ 1,000 mg/g = 42.37 g. Now add: 5.12 g + 42.37 g = 47.49 grams.

1200. b. Every 908 kilograms is equivalent to 1 ton, so divide: 5,448 kg $\div$ 908 kg/t. = 6 tons.

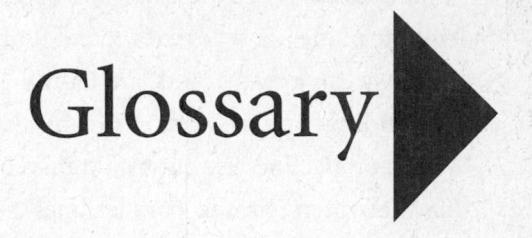

Glossary

acute angle   an angle that measures less than 90 degrees

acute triangle   a triangle with three angles that all measure less than 90 degrees

addend   one of the numbers being added in an addition sentence. In the number sentence $3 + 2 = 5$, 3 and 2 are the addends.

adjacent angles   two angles that are side-by-side, sharing a vertex and a side

algebraic equation   an equation that contains one or more variables. For example, $4 + x = 7$ is an algebraic equation.

algebraic expression   an operation that contains one or more variables. For example, $4 + x$ is an algebraic expression.

arc   a set of consecutive points on a circle

area   the amount of surface, in square units, that an object covers

base   In a term, the base is the variable, which may or may not be raised to an exponent. In $3x^3$, 3 is the coefficient and x^3 is the base.

binomial   a polynomial with only two terms. For example, $3x + 3$ is a binomial.

central angle   an angle formed by two radii; an angle whose vertex is at the center of a circle

chord   a line within a circle that connects two points on the circle

circle   a closed figure with no sides; a set of points that are equidistant from a single point, the center of the circle

circumference   the distance around a circle

coefficient   the number that is multiplied by the base of a term. The coefficient in the term $6x$ is 6, because 6 is multiplied by the base, x.

complementary angles   two angles whose measures total 90 degrees

concave polygon   a polygon in which one or more interior angles are greater than 180 degrees

congruent   equal in size or measure. Two angles or two line segments can be congruent.

constant   a real number (not a variable), such as 7 or –5

convex polygon   a polygon in which all of the interior angles are no greater than 180 degrees

coordinate plane   a set of axes, x and y, on and from which we can graph points and lines

corresponding angles   sets of congruent angles. When parallel lines are cut by a transversal (a third line that intersects the parallel lines), alternating angles are corresponding angles.

decimal   a number that is written using one or more of ten symbols. The ten decimal symbols are 0, 1, 2, 3, 4, 5, 6, 7, 8, and 9.

delta   a Greek letter, Δ, that represents change

denominator   the bottom number of a fraction. In the fraction $\frac{2}{8}$, the denominator is 8. Because a fraction is a division statement, the denominator is the divisor of that statement.

diagonal   a line drawn within a polygon connecting two nonconsecutive vertices

diameter   a line within a circle, through the center of the circle, that connects two points on the circle

difference   the result of subtracting one number from another. In the number sentence $5 - 3 = 2$, 2 is the difference.

dividend   the number being divided in a division problem (the numerator of a fraction). For example, in the number sentence $4.5 \div 9 = 0.5$, 4.5 is the dividend.

divisor   the number by which the dividend is divided in a division problem (the denominator of a fraction). For example, in the number sentence 4.5 9 = 0.5, 9 is the divisor.

equilateral triangle   a triangle that has three congruent sides and three congruent angles

exponent   Also called a power, it is the number of times a base must be multiplied by itself. In 4^3, 3 is the exponent, which tells us that three 4s must be multiplied: $4^3 = 4 \times 4 \times 4 = 64$.

exterior angle   an angle formed by a side of a polygon and a line that extends beyond the polygon

factor   each number being multiplied in a multiplication problem. For example, in the number sentence $1.2 \times 6 = 7.2$, 1.2 and 6 are factors.

fraction   a number that represents a part of a whole. A fraction itself is a division statement; $\frac{2}{8}$ is an example of a fraction in which 2 is divided by 8.

greater than sign   >, used when the first of two numbers is larger than the second of those two numbers. For example, "5 is greater than 4" can be represented with $5 > 4$.

greatest common factor   the largest number that divides evenly into two or more numbers. For example, the greatest common factor of 12 and 18 is 6. Both 12 and 18 can be divided evenly by 6. Other numbers (1, 2, and 3) are also factors of both 12 and 18, but 6 is the greatest common factor.

improper fraction   a fraction that has a value greater than or equal to 1, or less than or equal to –1. For example, $\frac{3}{2}$ and $-\frac{3}{2}$ are both improper fractions.

inequality   a number sentence in which the quantity on the left side is greater than, greater than or equal to, less than, or less than or equal to the quantity on the right side of the equation. $5x + 1 < 11$ is an inequality.

integer   The set of integers is made up of the whole numbers, their negatives, and zero. This set is written as $\{\ldots, -3, -2, -1, 0, 1, 2, 3, \ldots\}$.

interior angle   an angle found within a polygon

isosceles trapezoid   a trapezoid with congruent legs, base angles, and diagonals

isosceles triangle   a triangle that has two congruent sides and two congruent angles

least common multiple   a common multiple for two or more numbers that is the smallest value that is divisible by both numbers. For example, 30 is a common multiple of 3 and 5, because 30 is divisible by both 3 and 5; but 15 is the least common multiple of 3 and 5, because 15 is divisible by both numbers, and there is no smaller value that is divisible by both 3 and 5.

less than sign <, used when the first of two numbers is smaller than the second of those two numbers. For example, "4 is less than 5" can be represented with $4 < 5$.

like fractions two or more fractions with the same denominator. For example, $\frac{1}{3}$ and $\frac{2}{3}$ are like fractions, because they both have a denominator of 3.

like terms two terms that have the same base. $6y$ and $9y$ are like terms, because they both have a base of y.

major arc an arc that measures greater than 180 degrees

minor arc an arc that measures less than 180 degrees

minuend in a subtraction sentence, the number from which you are subtracting. In the number sentence $5 - 3 = 2$, 5 is the minuend.

mixed number a number that has a whole number part and a fraction part. The number $1\frac{1}{2}$ is a mixed number. Mixed numbers, like improper fractions, have a value that is greater than or equal to 1, or less than or equal to negative 1. The improper fraction $\frac{3}{2}$ is equal to the mixed number $1\frac{1}{2}$.

monomial an expression with a single variable and/or constant; a polynomial with only one term. For example, $3x$ is a monomial.

numerator the top number of a fraction. In the fraction $\frac{2}{8}$, the numerator is 2. Because a fraction is a division statement, the numerator is the dividend of that statement.

obtuse angle an angle that measures greater than 90 degrees

obtuse triangle a triangle with one angle that measures more than 90 degrees

ordered pair the x-value followed by the y-value of a point. For example, $(5, -4)$ is an ordered pair that describes the location of a point five units to the right of the y-axis and four units below the x-axis.

origin the point at which the y-axis and x-axis intersect on the coordinate plane. This point has the coordinates $(0,0)$.

parallelogram a quadrilateral whose opposite sides are parallel and congruent. Opposite angles are also congruent, consecutive angles are supplementary, and the diagonals bisect each other.

percent a ratio that represents a part to a whole as a number out of 100. For example, $0.25 = \frac{25}{100} = 25\%$.

percent decrease the difference between an original value and a new value divided by the original value. The new value is subtracted from the original value, because the original value is larger than the new value. Percent decrease is used to show the decline from an original value to a new value.

percent increase the difference between an original value and a new value divided by the original value. The original value is subtracted from the new value, because the new value is larger than the original value. Percent increase is used to show the growth from an original value to a new value.

perimeter the distance around an object, in linear units

polygon any closed figure made up of line segments. For example, a hexagon is a polygon.

polynomial an expression with one or more terms made up of variables and/or constants. For example, $3w^4 + 2x^2 + z + 6$ is a polynomial.

product the result of multiplication. For example, in the number sentence $1.2 \times 6 = 7.2$, 7.2 is the product.

proper fraction a fraction that has a value less than 1 and greater than negative 1. For example, $\frac{2}{3}$ is a proper fraction, as is $-\frac{2}{3}$.

proportion a relationship that shows two equivalent fractions, or ratios. For example, the fractions $\frac{3}{4}$ and $\frac{6}{8}$ are in proportion to each other because $\frac{3}{4} = \frac{6}{8}$.

Pythagorean theorem a theorem that states that the sum of the squares of the bases (legs) of a right triangle is equal to the square of the hypotenuse of the right triangle. Simply put, $a^2 + b^2 = c^2$, where a and b are the bases of the right triangle and c is the hypotenuse.

quadratic equation an equation that is written in the form $ax^2 + bx + c = 0$. The equation has one variable that is raised to the second power, and could have one variable that is raised to the first power (if b is not zero), and one constant (if c is not zero).

quadratic formula the formula used to find the roots of a quadratic equation: $x = \frac{-b \pm \sqrt{b^2 - 4ac}}{2a}$. Once a quadratic equation is in the form $ax^2 + bx + c = 0$, the values of a, b, and c can be substituted into the formula.

quadrilateral a four-sided polygon whose interior angles total 360 degrees

quotient the result of a division problem. For example, in the number sentence $4.5 \div 9 = 0.5$, 0.5 is the quotient.

radical the nth root of a radicand; a value that, when multiplied by itself according to the power of the radical (n), is equal to the radicand. For example, $\sqrt[3]{27} = 3$, because when 3 is multiplied by itself three times, the result is 27.

radicand the value under the radical symbol. In $\sqrt{81}$, 81 is the radicand.

radius the distance from the center of a circle to a point on the circle.

ratio a comparison, or relationship, between two numbers. Ratios are often shown with a colon, but they can also be expressed as fractions. The ratio 2 to 1 can be written as 2:1 or $\frac{2}{1}$.

reciprocal The reciprocal of a fraction is the multiplicative inverse of the fraction. To find the recip-rocal of a fraction, switch the numerator and the denominator. The reciprocal of $\frac{3}{4}$ is $\frac{4}{3}$.

rectangle a parallelogram whose angles are all congruent and whose diagonals are congruent

reflex angle an angle that measures greater than 180 degrees

regular polygon a polygon whose sides are all congruent and whose angles are all congruent

rhombus a parallelogram with four congruent sides and perpendicular diagonals that bisect its angles

right angle an angle that measures exactly 90 degrees

right triangle a triangle with one angle that measures exactly 90 degrees

rounding the process of taking a number and making it less precise by removing one or more digits from the end of the number, replacing those digits with zeros if necessary

scalene triangle a triangle that has no congruent sides or angles

secant a ray or line segment that touches a circle at two points

semicircle an arc that measures 180 degrees; an arc that connects the ends of a diameter

similar triangles two triangles that have identical angles. Similar triangles are not necessarily congruent.

slope the change in y-values divided by the change in x-values between two points on a line, written formally as $\frac{y_2 - y_1}{x_2 - x_1}$ or $\frac{\Delta y}{\Delta x}$

square a quadrilateral with four congruent sides; four congruent, 90-degree angles; two pairs of parallel sides; consecutive angles that are supplementary; and diagonals that are congruent, are perpendicular, bisect each other, and bisect their angles

straight angle an angle that measures exactly 180 degrees

strictly convex polygon a polygon in which all of the interior angles are less than 180 degrees

subtrahend   in a subtraction sentence, the number that you are subtracting. In the number sentence 5 − 3 = 2, 3 is the subtrahend.

supplementary angles   two angles whose measures total 180 degrees

sum   the result of adding two numbers. In the number sentence 3 + 2 = 5, 5 is the sum.

surface area   the total area of every face of a solid. Surface area is measured in square units.

system of equations   a pair or set of related equations in which the values of each unique variable are the same in each equation

tangent   a ray or line segment that touches a circle at exactly one point

term   an element in an expression that is subtracted, added, multiplied, or divided (or, in a single-term expression, an element that stands alone). In the expression $x - 6$, x and 6 are both terms.

trapezoid   a four-sided figure that has one pair of parallel sides

triangle   a polygon with three sides and three angles that total 180 degrees

trinomial   a polynomial with three terms. For example, $3x + 3y - 5$ is a trinomial.

unlike fractions   two or more fractions with different denominators. For example, $\frac{1}{3}$ and $\frac{1}{4}$ are unlike fractions, because they do not have the same denominator.

unlike terms   two terms that have different bases. For example, $6y$ and $9y^2$ are unlike terms, because they have different bases.

variable   a letter or symbol that takes the place of a number in an expression or an equation. In the expression $4 + x$, x is a variable.

vertical angles   Also known as opposite angles, these are angles that are not adjacent, but made by the intersection of two lines. Vertical angles are congruent.

volume   the amount of space taken up by an object. Volume is measured in cubic units, such as cm^3 or $in.^3$.

whole number   a positive number that has no fractional part—no digits to the right of the decimal point. For example, 45 is a whole number. This set is written as $\{0, 1, 2, 3, \ldots\}$.

***x*-axis**   the horizontal axis on the coordinate plane

***y*-axis**   the vertical axis on the coordinate plane

***y*-intercept**   the y-value of the point where a line crosses the y-axis